T0357919

Author photograph: Mary Adams

George Adams is an architect, avid birdwatcher, landscape designer, wildlife artist and photographer. *The Complete Guide to Australian Birds* is a photographic celebration of the beauty and rich diversity of Australia's birdlife; the result of several years' work in collaboration with his wife, Dianne, and a natural extension of their interest in the relationship between native plants and birds. Drawing on a lifetime's interest in natural science and concern for the damage inflicted by the built environment on the natural world, George developed a concept for landscaping in which creating habitat for birds and other wildlife is a vital part of the garden's design. This led to the publication of his first book, *Birdscaping Your Garden* (Rigby Publications, 1980). Continuously updated and revised, the new edition of the book is titled *Birdscaping Australian Gardens* (Penguin, 2015). His other books include *Gardening for the Birds* (Timber Press, 2018) and *Foliage Birds* (AH & AW Reed, 1980).

The Complete Guide to AUSTRALIAN BIRDS

Featuring 'in the field' images from
Australia's best wildlife photographers

GEORGE ADAMS

VIKING
an imprint of
PENGUIN BOOKS

VIKING

UK | USA | Canada | Ireland | Australia
India | New Zealand | South Africa | China

Penguin Books is part of the Penguin Random House group of companies
whose addresses can be found at global.penguinrandomhouse.com.

First published by Penguin Random House Australia Pty Ltd 2018

Cover design by Adam Laszczuk © Penguin Random House Australia Pty Ltd
Cover images: *(front)* the beautiful male Orange Chat alighting on the male plant of Silver Saltbush. Photographed at Wooleen Station in Western Australia, a conservationist haven and an excellent birding spot. The Orange Chat is found across Australia. Photograph © Shelley Pearson; *(spine)* Australian Reed-Warbler © John Anderson; *(back)* Soft-plumaged petrel © Todd Burrows
Text design by Post Prepress Group, Australia
Typeset in Arial Regular by Post Prepress Group
Printed and bound in China by RR Donnelley

A catalogue record for this
book is available from the
National Library of Australia

ISBN 978 0 14378 708 2

penguin.com.au

This book is dedicated to the enthusiasm and generosity of wildlife photographers and, in particular, to those who photograph birds. They brave the heat of the desert or a frosty, chilly morning on a mountaintop, a heaving boat in heavy Antarctic seas, and patiently wait for that perfect shot. Their images capture the behaviour, the quirkiness, the environment and reveal details that the human eye is not quick enough to see. These men and women contribute enormously towards our appreciation of the wonders of nature and our understanding of natural science.

The book is also dedicated to my wife, Dianne, who should be included as co-author but graciously declined. I would rather she had accepted. If it were not for her tireless research, design and dedication, this book would not have been possible.

CONTENTS

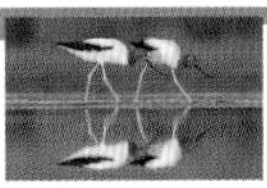

Introduction

Throughout the millennia, humans have had a continuing fascination with birds. The earliest known attempt to produce a comprehensive study of birds was by Aristotle (384–322 BC) and later, in Ancient Rome, Pliny the Elder (23–79 AD) introduced a classification system for birds based on the structure of their feet. His 36-book encyclopaedia entitled *Historia Naturalis* (Book 10 relates to the study of birds) remained in use, in various forms, for over 1500 years.

The first modern attempt to establish a natural classification of birds was in the 1750s when the Swedish scientist Carolus Linnaeus published his systematic biological classification of species, naming, ranking and classifying organisms into orders, families, genera and species based on body structure, bill, feet, colour and plumage structure. Once described and placed into his system it was thought that all species remained unchanged.

In 1859, Charles Darwin published *On the Origin of Species*, describing how, as a result of 'heritable behavioural and physical traits', organisms change over time. Charles Darwin's theory of natural selection soon began to be applied to the classification of birds.

The theory of continental drift had first been proposed in the 1500s but it was not until 1968 that a geoscientist, Jack Oliver, published a seminal paper 'Seismology and the New Global Tectonics' in the *Journal of Geophysical Research*, proving that 'Earth's crust is slowly shifting and moving'. Plate tectonics now forms the basis of all of our understanding 'of how the Earth works'. It also helped provide a timescale for the divergence of all bird groups on Earth.

The Australia–New Guinea land mass began breaking away from Gondwana about 55 million years ago. Drifting on its own tectonic plate like an ancient arc it carried the Gondwanan ancestors of emus, cassowaries and the huge parrot family, and the corvine ancestors of the passerines, plus areas of Gondwanan rainforests.

For 20–30 million years, as Australia drifted northward isolated from other landmasses, these ancient ancestor birds and rainforest plants continued to co-evolve and adapt. As the continent warmed and dried out, birds and plants continued to co-evolve. Over tens of millions of years the Australian corvine ancestors evolved into today's wrens, robins, magpies, thornbills, pardalotes, the huge honeyeater family, treecreepers, lyrebirds, birds-of-paradise and bowerbirds. They developed birdsong to advertise their occupation of a breeding or food territory, and to maintain contact with each other in evergreen, densely vegetated areas.

Australian Aboriginals passed down their knowledge of bird behaviour, plumage, calls, habitat, food and seasonal movement through oral tradition, dance and art. In 2010, in a narrow rock shelter in Arnhem Land, a red ochre painting of an extinct 'thunder bird', *Genyornis newtoni*, a heavily built, 2-metre-high, flightless bird, was discovered. This red ochre rock painting was recently dated at 40,000 years old, making it the world's oldest recorded bird painting.

When the First Fleet arrived in Port Jackson in 1788, the convicts, marines, seamen and settlers began naming the local birds after similar-looking British birds. Fairy wrens have cocked tails and a twittering call and were thought to be related to the northern hemisphere wrens; Australian robins looked and acted like northern robins and were assumed to be related to them and misnamed after them. Throughout the 18th and 19th centuries the belief among ornithologists was that Australian birds originated in the northern hemisphere.

In 1990 at Yale University the American ornithologists Sibley and Ahiquist, using DNA-DNA hybridisation techniques to analyse avian relationships, stunned the world by announcing that the world's 4500 songbirds (including the iconic northern hemisphere jays, mockingbirds, robins and thrushes) all had ancestral links back to Australia.

Sibley and Ahiquist also showed that a large number of Australian passerines that were thought to be related to northern hemisphere birds, such as wrens, robins and nuthatches (sittellas), were the result of 'convergent evolution' rather than genetic relatedness. They were the product of an independent 'radiation' evolving from the Gondwana ancestors of the crow family, into a variety of new forms and evolving similar traits as a result of adapting to similar ecological niches. Proof of the continental drift theory allowed Sibley and Ahiquist to estimate a timescale of divergence for all bird groups.

A new era for the study of avian classification had begun.

In Australia, Les Christidis and Walter Boles confirmed Sibley and Ahiquist's results. Christidis reviewed early records of DNA tests comparing protein with various bird species. Australia was the birthplace of, the parrot and pigeon families, and the birthplace of the world's songbirds, 'the land where song began'. When sceptics complained about the lack of fossil evidence in Australia, Boles and colleagues working at the Riversleigh World Heritage fossil site in Queensland in 1993, uncovered a 54-million-year-old fragment of a finch-sized bird. This was the earliest record of a songbird ever found, eclipsing the existing world record by 25 million years.

In 2008 Les Christidis and Walter Boles published *Systematics and Taxonomy of Australian Birds*, an up-to-date classification of Australian birds.

Australia is one of the world's ten mega-diverse countries and is fortunate to have a rich diversity of birds and an unusually high number of endemic bird species found across its many, equally diverse and beautiful landscapes. The jabbering of parrots, the laughter of kookaburras, the song of the magpie or the trilling warble of fairy-wrens all bestow a real sense of 'place' that is uniquely Australian.

Unfortunately today, with all the scientific knowledge we have at our disposal, thoughtless destruction of habitat, pollution and disrupted ecosystems are all contributing to an unnatural decline in birdlife and number of species. The Royal Australian Ornithologists Union (RAOU) lists Australia as having 793 birds, with 32 introduced species and 21 presumed extinct, leaving 740 native bird species surviving, 45 per cent of which are endemic. The Australian Government, Department of Environment and Energy, currently lists 16 birds as critically endangered; 55 birds as endangered and a further 62 birds as vulnerable. We have proved to be poor custodians of our amazing natural heritage our governments deciding to place little value on our birdlife, which represents a priceless resource not found in any other place on Earth.

It is hoped that this book will increase understanding, awareness and enjoyment of Australia's birds and underline the importance of preserving their habitats so that future generations can enjoy the sights and sounds that are quintessentially Australian.

Using this book

The photographs have been taken in the 'field', where the birds live, in every diverse habitat throughout Australia and are a window into the bird's everyday life. They illustrate what birders call the 'giss' of the bird – what Simpson and Day call 'an all-embracing term, the entire 'character', the 'essence', of the bird in the field'. We see glimpses into the bird's family life, birds at play, at the nest, dancing in display rituals, birds in flight, nest building, feeding, hunting or showing young how to find food. In this book each photograph conveys the absolute beauty and joy of Australia's unique birdlife and treats each species with reverence, as one of nature's own *objets d'art*.

Presenting birds in photographic format will complement the many hand-drawn field guides available and using the two formats together can provide the most complete overview of the bird. Most species have two photographs, illustrating male and female or photographs taken from a different angle where the sexes are alike or only have minor variations between sexes. Where noticeable geographic variations occur for a specie the various 'races' are described and illustrated. Immature birds are described and illustrated where they differ markedly from the adult birds.

The photographs appear directly alongside concise information including the bird's scientific name, common name, size, description, behaviour, preferred habitat, feeding habits, voice, status, breeding patterns and distribution map. Bird families are grouped in 27 chapters containing birds of the same or related families or birds of similar habits or habitat. Chapter introductions contain general interest information such as nesting and behaviour traits on each of the families included in the chapter.

The bird names, both scientific and common, generally follow the naming conventions of Christidis and Boles' *Systematics and Taxonomy of Australian Birds* (2008). In sequence of families the book also follows the current taxonomic associations as set out by Christidis and Boles, except where similarity of species dictates placing them with birds of similar appearance to save confusion in identification.

All wild birds that have been regularly recorded on the Australian mainland, Tasmania and offshore continental islands and oceans have been included. Non-breeding summer migrants and introduced species that have established viable feral populations are included. The Cocos (Keeling) islands and Christmas Island are not included, as they are closer to Java than Australia, and avian fauna is more Indonesian than Australian. Rare vagrants and accidental arrivals are not included. The birds of Australian dependencies are generally not included as their birds generally show affinities with other closer land masses. For example, Christmas Island is just 350 km south of Java; Lord Howe and Norfolk Islands are closer to New Zealand than Australia.

The binocular icon indicates areas where the bird is regularly found.

Abbreviations: CP. Conservation Park; CR. Conservation Reserve; NP. National Park;
NR. Nature Reserve; RP. Regional Park; GR. Game Reserve ; TP. Treatment Plant

Identifying birds

When attempting to identify a bird, firstly get an overall impression of the giss of the bird – its size and shape, overall colouration and movement. Each specie usually has a 'shape' that is characteristic of that group. Take into account the locality, habitat and any feeding or flight behaviour.

Take notice of the bird's head – often the head marking will provide the salient information to confirm identity. The size, shape and position of the tail – cocked, trailing, fanned, still or flicking – should be appraised when taking in a view of 'the whole bird'. Markings on the tail and wings are also important indicators to the bird's identification.

Plumage can vary and look different in different light conditions. For example, blue hues can often appear black. The plumage of some species, particularly sea birds, can change from crisp and new to tattered and worn by the end of the season before the birds begin their moult. During the moulting phase some birds, such as the male fairy-wren look quite unusual, halfway between breeding and non-breeding plumage, often referred to as 'eclipse' plumage. Different subspecies may also interbreed where their ranges overlap giving rise to birds that don't quite match the general description. Some species also have colour morphs, light to dark and intermediate. Where this occurs it is indicated in the text. Subspecies (races) are included where their differences can be distinguished in the field. Variations in plumage such as display or breeding are also noted.

Principal external features of a bird

Reference to the drawing below will explain the terminologies used in this book to describe a bird's crucial features. Familiarity with a bird's distinguishing features will provide clues to help in identification.

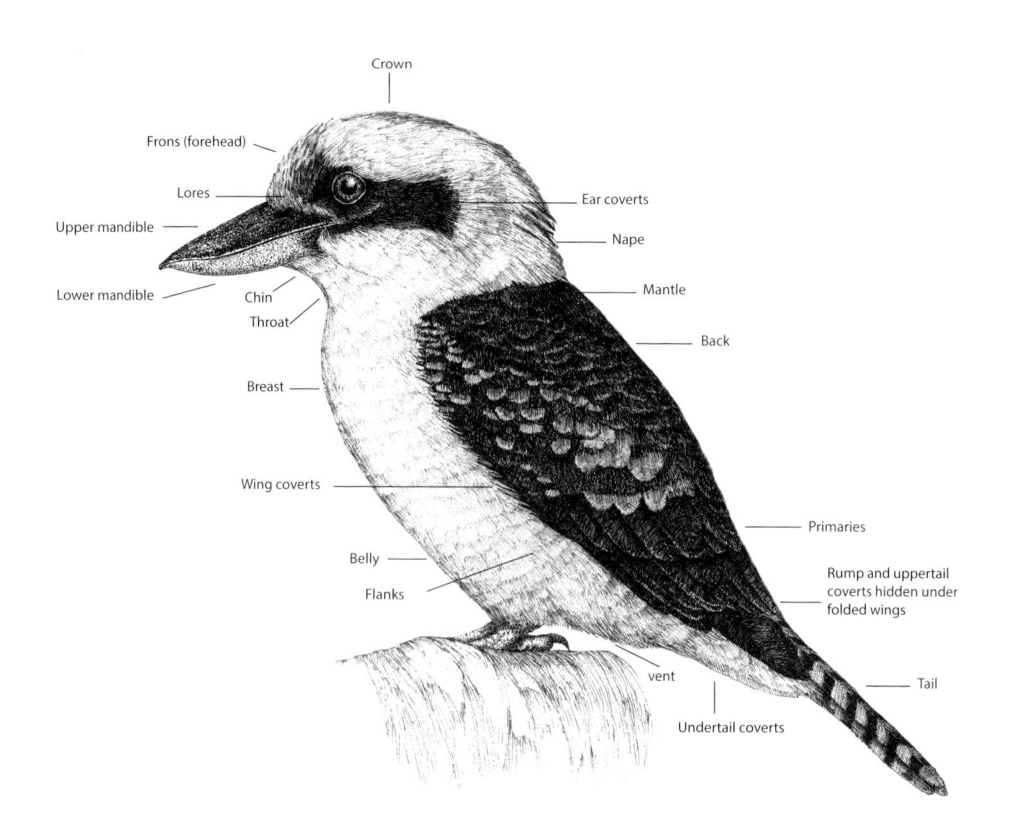

A bird’s bill

Each species of bird has its own food preferences, based on the bird’s behaviour and physical characteristics, so different species can live in the same territory without competing for food. A bird’s bill is its most important implement for gathering food, and the size and shape of the bill is a good indication of the type of food eaten by a species.

Insectivores, such as the Spotted Pardalote, have strong, stout bills for gleaning insects and insect larvae from foliage.

Honeyeaters, such as the Western Spinebill, have long bills for probing tubular-shaped flowers.

Meat-eaters, such as the Grey Butcherbird, have large, strong, hooked bills for grasping and tearing apart prey.

Parrots, such as the Sulphur-crested Cockatoo, have powerful bills for cracking hard nuts and seeds.

Distribution maps

The distribution maps were generated from a wide range of ornithological records, including the *Atlas of Living Australia*, historical records and birding tour reports, supplemented with vegetation surveys and maps and biological surveys to determine suitable habitat in more remote areas where particular species should be present. The maps are generally in three tones. Dark grey indicates areas where the birds will most frequently be seen. Mid-grey indicates areas where the birds can generally be seen, or where they may occur in good seasons. Palest grey indicates where the birds may be sparsely distributed. Broken lines denote the range of subspecies. Keep in mind that birds are wild creatures of opportunity, notorious for turning up where they shouldn’t be.

Birdscaping your garden

You can make a difference in your own garden by creating suitable habitats for the birds in your area. A garden landscaped to attract birds needs abundant natural food sources such as flower nectar, grass seed heads, fruits, berries and a diversity of plants to attract insects, because many birds are insectivores. A birdbath or small pond will bring thirsty birds as well as an interesting assortment of wildlife, including frogs, dragonflies and butterflies. You can attract a greater variety of birds to nest in your garden by providing a diversity of plant habitats, such as clumps of shrubbery, tangled thickets and taller trees.

Providing cover and shelter

Australian native plants provide effective shelter for native birds, whose colour and markings have adapted to blend with the indigenous flora. Effective shelter can be provided in a relatively small area by plants that provide a solid mass of foliage from ground level to 3m or higher. A row of trees or shrubs can also be a pathway for birds across a garden and will attract smaller birds that would otherwise not cross the open space for fear of attack from predators. Select plants to provide sequential flowering for year-round flowers and fruit.

Providing nesting sites

Birds choose their nesting sites carefully and guard their territory. By keeping their nest site and breeding territory off limits to members of the same species that compete for the same food, the birds ensure sufficient food for themselves and their nestlings. Nesting sites must have some protective cover. The eggs and nestlings are most vulnerable to predators and
the degree of exposure often determines the nesting location. Songbirds usually seek protective shelter in dense foliage or prickly shrubs.

Many Australian birds, such as parrots, lorikeets, Kookaburras and treecreepers, as well as many mammals, nest in hollows found in the dead limbs of mature trees. In undisturbed bushland, mature eucalypts provide numerous nesting hollows. Most eucalypts need to be 50 to 60 years old before adequately sized nest hollows are available for wildlife. This age usually coincides with their optimum commercial value and they are often selected for logging.

In urban areas the removal of old, decayed or dead trees, the removal of tree limbs, patching up cavities in established trees and competition from introduced starlings, including the Common Starling, Common Myna and Eurasian Tree and House Sparrow, have caused many cavity-nesting birds and mammals to be deprived of nesting opportunities, resulting in local extinctions.

In rural areas, excessive land clearing has resulted in a shortage of nesting sites and depletion of habitat. (La Trobe University has estimated that 90 per cent of Victoria's natural vegetation, 'mainly the woodland ecosystems', has been lost since European settlement.)

You can provide homes for birds by purchasing or constructing a simple nest box. The diameter of the entrance hole will determine which species of bird can use the nest box.

Using water to attract birds

A supply of fresh, clean water is an important element for attracting birds to your garden. Throughout the year, they need water for bathing and drinking. The simplest way to provide water is to set out a birdbath or create a beautiful garden pond in a quiet area of the garden.

Birds bathe and preen in order to keep their feathers in perfect condition for flying and to maintain the waterproofing and insulating properties. When bathing, most birds scoop water up in their bills and splash it over their backs, flapping their wings and ruffling their feathers in a manner that could attract the attention of cats. Wet feathers will hinder the birds' ability to fly and escape a cat or other predator, so the position and design of the birdbath is critical for the birds' survival. Place your birdbath near protective shrubbery – close enough to allow the birds to escape to safety, but not so close that a cat can hide at close range and easily pounce on the unwary birds.

Australia's principal vegetation habitats

The ancient ancestry of Australia's birds and plants, co-evolving and adapting to their environment, has created the amazing diversity of unique endemic bird and plant habitat species we have today.

We can follow this co-evolution back to around 55 million years ago, when the Australia–New Guinea landmass began to drift away from Gondwana on its own tectonic plate. Like a gigantic 'ark', it held a small group of corvine, or crow-like, ancestors and rainforest vegetation, which had been present on Gondwana before the split.

Drifting in isolation for 20–30 million years, the corvine ancestors began evolving with the rainforest plants, and then radiated across the landscape as they evolved to live with the emerging hard-leafed woodland vegetation, exploiting each ecological niche.

Compared with birds elsewhere, Australian birds exert more ecologically powerful influences on our native vegetation and the shape of our forest environments. For example, the Superb Lyrebird is an important ecosystem engineer, able to influence the germination of plants and determine the forest structure. Other birds pollinate, help germinate and protect plants from insect pests.

A relic of this ancient past, the Antarctic Beech Tree is an important Gondwana relic that can still be seen growing today, in the lush prehistoric landscapes of the cool temperate rainforests of New South Wales and Queensland. DNA studies have suggested that the Noisy Scrub-birds are living fossils that evolved 97 million to 65 million years ago. These studies have also shown that the Superb Lyrebird is the closest surviving relative of the very first songbird to have ever evolved. It is now believed Australia was the birthplace of songbirds, parrots and pigeons, and the land where song began.

Throughout the book, reference is made to the habitat that particular birds prefer. Generally, each vegetation habitat is the result of the interaction of species within that area, including the endemic birdlife species. The plumage pattern and colours of these birds have evolved with the colours and textures of the local native vegetation to provide effective camouflage.

Where the book lists parks, reserves and gardens under 'Habitat', availability of suitable native vegetation is implied.

Habitats are complex, overlap and merge. They are broadly categorised in this book as follows:

Closed forests or rainforests

Rainforests are characterised by high-rainfall-dependent, tall, dense trees that form a continuous canopy, with shade-tolerant understorey plants and vines below. The vines, referred to as liana vines, are rooted in the soil, grow vigorously up trees to the canopy and depend on the trees for support. They are long, usually woody and typical of rainforest habitat.

The rainforests of eastern Australia support 140 species of birds, with over 60 species found only in rainforest habitat.

In Australia there are four main types of rainforest: wet tropical, dry, subtropical and temperate.

Wet tropical rainforest (Row 1)

Found from eastern Cape York, Qld, southward to near Sarina, Qld. Typically trees 30–40m tall with three layers of vegetation and dense liana vines. The oldest tropical rainforest on Earth. Many of its species originated when Australia was still part of Gondwana. Provides habitat for numerous rare species of plants and animals, with 370 bird species having been recorded, 11 species being endemic. Fruit-eating birds such as the Cassowary, Fig-parrot and fruit-doves are prominent, feeding on the abundant rainforest fruits. Many of the bird species are New Guinean in origin and are found nowhere else; they include the Eclectus Parrot, Red-cheeked Parrot, Trumpet Manucode and Magnificent Riflebird.

Dry rainforest

Dry monsoon rainforest (Row 2)
Outlying pockets of tropical rainforest, also referred to as 'monsoon vine thicket', where rain is seasonally scarce in the dry season. Found in isolated pockets in Arnhem Land and throughout the Kimberley region. Usually stunted trees 5–10m tall, with fewer plant species, dominated by liana vines, and mainly leaf litter forest floor. Emergent trees include Hoop Pine and Lacebark.

Monsoon rainforest (Row 3)
Typically found around permanent water with 20–30m tall trees, and dense understorey of small trees, ferns and few vines.

Subtropical rainforest (Row 4)

Found from Cooktown, Qld, south to the Illawarra district of NSW. The most widespread rainforest, with trees 20–30m tall, two layers of vegetation and fewer vines. Laurels, myrtles and Sassafras are the dominant trees. Distinctive birds include the Topknot Pigeon, Australian Logrunner, Sooty Owl, Regent Bowerbird, Rufous Scrub-bird, Large-billed Scrubwren and Yellow-throated Scrubwren.

Temperate rainforest

Warm temperate rainforest (Row 1)
The main type of rainforest south of Coffs Harbour, NSW, generally with low soil fertility and wet, cool winters. These forests typically receive over 1300mm of rainfall annually and are characterised by a few species of slender-trunked trees and a sparse fern understorey. Coachwood and Sassafras are the dominant tree species. Typical birds include the Superb Lyrebird, Brown Cuckoo-dove, Wonga Pigeon, Red-tailed Black Cockatoo, Yellow-tailed Black Cockatoo, Australian King Parrot, Blue-winged Parrot, Koel, Sooty Owl, Marbled Frogmouth and Noisy Pitta.

Cool temperate rainforest (Row 2)
Mostly found in Tasmania, also in wet montane southeastern Australia and isolated pockets on mountaintops, north to the Gold Coast hinterland. Often shrouded in mist, these forests typically receive over 1750mm of rainfall annually and have two layers of vegetation, dominated by Antarctic Beech and Sassafras, with dense ferns, tree ferns and mosses to the forest floor and few vines. Typical birds include the Rose, Pink and Scarlet Robins, Logrunner, Eastern Whipbird, Golden Whistler, Rufous Fantail and Satin Bowerbird.

Open Forests

Open forests are also called ‘sclerophyll forests’ and are characteristically tall eucalypt forests with an open canopy and an understorey of grasses and shrubs. They provide habitat for 140 bird species.

Wet eucalypt forest (Row 3)
Includes both mixed forests and wet sclerophyll forests. Tall open forests with stands of eucalypts generally 30m tall, but as high as 70m in the Mountain Ash forests of southeastern Australia and the Karri forests of southwestern Western Australia. Characterised by a dense understorey with dense, tall shrubs, small trees and tree ferns in moist areas. The Mountain Ash is the world’s tallest flowering plant and one of the tallest trees in the world. Lowland wet eucalypt forest understorey can be dominated by rainforest species and is called ‘mixed forest’, or it can have a variety of tall broad-leafed shrubs and small trees and is called ‘wet sclerophyll forest’.

Dry eucalypt forest (Row 4)
Identified by the dominance of eucalypts. Found on less fertile soils, with medium-sized trees growing to less than 30m tall and a more open canopy and well-developed multilayered understorey of shrubs adapted to dry conditions, including a rich diversity of wattles, sheoaks and Native Cherry. Grevilleas, hakeas, boronias and other flowering shrubs can also be present.

Woodland

Woodlands support over half of Australia's land bird species. Characterised by scattered medium to tall trees, usually eucalypts with fully developed crowns, and a grassy understorey. When there are fewer trees and the understorey is predominately grass, it is termed 'savannah'.

Tropical eucalypt woodland (Row 1)
Found in northern Australia. This habitat supports 170 species of birds. Contains the tall bunchgrass savannah and related eucalypt woodland of far northern Western Australia, the Northern Territory and Far North Queensland, including Cape York. Characterised by 1–3m tall annual sorghum grass in the wet season.

Temperate eucalypt woodland (Row 2)
Found in southeast and southwestern Australia, mainly west of the Great Dividing Range from southern Queensland to Tasmania. Characterised by an understorey of grasses and herbs up to 1m tall. It supports 170 species of birds. In southwestern Western Australia it has mostly disappeared due to land clearing for farming and grazing in the 'wheat–sheep belt'.

Arid and semi-arid woodland (Row 3)
Open eucalypt savannah woodland 10–15m tall. Characterised by communities of Mallee (shrubby eucalypts) and Mulga and perennial grass and herb understorey with scattered shrubs. The Bush Stone-curlew, Glossy Black Cockatoo, Pied Honeyeater and Varied Sittella are some of the birds present; many species are listed as vulnerable.

Mallee (Row 4)

Mallee supports approximately 160 species of birds. It occurs in vast arid and semi-arid regions of southern Australia and on the southern edges of the inland deserts dominated by 'mallee trees'. These are shrubby eucalypts with multiple stems, often with spectacular flowers, that provide a uniform scattered canopy for an understorey of shrubs and groundcovers. In wetter areas with 300–500mm annual rainfall, mallee habitat has an understorey of melaleuca species. In dryer areas saltbush species are the dominant understorey plant.

Mulga (Row 1)

Approximately 80–90 bird species occur in mulga habitat, with five species found there exclusively. Often called acacia scrubland, 'mulga' refers to the habitat's dominant plant, *Acacia aneura.* Covering vast areas of southern Australia, mulga is found scattered through the tablelands and alluvial flats of inland Australia. It meets the edge of the mallee regions in the south and the edge of the inland deserts in the north, in areas with less than 300mm annual rainfall. Characterised by 2–9m tall trees, usually exclusively acacia, including Mulga, with a sparse understorey of emu-bushes, *Dodonaea* and *Senna* species. In sandy soils, the understorey is spinifex; in loamy soils, the understorey is annual grasses.

Lowland shrub (Row 2)

Also called 'shrub steppes'. These treeless areas are mostly dominated by well-separated saltbush species growing up to 2m tall in the red loam inland soils. Lowland shrub habitats include, for example, the Nullarbor Plain and the gibber plains or 'stony desert'. These habitats have an annual rainfall of less than 500mm, which usually falls in winter, and support over 50 bird species.

Heathland

There are several types of heath in Australia; all are generally treeless communities of shrubs under 2m in height, often with spectacular colourful plants including epacris, banksias, grevilleas and bottlebrush. Heathlands support about 80 species of birds, with seven species found exclusively in this habitat.

Inland heathland (Row 3)
Found in southwestern Australia and the South Australia–Victoria border desert regions on low-nutrient sand plains. It consists of stunted mallees, sheoaks and a variety of nectar-rich, woody shrubs, which form heathland that attracts similar bird species as those found in coastal heath, but also includes other inland birds.

Coastal heathland (Row 4)
The most widespread heathland found in Australia. Coastal heathland forms where soil and wind conditions prevent tall trees from growing, such as areas that are windswept and salt-sprayed. It occurs on nutrient-poor acid soils in temperate and tropical regions. Heathland plants consist of stunted trees and shrubs shorter than 2m high and cover at least one third of the land area. Plants such as banksias, *Angophora* and *Xanthorrhoea* species have nectar-rich flowers attractive to pollinators such as honeyeaters. Smaller colourfully flowered plants that form dense, prickly thickets provide cover for smaller birds to nest and shelter.

Alpine heathland (Row 1)
Dominates the high alpine regions of the southeastern highlands of Australia, including Tasmania. They are the least diverse of the heathlands but contain a large number of endemic plants with restricted ranges, including shrub-like grevillea, epacris and *Kunzea* species. Alpine heathlands generally have a low diversity of birdlife. The Nankeen Kestrel and Australian Magpie occupy the alpine areas during summer, retreating to lower elevations in autumn. Some species, including the Flame Robin, brave the harsh winter climate of the higher altitudes, and the Gang-gang Cockatoo inhabits the subalpine snow gums in winter.

Wetlands

Wetlands are areas where water covers the soil all year or at certain times of the year. Australian wetlands support about 110 species of birds, with over 50 species found exclusively in this environment.

Australia presently has 65 Ramsar Wetlands of International Importance for conserving biological diversity, which cover 8.3 million hectares, plus 900 wetlands of national importance.

Far Northern Australian wetlands (Row 2)
These wetlands evaporate in the dry season and fill again in the wet season, and mostly attract seasonally migratory birds. Magpie Geese are the iconic bird of these wetlands.

Wetlands of northeastern Australia (Row 3)
These wetlands feature the vast coastal floodplains, which are a far more reliable year-round resource due to regular rainfall, as are the less extensive wetlands of southeastern Australia.

The inland wetland system (Row 4)
Dominated by the Murray–Darling Basin, the only inland system consistently filled by summer and winter rains. These wetlands are vitally important to a large range of species, including many endangered Australian migratory birds and international non-breeding migratory summer visitors.

Grasslands

Grasslands support over 100 bird species, with about 18 species found exclusively in this habitat.

Most grassland occurs in arid or semi-arid regions. The variety of grassland habitats includes spinifex and arid tussock, or Mitchell grasslands and the alpine grasslands of southeast Australia. Other types of grassland include swampy grasslands, floodplain grasslands of tropical coastal rivers, and tropical grasslands found inland from the Gulf of Carpentaria.

Spinifex grasslands (Row 1)
Occur in arid, sandy and sandstone areas and cover approximately one quarter of Australia. Spinifex grows in hummocks 900–1500mm in diameter and 300–600mm tall, separated by bare ground. After rain the 'bare ground' comes alive with a rich variety of wildflowers. Trees and shrubs are widely spaced.

Arid tussock, or Mitchell grasslands (Row 2)
Dominated by *Astrebla* species, growing in tussocks up to 900mm tall and 230mm in diameter, separated by bare ground on mostly treeless plains. Found in western Queensland and eastern Northern Territory.

Mangroves

Australia has 45 species of mangroves, which make up half the world's total species. Mangroves perform vital ecological roles, including protecting coastal ecosystems against storm surge, and storing up to ten times more carbon than an equivalent area of terrestrial forest. Mangrove roots provide protective nursery habitats for breeding fish and shellfish.

Mangrove forests attract about 70 species of birds, with 13 species found exclusively in this environment and a further 20 species typically found here. They include the Mangrove Honeyeater, Mangrove Robin, Mangrove Grey Fantail and Broad-billed Flycatcher.

Mangroves line the muddy shores of estuaries, bays and the coastline of northern Australia, growing down the east coast to Westernport in Victoria and down the west coast to Carnarvon in Western Australia. Most species grow across northern Australia, with 36 species occurring in Darwin Harbour. Twelve varieties grow to over 20m high, forming a continuous canopy. (Row 3)

In the far south they are represented by a single species, the Grey Mangrove (*Avicennia marina*), which grows about 3m tall. (Row 4)

Typical nest types

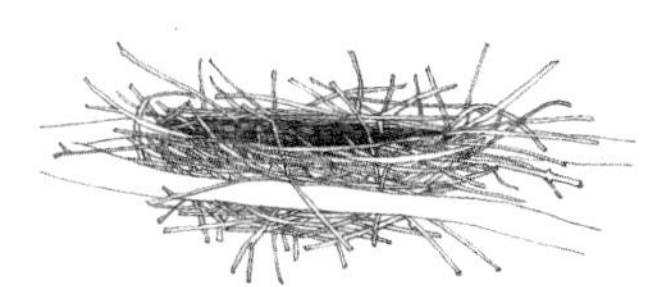

Stick platform nest typical of native pigeons, frogmouths, bitterns and birds of prey.

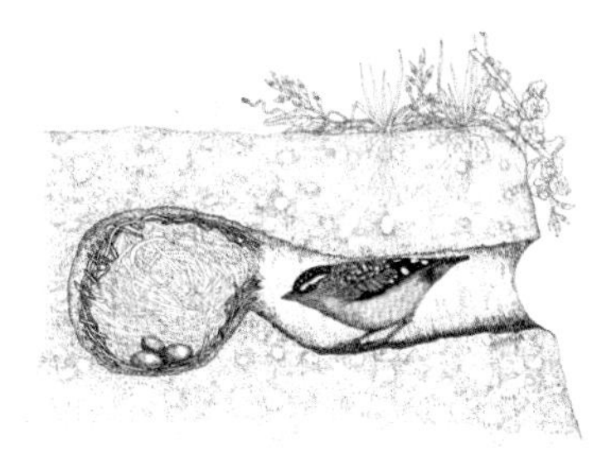

Excavated underground nest typical of pardalotes, kingfishers, Rainbow Bee-eater, White-backed Swallow, Little Penguin and Short-tailed Shearwater.

Stick bowl nest typical of Black-necked Stork, butcherbirds, wattlebirds, Australian Magpie and currawongs.

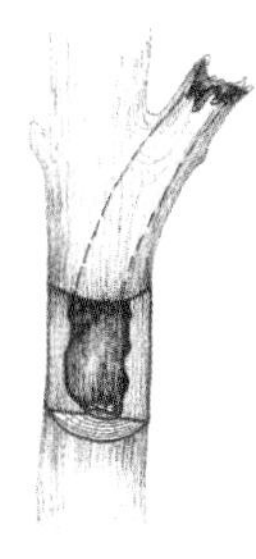

Tree hollow nest typical of cockatoos, corellas, lorikeets, parrots, owls, treecreepers, kookaburras, Australian Wood Duck, Gouldian Finch and Dollarbird.

Cup nest of bark, grass and spider web, typical of robins, flycatchers, sittellas, monarchs and trillers.

Bowl nest of mud and grass built on a horizontal branch or shelf, typical of swallows, White-winged Chough, Apostlebird and Magpie-lark.

Cup nest with tail, typical of fantails.

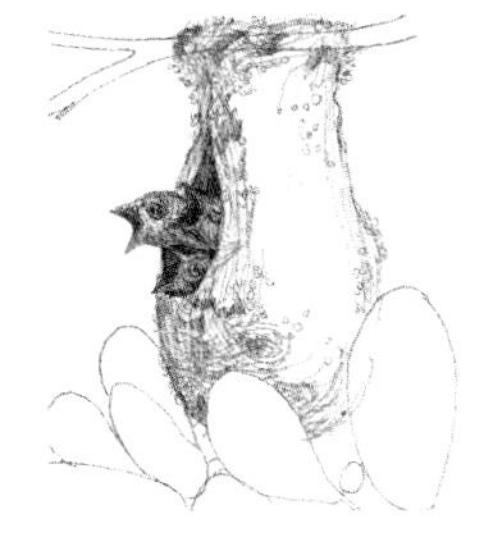

Pendulous nest of plant fibre and spider web with side entrance, typical of Mistletoebird, Weebill, gerygones, thornbills, Olive-backed Sunbird and Metallic Starling.

Sphere nest of grass with side entrance spout, typical of finches.

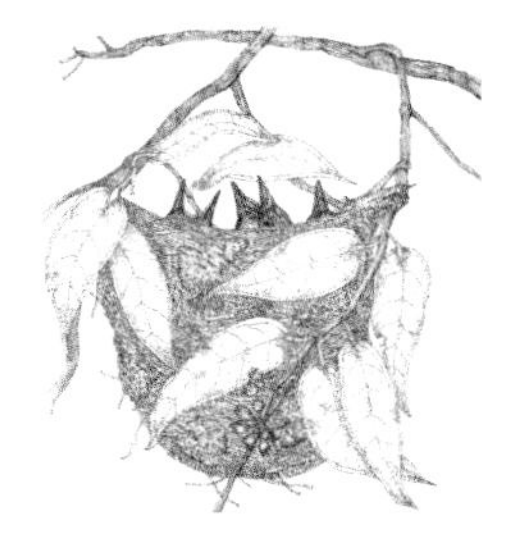

Suspended cup nest of fibrous plant material and spider web, typical of honeyeaters, Silvereye, white-eyes, Figbird, orioles and Yellow-breasted Boatbill.

Domed grass nest with side entrance, typical of fairy-wrens, emu-wrens, grasswrens, cisticolas, pittas and lyrebirds.

Family Casuariidae, Megapodiidae, Phasianidae

Cassowary

Emu

Brush-turkey

Scrubfowl

Malleefowl

Quails

A grouping of generally large and flightless birds and native quails.

Family Casuariidae includes the flightless Cassowary and Emu.

The Cassowary is a keystone species in the wet tropical rainforest ecology, eating up to 150 rainforest fruit species and passing the seed around ten hours later in a ready-made fertiliser up to a kilometre away from the parent plant. Approximately 100 rainforest plant species totally rely on the Cassowary for dispersal of their seed. In the breeding season the female lays four eggs. The male Cassowary incubates the eggs and raises the young, while the female finds another partner or joins a non-breeding, wandering flock.

The Emu, Australia's largest bird, is a widespread nomadic wanderer. Females lay up to 20 eggs then leave the male to care for the young. Flightless birds, they are able to run at 48km/h with a characteristic swaying action.

Family Megapodiidae includes the three Australian species of 'mound builders' or megapodes, the Australian Brush-turkey, Malleefowl and Scrubfowl.

Incubating mounds are constructed from fallen debris and rely on the heat generated by decaying matter to incubate the eggs. Up to 25 eggs are laid separately over several weeks and must remain at a constant temperature for the incubation period. The male regulates the temperature and humidity by digging shafts and removing or adding material as required. The young break out of the egg with open eyes and developed feathers and are strong enough to dig their way out of the mound, run for cover and raise themselves. The young are the most mature of any bird species, some being able to fly 24 hours after hatching.

Mounds are of varying sizes; for example, the Australian Brush-turkey builds a 4m dia. mound, reaching over 1m high. The Malleefowl has a mound 5m dia. and 1.5m high, while Australia's smallest mound-building bird, the Orange-footed Scrubfowl, builds the largest mound, moving tons of sandy soil and vegetable matter from up to 150m away from the mound. This mound can reach 6–7m dia. and 3m high and occasionally reaches 15m dia. and 4m high. Mounds are tended year-round and birds can often be seen at or near their mound.

Family Phasianidae includes the native Stubble Quail, Brown Quail and King Quail.

These birds are seldom seen because their mottled, cryptic plumage provides excellent camouflage and their habit of squatting motionless when disturbed makes them difficult to observe. When almost trodden on, they burst from cover with a whirring of wings and low flight. This habit makes Stubble Quail and Brown Quail choice targets for shooters.

The introduced Indian Peafowl, Wild Turkey, Red Junglefowl (feral chicken) and Common Pheasant are also included in family Phasianidae. A wild colony of Indian Peafowl or Peacocks (*Pavo cristatus*) persist on Rottnest Is., Western Australia.

Opposite: Female Emu in breeding plumage.

Southern Cassowary

Casuarius casuarius

Endangered

1700–1800mm

Thickset, flightless bird with a brownish grey casque about 17cm high. Head, neck: Bright blue bare skin, purple-pink along back of neck with long deep pink wattles. Body: Black, shaggy, bristle-like plumage. Legs: Short and powerful.
Female: Similar, larger, brighter.
Hatchlings: Striped.
Immature: Small casque, wattles and browner plumage.

Runs at great speed, head down when disturbed. Uses its bony casque as a battering ram through dense undergrowth. Aggressive. 3 short toes, innermost toe has a 120mm-long 'nail' spike that can be used in defence.

Habitat: Dense, tropical rainforests and fringes near watercourses.

Feeding: Fallen native rainforest fruits and berries, also fungi, snails, carrion.

Voice: Resonant booming, rumbling and hissing sounds. Foot stamping.

Status: Endangered. Sedentary.

Breeding: June–Oct. Nest is a scrape on the ground / 4 eggs.

Mission Beach, Qld.

Left: Adults. Right: Adult with chick.

Emu

Dromaius novaehollandiae

Endemic

1500–1900mm

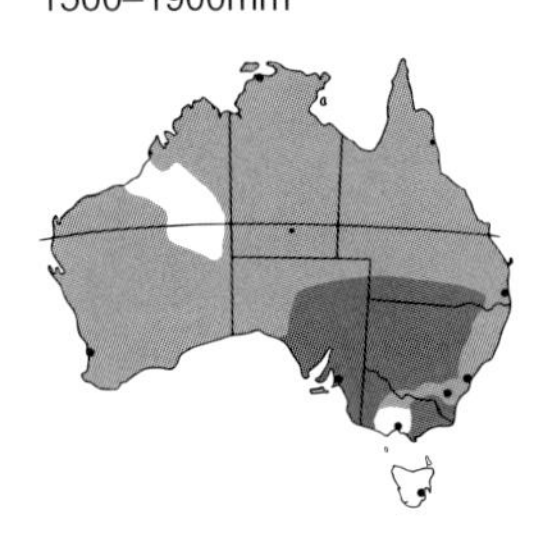

Shaggy, hair-like grey-brown to dusky brown plumage. Head, neck: Bare bluish-grey skin. Legs: Dark brown, powerful, 3-toed feet. Eyes: Yellow, grey or red. Birds are darker in the west and paler in the north. Southeastern birds have a pale collar when breeding.
Female: Similar, slightly larger than male. Breeding season plumage darkens slightly and patches of bare facial skin change to turquoise blue.
Immature: Greyer. Hatchlings: Downy, dark brown with cream stripes.

Highly nomadic, especially during droughts when they often travel in flocks of several hundred birds. Runs swiftly.

Habitat: Woodlands, scrub, semi-arid grasslands, plains, alpine grasslands, agricultural areas. Avoids rainforests and deserts.

Feeding: Omnivorous. Mostly fruit, seeds and insects, including grasshoppers. Also flowers, leaves, herbs, grass.

Voice: Usually silent. Occasionally deep guttural grunts by both sexes and dull resonant booming by female.

Status: Locally common.

Breeding: Apr.–Oct. Male makes a scrape on ground, loosely lined with plant litter / 7–11 eggs.

Wyperfeld NP, Vic.

Left: Male with chicks.
Right: Young female birds.

Australian Brush-turkey

Alectura lathami

Endemic

700mm

Sexes similar. Bare red head and neck sparsely covered with black, hair-like bristles. Large, pendulous bright yellow wattles, most prominent in the breeding male. Body: Dull black. Breast, belly: Black and scalloped. Bill: Black. Eyes: Brown. Tail: Side-on flattened fan shape.
Female: Smaller wattles.
Immature: Duller, more feathered neck.
Race *purpureicollis*, Cape York, Qld. Wattles: Bluish white.

If disturbed, birds usually run into undergrowth or, occasionally, bursting-leaping, heavy flight into tree cover. Pairs often seen at their large incubating mound.

Habitat: Damp, warm coastal forests, woodland, parks and gardens.

Feeding: Ground foraging for insects, native fruits and seeds.

Voice: Harsh grunts. Territorial booming at nest mound.

Status: Locally common. Sedentary.

Breeding: Aug.–Dec. Incubation mound 4m dia. x 1.5m high, maintained by male / 18–24 eggs.

Lamington NP, Qld.

Left: Male showing 'flattened' tail.
Right: Base of neck flared in ruff of yellow wattles.

Orange-footed Scrubfowl

Megapodius reinwardt

Endemic

420–470mm

Sexes alike, female slightly smaller. Small, pointed reddish brown crest. Slate grey head, neck and underparts. Upperparts, wings: Dark chestnut or olive-brown. Bill: Reddish brown. Eyes: Brown encircled by greyish bare skin. Legs: Bright orange.
Immature: Similar, duller.

Pairs form permanent bonds and are usually seen near their large incubation mound. Occasional flight with heavy wingbeats to a low branch or for short distances.

Habitat: Dense coastal tropical rainforests, monsoon rainforests.

Feeding: Forest floor for fallen fruit, seeds, snails, grubs and larvae.

Voice: Screams, chuckling, chortling. Occasional duets.

Status: Locally common. Sedentary.

Breeding: Aug.–Mar. Incubation mound 3m wide x 1.5m high, with eggs laid in a hole 2m deep / 18–24 eggs.

Centenary Lakes, Cairns, Qld. Botanical gardens, Darwin, NT.

Left: Dust bathing.
Right: Scratching leaf litter at mound.

Malleefowl

Leipoa ocellata

Endemic Endangered

600mm

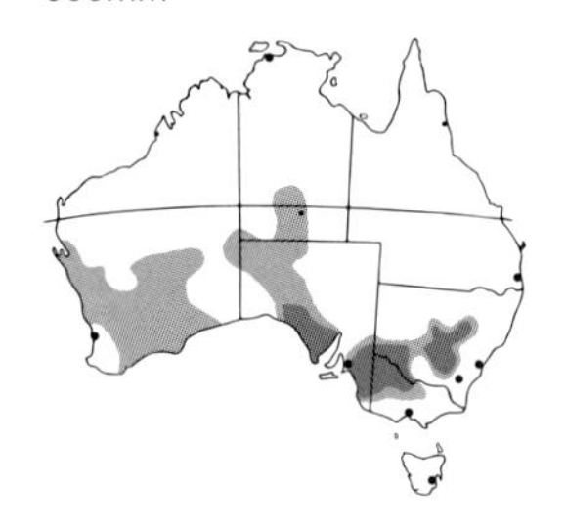

Sexes alike. Heavily built bird with a buffy grey head, white line below eyes and large earhole. Bill, lores: Grey. Throat: Rich cream. Neck, breast: Buffy grey with blackish centre stripe. Upperparts: Barred black, white and grey cryptic plumage. Underparts: Pale fawn. Immature: Blue eyeline. Hatchlings: Grey-brown with black and brown speckles to upperparts. Chicks run on hatching and fend for themselves.

Wary, freezes when disturbed then quietly moves away. Rarely flies. Heavy flapping wingbeats. Pairs often seen at their large sandy incubation mound. Dust-bathes. Roosts low in trees. Territorial area 40–70 hectares.

Habitat: Dry inland mallee, acacia scrub, open dry woodlands in semi-arid zones.

Feeding: Fruit, flowers, ants, lerp and seeds, particularly acacia and cassia seeds.

Voice: Territorial booming call. Alarm grunt and soft clucking contact call.

Status: Vulnerable to critically endangered.

Breeding: Sept.–Apr. Incubation mound 2–5m dia., 1–1.5m high / 5–30 eggs. Male digs nest mound to keep temperature of egg chamber at 33°C.

Little Desert NP, Vic.

Left and right: Malleefowl 'working' the nest mound, right showing breast stripe.
Below: Typical Malleefowl mound.

Stubble Quail

Coturnix pectoralis

Endemic

180–190mm

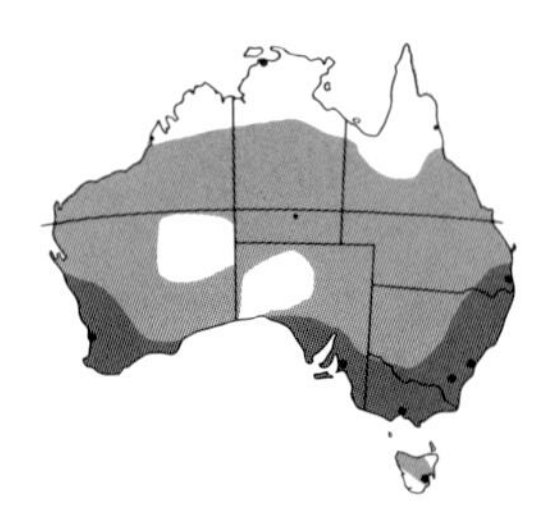

Plump, round-winged bird with a stumpy tail. Upperparts: Heavily mottled buff, brown, black. Crown: Streaky grey-white with long white eyebrow. Bill: Grey. Eyes: Red. Face, chin, throat: Orange-tan. Breast: Black centre patch. Underparts: White, strongly streaked black.
Female: Throat: White. Underparts: Paler, finely speckled breast, flanks. Lacks black breast patch.
Immature: Duller.

Usually runs quickly but will occasionally fly low and drop quickly into cover. Wings whirring on take-off.

Habitat: Dense areas of dry open grasslands, swamp margins, crop stubble, spinifex and saltbush shrubland.

Feeding: Weed seeds, spilt grain in crop stubble and insects.

Voice: Clear, piping 3-note whistle. Deep purring.

Status: Common and highly nomadic, following ripening of seeds.

Breeding: Oct.–Feb. Year-round in north in periods of continuous rain. Scrape in the ground, grass-lined nest, under cover / 7–14 eggs.

Left: Male. Right: Female.

Brown Quail

Coturnix ypsilophora

180–200mm

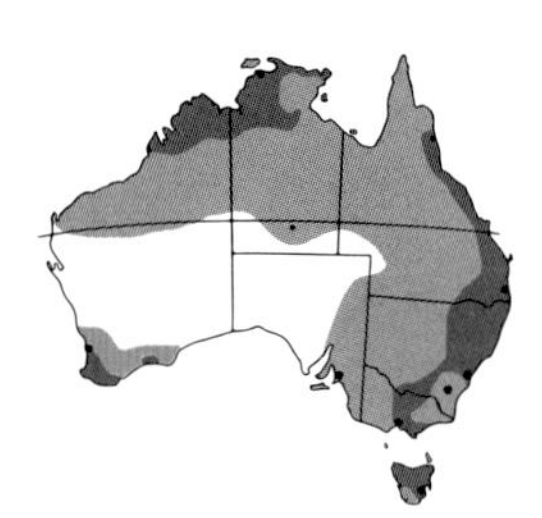

Largest native quail. Overall plumage varies from buff-brown to rufous-brown. Upperparts: Finely streaked silvery white. Underparts: Brownish with fine white-grey V-shaped barring. Face: Plain with dark ear-spot.
Nominate race in Tas. Larger with yellow eyes.
Race *australis*, mainland. Red eyes.
Female: Both races, less rufous, heavier striations.
Immature: Duller.

Usually seen in pairs or in a covey of 10–30 birds. Explosive whistling take-off flight that is short and fast before diving into cover.

Habitat: Dense areas of damp grasslands, tall tropical grasslands, swampy coastal heaths, melaleuca and banksia thickets, spinifex savannah.

Feeding: Seeds and insects.

Voice: Loud, rising whistling.

Status: Common.

Breeding: Aug.–Mar. in south. Oct.–May in north. Nest is a grass-lined scrape, in dense cover and near water / 7–11 eggs.

Left: Female. Right: Male.

King Quail

Excalfactoria chinensis

120–150mm

Distinguishing deep blue-grey chest, body and sides of head and yellow legs. Eyes: Red. Lores to upper throat: Black and white pattern. Belly, vent: Deep chestnut.
Female: Rufous face, buff throat. Upperparts: Dark brown, pale streaks. Eyes: Brown. Underparts: Buff, black barring.
Immature: Similar to female, spotted below.

Usually seen in pairs or coveys of 5–60 birds. Small, shy and elusive. Feeble flight.

Habitat: Tropical grasslands, swampy heaths, weedy areas, crop stubble, forest fringes.

Feeding: Seasonal grasses, seeds and insects.

Voice: High-pitched, monotonous, 3 whistled notes, usually near sunset.

Status: Uncommon. Patchy distribution. Discovered in Kimberley area in 1974.

Breeding: Sept.–Mar. in south, most months in north. Nests on the ground in dense swampy areas and grasslands / 4–9 eggs.

Great Sandy NP, Qld.

Left: Male. Right: Female with chick.

2

Family Anseranatidae, Anatidae, Podicipedidae, Spheniscidae

Magpie Goose

Geese

Swans

Ducks

Grebes

Penguins

A grouping of waterbirds.

Family Anseranatidae includes the Magpie Goose, the sole member of this ancient family. It is well adapted to the cycle of wet and dry seasons of Australia's coastal wetlands. It is unusual in that it has partially webbed feet and strongly clawed toes and the ability to fly at all times, even with progressively moulting flight feathers. The distinctive knob on its head increases in size, indicating the bird's age. Once abundant in the south of their range, they were on the brink of extinction due to over-hunting and drainage of wetlands. They are now fully protected.

Family Anatidae includes swans, geese and ducks. This family group all have bills adapted for feeding on aquatic plant matter and filtering small organisms and plant matter from pond water.

The Black Swan is endemic to Australia and the only all-black swan in the world. Although highly mobile and nomadic, during September to February their flight feathers moult and they unable to fly. During this time they often gather in groups of thousands on freshwater or briny swamps, lakes or estuaries. The Cape Barren Goose is unique to southern Australia. Once hunted to near extinction, they are now protected but remain one of the world's rarest geese. Australia's two native pygmy-geese species are tiny perching ducks.

Ducks are divided into dabbling ducks, which feed on the surface or tip headfirst in shallow water to feed on aquatic vegetation that is sieved through a specialised bill fringed with fine lamellae (grooves), and diving ducks, which feed on the bottom of ponds. Most ducks also forage on land for seeds and insects. Adapted to periodic drought, most ducks can fly great distances in search of suitable habitat and can breed rapidly in response to floodwaters.

Family Podicipedidae includes grebes. These aquatic birds are rarely seen on land. They dive for prey and swim long distances underwater, often staying under for up to a minute. They rarely fly, except during times of drought, when they can undertake long migrations, usually at night, in search of suitable habitat. During the breeding season grebes eat their feathers and feed them to their young, apparently to block the young bird's stomach and prevent sharp fish bones from entering the intestine.

Family Spheniscidae includes penguins. Penguins have evolved from small petrel-like ancestors. Their wings developed into flippers for 'flying' underwater and their plumage into a scale-like covering over a thick waterproof down that protects them from cold. Penguins moult each year, and some Antarctic species can be found on southern Australian beaches during the moulting phase. Moulting needs to be completed on shore as the bird's plumage is not waterproof during this phase. Birds also fast for about three weeks over this period, before returning to the sea to feed and rebuild their condition.

Australia's sole resident penguin, the Little Penguin, is the smallest of all penguins. It commonly nests in burrows and approaches the nest in groups, after dark, for protection against aerial predators.

Opposite: Magpie Geese, Mamukala Wetlands, NT.

Magpie Goose

Anseranas semipalmata

Male: 750–920mm
Female: 710–815mm

Sexes similar. Overall black and white bird with pink facial skin and a distinctive knob that becomes larger with age. Distinctive, orange, partially webbed feet with large clawed toes and long orange legs. Bill: Slender, hooked, pink. Eyes: Brown. Neck, wings, tail, thighs: Black. Underparts and back: White.
Female: smaller with smaller 'knob'. Immature: Similar, but mottled brown-grey to white.

Usually seen in large flocks of several thousand flying with neck outstretched and strong, laboured flight, with pronounced black 'fingered' wing tips. Roosts in tall vegetation.

Habitat: Freshwater swamps, foodplains, margins of rivers, lagoons, billabongs.

Feeding: Aquatic vegetation, wild grasses including Bulkuru Sedge and Wild Rice. Uses strong, clawed feet to dig for corms.

Voice: Loud, repetitive honking.

Status: Locally abundant. Nomadic.

Breeding: Wet season. Roughly cup-shaped nest, floating over deep water / 6–9 eggs.

Kakadu NP, NT.

Left: Male (foreground) and female. Right: Male in flight.

Cape Barren Goose

Cereopsis novaehollandiae

Endemic

750–1000mm

Sexes alike. Overall pale grey with black spot markings to wing coverts and shoulders. Crown: White. Bill: Short, black with greenish yellow cere that covers the base of the bill. Eyes: Hazel. Legs: Pink. Feet: Black. Immature: Duller with paler cere.

Commonly seen grazing grasslands on offshore islands and the adjacent mainland. Noted for belligerent defence of the nest. Flight is strong with rapid wingbeats and gliding. Pairs form a permanent bond.

Habitat: Offshore islands with tussock grass, heathlands, swamp edges, pasture.

Feeding: Herbivorous, foraging on heathland grasses, herbage, seeds, including Blue Tussock Grass and the fruits of Common Boobialla. Able to drink brackish and salt water.

Voice: Grunting, honking.

Status: Locally, moderately common. Introduced to Kangaroo Is., SA.

Breeding: May–Aug. Cup-shaped nest lined with down, often in coastal saltbush. Both parents tend the chicks / 4–5 eggs.

Eyre Peninsula, SA.

Left: Group grazing. Right: Rear view of adult.

Black Swan

Cygnus atratus

Endemic

1200–1300mm

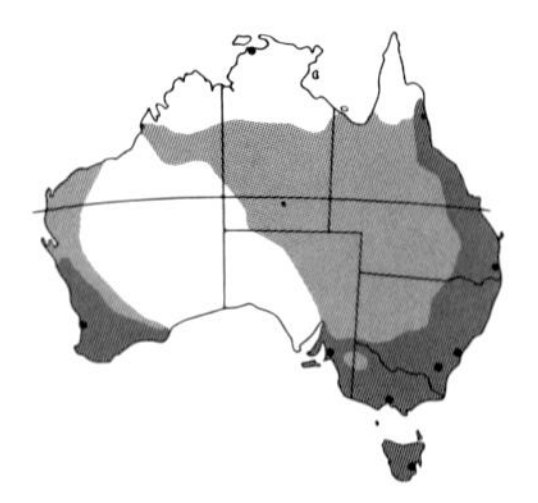

Sexes similar. Overall black with white outer flight feathers that are visible in flight. Bill: Red with white bar near tip. Eyes: Red. Legs: Dark grey.
Female: Smaller with paler bill and eyes. Immature: Grey-brown. Eyes: Dark.

Seen in large flocks of thousands after the breeding season. May fly long distances, usually at night in 'V' formation with necks outstretched. Pairs form a permanent bond.

Habitat: Fresh or brackish swamps or shallow open water where they can reach the bottom to feed.

Feeding: Herbivorous. Aquatic vegetation and surface filter-feeding. Uses long neck and up-ends to reach deeper vegetation.

Voice: High-pitched musical bugle notes. Often at night during flight.

Status: Common. Usually sedentary, but nomadic when food is scarce.

Breeding: Feb.–Sept. Nest is a mound of vegetation / 4–6 eggs.

Left: Black -Swans feeding. Right: Adult and juvenile.

Spotted Whistling-duck

Dendrocygna guttata

430–500mm

New Guinea species that has recently colonised the far north of Australia. Related to the Wandering and Plumed Whistling-ducks, but lacks flank plumes. Sexes alike. Overall dark chocolate brown with buff scalloped plumage and dark crown. Face and throat: Grey with dark grey eye patch. Flanks and undertail: Chestnut to buff, boldly spotted white. Bill: Dark red. Legs, feet: Pinkish brown.

Gregarious. Small flocks or family groups. Wings 'whistle' in flight.

Habitat: Freshwater wetlands, marshes, mangroves, lagoons.

Feeding: Mostly nocturnal. Grass seeds, aquatic vegetable matter and small snails.

Voice: Low whistling sounds. Nasal 'wheeow' often repeated in quick succession and piping 'pi-pi'.

Status: Small populations regularly breed at Weipa and Coen, Qld. Increasing in numbers and extending their range southward, with regular sightings as far south as Cairns, Qld, and Darwin and Melville Is., NT.

Breeding: From Sept. Nest in a tall tree hollow / 10–12 eggs.

Opposite: Small flock at Centenary Lakes, Cairns, Qld.

Plumed Whistling-duck

Dendrocygna eytoni

Endemic

Male: 435–615mm
Female: 415–560mm

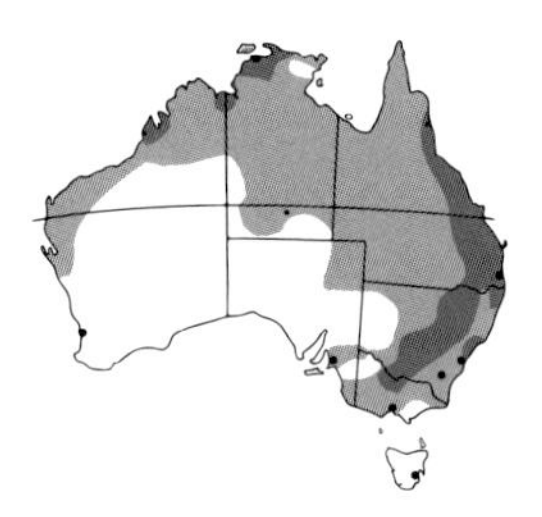

Sexes alike. Tall duck with distinctive long, lanceolate, off-white plumes edged in black extending over back. Sides are rich chestnut barred black. Crown, nape, hind neck: Brown-buff. Back: Brown, scalloped black. Wings: Brown. Underwings: Pale brown. Tail, rump: Dark brown. Bill: Pink, mottled black. Eyes: Orange-yellow. Feet, legs: Pink. Immature: Duller, paler.

Usually seen in large flocks. Rarely dives or swims. Wings 'whistle' in flight. Pairs are monogamous.

Habitat: Wetland edges, tropical grasslands near water with short grasses for grazing.

Feeding: Mostly after dusk on grazing grassland and wetland edges or dabbles on water surface. Often flies long distances at night to feed. Communal, daytime roosting camps on wetland banks.

Voice: Shrill, high-pitched whistle, twittering.

Status: Locally abundant. Nomadic, moving inland in wet season.

Breeding: Feb.–Apr. Nest is a grass-lined scrape on ground in long grass / 8–14 eggs.

Kakadu NP, NT.

Left: Adults. Right: Adults in flight.

Wandering Whistling-duck

Dendrocygna arcuata

Size: 550–575mm

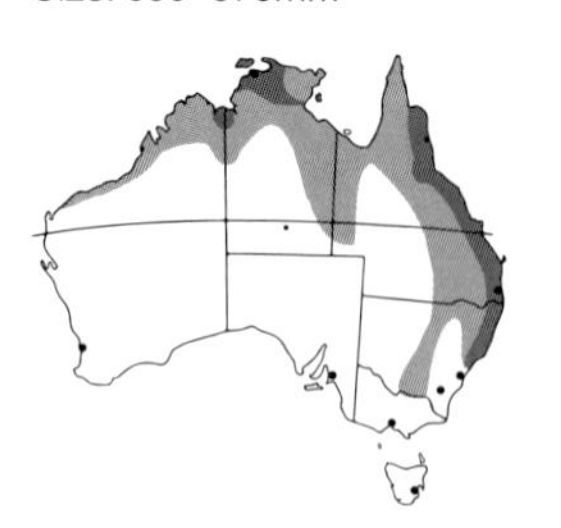

Sexes alike. Overall rich chestnut plumage. Face, neck, upper breast: Buff. Crown and down hind neck: Black stripe. Flanks: Distinctive, elongated, chestnut-edged, off-white plumes. Wing coverts: Chestnut. Flight feathers: Dark brown. Bill, legs: Black. Eyes: Reddish brown. Immature: Duller.

Usually seen in large flocks, often with Plumed Whistling-ducks, either on or close to water. Wings 'whistle' in flight.

Habitat: Deep tropical lagoons, wetlands, swamps, sewerage farms.

Feeding: Deep-diving for aquatic grasses, waterlily bulbs, submerged plants, seeds and insects.

Voice: High-pitched, loud whistling, twittering, especially in flight.

Status: Locally abundant. Nomadic, usually in flocks that move inland in the wet season.

Breeding: Rain dependent, usually Jan.–July. Nests in long grass / 6–15 eggs.

Kakadu NP, NT.

Left: Plumage detail. Right: Adult pair.

Musk Duck

Biziura lobata

Endemic

Male: 600–730mm
Female: 470–600mm

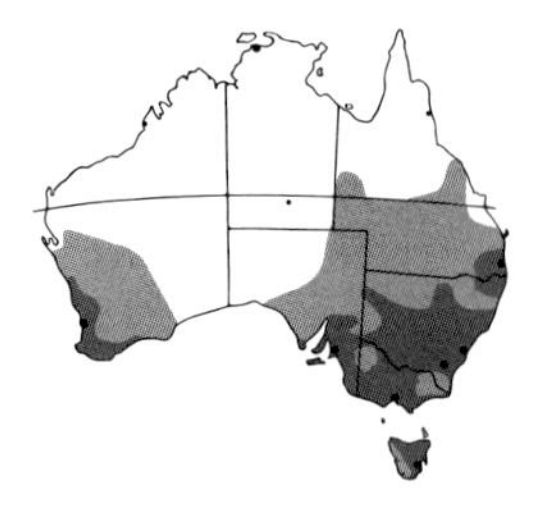

Overall sooty brown, finely barred paler brown and finely spotted head. Distinctive pendulous lobe under a broad, black bill. Tail: Stiff feathers, erect when raised. Eyes: Dark brown. Legs, feet: Dark grey.
Female: Similar, smaller, lacks lobe under bill. Immature: Similar to female.

In courtship display, the lobe is inflated, together with raising the tail feathers into a fan and kicking up 2m sprays of water. Rarely seen on land. Dives for cover when alarmed. Flight is laboured with 'crash' landings. Emits a strong musk odour from rump gland to attract female.

Habitat: Large and permanent lakes, swamps with dense reed beds and open water.

Feeding: Diving deep, remaining submerged for at least a minute. Takes insects, small aquatic creatures and occasionally aquatic plants and seeds.

Voice: Rumbling grunts and whistles during flamboyant water courtship display. Usually silent.

Status: Uncommon. Nomadic.

Breeding: Sept.–Oct. Nests in a clump of reeds in water or low-hanging tea-tree branches / 1–3 eggs.

Left: Female. Right: Male in courtship display.

Freckled Duck

Stictonetta naevosa

Endemic

Male: 520–590mm
Female: 480–540mm

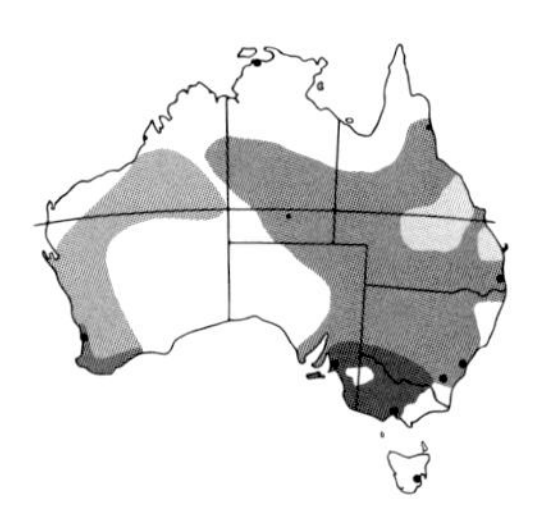

Dabbling duck, overall dark brown with finely freckled white-buff plumage. Head: Darker with distinctive small peak. Bill: Mid-grey, dish-shaped, upward-hooked tip with a patch at the base that turns bright red during the breeding season. Wings: Dark brown. Underwings: Dull white with dark primaries. Non-breeding: Lacks red patch on base of bill.
Female: Paler, finely freckled white-buff. Immature: Similar to female.

Usually seen in family groups. Roosts in deep cover during the day, feeding at dusk.

Habitat: Prefers permanent freshwater swamps and creeks with bulrushes or tea-tree cover.

Feeding: In shallow water at dawn and dusk, taking crustaceans, zooplankton, algae, seeds and aquatic plants.

Voice: Seldom heard, grunts and fluting whistles.

Status: Rare. Usually sedentary. Patchy distribution across Qld, NT, WA, SA. Endangered Vic. Vulnerable SA., NSW.

Breeding: Sept.–Nov. or depending on rainfall. Shallow cup nest on the ground under shrubby cover / 5–14 eggs.

Left: Breeding male. Right: Non-breeding.

Pink-eared Duck

Malacorhynchus membranaceus

Endemic

Male: 380–450mm
Female: 400mm

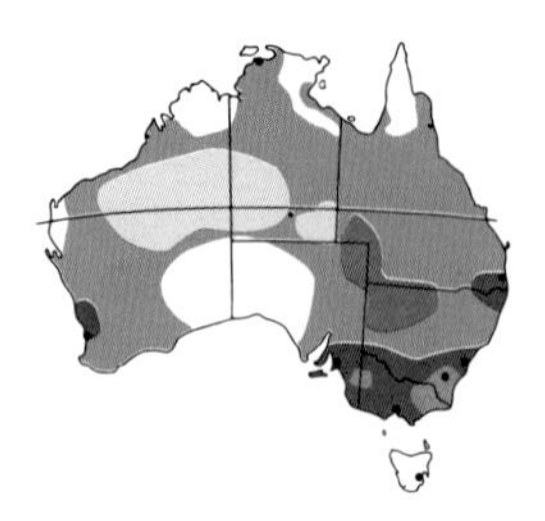

Sexes alike. Small duck with a large, grey, spatulate bill and distinctive strongly barred brown flanks. Face: White, finely barred with large brown eye patch and small deep pink ear patch. Crown, hind neck, back: Brown. Upperwings: Brown with white trailing edge. Rump, tail tip: White. Underwings: Light grey, finely barred brown. Feet, legs: Mid-grey.
Immature: Paler. Lacks pink ear patch.

Often seen on open shallow water in huge flocks.

Habitat: Prefers shallow floodwaters, fresh or brackish water, sewerage ponds, wetlands and coastal mangroves.

Feeding: Paddling in shallows stirring up aquatic organisms, algae, plankton and floating seeds, which are sieved through submerged bill.

Voice: Musical chirruping notes.

Status: Uncommon. Nomadic following rainfall and floods.

Breeding: Anytime, depending on rainfall and height of water. Nests in a shrub or tree hollow / 5–8 eggs.

Werribee TP, Vic.

Left: Plumage detail. Right: Adult preening.

Radjah Shelduck

Tadorna radjah

500–550mm

Sexes similar. Head, neck, breast, belly: White with chestnut upper-breast band. Upper back: Chestnut. Lower back, rump, tail: Black. Upperwing coverts: White. Underwings: Dark tipped, white covert panels and deep green speculum. Eyes: White. Bill, legs: Pink.
Female: Narrower upper-breast band. Immature: Duller with incomplete breast band.

Usually in pairs or small family groups roosting in cover on the water's edge during the day. Flight is slow. Pairs are monogamous.

Habitat: Brackish waterways, tropical coastal wetlands, mudflats, saltmarshes, mangrove and paperbark swamps.

Feeding: Along water's edge, sieving for worms, molluscs, insects, algae and sedges.

Voice: Harsh rattling, whistling commonly in flight.

Status: Moderately common. Nomadic. Tame, vulnerable to shooting. Decimated by hunting in the east.

Breeding: May–June. Nests in a tree hollow close to water / 6–12 eggs.

Kakadu NP, NT.

Left: Radjah Shelduck. Above Right: Male in flight. Below Right: Female in flight.

Australian Shelduck

Tadorna tadornoides

Endemic

590–720mm

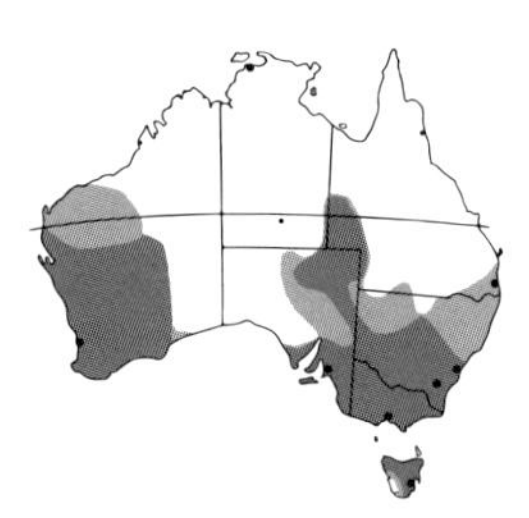

Bright cinnamon breast and white neck ring. Head: Sooty black with green sheen. Upperparts: Black. Wings: Overall black with rufous and glossy green speculum and white upperwing coverts. Underwing: White, broadly edged black. Non-breeding: Breast: Yellowish brown. Neck ring: Less defined. Female: Similar, with chestnut breast, white around eyes and bill, and narrower white neck ring. Immature: Duller. Lacks neck ring. Head: White flecked.

Wary, usually seen well away from the shore. After breeding, during moulting phase, thousands of birds flock together forming rafts. Strong, direct flight in long lines or 'V' formation.

Habitat: Marshes, pastures, brackish or freshwater shallow lakes, open woodlands, tidal flats, saltmarshes near fresh water.

Feeding: Resting at water's edge by day, feeding in the late afternoon. Aquatic plants, grasses, insects and molluscs. Rarely dives.

Voice: Loud, deep honking, commonly in flight. Female: Higher-pitched buzzing call.

Status: Locally abundant.

Breeding: Commonly July–Aug. Nest is usually in a hollow limb in tall tree 2–25m high, occasionally on ground / 5–15 eggs.

Left: Male (foreground) and female. Right: Female.

Australian Wood Duck

Chenonetta jubata

Endemic

Male: 420–590mm
Female: 420–550mm

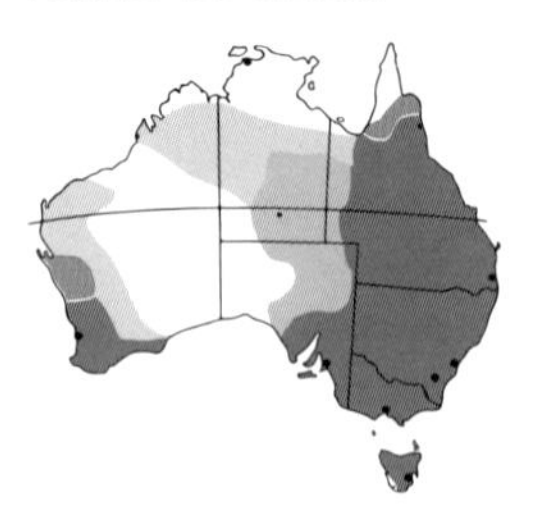

Distinctive black feather 'mane'. Head, neck: Dark brown. Breast: Mottled brown-grey. Upperparts: Grey, finely barred with 2 stripes along back. Lower belly, undertail: Black. Wings: Grey with iridescent green-black speculum and white rear edge of secondaries. Bill: Stubby, olive-brown. Eyes: Brown. Legs, feet: Dark grey with claws for climbing trees.
Non-breeding: Some white feathers around eye, chin, belly. Female: Head: Greenish brown, white line above and below eye. Body: Mottled brown. Underbelly, undertail: White. Immature: Similar to female, paler.

Best adapted for walking. Usually seen grazing, often at night.

Habitat: Margins of swamps, lagoons and rivers.

Feeding: Generalised grazers on grasses and herbs and occasionally insects.

Voice: Female: Drawn-out nasal 'mew'. Male: Higher pitched, shorter note.

Status: Locally abundant. Nomadic.

Breeding: Sept.–Nov. in south. Inland after heavy rain. Nests in hollow tree limbs / 9–12 eggs.

Left: Breeding male (foreground) with chicks and female. Right: Breeding male.

Blue-billed Duck

Oxyura australis

Endemic

370–440mm

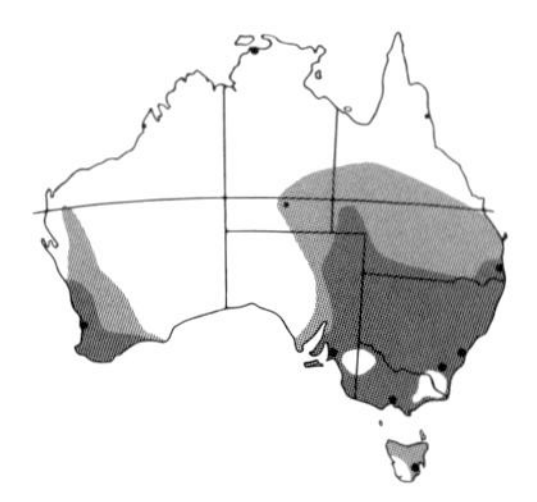

Bright blue bill when breeding. Head, neck: Glossy black. Upper breast, flanks, back: Rich chestnut. Belly: Brown, flecked black. Undertail: Black. Wings: Dark brown. Eyes: Dark brown. Non-breeding: Grey plumage, feathers edged pale brown. Bill: Slate grey. Head, chin: Speckled black-grey.
Female: Blackish brown, finely barred light brown. Chin, throat: Brown, speckled black. Breast, belly: Mottled light brown–black. Bill, feet: Grey brown. Immature: Similar to female. Paler with grey-green bill.

Completely aquatic diving duck almost unable to walk on land. Remains well out from shore. If alarmed, usually dives, resurfacing up to 50m away.

Habitat: Deep, permanent, heavily vegetated freshwater swamps. Open lakes in winter.

Feeding: Bottom-feeding for vegetable matter and insects.

Voice: Mostly silent. Male: Rattling sounds in display. Female: Soft quacking.

Status: Moderately common.

Breeding: Sept.–Mar. Cup-shaped nest in dense reeds / 4–12 eggs.

Werribee TP, Vic. Bool Lagoon, SA.

Left: Male. Right: Female.

Cotton Pygmy-goose

Nettapus coromandelianus

Male: 350–380mm
Female: 330–380mm

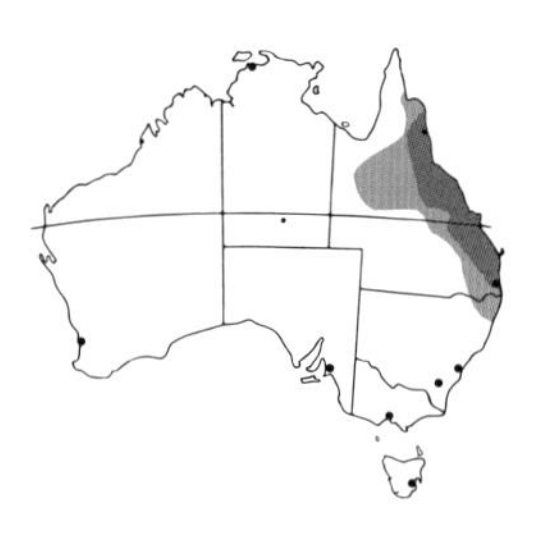

Race *albipennis* present in Australia. Small perching duck with a goose-like bill, white face, head, neck and underparts. Upperparts, crown: Dark glossy green. Breast: White with tapering dark breast band. Underwing: Green with white wing bar, white trailing edge and black tips. Flanks: White with fine buff barring. Eyes: Red. Feet: Olive green. Bill: Short, stubby.

Female: Duller, dark stripe through eyes and a white eyebrow. Immature: Similar to female, browner.

Completely aquatic. Seldom leaving the water, it seeks cover in waterlilies by day and feeds at night. Flight is low.

Habitat: Deep, permanent, freshwater lagoons, lakes, swamps with waterlilies.

Feeding: Cruises among waterlilies feeding on seeds, other aquatic plants and weeds and sometimes insects.

Voice: Loud staccato cackles and quacks. Trumpets in flight.

Status: Uncommon. Rare in south of range. Endangered NSW.

Breeding: Sept.–Nov. in south. Jan.–Mar. in north. Nests in a tree hollow up to 10m high / 6–9 eggs.

Townsville Common CP, Qld.

Left: Male (centre) with females. Right: Male.

Green Pygmy-goose

Nettapus pulchellus

340–380mm

Glossy green-black head, neck and upperparts. Face: Bright white cheek patch and small white patch forward of the eye. Eyes: Dark brown. Bill: Short, green-grey above, pink below and tip. Breast, belly, flanks: Grey, finely barred green-brown. Upperwings: Grey green with iridescent green highlights and white panel through secondaries. Legs, feet: Green-grey.
Female: Similar, paler, white eyebrow, less green on neck. Immature: Similar to female.

Usually in pairs or small groups. Aggressive. Almost completely aquatic. Flight is low, swift, direct, with whistling wingbeats.

Habitat: Tropical lakes, lagoons with aquatic vegetation including Blue Waterlily.

Feeding: Primary food is seeds and buds of waterlilies. Dives for pond weeds and aquatic plants.

Voice: Musical trills. Whistling.

Status: Common, sedentary.

Breeding: Dec.–Mar. Nest is in a hollow, up to 10m high in a tree growing in water or swampy vegetation / 8–12 eggs.

Kununurra, WA. Cairns, Qld.

Opposite: Male (background) and female.

Australasian Shoveler

Anas rhynchotis

460–530mm

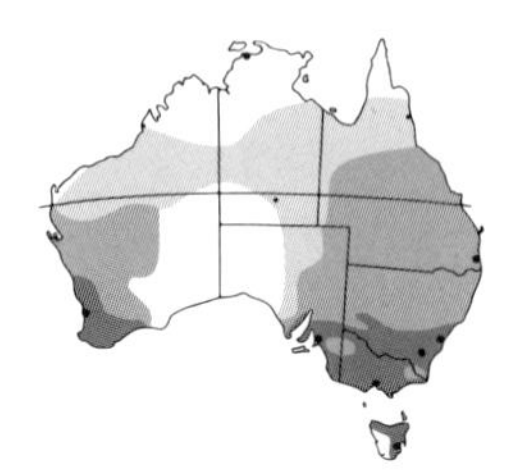

Dark blue-grey head with green sheen and vertical white crescent between the yellow eyes and the large blue-grey 'shovel-like' bill. Neck, breast: Speckled dark brown. Flanks: Chestnut with large white patch. Wings: Blue-grey, black-white scapulars; dark green speculum edged above with tapering white band. Underwings: Dark brown, white coverts. Legs, feet: Orange. Non-breeding: Like female. Legs, feet: Orange-brown.
Female: Duller. Lacks face crescent. Crown: Dark brown. Nape, upperparts: Scalloped pale brown. Underparts: Pale chestnut, mottled brown. Eyes: Brown. Legs: Grey-green.

Immature: Like female.

Often seen well out from shore.

Habitat: Thickly vegetated fresh or saline deep lakes, ponds, farm dams, tidal flats, swamps.

Feeding: Dabbler, surface feeder, sifting water or mud for aquatic animals, zooplankton.

Voice: Usually silent. Male: Soft, courting 'took-took'. Soft chattering in flight.

Status: Moderately common. Usually sedentary.

Breeding: Aug.–Dec. or after floods. Nests in tall grass near water's edge / 9–11 eggs.

Left: Male. Right: Female

Grey Teal

Anas gracilis

Male: 410–480mm
Female: 370–440mm

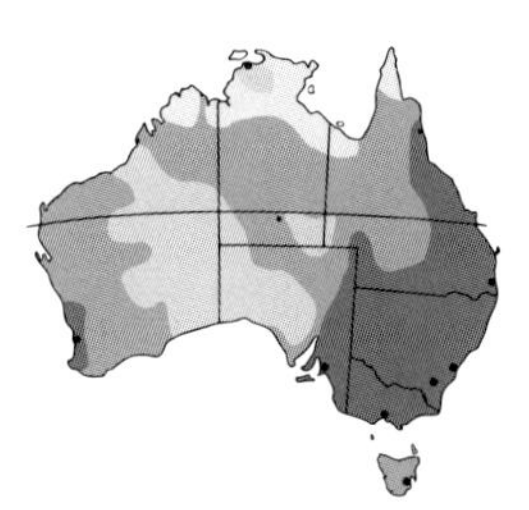

Sexes alike. Small duck, mostly mottled grey-brown. Each body feather is buff-edged except for rump. Face, chin, throat: Whitish. Upperwings: Dark-brown, thin white bar and iridescent green speculum. Underwings: Grey-brown with white wedge on lower coverts seen in flight. Bill: Dark green. Eyes: Red.
Immature: Paler.

Dabbling duck well adapted to Australia's erratic rainfall.

Habitat: Sheltered fresh or brackish water, or salt water, tree-lined waterways, inland swamps.

Feeding: Seeds from aquatic plants and grasses. Aquatic insects, crustaceans.

Voice: Female: Loud repetitive quacks. Males: Sharp whistles, low grunts.

Status: Abundant. Nomadic. Migrates to coastal areas as inland wetlands dry out in summer.

Breeding: July–Dec. or anytime in favourable conditions. Nests on the ground, in rock crevices or tree hollows, nest boxes / 4–14 eggs.

Opposite: Adult with chicks.

Chestnut Teal

Anas castanea

Endemic

Male: 410–480mm
Female: 380–460mm

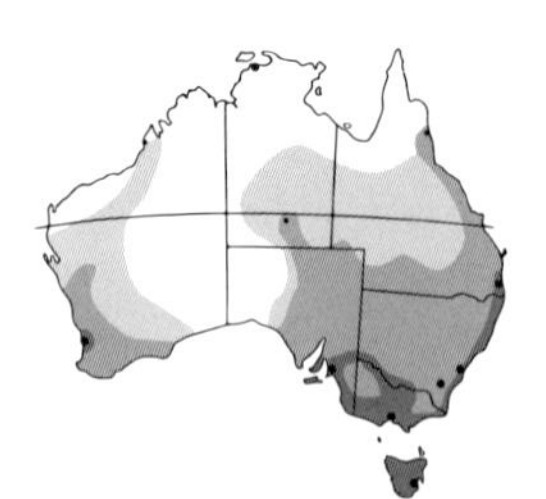

Glossy iridescent green head and neck. Underparts, breast: Chestnut with dusky brown central marking to each feather. Upperparts: Dark brown. Wings: Glossy green to purple speculum edged white above, narrower white band below. Rump, undertail: Black. Flanks: Prominent white patch. Underwings: Dark grey-brown with white wedge shape to lower coverts. Bill: Grey. Eyes: Red. Non-breeding: Resembles female.
Female: Mottled dark brown with grey-edged feathers. Crown, face: Dark streaked. Underparts: Paler brown. Immature: Speckled. Similar to female.

Usually in pairs or small groups.

Habitat: Brackish or fresh water, lagoons, saltwater estuaries, creeks with mangroves.

Feeding: Dabbling duck. Aquatic plants and insects.

Voice: Female: Loud, descending quacks, very similar to Grey Teal. Male: Low grunts, whistles.

Status: Locally common. Most abundant in Tas. Mostly sedentary.

Breeding: June–Dec. Nests in rushes or in tree hollows near water, nest boxes / 7–10 eggs.

Left: Male (foreground) with female and chicks. Right: Female nesting in a tree hollow.

Hardhead

Aythya australis

Male: 415–490mm
Female: 420–545mm

Mostly dark brown duck with a distinctive white eye. Throat, neck, upperparts: Mottled rich dark brown. Breast: Dark rufous brown. Lower breast, belly: White, mottled brown. Undertail coverts: White. Wings: Brown with broad, white band seen in flight. Underwings: White, thin white edge. Bill: Black, slate blue tip.
Female: Paler. Eyes: Brown.
Immature: Similar to female, yellowish brown plumage.

Australia's only true diving duck. Stays well out from shore, usually in pairs or large groups. Rarely comes ashore. Flight is fast. Rises vertically with loudly whirring wings if alarmed.

Habitat: Large, deep-water swamps, lagoons and flooding inland rivers.

Feeding: Dives deep for small aquatic creatures and some plant material. May travel 30–40m before resurfacing.

Voice: Wheezy whistling. Female: Soft croaking. Usually silent.

Status: Common. Usually sedentary where permanent deep swamps, streams occur.

Breeding: Oct.–Dec. Nests in dense reeds or tea-tree growing in 1m-deep water / 9–12 eggs.

Left: Male. Right: Female.

Pacific Black Duck

Anas superciliosa

Male: 500–610mm
Female: 470–580mm

Sexes similar. Overall dark brown with buff-edged feathers. Head, face: Buff with distinctive blackish stripe on crown, through eye, from under chin to cheeks. Eyes: Brown. Bill: Leaden grey. Upperwings: Dark brown with iridescent purple to dark green speculum. Underwings: White with brown trailing edge, tips. Legs: Grey.
Female: Generally smaller.
Immature: Upperparts: Black. Underparts: Yellow.

Most abundant of Australia's ducks, usually in pairs or flocks. Flight is swift, direct.

Habitat: Open water, wetlands, urban parks, dams, lagoons.

Feeding: Dabbling duck. Seeds from aquatic or ditch grasses. Aquatic insects, crustaceans.

Voice: Female: Quacks loudly. Male: 'Peeps' in courtship.

Status: Abundant, sedentary in permanent water. Tame in urban parks. Nomadic following floods or drought. Some northward migration in winter.

Breeding: Mar.–May in north. July–Oct. in south. Nests on ground or in a tree hollow / 6–16 eggs.

Left: Displaying iridescent speculum.
Right: Breeding pair.

Australasian Grebe

Tachybaptus novaehollandiae

250–270mm

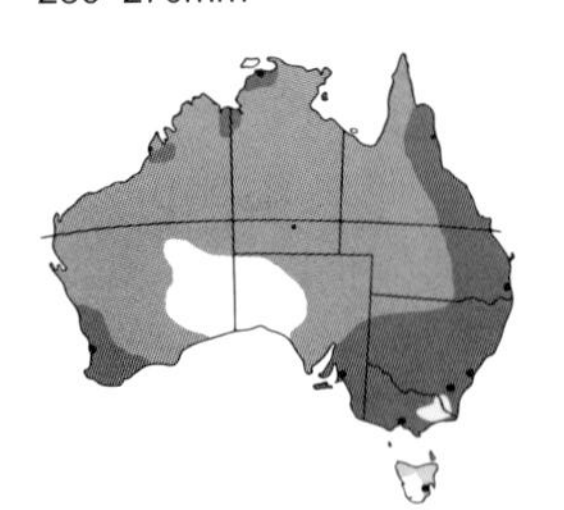

Sexes similar. The smallest grebe, noted for its 2 distinct plumage phases. Breeding: Dark brown upperparts with pale chestnut flanks. Head, throat: Glossy black with prominent chestnut patch from behind the eyes to upper neck and bright yellow facial patch at gape. Eyes: Bright orange-yellow. Bill: Black, tipped yellow. Underparts: Silver-grey.
Non-breeding: Upperparts: Duller grey-brown. Throat, face below eye, foreneck: Whitish. Bill: Whitish.
Female: Smaller, with shorter bill. Immature: Similar to non-breeding adult.

Solitary and shy. Dives deep and swims away when disturbed. Poor flyer. Travels short distances with rapid wingbeats, usually at night.

Habitat: Sheltered fresh water with emergent vegetation.

Feeding: Small fish, crustaceans. Dives deep, chases prey.

Voice: Shrill rattling trill.

Status: Moderately common. Usually sedentary, southern birds may migrate north for winter, often at night.

Breeding: Sept.–Mar. Floating nest anchored in rush cover / 4–7 eggs.

Left: Breeding plumage, at nest.
Right: Non-breeding plumage.

Hoary-headed Grebe

Poliocephalus poliocephalus

Endemic

270–300mm

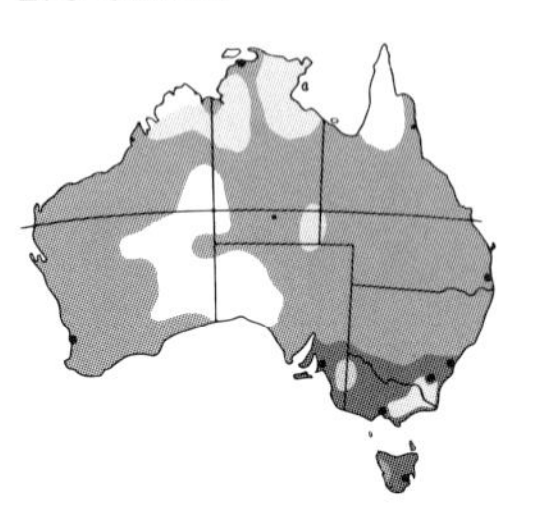

Sexes alike. Named for the long silvery breeding plumage on its black head and nape. Distinctive short pointed bill. Upperparts: Mid-grey. Underparts: White. Upperbreast: Pale buff. Eyes: Yellow, female eyes paler. Non-breeding: Dull grey crown. Lower face, throat, and neck: Off-white. Bill: Pale. Immature: Similar to non-breeding adults.

Wary. Usually in pairs or groups. Unlike other grebes, they most often take flight when disturbed, flying close to the water surface.

Habitat: Open freshwater ponds with patches of reed cover.

Feeding: Dives for aquatic insects, crustaceans and plankton.

Voice: Mostly silent. Occasional short, guttural note.

Status: Locally abundant, sedentary or nomadic. Established in NZ.

Breeding: Nov.–Jan. or after good rain. Floating nest in shallow water in emergent plants; several pairs may nest communally / 5–6 eggs.

Werribee TP, Vic.

Left: Non-breeding. Right: Breeding.

Great Crested Grebe

Podiceps cristatus

500mm

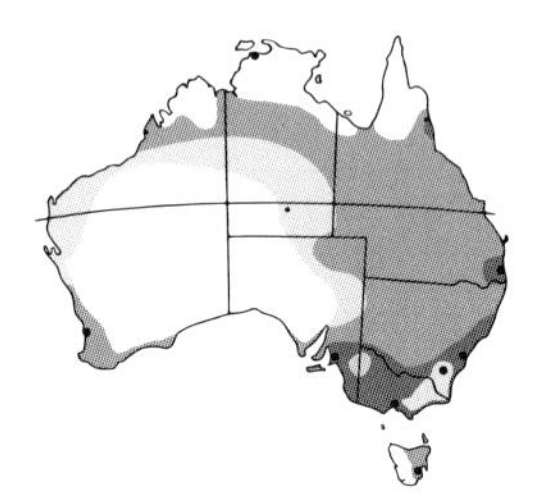

Sexes similar. Largest grebe, with a long, slender neck, distinctive upper ruff (more pronounced in breeding) and black, chestnut-tipped double crest on each side of the head. The face is sharply divided, white below the eye, black above. Eyes: Red. Bill: Long, brown. Upperparts: Grey-brown. Underparts: Silky white. Underwings: Dark grey with white panels visible in flight. Female: Slightly smaller. Immature: Lacks ruff and crest.

Cosmopolitan bird noted for its elaborate courtship dance, rising from the water, shaking head and tail and carrying a reed. Rapid flight. Shallow wingbeats.

Habitat: Aquatic, deep lakes, swamps. Open fresh or brackish water with emergent vegetation.

Feeding: Dives for fish and crustaceans.

Voice: Rattling and croaking.

Status: Moderately common.

Breeding: Nov.–Mar. Nest is a floating mat of reeds / 3–7 eggs.

Lake Barrine NP, Qld. Hattah-Kulkyne NP, Vic.

Opposite: Adults carry the young on their back then submerge and resurface nearby, leaving chicks to float, thus teaching them how to swim and dive.

Northern Mallard

Anas platyrhynchos

Introduced

Male: 580–680mm
Female: 520–580mm

Large dabbling duck. Iridescent green head, distinctive white neck ring and chestnut, tinged purple breast. Upperparts: Grey-brown. Rump, tail: Black. Wings: Grey-brown with blue-violet iridescent panel. Underparts: White. Bill: Yellow-green. Non-breeding: Dull green bill. Female: Smaller, overall dusky brown streaked, with orange-grey bill and dull eye stripe. Immature: Similar to female. Usually mate for life. Male becomes aggressive around nest.

Habitat: Urban parks, sometimes in the wild. Potentially endangers the Pacific Black Duck through interbreeding.

Feeding: Aquatic or ditch grass seeds, insects, crustaceans.

Voice: Female: Quacks loudly. Male: 'Peeps' in courtship.

Status: Moderately common.

Breeding: July–Dec. Grass bowl nest well hidden on the ground / 8–12 eggs.

Opposite: Male (foreground) and female.

Mute Swan

Cygnus olor

Introduced

1300–1600mm

Sexes alike. Large, white swan with orange-red bill with black base and tip and a knob at the top. Face: Black. Breeding male: Enlarged knob at base of bill. Immature: Grey with grey bill.

Habitat: Urban ponds, lakes.

Feeding: Mostly aquatic plants, occasionally frogs, fish, insects.

Status: Common. Sedentary.

Breeding: Late winter. Bowl-shaped nest of aquatic plant material placed on the ground / 2–5 eggs.

Opposite: Breeding pair feeding.

Little Penguin

Eudyptula minor

300mm

Sexes alike. Upperparts: Deep blue. Underparts: White, silver sheen. Feet: Flesh pink with black sole. Bill: Black.
Moult phase: Duller, rust-tinged.
Immature: Bluer.
Smallest and only penguin to breed in Australia. Roosts and nests in short burrows. Leaves roosting colony before dawn, returning at nightfall. When preening they take a small drop of oil from an oil gland above the tail to keep feathers waterproof.

Habitat: Coastal dunes and offshore islands.

Feeding: Hunts by concentrating small shoals of fish. Swallows prey underwater.

Voice: Sharp yapping at sea. Onshore braying, yaps, grunts.

Status: Moderately common.

Breeding: Aug.–Feb. in the east. Apr.–Dec. in the west. Burrow or crevice nest / 1–3 eggs.

Phillip Is., Vic.

Left: Moult phase. Right: Adult leaving nesting burrow.

Rockhopper Penguin

Eudyptes chrysocome

450–600mm

Sexes similar. Distinguishing bright yellow up-curved eyebrow starting at bill and finishing in drooping, fanned plumes. Upperparts, throat: Blue-black. Bill: Large, red-brown with a fleshy pink membrane around base. Eyes: Bright red. Feet: Pinkish, soles black.
Female: Lacks fleshy membrane.
Immature: Smaller crest. Dull brown bill. Whitish grey throat. Race *moseleyi* most common visitor to southern Australia.

Jumps along on land and jumps feet-first into sea, unlike other penguins that dive.

Habitat: Visits the waters and beaches of southern Australia.

Feeding: Porpoise-like diving. Prefers krill, crustaceans, squid.

Voice: At sea, barks, croaks contact calls. Harsh and noisy in roosting colonies.

Status: Occasional winter visitor.

Breeding: Rocky islands in the Southern Ocean.

Opposite: Entering water feet-first.

Fiordland Penguin

Eudyptes pachyrhynchus

520–700mm

Sexes alike. Distinguished by their broad, yellow, spiky eyebrow which splays and droops down the neck. Upperparts, throat: Blue-black. Cheeks: White flecks. Flippers: Upperpart blue-black with outside white edge; underside is white. Underparts: White. Eyes: Red. Bill: Orange-pink.
Immature: Bill: Grey-brown. Crest: Short. Throat: Whitish.

Timid. Communicates at sea with barking calls and on land uses visual and vocal displays.

Habitat: Southern oceans.

Feeding: On or near the water surface in groups or solitary.

Voice: Guttural, harsh barking, usually quiet.

Status: Uncommon, but regular. Usually immature birds, winter–spring stragglers, seen from southern WA, SA, Tas., Vic., southern NSW.

Breeding: NZ South Is. and offshore islands.

Left: Adults showing plumage.
Right: Displaying peculiar gait.

Macaroni Penguin

Eudyptes chrysolophus

700mm

Sexes similar. Large crested penguin distinguished by its spiky orange feather plumes rising from the centre of its forehead and extending horizontally back towards the nape. Bill: Massive, pink or red-brown, with deep pink skin around gape. Upperparts: Black. Face, throat, underparts: White. Frons: Long golden plumes.
Female: Slightly smaller.
Immature: Duller, shorter plumes.
Most abundant penguin species.

Habitat: Oceans and coast of Antarctica, often seen feeding in southern Australian waters, mostly Tas.

Feeding: Mostly krill and fish.

Voice: Loud, trumpeting.

Status: Non-breeding, uncommon visitor, mostly Tas.

Breeding: Mostly on Heard and Macquarie Is.

Opposite: Group showing long yellow plumage.

3

Family Hydrobatidae, Diomedeidae, Procellariidae, Pelecanoididae

Storm-petrels

Albatross

Petrels

Fulmar

Prions

Shearwaters

Diving-petrels

A group of mostly oceanic seabirds made up of four families.

Family Hydrobatidae contains the smaller storm-petrels. Storm-petrels are noted for their dainty flight, fluttering over waves with feet dangling in the water while picking small morsels of food from the surface. Southern storm-petrels belong to the sub-family Oceanitinae.

Family Diomedeidae contains the Albatross.

Albatross travel enormous distances. They nest on offshore islands and build a large bowl-shaped nest from mud and plant material. Mollymawks are medium-sized albatross, found only in the southern hemisphere, where they are the most common of the albatross. Albatross take up to ten years before reaching maturity and only lay one egg every two years. Their greatest threat is from long-line fishing.

Family Procellariidae is a variable group, containing petrels, fulmars, shearwaters and prions.

Petrels have a highly developed sense of smell, which allows them to locate prey at great distances. They emit a musky odour, and the cumulative musky scent emitted by the breeding colony allows returning birds to locate the colony at night. Petrels are strictly pelagic outside the breeding season.

Giant-petrels and fulmars have prominent nasal tubes and a stiff-winged flight. They excrete salt from a gland above the nasal passage, which desalinates their bodies.They have a foul-smelling and oily stomach fluid that they either feed to chicks; use for sustenance on long journeys; or spit at intruders, causing matting of the intruders' plumage and possible death. They largely feed on carrion and often attack other seabirds, killing them by battering or drowning.

Pelagic gadfly petrels pick prey from the ocean surface, seemingly walking on water. They have a fast-weaving, arching flight and are distinguished from shearwaters by their short nostril tube and narrow bill.

Shearwaters are mid-sized, long-lived, long-winged birds, common in temperate and cold waters. They are pelagic outside the breeding season. Noted for their stiff-winged 'shearing' flight technique, they move across the wave fronts with a minimum of wing movement. Many are long distance migrants. The Short-tailed Shearwater journeys 20,000 nautical miles across the Pacific Ocean to the Aleutian Islands in Alaska then returns to Tasmania, the world's longest migration journey.

Prions are also known as 'whalebirds' for their habit of feeding with whales, and like whales, their bills are edged with filaments (lamellae) for sieving plankton. They are commonly found washed up on the beaches of southern Australia.

Family Pelecanoididae contains the diving-petrels.

Diving-petrels are small auk-like petrels that dive for prey, using their flapping wings underwater in pursuit to depths down to 60m. They fly low over water and can fly straight through waves without interruption to their flight path.

Opposite: Albatross, Giant-petrels and gulls.

Wilson's Storm-petrel

Oceanites oceanicus

180mm
Wingspan: 380–420mm

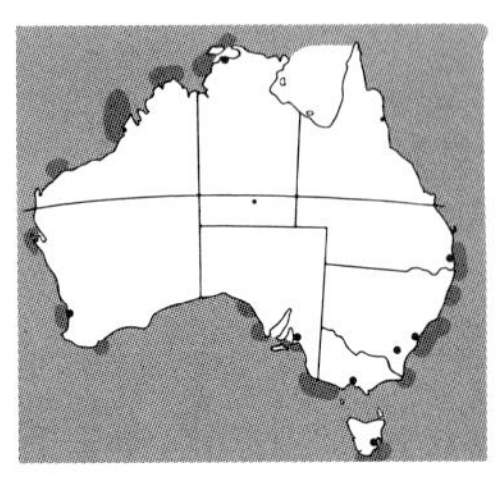

Sexes, immature: Alike. Overall sooty brown with steep, high frons and broad white crescent across rump, spreading to slightly under tail. Upperwings: Pale diagonal bar. Tail: Square-cut. Bill: Short, black with small nasal tube. Chin: Small, white patch. Legs: Long, thin, black. Feet: Black, bright yellow webs.

Flying with legs trailing, rapid wingbeats, short glides or fluttering wings while hovering. Follows trawlers.

Habitat: Common in all Australian oceanic and inshore waters.

Feeding: Runs along the water, flutters and hovers eating small fish, crustaceans, carrion.

Voice: Chattering while feeding.

Status: Moderately common Apr.–June.

Breeding: Antarctic coast.

Green Cape, NSW. Point Lonsdale, Vic.

Left: Surface feeding. Right: Inflight showing trailing legs.

Grey-backed Storm-petrel

Garrodia nereis

170–180mm
Wingspan: 390mm

Sexes, immature: Alike. Black head, with steep, high frons. Underparts: White, sharply defined. Upperparts: Lightly scalloped ash grey with faint diagonal bar to wings. Underwings: White with broad, black leading edge and grey trailing edge. Tail: Square edged. Legs, feet: Long, trailing. Eyes: Brown. Bill: Black.

A little-known species. Follows trawlers, attracted to lighthouses. Flight is fast, erratic or direct. Fluttering wingbeats followed by long glides.

Habitat: Oceanic. Usually seen in Bass Strait and Tasman Sea on edge of continental shelf.

Feeding: Plankton, crustaceans, squid, fish, picked from waves.

Voice: High-pitched twittering.

Status: Uncommon.

Breeding: Oct.–May. Islands southeast of NZ. South Atlantic and Indian Oceans.

Left, right: Running on water feeding.

White-faced Storm-petrel

Pelagodroma marina

200mm
Wingspan: 420–430mm

Sexes alike. Distinctive white face. Broad, long white brow. Large dark-grey lower eye patch. Crown: Grey. Nape, back: Grey, aging to brownish. Wings: Grey, long, rounded. Rump: Pale grey. Tail: Dark grey, square tip. Legs: Grey, long, trailing. Underparts: White. Underwings: White with a thin, dark-grey leading edge and grey flight feathers.
Immature: Paler.

Rarely follows ships. Erratic flight, weaving or low glides with intermittent wingbeats or 'bouncy' pattering over the water with outstretched, upright wings held into the wind.

Habitat: Oceanic, beyond continental shelf.

Feeding: In groups, surface-seizing plankton, small fish.

Voice: Repetitive 'woo-woo'. Silent at sea.

Status: Locally common.

Breeding: Oct.–Dec. Southern Australia, north to Broughton Is., NSW. Nests in rock crevices close to water / 1 egg.

Left: Inflight. Right: Surface feeding.

Black-bellied Storm-petrel

Fregetta tropica

180–200mm
Wingspan: 460mm

Sexes, immature: Alike. White, crescent-shaped patch on rump and tail coverts extending to underparts. Upperparts: Blackish brown. Head, breast: Black, small white patch on chin. Underparts: White, variable, streaky black centre line extending from breast to undertail coverts, sometimes absent. Underwings: White, dusky margins. Tail: Black, square. Legs: Black, trailing. Eyes, bill: Black.

Singular or at times pairs. Follows ships. Flapping, gliding, flight follows wave contours.

Habitat: Oceanic, subtropical.

Feeding: Surface-seizing crustaceans and squid.

Voice: Silent at sea. Long, high-pitched, piping whistle at breeding colonies.

Status: Uncommon. Solitary Circumpolar distribution.

Breeding: Subantarctic islands.

Left: Upper parts. Right: Underparts.

White-bellied Storm-petrel

Fregetta grallaria

180–200mm
Wingspan: 460–480mm

Sexes, immature: Alike. Light, dark and intermediate morphs. Light morph is the most common. Head: Black hood with steep, high frons. Back, rump: Brownish black lightly scalloped white. Tail covert: Broad, white crescent. Wings: Long, dusky black, paler shoulder patch. Underwings: White central panel, broad blackish flight feathers and leading trailing edges. Underparts: White. Dark morph: Blackish brown. Mottled white belly and upper tail covert.

Erratic, weaving flight into the wind on uplifted wings, feet dangling. Hugs wave surface.

Habitat: Southern hemisphere, tropical, subtropical oceanic. Rarely inshore waters.

Feeding: Surface-seizing crustaceans and squid.

Voice: Silent at sea and in flight. Shrill whistles in burrows.

Status: Rare winter visitor.

Breeding: Lord Howe group. In rock crevices or burrows / 1 egg.

Ball's Pyramid off Lord Howe Is.
Left: Upperparts. Right: Underparts.

Wandering Albatross

Diomedea exulans

Up to 1350mm
Wingspan: Up to 3500mm

Upper body, crown: White. Upperwings: White with black tips. Underparts, underwings: White, black-tipped. Plumage varies with age, 8–9 year old mature males are whitest. Bill: Pink, yellow tipped.
Female: Similar but smaller and duller with faint breast barring.
Immature: Brown with white underwings and face. Upperwings: Brown, whitening with age.

Largest flying bird existing today, travels great distances circling the world's oceans. Gliding, soaring flight on updrafts from waves. Follows ships for days.

Habitat: Open ocean and coastal waters.

Feeding: Fish, squid, crustaceans and carrion.

Voice: Occasional, throaty gurgling at sea. Silent in flight.

Status: Visits southern Australian waters June–Sept. Moderately common to endangered. Many are drowned in long line fishing operations.

Breeding: Sept. Antarctic and subantarctic islands. Nest is a mud bowl on ground / 1 egg.

Left: Immature. Centre and right: Adult.

Antipodean Albatross

Diomedea antipodensis

Up to 1200mm
Wingspan: Up to 3000mm

Smaller, slighter and darker than the Wandering Albatross. Variable plumage, whitening with age. Mostly white body with close, wavy, dark brown markings. Upperwings: Brown. Underwings: White with narrow trailing edge. Head, throat: White, sometimes with a dark crown. Neck: White with variable brownish blotching. Uppertail: Black, white tipped. Bill: Pale pink, horn-coloured tip. Eyes: Dark brown. Legs, feet: Grey to pinkish.
Female: Darker than male. Face, chin, throat: White. Belly: Whitish with brown close wavy pattern on sides and flanks. Undertail coverts: Brown. Breeding: Dark brown upperparts and 'wave'-patterned breast band and back.
Immature: Difficult to separate from immature Wandering Albatross. Brown with white underwings and face. Upperwings: Brown.

Flight is gliding, soaring on updrafts from waves. Elaborate courtship dances and sequence of calls. Form permanent bonds.

Habitat: South Pacific Ocean and coastal waters, south of the Tropic of Capricorn.

Feeding: Fish, squid, crustaceans and carrion.

Voice: Occasional, throaty gurgling sounds at sea. Silent in flight.

Status: Moderately common to endangered. June–Sept. visitor.

Breeding: Biennially in colonies. Antipodes and Campbell Is., NZ. Nest is an earth platform on ground / 1 egg.

Left: Female (foreground) and male.
Right: Female in flight.

Southern Royal Albatross

Diomedea epomophora

1150–1250mm
Wingspan: Up to 3280mm

Sexes similar. Similar to the Wandering Albatross, but with a distinctive black cutting edge to upper bill. Bill: Pink. Upperwings: Brownish black. Underwings: White with black trailing edge and tips. Female: Smaller. Immature: Dark speckled crown, rump. Upperwings: Black, speckled white, whiter with age.

Visits fishing boats. Rarely follows ships. Uses updraft off waves to gracefully glide, soar and bank, only occasionally flapping wings.

Habitat: Southern oceans.

Feeding: Fish, squid, crustaceans, carrion.

Voice: Harsh, squawking, squabbling. Gobbling sounds.

Status: Moderately common. Winter visitor to offshore waters NSW, Vic., Tas., SA and casual visitor WA.

Breeding: Nov.–Dec. every 2 years. NZ islands. Mud nest / 1 egg.

Top left: Southern Royal Albatross adult.

Northern Royal Albatross

Diomedea sanfordi

Endangered

1150–1250mm
Wingspan: Up to 3280mm
See map for Southern Royal Albatross

Very similar to the Southern Royal Albatross but upperwings are all black at all ages. Underwings: White with black trailing edge and leading edge to elbow and tips.

Winter range as for Southern Royal Albatross.

Immature: Dark speckled crown and rump.

Status: Moderately common in small breeding area.

Breeding: Chatham Is., NZ.

Top centre and right: Northern Royal Albatross.

Black-browed Albatross

Thalassarche melanophris

840mm
Wingspan: 2200mm

Sexes similar. White body. Back, upperwings: Black. Underwings: Wide, black margins, white panel in centre. Bill: Yellow. Eyes: Black with thin black eyebrow. Underwings: White, wide black leading edge.
Immature: Upper- and underwings: Dark brown. Crown, collar: Grey. Bill: Dark brown.

Most abundant mollymawk, commonly seen from land in autumn to spring in southern Australian waters. Follows ships. Gathers around trawlers. Graceful, effortless flight.

Habitat: Oceanic and often coastal waters.

Feeding: Forages inshore in bays or harbours in May–Nov., taking plankton, fish, squid, carrion.

Voice: Grunting noises and whistled braying. Silent in flight.

Status: Circumpolar, abundant. Winter–spring visitor, mostly over shelf break.

Breeding: Sept.–Apr. Heard, Macquarie and subantarctic islands. Mud nest / 1 egg.

Left and centre: Black-browed Albatross.

Campbell Albatross

Thalassarche impavida

860mm
Wingspan: 2200mm
See map for Black-browed Albatross

Very similar to the Black-browed Albatross in range, appearance, behaviour, but has heavier black brow line, broader wing margins and honey-coloured eyes.

Winter visitor to southern Australian coastal waters from Geralton, WA, south to central Qld.

Breeding: Campbell Is., NZ.

Far right: Campbell Albatross.

Shy Albatross

Thalassarche cauta

Endemic

900–1000mm
Wingspan: 2250mm

Sexes similar, female smaller. Overall white with a pale grey face and dark eyebrow. Bill: Grey with pale yellow tip and yellow base on upper ridge. Upperwings, back: Ash grey. Underwings: White with narrow black edging and tips. Underparts: White. Feet: Blue-grey.
Immature: Similar. Bill: Grey, dark tipped. Feet: Darker.

Only albatross to breed in Australia. Occasionally follows ships or gathers around trawlers. Flight is laboured in soft breezes.

Effortlessly soars high in windy conditions.

Habitat: Oceanic and coastal waters.

Feeding: Fish, squid, cuttlefish, carrion.

Voice: Hoarse gurgling and cackling sounds.

Status: Common, abundant, but vulnerable.

Breeding: Sept.–Oct. Offshore Tas. islands. Nest is a mud bowl / 1 egg.

Left: Pair following trawler.
Right: Adult in flight.

Salvin's Albatross

Thalassarche salvini

900–1000mm
Wingspan: 2200mm

Sexes similar. Similar to the Shy Albatross but with a grey hood. Pale grey face and dark eyebrow. Face, throat and uppermantle: Mid-grey. Crown, chin: Light grey. Upperwings, tail, and back: Dark grey, palest on back. Rump: White. Underwings: White with narrow black edging and tips. Underparts: White. Bill: Grey with pale yellow tip and yellow base on upper ridge. Feet: Blue-grey.
Female: Smaller. Immature: Grey darker over crown and mantle. Bill: Darkish blue-grey, dark tipped. Feet: Darker.

Flight is laboured in soft breezes. Effortlessly soars high in windy conditions.

Habitat: Oceanic and coastal waters.

Feeding: Fish, squid, cuttlefish, carrion.

Voice: Hoarse gurgling and cackling sounds.

Status: Abundant, but vulnerable. Recorded NSW to southern WA.

Breeding: Annually. Snares and Bounty Is., NZ, and Crozet Is., Indian Ocean, in dense colonies. Nest is a mud platform / 1 egg.

Left: Underparts. Centre: In flight showing upperparts. Right: Bill detail.

Grey-headed Albatross

Thalassarche chrysostoma

910mm
Wingspan: 2150mm

Sexes similar, female smaller. Head and neck darkish blue-grey. Bill: Black with golden stripe along upper and lower ridge, faintly tipped pinkish red. Underwings: White with broad black band to leading edge and narrow black band to trailing edge. Upperwings, back, mantle: Black. Feet: Light blue-grey. Immature: Light brownish grey. Dark eyebrow. White cheek patch. Bill: Grey-brown.

Follows ships and trawlers at a distance. Wheels, glides in wide arcs using updrafts off waves. Often rests on surface in calm conditions. Usually solitary.

Habitat: Oceanic.

Feeding: Fish, squid, crustaceans, carrion.

Voice: Braying calls. Silent in flight.

Status: Vulnerable. Uncommon but regular winter visitor to southern coastal waters in May–Nov.

Breeding: Sept.–Feb. Alternate years. Subantarctic islands including Macquarie Is. Mud nest / 1 egg.

Off Portland Port, Port Fairy, Vic.

Left and centre: Flight details.
Right: Bill detail.

Indian Yellow-nosed Albatross

Thalassarche carteri

700–820mm
Wingspan: 1800–2000mm

Sexes alike. Smallest mollymawk. Distinctive slender, glossy black bill with bright yellow stripe to upper ridge and pink tip. Eyes: Brown with grey patch below. Head, neck: White. Upperparts, wings: Grey. Rump, uppertail coverts: White. Immature: Lacks grey eye patch. Bill: Black.

Follows fishing boats. Flight is low to mid-range, using updrafts off waves, strong glides with few wingbeats.

Habitat: Subtropical, warmer subantarctic waters. Inshore and offshore waters.

Feeding: Gregarious, feeding with other albatross. Mostly fish, squid and trawler waste.

Voice: Throaty grunts when squabbling over food.

Status: Vulnerable. Abundant in southwest WA, SA, in Mar.–May and Tasman waters, May–June.

Breeding: Islands in the mid-Atlantic Ocean.

Left: Flight detail. Right: Landing on water.

Buller's Albatross

Thalassarche bulleri

760–810mm
Wingspan: 2130mm

Sexes similar, female smaller. Similar in appearance to the Grey-headed Albatross but smaller and more slender with broader yellow stripes on its long slender bill, more distinct dark area in front of eyes, and wider black leading edge to underwings. Head: Dark grey hood with small silvery cap. Feet: Flesh pink.
Immature: Light brown. Bill: Grey-brown.

Graceful, effortless flight. Gliding between wave troughs or soaring and wheeling in wide arcs.

Habitat: Oceanic and inshore. Prefers warmer waters, warm currents of southern seas.

Feeding: Fish, squid and crustaceans.

Voice: Nasal croaking. Silent at sea.

Status: Regular but uncommon visitor to southeastern Australia, Apr.–Jul.

Breeding: NZ offshore islands and Eaglehawk Neck, Tas.

Left: Flight study. Right: Bill detail.

Sooty Albatross

Phoebetria fusca

850mm
Wingspan: 2000mm

Sexes similar, female smaller. Overall sooty brown. Head: Dark brown. Eyes: Dark brown with partial white hind eye ring. Tail: Distinctive, long, pointed, wedge-shaped. Bill: Slender black, with dull yellow line on lower mandible.
Immature: Similar, pale yellow-grey bill stripe and mottled buff collar.

Graceful flight with aerobatic gliding and rarely flapping wingbeats. Regularly follows ships for long periods.

Habitat: Oceanic.

Feeding: Fish, squid and crustaceans.

Voice: Harsh, threatening call when feeding. Silent other times at sea.

Status: Circumpolar. Uncommon, but regular visitor to southern Australian open ocean, Jan.–Nov.

Breeding: Circumpolar subantarctic islands.

Left: In flight showing underparts. Right: Upperparts showing distinctive tail.

Light-mantled Sooty Albatross

Phoebetria palpebrata

800mm
Wingspan: 2000mm

Sexes similar. Head, wings, tail: Sooty dark brown. Mantle, back, underparts: Pearly grey. Bill: Glossy black with pale blue line to lower mandible. Eyes: Dark brown with partial white hind eye ring.
Immature: Similar, mantle mottled light grey.

Similar to the Sooty Albatross, but prefers colder seas. Flight is masterful, effortless, graceful, seemingly suspended in air. Follows close to ships.

Habitat: Oceanic, occasionally inshore waters.

Feeding: Mostly squid, also fish, crustaceans.

Voice: Harsh trumpeting at sea.

Status: Uncommon but regular winter–spring visitor to southeastern Australia. Vulnerable.

Breeding: Nov.–Jan. Subantarctic islands, including Macquarie Is. Cup-shaped mud nest / 1 egg.

Left: Underparts seen in flight. Right: Flight detail.

Southern Giant-petrel

Macronectes giganteus

900mm
Wingspan: 1500–2100mm

Sexes alike. Largest petrel. Massive, pale yellowish bill with diagnostic green-tinged tip and conspicuous long nasal tube. Head, neck, breast: Mottled greyish white. Upperparts, wings: Dark brown. Underparts, wings: Dull white with smudgy brown markings. Feet: Dark grey. White morph: Overall white with some dark flecks. Feet: Pink. Immature: Overall glossy, dark brown. Bill: Greenish tip. Face: Whitens with age. White morphs: Mostly white.
Laboured flight, 'humped-back' appearance. Follows ships.

Habitat: Oceans and bays.

Feeding: Mostly from surface. Fish, squid, crustaceans, carrion.

Voice: Squawking, raucous, groans. Usually silent at sea.

Status: In Australian waters immature are moderately common; adults are rare.

Breeding: Oct.–Nov. Subantarctic islands. On grass nest / 1 egg.

Left: White morph. Centre: Adult. Right: Immature.

Northern Giant-petrel

Macronectes halli

900mm
Wingspan: 1500–2100mm

Sexes alike. Overall sooty brown or dark grey. Head, neck, breast: Mottled whitish brown, face whitening with age. Eyes: Pale blue. Bill: Massive, pale pinkish with diagnostic red-tinged tip and conspicuous long nasal tube. Underwings: Dark leading edge. Immature: Overall sooty brown with dark eyes and dull reddish-tipped bill.

Follows ships. Similar to Southern Giant-petrel. Both species interbreed, producing hybrids that share both features. 'Humped-back' appearance in flight. Rigid wing glides and laboured wingbeats.

Habitat: Oceans and bays.

Feeding: Fish, squid.

Voice: Usually silent. Throaty raucous calls.

Status: Moderately common. Vulnerable.

Breeding: July–Feb. Subantarctic islands including Macquarie and Heard Is. Grassy depression nest / 1 egg.

Left: Adult. Right: Immature.

Southern Fulmar

Fulmarus glacialoides

450mm
Wingspan: 1000mm

Sexes, immature alike. Thickset, gull-like bird with pale blue-grey upperparts. Upperwings: Darker blue-grey, black tipped with white patches. Underwings: White, grey tipped, grey patches. Head, neck, underparts: White. Bill: Pinkish or grey, black tipped with long, dark blue-grey nasal tube. Feet: Pinkish blue with pink webs.

Follows trawlers. Wheeling, gliding flight.

Habitat: Southern oceans.

Feeding: Small squid, fish, plankton, carrion, scavenging. Surface feeding; snatching or plunging from a graceful, low glide.

Voice: Noisy cackling when quarrelling over food.

Status: Uncommon. Occasional visitor, usually May–Oct.

Breeding: Antarctic coast and surrounding islands.

Left: Adult. Right: Upperparts.

Salvin's Prion

Pachyptila salvini

280–300mm
Wingspan: 580mm

Sexes, immature: Alike. Similar to Antarctic Prion. Upperparts, head: Blue-grey dark eye, white eyebrow and frons. Bill: Dark blue-grey, wide with bowed sides and visible side lamellae. Upperwings: Blue-grey with wide black 'M' pattern across wings. Tail: Broad, wedge-shaped, broadly tipped black. Underparts, wings: White. Collar: Blue-grey, incomplete.

Gregarious. Avoids ships. Gliding, or with slow wingbeats, skims low across the water.

Habitat: Oceanic.

Feeding: In flocks. Skims water, scooping up krill, squid, plankton, or swims and dives for prey.

Voice: Unknown.

Status: Uncommon. Offshore, winter–spring non-breeding visitor to southern Australia.

Breeding: Subantarctic islands in southern Indian Ocean.

Left: Adult. Right: Upperparts.

Broad-billed Prion

Pachyptila vittata

280–300mm
Wingspan: 600mm

Sexes alike. Largest and most distinctive prion with its strikingly wide and massive, boat-shaped bill with visible lamellae and broad head with high, steep frons. Eyes: Dark brown with dark stripe through eyes and thin white eyebrow. Lores: White. Upperparts: Blue-grey, with a dark 'M' pattern across wings. Tail: Wedged-shaped, narrow black tip. Underparts, underwings: White with pale half-collar.

Rarely in Australian waters. Flight is fast, erratic, usually in flocks. Glides, banks, flutters on water.

Habitat: Oceanic.

Feeding: Krill, squid, plankton.

Voice: Rasping calls, clucking.

Status: Uncommon, non-breeding visitor to southern Australian coast.

Breeding: Subantarctic islands and NZ.

Left and right: Adult in flight.

Antarctic Prion

Pachyptila desolata

270–290mm
Wingspan: 640mm

Sexes, immature: Alike. Smallest whalebird. Often indistinguishable from Salvin's Prion, the Antarctic Prion has a darker blue-grey bill, although not quite as wide, with straight sides and lamellae not visible when bill is closed. Collar: Incomplete, darker blue-grey, slightly longer.

Flight is fast and erratic, usually high above water.

Habitat: Oceanic.

Feeding: Usually in flocks. Skims water with broad bill, scooping mostly krill, squid, plankton, or swims and dives for prey.

Voice: Silent at sea. Dove-like cooing at nest.

Status: Moderately common in southern Australian waters.

Breeding: Oct.–May. Subantarctic islands including Macquarie Is. Nest burrow / 1 egg.

Left: Underparts. Centre and right: Upperparts.

Slender-billed Prion

Pachyptila belcheri

250–270mm
Wingspan: 580mm

Sexes, immature: Alike. Upperparts: Light blue-grey with pale 'M' across wings, becoming indistinct and paler with age. Eyes: Long, dark-grey eye stripe and long white eyebrow. Bill: Bluish, slender, long, straight sides, lamellae not visible when bill is closed. Face, lores, throat: White. Underparts: White with faint, short, blue-grey collar. Tail: Wedge-shaped, black tipped, grey outer margins.

Flight is fast, erratic, close to surface.

Habitat: Oceanic.

Feeding: Mostly small crustaceans.

Voice: Silent at sea.

Status: Uncommon, non-breeding winter visitor to southwest coastal Australia. Occasionally southeast coast.

Breeding: Southern Indian, southwest Atlantic and southeast Pacific Oceans.

Left and right: Adult in flight.

Fairy Prion

Pachyptila turtur

230mm
Wingspan: 560mm

Sexes, immature: Alike. Smallest prion. Upperparts: Blue-grey. Upperwings: Blue-grey, with pale 'M' across wings. Eyes: Dark-grey eye stripe and white eyebrow. Underparts: White, with very short, faint collar. Tail: Wedge-shaped, black tipped, white outer margins.

Usually in flocks. Rarely follows boats. Swift, erratic flight, close to waves.

Habitat: Oceanic, rarely close to coast.

Feeding: Usually in large flocks. Surface-seizing, diving for crustaceans, squid.

Voice: Cooing courtship call.

Status: Vulnerable. Common in southeastern Australia. Only prion breeding in Australian waters.

Breeding: Sept.–Mar. Offshore islands of Tas. and southeastern Australia. Nests in colonies. Burrow or rock crevice / 1 egg.

Left: Adult in flight. Right: Upper wing detail.

Cape Petrel

Daption capense

400mm
Wingspan: 800–900mm

Sexes, immature: Alike. Distinctive black and white patchwork to upperparts with underparts white. Tail: White, broadly tipped black. Head, throat, mantle: Black, brownish with age. Underwings: White edged black and black scalloping. Bill: Short, black. Eyes: Dark. Feet: Black. Race *australe*, Australia, NZ. Upper body overall darker.

Gather in large quarrelsome flocks. Follow ships for scraps. Brief periods of gliding followed by rapid wingbeat flight.

Habitat: Oceans, bays, but mostly well offshore.

Feeding: Scavengers. Surface feeds or dives, mostly krill, squid.

Voice: Noisy 'cak-cak-cak'.

Status: Common. Regular winter–spring visitor.

Breeding: Antarctica and subantarctic islands.

Left: Underparts. Centre, right: Upperparts.

Blue Petrel

Halobaena caerulea

300mm
Wingspan: 650mm

Sexes, immature: Alike. Only petrel with white tail tip. Plain blue-grey upperparts and tail. Upperwings: Blue-grey with dark 'M' pattern across wings. Crown, nape, eye patch: Grey-black. Frons: White mottled dark grey. Eyes: Brown. Bill: Slender, black, sides tinged blue. Underparts: White with pale grey across breast. Tail: White tipped, square cut.

Often beach-washed in southern Australia. Flocks form at sea. Follow ships and whales. Dipping and rising between waves. Glides, skims close to surface.

Habitat: Oceanic.

Feeding: Swims, dives and dips, mostly taking krill, small squid, occasionally small fish.

Voice: Silent at sea.

Status: Regular visitor to southern Australian seas in winter–spring.

Breeding: Antarctica islands.

Opposite: Adults in flight.

White-chinned Petrel

Procellaria aequinoctialis

500mm
Wingspan: 1400mm

Sexes, immature: Similar. Heavily built. Overall blackish brown with white chin patch variable in size or sometimes absent. Bill: Variable pale colour with large hook. Feet: Black.

Follows ships, gathers around trawlers aggressively feeding on scraps. Flight is graceful, usually low over water, but also soars high. Slow wingbeats and long glides.

Habitat: Oceanic.

Feeding: Squid, fish, crustaceans.

Voice: Silent at sea. Noisy feeders around trawlers.

Status: Vulnerable. Uncommon visitor to Australian waters, mostly seen in summer off Tas.

Breeding: Antarctic and subantarctic islands.

Left: Adult landing. Right: Adult feeding.

Westland Petrel

Procellaria westlandica

450–460mm
Wingspan: 1400mm

Sexes alike. Similar to Black Petrel. Distinctive dark-tipped pale ivory bill. Thickset, bulky, overall blackish body aging to a mottled brown. Underwings: Flight feathers have paler sheen. Tail: Wedge-shaped. Legs: Black, trailing.
Immature: Black bill.

Gathers around fishing boats. Flight is laboured in light winds, becoming more vigorous as wind strengthens. Glides close to water between wave crests.

Habitat: Oceanic, coastal.

Feeding: Fisheries waste, small fish and squid.

Voice: Silent at sea.

Status: Sporadic summer visitor to eastern Australian waters. Uncommon. Vulnerable.

Breeding: South Is., NZ.

Left: Adult feeding. Right: Adult in flight.

Black Petrel

Procellaria parkinsoni

460mm
Wingspan: 1100mm

Sexes alike. Gracefully proportioned, overall sooty black becoming browner with age. Bill: Small, horn-coloured with greenish yellow tinge and black tip, upper-ridge and cutting edge. Tail: Wedge-shaped, feet protude slightly in flight. Immature: Similar. Bill: Whiter.

Rare, but possibly regular visitor to eastern Australian waters. Seldom follows ships, but gathers around trawlers. Flight is graceful and buoyant. Glides with stiff, backswept outer wings.

Habitat: Oceanic, coastal.

Feeding: Mostly squid, also fish, crustaceans.

Voice: Usually silent at sea.

Status: Summer visitor to Tasman Sea and eastern Australia.

Breeding: NZ islands.

Opposite: Adult feeding and in flight.

Bulwer's Petrel

Bulweria bulwerii

250–290mm
Wingspan: 780–900mm

Sexes, immature: Alike. Slender bodied and overall sooty brown petrel with a short bill, head and neck. Pale grayish wing bars and long, wedge-shaped tail in flight. Legs: Pale, short.

Gadfly petrel, occasionally seen in waters off northern Australia. Flight is erratic, buoyant and close to waves with bursts of rapid wingbeats. Circles when feeding.

Habitat: Oceanic.

Feeding: Usually solitary. Surface-seizing or resting on water plucking prey, mostly squid, fish.

Voice: Silent at sea.

Status: Summer visitor. Most common off northwestern Australia. Rare elsewhere.

Breeding: Northwest and central Pacific Ocean.

Left: Adult at breeding area.
Right: Adult, note long pointed wings.

Wedge-tailed Shearwater

Ardenna pacifica

420–440mm
Wingspan: 1000mm

Sexes, immature: Alike. Common 'muttonbird' found in warm coastal waters. Overall dark brown with broad wings and blackish flight feathers. Underwings: Pale base to primaries and secondaries. Bill: Leaden grey. Feet: Pinkish, black toe nails. Tail: Long, wedge-shaped, pointy. Light morph: Whitish underparts.

Distinctive, drifting, buoyant flight. Low over waves. Flocks around trawlers.

Habitat: Oceanic and tropical, coastal waters.

Feeding: Solitary or small groups at sea, or flocking at food source, for squid, small fish.

Voice: Silent at sea. Slow wailing song at night.

Status: Common, abundant. Migrates north after breeding.

Breeding: Oct.– Apr. Offshore islands of eastern and western Australia. Nest in burrow / 1 egg.

Off North Stradbroke Is., Qld.

Left: Adult. Right: Light morph.

Buller's Shearwater

Ardenna bulleri

450–460mm
Wingspan: 1000mm

Sexes, immature: Alike. Elongated neck, bill, wings and tail. Crown: Flat, sooty brown. Upperparts: Mid-grey. Upperwings: Mid-grey with black tips; dark 'M' pattern broken across rump; pale crescents near body. Face, underparts, underwings: White. Bill: Leaden blue-grey. Tail: Wedge-shaped, black.

Ignores ships. Flight is low, unhurried, drifting, buoyant.

Habitat: Oceanic, coastal.

Feeding: Solitary or flocks, swims and dives for prey, squid, crustaceans.

Voice: Silent at sea. Noisy, wailing, howling at nesting colonies.

Status: Moderately uncommon but regular spring–summer visitor to the southeastern coast of Australia.

Breeding: Sept.–May. Poor Knight Is., NZ.

Left: Upperparts. Right: Underparts.

Flesh-footed Shearwater

Ardenna carneipes

450–500mm
Wingspan: 1300mm

Sexes, immature: Alike. Heavily built with a large head, broad wings and overall blackish brown. Eyes: Dark brown. Bill: Large, pale horn and blackish tip. Legs: Pinkish.

Largest Australian shearwater. Gathers around trawlers. Glides close to water. Usually solitary at sea. Flocks form at food source.

Habitat: Pelagic birds, most often found over continental shelves and slopes.

Feeding: Surface-seizing, diving, and swimming after prey, taking small fish, squid, crustaceans, carrion.

Voice: High-pitched sharp call. Food squabbling calls.

Status: Common, abundant in southern waters in summer.

Breeding: Sept.–May on islands off the east coast including Lord Howe Is. Aug.–May on southwestern coast of WA. Nests and roosts in burrows / 1 egg.

Left: Adult showing its pink legs.
Right: Flight.

Sooty Shearwater

Ardenna grisea

450–460mm
Wingspan: 1000mm

Sexes, immature: Alike. Overall dark brown with pale, silvery patch on underwings. Wings: Slender, straight. Bill: Long, pale lead grey, dark tip. Eyes: Brown. Tail: Short, rounded. Legs, feet: Pinkish. Moult phase in summer.

Noted for its autumn migration in large flocks to the cold waters of the northern hemisphere. Follows fishing boats and whales. Flies close to water. Shallow wingbeats.

Habitat: Oceanic, occasionally coastal.

Feeding: Usually surface-seizing but can dive in excess of 60m for fish, squid.

Voice: Quiet at sea. Cooing and wailing at nesting site.

Status: Near threatened.

Breeding: Sept.–Apr. NZ, southern Australia, Tas. and offshore islands, Macquarie Is. Nests in burrows / 1 egg.

Opposit: Adults.

Short-tailed Shearwater

Ardenna tenuirostris

400mm
Wingspan: 1000mm

Sexes, immature: Alike. Also called 'Tasmanian Muttonbird'. Mid grey-brown body with some reflective, glossy flight feathers. Underwings and throat patch appear paler. Tail: Short, rounded. Bill: Lead grey, slender.

Strong, direct flight with stiff, straight wings and few wingbeats. Dips close to water. Follows wave contours. Only petrel to breed solely in Australia.

Habitat: Oceanic and coastal waters.

Feeding: Huge floating flocks feed on plankton, crustaceans, fish, squid. Returns to breeding area at night, leaves before dawn.

Voice: Noisy wailing, wheezing.

Status: Common, abundant. Anual migration to northern Pacific waters.

Breeding: Oct.–Apr. Offshore islands. Nests in burrows in huge colonies / 1 egg.

Bruny Is., Tas. Phillip Is., Vic.

Opposite: Adult birds.

Streaked Shearwater

Calonectris leucomelas

450–480mm
Wingspan: 1200mm

Sexes, immature: Alike. Large shearwater with a distinctive, large, pale grey, dark-tipped bill. Crown, hind neck: Streaky brown and white. Frons, face: White. Eyes: Dark with white eyebrow. Upperparts: Dark grey-brown feathers with pale margins, appear scalloped; indistinct 'M' pattern across wings and rump. Underparts: White. Underwings: White with broad brown tips and trailing edge.

Follows trawlers warily. Flight is low, graceful and effortless with long glides between wingbeats.

Habitat: Oceanic, coastal.

Feeding: Surface-seizing, shallow plunges. Mostly small fish, squid.

Voice: Silent at sea.

Status: Regular, common visitor to north, northwestern and eastern Australia. Nov.–May.

Breeding: Asia.

Off North Stradbroke Is., Qld.

Left: Underparts. Right: Upperparts.

Fluttering Shearwater

Puffinus gavia

320–330mm
Wingspan: 750mm

Sexes, immature: Alike. Upperparts: Dark brown. Underparts: White with slight collar and black thigh patch. Underwings: White.

Flight is close to water with rapid wingbeats and short glides. Does not gather at trawlers, but often seen gathering around schools of small fish. Forms rafts in large numbers to forage, dipping head underwater seeking prey, before diving.

Habitat: Usually coastal, frequently in estuaries.

Feeding: In large flocks. Dives for prey.

Voice: Cackling calls. Staccato flight calls.

Status: Locally abundant, common visitor to southeastern coasts in winter.

Breeding: Coastal islands, NZ.

Left: Upperparts. Centre: Underparts. Right: At breeding ground.

Hutton's Shearwater

Puffinus huttoni

350mm
Wingspan: 780mm

Sexes, immature: Alike. Similar to Fluttering Shearwater, but larger. Upperparts: Dark brown. Underparts: White with dark, long, incomplete collar and large, dark thigh patch. Underwings: White with broad dark brown trailing edge and tips, dark brown smudge close to body. Tail: Short, wedge-shaped. Bill: Dark grey. Legs: Pink.

Regular winter visitor to waters off eastern and southern Australia, often in flocks. Flight is fast, close to water with rapid wingbeats and short glides.

Habitat: Oceanic, coastal, occasionally estuaries.

Feeding: In flocks. Dives and swims for shoaling fish, krill, crustaceans.

Voice: Repetitive cackling sounds. Silent at sea.

Status: Regular, but uncommon winter visitor. Usually young birds.

Breeding: Sept.–Apr. NZ.

Opposite: Adults in flight.

Little Shearwater

Puffinus assimilis

280–300mm
Wingspan: 630mm

Sexes, immature: Similar. Smallest Australian shearwater. Upperparts, wings: Blue-black. Bill: Short, grey. Eyes: Brown. Dark cap to above eyes. Face: White. Underparts: White with short collar. Underwings: White with narrow black leading edge, broader trailing edge and tips. Undertail: Dark edged. Legs, feet: Bright blue, flesh webs.

Flying close to water surface on small, short, stiff wings with distinctive rapid wingbeats and short glides. Follows ships.

Habitat: Oceanic, coastal.

Feeding: In flocks. Dives, swims, surface-seizes, mostly squid, also fish, crustaceans.

Voice: High-pitched wheezing whistles.

Status: Common in southeast and southwest Australia.

Breeding: Feb.–Nov. Nominate race breeds on Lord Howe Is. Race *elegans* breeds on offshore islands from Recherche Archipelago to Abrolhos, WA. Nest in burrows / 1 egg.

Left: Take-off. Right: Underparts.

Grey Petrel

Procellaria cinerea

450–500mm
Wingspan: 1200mm

Sexes alike. Heavy body with small head and slender grey-black wings. Upperparts: Silvery grey wearing to brownish grey. Underparts: White. Underwings, undertail coverts: Grey. Bill: Yellow-green. Tail: Wedge-shaped.
Immature: Plain grey.

High wheeling flight. Gliding in high winds. Follows ships at a distance. Scavenges at fishing trawlers. Solitary or small groups.

Habitat: Mostly pelagic.

Feeding: Shallow dives down to 10m, swimming underwater for small fish and marine prey.

Voice: Silent at sea. Cackles, resonating alarm calls at nest.

Status: Rare. Vulnerable.

Breeding: Feb.–Sept. Islands of Southern Ocean including Macquarie Is. Nest in burrows / 1 egg.

Opposite: Adults in flight.

Tahiti Petrel

Pseudobulweria rostrata

380–400mm
Wingspan: 950mm

Sexes, immature: Alike. Tropical gadfly petrel with a black, bulbous bill. Head, breast, upperparts, tail: Glossy sooty brown. Breast, underparts: White. Eyes: Dark brown. Legs, feet: Pink, black.

Flies low over ocean with banking glides and easy wingbeats.

Habitat: Oceanic.

Feeding: Crustaceans, squid, small fish plucked from the surface in flight.

Voice: Noisy, squabbling trills.

Status: Regular but uncommon summer–autumn visitor.

Breeding: Pacific islands.

Off Eaglehawk Neck, Tas.

Opposite: Adults in flight.

Kerguelen Petrel

Lugensa brevirostris

330–360mm
Wingspan: 800mm

Sexes, immature: Alike. Gadfly petrel with large, distinctive, high and bulky frons. Overall glossy, reflective grey. Head: Large with distinctive, high, bulky frons. Bill: Black, stubby, tube nostrils on basal quarter giving a 'saddle' profile. Underwings: Dull white leading edge close to body and silvery flashes from reflective flight feathers. Tail: Long, pointy wedge-shaped.

Swift-like flight, soars and glides high. Rapid wingbeats.

Habitat: Oceanic.

Feeding: Usually feeds at night on squid, crustaceans, occasionally fish.

Voice: High-pitched alarm calls.

Status: Solitary. Visitor to offshore waters, more common in southern and western Australia. Rare in the east.

Breeding: Subantarctic islands, south Indian and Atlantic Oceans.

Opposite: Adults in flight.

Kermadec Petrel

Pterodroma neglecta

380mm
Wingspan: 900mm

Sexes, immature: Alike. Gadfly petrel. Three main colour morphs. All have brown wings with distinctive white base to primaries, more prominent on underwings. Bill: Black. Feet: Pink, bluish extremities. Eyes: Dark brown. Tail: Short, squared. Dark morph: Brown, dark grey. Intermediate morph: Variable. Underparts: Whitish. Upperparts: Mid-grey with lighter mottling around face. Pale morph: Overall whitish with pale buff tint. Ignores ships. Deep wingbeats with slow glides.

Habitat: Tropical and subtropical South Pacific.

Feeding: Plucks crustaceans, squid from surface, rarely dives.

Voice: Silent at sea.

Status: Vulnerable. Rare summer–winter visitor to Australian waters.

Breeding: Jan.–July. Ball's Pyramid, Norfork Is., and Juan Fernandez Is. Nest is lined depression / 1 egg.

Left: Dark morph. Right: Pale morph.

Herald Petrel

Pterodroma heraldica

340–390mm
Wingspan: 950mm

Sexes, immature: Alike. Gadfly petrel. Dark and light morphs. Pale morph is most common to Australian waters. Upperparts: Dark grey-brown with smudged 'M' pattern across wings. Underwings: Distinctive, dark 'M' pattern. Bill: Black, hooked. Side of face: White patch. Throat: Whitish with grey collar. Underparts: White.

Glides close to water. Slow wingbeats with wheeling, banking flight.

Habitat: Tropical and subtropical oceanic waters.

Feeding: Crustaceans, squid from surface.

Voice: High-pitched chattering.

Status: Uncommon. Vulnerable. Usually sedentary around its breeding ground.

Breeding: Sporadically all year. Recently found breeding on Raine Is., Qld. Nests on ground or crevice / 1 egg.

Opposite: Adults in flight.

Great-winged Petrel

Pterodroma macroptera

400–420mm Wingspan: 1000mm

Sexes, immature: Alike. Overall, dark brown with a short neck and long wings that extend past the tail. Bill: Heavy, large, black with small area of pale grey around the base.

Solitary at sea. Ignores ships. Flight is fast with vertical flight to 30m or more and then diving vertically into the water.

Habitat: Oceans, coastal waters.

Feeding: Silently at night on krill, squid, crustaceans.

Voice: Silent at sea. Sharp, squeaky calls.

Status: Common. Solitary. Common in offshore waters.

Breeding: May–Aug. Subantarctic islands, NZ. Usually nests in burrows. 1 egg.

Left Great Winged Petrel.

Grey-faced Petrel

Pterodroma gouldi

Similar to the Great-winged Petrel, but with larger pale-grey face mask. (Former sub-species). Endemic NZ, regular visitor to southern and eastern Australia. Oct to Apr. Usually nexts in burrows.

Above Right: Grey-faced Petrel

Soft-plumaged Petrel

Pterodroma mollis

320–370mm Wingspan: 900mm

Sexes, immature: Alike. Typical gadfly petrel. Dark and light colour morphs. Upperparts: Slate grey. Upperwings: Dark brown indistinct 'M' pattern. Breast: Soft grey grading to white belly with complete or incomplete grey breast band. Frons: Freckled white and dark grey. Underwings: Dark grey, broad, dark diagonal, indistinct on dark morphs and visible on light morphs. Eyes: Brown with white eyebrow and smudgy black eye patch.

Distinctive rapid wingbeats, long glides in large arcs. Usually ignores ships.

Habitat: Oceanic.

Feeding: Mostly squid.

Voice: Shrill whistle.

Status: Vulnerable. Common, non-breeding visitor.

Breeding: Subantarctic islands, southern Indian and southern Atlantic Oceans.

Opposite: Adults in flight.

White-headed Petrel

Pterodroma lessonii

400–420mm Wingspan: 1000mm

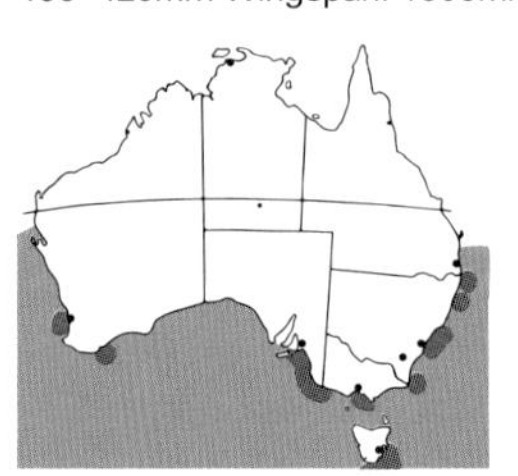

Sexes, immature: Alike. Stocky bird with long wings. Head: White with pearly pale-grey crown and collar. Bill: Black, sturdy. Eyes: Dark, with dark-grey smudgy eye patch. Nape, rump: Pearly grey. Back: Mottled, scalloped greys. Upperwings: Dark grey with dusky black flight feathers. Underwings: Dark grey appearing black at sea. Underparts: White. Tail: White, pointed, wedge-shaped when fanned. Feet: Pink. There are several colour morphs.

Mostly solitary in late summer through winter. Glides low over waves. Steady wingbeats. In wind, glides high in arcing flight.

Habitat: Southern seas.

Feeding: Mostly at night. Prefers squid.

Voice: 'Wil-wik' variations at sea.

Status: Uncommon visitor.

Breeding: Nov.–Dec. Subantarctic islands. Nests in colonies, in burrows / 1 egg.

L, R: Upperparts. Centre: Flight detail.

Providence Petrel

Pterodroma solandri

400–420mm Wingspan: 1000mm

Sexes, immature: Alike. Heavily built gadfly petrel. Head: Dark grey-brown and scaly white around base of the stout, black bill. Eyes: Dark brown. Upperwings, back, tail: Mottled dark grey. Underparts: Dark grey, belly paler with age and wear. Underwings: Dark grey with white coverts and primaries.

Effortless flight. Slow wingbeats. Soars fast in high arcs in wind.

Habitat: Tropical oceans.

Feeding: Krill, small fish.

Voice: Loud screeching.

Status: Vulnerable. Migratory. Common off continental shelf of NSW in Apr.–Oct.

Breeding: Feb.–Nov. Lord Howe and nearby islands. Nests in colonies, in burrows / 1 egg.

Left: Adult. Right: Landing.

White-necked Petrel

Pterodroma cervicalis

400–430mm
Wingspan: 1000mm

Sexes, immature: Alike. Gadfly petrel with a white face and body and black cap extending to below eyes. Back, wings, tail: Blue-grey with broad black 'M' band across wings and white rear neck. Underparts: White with dark base to primary feathers. Underwings: White with dark trailing edge. Tail: Dark, longish, wedge-shaped in flight. Legs: Flesh pink. Eye, bill: Black.

Usually solitary, sometimes loose flocks. Graceful, effortless flight. Strong, rapid wingbeats followed by short glides.

Habitat: Oceanic, tropical or subtropical.

Feeding: Dipping and picking fish and squid from surface.

Voice: Long, high-pitched, loud calls at breeding ground. Silent at sea.

Status: Uncommon. Solitary.

Breeding: Nov.–May Kermadec Is., NZ.

Opposite: Adult in flight.

Common Diving-petrel

Pelecanoides urinatrix

200–250mm
Wingspan: 400mm

Sexes, immature: Alike. Stubby body, stubby bill and rounded wings. Upperparts: Blackish, becoming browner with age. Underparts: Dusky white. Legs, feet: Short, deep blue, brighter in the breeding season.

Flight is low, rapid whirring wingbeats, short glides, diving and swimming using wings like flippers.
Returns to breeding area at night, leaves before dawn.

Habitat: Oceanic, coastal.

Feeding: In flocks, mostly crustaceans.

Voice: Calls in flight and ashore.

Status: Common.

Breeding: July–Dec. Southeastern mainland, Tas. offshore islands. Nests in a burrow or rock crevice / 1 egg.

Left: Upperparts. Right: Underparts.

Gould's Petrel

Pterodroma leucoptera

300mm
Wingspan: 700mm

Sexes, immature: Alike. Small gadfly petrel. Upperparts: Sooty brown and grey. Head: Sooty brown crown extending to below eye, sides of face and over neck. Frons: Strongly white speckled. Upperwings: Broad black diagonals meet across rump forming 'M' pattern. Underparts: White from face to undertail, with dusky partial neck band. Underwings: White, black leading edge to outer wing and black line to centre wing, making an indistinct 'M'. Black trailing edge. Tail: Short, rounded, dark.

Relatively slow flight. Glides close to water.

Habitat: Oceanic.

Feeding: Small squid, fish skimmed from the surface.

Voice: Silent at sea. High-pitched squeaky notes ashore.

Status: Rare. Vulnerable.

Breeding: Oct.–Apr. New Caledonia and Cabbage Tree Is. and islands off Port Stephens, NSW. In colonies, on ground / 1 egg.

Opposite: Adult in flight.

Black-winged Petrel

Pterodroma nigripennis

280–300mm
Wingspan: 670mm

Sexes, immature: Alike. Gadfly petrel. Frons, underparts: White. Crown, nape, incomplete collar, back: Blue-grey. Eyes: Brown with smudgy dark grey eye-patch. Underwings: White with black leading edge to outer wing and black line to centre wing, making an indistinct 'M'. Black trailing edge. Upperwings: Broad, black diagonals meet across rump forming 'M' pattern. Tail: Wedge-shaped. Feet: Pinkish.

Fast flight, swooping, weaving in wide arcs with sudden changes in elevation.

Habitat: Tasman Sea and central Pacific Ocean.

Feeding: Shrimps, fish, squid.

Voice: Shrill 'pee-pee'.

Status: Rare, locally common.

Breeding: Dec.–Jan. Norfolk Is., Lord Howe Is., Capricorn group, Qld, NZ, New Caledonia. Burrow / 1 egg.

Left: Underparts. Right: Upperparts.

4 Family Pelecanidae, Sulidae, Anhingidae, Phalacrocoracidae, Fregatidae, Phaethontidae

Pelican

Gannets

Boobies

Darter

Cormorants

Frigatebirds

Tropicbirds

A group of six families belonging to the Order Pelecaniformes, the Pelican Order.

Family Pelecanidae includes the Australian Pelican.
The Australian Pelican has survived virtually unchanged in Australia for 30–40 million years, completely adapted to its environment. The large expandable skin throat pouch is used as a scoop-net to capture small fish and crustaceans and also in courtship, where it takes on a reddish courtship hue that dissipates after pairing. Chicks leave the nest when able to walk, joining crèches of approximately 100 birds.

Family Sulidae includes gannets and boobies.
Gannets patrol the coastal temperate waters, plunging vertically from up to 10m for shoaling fish. In tropical and subtropical areas the related boobies replace them. Boobies have coloured facial skin and throat pouches that help to regulate their temperature when nesting on treeless tropical islands.

Family Anhingidae has a single Australian representative, the Australasian Darter.
The Australasian Darter floats with its body submerged and its snake-like neck and head above the water. It sinks to hunt prey, often stalking or waiting for prey to come into range, where it is speared by the bill and springing action of the neck. The bill edges have reverse serrations to secure the prey. To reduce buoyancy and allow the bird to stay underwater for extended periods, the plumage is not fully waterproof. Often seen with their wings hung out to dry, they first use their beaks to squeeze out the excess water.

Family Phalacrocoracidae includes the cormorants.
Cormorants feed by leap-diving from the surface of the water in underwater pursuit of prey. They use their webbed feet for propulsion and their wings for braking and turning. Prey are snatched with the hooked bill and taken to the surface to be consumed. Cormorants' plumage and drying habits are similar to those of the Australasian Darter.

Family Fregatidae includes frigatebirds.
Frigatebirds are masters of the air, unable to walk or swim. They drink in flight, skimming over water, pick their food from the water surface, or snatch it from terns or boobies in mid-air. They can remain airborne for a week at a time. The males attract females by inflating their red balloon-like 'gular' pouch in the breeding season. Their miniature-webbed feet are clawed to allow them to perch.

Family Phaethontidae includes tropicbirds.
Tropicbirds have weak, almost useless webbed feet on land, and like frigatebirds cannot swim but are essentially aerial. When the young leave the nest they fly straight out to sea unaided and roam the tropics. Tropicbirds feed by hovering and then plunge diving to the water surface similar to the related gannets.

Opposite: Australian Pelicans

Australian Pelican

Pelecanus conspicillatus

1600–1800mm
Wingspan: 2300–2500mm

Sexes alike, males slightly larger. Large bird with a huge bill. Head, neck, body: White, with a grey stripe down back of head and neck. Shoulders, rump, tail: Black. Upperwings: Black with wide white bar pattern. Underwings: White with black tips, broad trailing edge and narrow centre strip. Eyes: Brown with pale-yellow eye skin. Bill: Bluish with pinkish ridge and expandable skin pouch. In courtship bill pouch turns deep pink with a cobalt blue base and a black line extending from base to tip. Colours dissipate after pairing. Immature: Browner.

Flying in loose 'V' formations, they can travel long distances. Riding updrafts of warm air in wide soaring circles, they can reach heights of 3000m.

Habitat: Sheltered estuaries, rivers, mudflats, inland lakes.

Feeding: Usually gregarious, co-operative, feeding in shallow waters on fish, crustaceans.

Voice: Usually silent. Gruff croaking squabbling over food.

Status: Common. Highly nomadic.

Breeding: Any time of year. Often in huge inland colonies. Incubation is shared and eggs are incubated on adults' feet. Scrape depression lined with sticks and seaweed / 2–3 eggs.

Opposite: Pelicans with courtship-coloured bills.

Cape Gannet

Morus capensis

900mm
Wingspan: 1800mm

Sexes alike. Similar to Australasian Gannet but has fine black lines down centre of throat and foreneck. Overall white with black tail, primaries and secondaries. Crown, hind neck: Yellow. Eyes: Blue with blue eye ring. Underwings: White with black trailing edge and tips. Immature: Speckled, streaked brown.

Strong flap and glide flight.

Habitat: Oceans and bays to 100km offshore.

Feeding: Plunge-dives for shoaling fish.

Voice: Silent at sea. Rasping call.

Status: Vulnerable.

Breeding: Apr.–Dec. Southern Africa, but since 2003 a new nesting colony has regularly occurred in southern Vic. Mounded seaweed nest / 1 egg.

Wedge Light platform at lower Port Phillip Bay, and Lawrence Rocks, off Portland, Vic.

Opposite: Nesting colony. Nesting colonies in southern Africa may contain over 140,000 birds.

Australasian Gannet

Morus serrator

840–920mm
Wingspan 1800mm

Sexes alike. Sleek, mostly white body. Crown, face, neck: Pale yellow tapering to white. Primary, secondary and inner tail feathers: Black. Eyes: Pale grey with blue eye ring. Bill: Long, straight, slender with backward-pointing serrations. Underwings: White with black tips and trailing edge. Immature: Speckled, streaked brown, attaining adult plumage over 2 years.

Often seen performing spectacular high plunge-diving off the coast. Effortless flight and glides, flying in lines close to waves, wheeling, soaring, high-diving and plunging after prey with folded-back wings from a height of up to 15m.

Habitat: Coastal and offshore waters, estuaries and bays.

Feeding: Herding and plunge-diving for fish, usually in groups.

Voice: Silent at sea. Loud, harsh cackling at colony.

Status: Common. Widespread in summer in southern Australian waters.

Breeding: Oct.–Dec. Bass Straight islands and Tas. Nest is an earth mound / 1 egg.

Point Danger, Portland, Vic.

Left: Adult in flight showing upperparts. Right: Underparts.

Masked Booby

Sula dactylatra

750–850mm
Wingspan: 1600mm

Sexes similar. Overall white. Flight feathers, tail: Dark brown. Eyes: Yellow. Face mask: Blue-black around eyes, base of bill and throat. Bill: Variable, but mostly yellow with blue-black base. Feet: Blue-grey or blue-green.
Female: Bill dull green-yellow.
Immature: Mottled brown upperparts, brown head. Underparts: White.

Habitat: Tropical seas and reefs.

Feeding: Deep, vertical plunge-dives to a depth of 35m or more. Bill used to 'spear' fish, squid.

Voice: Silent at sea. High-pitched whistles, noisy honks at nest.

Status: Locally abundant.

Breeding: All year, mostly July–Dec. Lord Howe, Norfolk and offshore islands, Great Barrier Reef and WA. Nest is a scrape in sand / 2 eggs.

Opposite: Adult in flight.

Red-footed Booby

Sula sula

710–790mm
Wingspan: 1400mm

Sexes similar, male is smaller. Smallest booby with distinctive bright red legs visible in flight. White or brown morphs and several intermediaries. Eyes: Grey with dark grey to purplish eye skin. Bill: Variable.
White morph is most common in Australia. Mostly white with black primary and secondary feathers and yellowish crown. Ash brown morph: Plumage is pale to dark grey; tail is usually white.
Immature: Dark brown, mottled, paler breast, belly. Feet: Pinkish.

Fast flight with long glides skimming the waves. Dives from heights of 8m or more. Often follows boats.

Habitat: Tropical and subtropical seas. Most pelagic booby.

Feeding: Plunge-dives for fish, squid.

Voice: Silent at sea. Harsh, repetitive 'kreck' call.

Status: Locally abundant near breeding colonies.

Breeding: Apr.–Oct. In colonies on offshore tropical islands. Stick nest in a shrub / 1 egg.

Christmas Is.

Left to right: Brown, intermediate and white morphs.

Brown Booby

Sula leucogaster

750–800mm
Wingspan: 1400mm

Sexes similar. Upperparts, rump, tail, throat: Dark brown. Underparts: Sharply defined white. Underwings: White, with dark tips, edges. Bill: Sharp, serrated, creamy white with blue base and blue facial skin. Legs, feet: Yellow. Eyes: Variable, cream, yellow, brown.
Female: Larger with yellow facial skin and small blue patch in front of eyes. Immature: Similar, but underparts mottled grey-brown.

Leisurely glides and wingbeats, close to waves, often low plunges. Least pelagic booby.

Habitat: Oceanic, bays, harbours, lagoons around islands.

Feeding: Dives, chases or plucks fish, squid from water.

Voice: Usually silent at sea. Male: High-pitched hissing. Female: Low-pitched honks, 'kruk' calls.

Status: Common.

Breeding: All year. Qld offshore islands and northwest WA. Scrape on ground nest / 2 eggs.

Lady Elliot Is.

Left: Male (left) and female. Right: Female.

Abbott's Booby

Papasula abbotti

Critically Endangered
800mm
Wingspan: 800mm

Sexes similar. Overall off-white with black eye patch and long pointed black tail and upperwings. Legs, feet: Blue with black tips. Flanks: Black marks. Underwings: White with black trailing edge and tips. Bill: Pale grey with black tip.
Female: Pink bill with black tip.
Immature: Similar to male, bluer around eyes.

Flight is slow, effortless. Glides with occasional wingbeats.

Habitat: Oceanic.

Feeding: Dives for fish, squid. Snatches flying fish.

Voice: Deep bellow.

Status: Critically endangered.
Breeding: Apr.–Dec. Christmas Is. Nest is cup-shaped stick platform high in rainforest trees / 1 egg.

Left: Male. Right: Male in flight.

Australasian Darter

Anhinga novaehollandiae

850–900mm
Wingspan: 1200mm

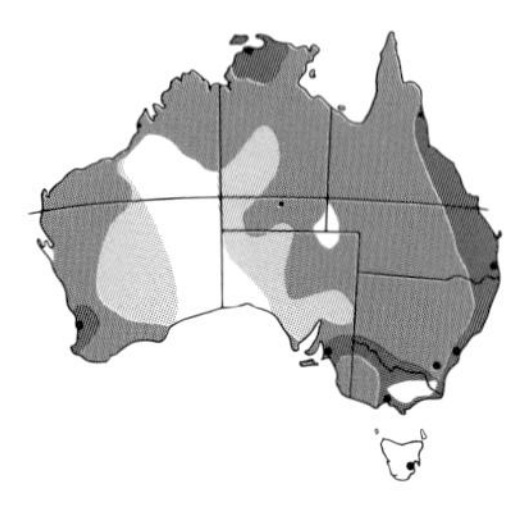

Also called 'Snake-Bird' as it swims with its body submerged, with only its snake-like head and long neck visible, writhing when disturbed. Overall black with long white streak from behind eyes down neck. Neck: Kinked with reddish patch on 'bend' brighter on breeding males. Upperwings: Black with white bars. Tail: Large, rounded. Female: Head, neck: Pale grey with white streak. Underparts: Creamy white. Immature: Similar to female, greyer.

Usually solitary or in small groups. Often preening or holding their wings to dry. Soars effortlessly on thermals. Rapid shallow wingbeats then glides in direct flight, in unison.

Habitat: Fresh or salt water. Sheltered bays, harbours and waterways. Swamps, wetlands, lakes with vegetated areas.

Feeding: Stalks prey. Dives for up to a minute and re-dives. Uses long, sharp, pointed bill to 'spear' fish or other aquatic prey.

Voice: Harsh rattling and cackling at nest.

Status: Moderately common. Solitary.

Breeding: All year, in widely spaced loose colonies. Rough nest in tree or ground / 4 eggs.

Left: Breeding male. Right: Female.

Little Pied Cormorant

Microcarbo melanoleucos

610mm
Wingspan: 1000mm

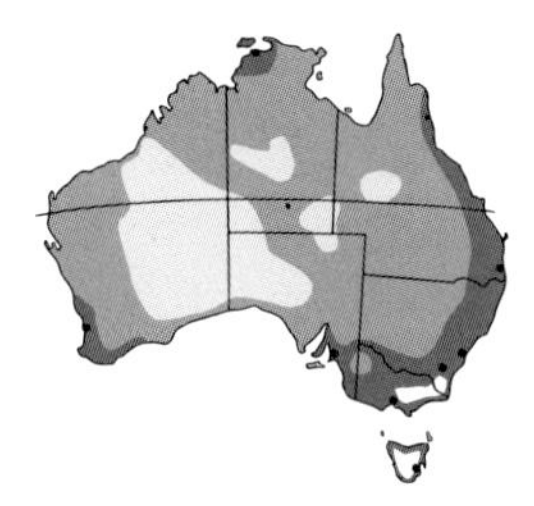

Sexes similar. Smallest and most commonly seen Australian cormorant. Upperparts are black with a bronze-green sheen. Underparts: White to above the eyeline. Bill: Yellow, bordered black. Breeding plumage: Fuller, spiky crest.
Immature: Bill: Grey brown. Upperparts: Blackish and black below the eye.

Flight is strong and direct with head slightly up. Often in urban parks. Roosts and nests communally.

Habitat: Coastal lagoons, sheltered harbours, dams, lakes. Fresh or salt water.

Feeding: In flocks in coastal areas or solitary inland. Aquatic animals, fish, crustaceans, insects. Yabbies inland.

Voice: Cooing and harsh calls at nest.

Status: Abundant.

Breeding: All months in southern Australia and NZ. In colonies. Rough stick platform nest in trees usually growing in water / 3–5 eggs.

Left: Adult upperparts. Centre: Underparts.Right: Swimming low in water.

Pied Cormorant

Phalacrocorax varius

660–840mm
Wingspan: 1500mm

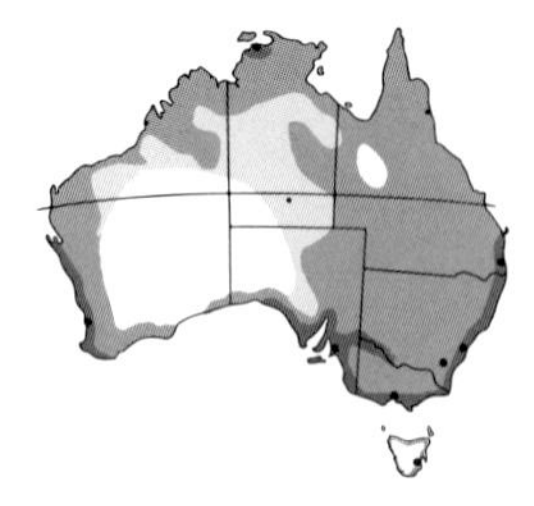

Sexes similar. Largest white-breasted Australian cormorant. Similar to Black-faced Cormorant but larger, with a black crown starting above the eye. Upperparts: Glossy black. Eye ring: Blue. Lores: Orange-yellow. Bill: Dark grey, hooked. Underparts: White with black streaks on flanks. Breeding: Bare chin patch turns purplish pink. Immature: Upperparts: Mottled brown. Underparts: White with brown patches.

Widespread, solitary, sometimes in huge flocks. Usually flying in 'V' formations, less frequently in a single line.

Habitat: Inshore waters including bays, estuaries, harbours and inland rivers and wetlands.

Feeding: Diving and underwater pursuit for shoaling fish, crustaceans and molluscs. Webbed feet used for propulsion.

Voice: Guttural cackling at nest.

Status: Moderately common to abundant.

Breeding: Any time, dependent on food supply. Colonial. Large, bulky nest of sticks and seaweed in trees over water, on ground or in small shubs on islands / 3 eggs.

Left: Non-breeding. Right: Breeding.

Great Cormorant

Phalacrocorax carbo

720–920mm
Wingspan: 1300–1500mm

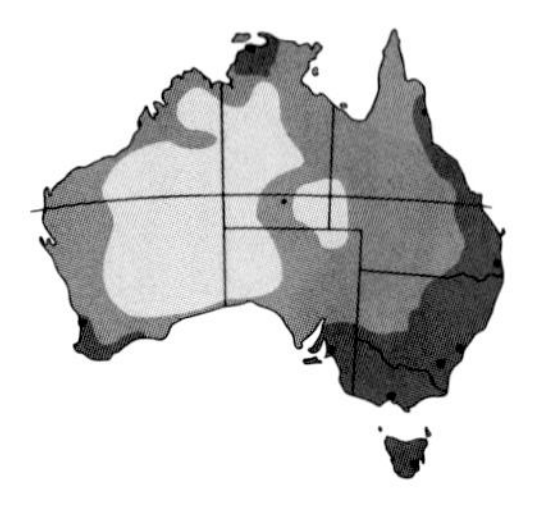

Sexes alike. Australia's largest cormorant. Overall black plumage with a slight greenish sheen. Bill: Dark grey. Facial skin, throat: Yellow. Legs, feet: Black. Breeding plumage: Orange-red facial skin and throat. Wings: Bronze patterned. Immature: Browner, smudgy white face.

Flies in 'V' formations. Rapid wingbeats, gliding. Often seen resting on sandbanks or poles drying their outstretched wings.

Habitat: Inland rivers, farm dams, coastal estuaries and sheltered inlets.

Feeding: In groups or solitary. Shallow underwater dives using feet to propel itself after prey.

Voice: Usually silent. Breeding male: Croaking stuttering groans. Female: Hissing noises.

Status: Widespread, abundant, nomadic.

Breeding: Any month. In colonies, large, stick-platform nest in treetops or sometimes on cliff or ground / 4 eggs.

Left: Adult in breeding plumage with chicks at nest. Right: Non-breeding plumage.

Little Black Cormorant

Phalacrocorax sulcirostris

580–640mm
Wingspan: 1000mm

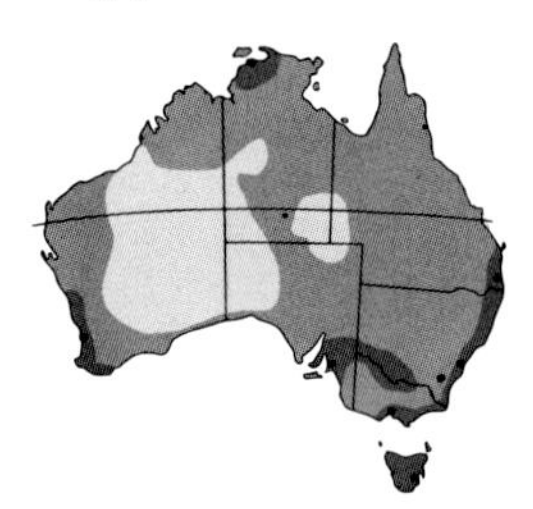

Sexes alike. Distinctively small, slim, overall black cormorant with a greenish sheen on the back, dark facial skin and emerald-green eyes. Bill: Grey, slender and hooked. Breeding plumage: Head, neck: Fine white flecks. Immature: Brownish.

Most commonly seen on inland waters. Flies in large flocks in V-shaped formation.

Habitat: Prefers inland freshwater wetlands. Less frequently sheltered coastal waterways.

Feeding: Dives and swims for fish, crustaceans, aquatic insects. Working in groups they corral schools of fish and feed together.

Voice: Silent outside breeding season. Croaking and ticking sounds when fishing communally, and at nest.

Status: Common. Nomadic. Sedentary in good conditions.

Breeding: Sept.–May. In colonies, large stick-platform nest in trees or on ground / 4–6 eggs.

Left: Adult, holding wings out to dry
Right: Breeding adult.

Black-faced Cormorant

Phalacrocorax fuscescens

Endemic

600–700mm
Wingspan: 1000mm

Sexes alike. Sea cormorant with white breast, naked black face and glossy black crown, hind neck and upperparts. Underparts: White. Flanks: Black. Eyes: Turquoise. Bill: Black, hooked. Legs, feet: Black. Breeding male: White 'nuptial' plumes to hind neck. Immature: Upperparts: Brown. Underparts: White with brown spots.

Confined to inshore marine habitats. Coastal waters, bays and offshore islands. Flight is straight, close to water with rapid shallow wingbeats. Glides with stiff wings held at right angles to body.

Habitat: Coastal waters, bays and offshore islands.

Feeding: Dives for fish, squid.

Voice: Gutteral croaking and hissing.

Status: Moderately common. Sedentary.

Breeding: Sept.–Jan. Nest on ground in seaweed, debris / 3–5 eggs.

Bruny Is. and Denison Canal, Dunalley, Tas.

Left: Adult. Note hooked bill. Right: Adult group with breeding males.

Great Frigatebird

Fregata minor

860mm–980mm
Wingspan: 2050–2500mm

Lightly built, mostly black with a red throat pouch, inflated in courtship. Deeply forked black tail. Wings: Black. Bill: Long, pale grey, hooked. Eyes: Brown. Eye ring: Black. Legs, feet: Red. Female: Similar, larger. Eye ring, bill: Dull pink. Throat, breast: White. Belly: Black centre. Underwings: Black. Immature: Upperparts: Dark brown. Underparts: White to streaked ginger. Head: Tinged rufous.

Wholly adapted to flight. Soars on thermals. Perches only. Neither walks nor swims.

Habitat: Tropical, subtropical seas, inshore islands.
Feeding: Fish taken in flight. Snatches prey from other birds.
Voice: Gobbling sounds.
Status: Abundant. Sedentary.
Breeding: Jan.–Oct. Tropical islands. Stick-platform nest in shrubs / 1 egg.

Lady Elliot Is., Qld.

Left: Male. Right: Mating pair.

Lesser Frigatebird

Fregata ariel

700–800mm
Wingspan 1900mm

Smallest frigatebird. In flight distinguished by its white 'armpits'. Underwings, belly: Black. Tail: Long, black, deeply forked. Male: Glossy black, red inflatable throat pouch, inflated during courtship. Bill: Grey-black. Eye ring: Black. Legs, feet: Pink. Female: Throat: Black. Breast, collar: White. Bill: Greyish pink. Eye ring: Pink. Immature: Ginger streaked head. Breast: Blackish. Belly: Off-white.

Soars effortlessly. Neither walks nor swims. Perches.

Habitat: Worldwide. Tropical.
Feeding: Surface-seizes fish.
Voice: Grunts, whistles at nest.
Status: Locally abundant. Sedentary. Immature birds dispersive.
Breeding: May–Dec. Coral Sea, Great Barrier Reef islands and coast. Stick-platform nest on ground or in trees / 1 egg.

Left: Female. Right: Male (foreground) with juvenile and female.

Red-tailed Tropicbird

Phaethon rubricauda

460–470mm / 500mm streamers
Wingspan: 1000mm

Sexes similar. Overall white plumage with a satiny sheen and delicate pink hue. Black crescent around eyes and black innermost flight feathers and flanks. Bill: Sturdy, red to orange-red. Tail: Bright red streamers, sometimes broken, giving a stumpy tail appearance. Legs, feet: Blue-grey.
Immature: Black bill. Underparts: Lightly barred. Lacks streamers.

Fast, high flight. Hovers. Fluttering wingbeats, alternating glides. Weak legs, shuffles along using breast and wings on land. Only comes ashore to breed. Usually solitary.

Habitat: Tropical and subtropical seas.
Feeding: Pelagic. Completely submerges from high plunges.
Voice: Harsh rattling calls in flight. Screams at nest.
Status: Widespread and locally abundant.
Breeding: Sept.–Dec. Across Indian and Pacific Oceans. Rowley Shoals Marine Park, WA. Loose colonies. Nest is a scrape in a shady place / 1 egg.

Left, right: Adult.

White-tailed Tropicbird

Phaethon lepturus

400mm / 300mm streamers
Wingspan: 950mm

Sexes similar. Small, graceful, overall white with white tail streamers. Black comma-shaped eye patch; black tertiary coverts and primaries. Bill: Yellow. 'Golden morph': Indian Ocean / Christmas Is. Overall golden tinge. Bill: Orange.
Immature: Lacks tail. Smaller, finely streaked, mottled black. Bill: Cream-green.

Follows ships. Fast flight. Flutters, soars, glides; hovers before diving. Usually solitary, at times in pairs.

Habitat: Tropical and subtropical seas.
Feeding: Poor swimmer. Plunges or vertical dives from 10–20m for prey. Flying fish caught in flight.
Voice: Harsh screaming, rattling at nest. Often silent at sea.
Status: Nomadic. Widespread.
Breeding: Jan.–Apr. Loose colonies. Nests in hollows or crevices on tropical islands / 1 egg.

Left: Adult. Right: Immature.

Bitterns

Herons

Egrets

Ibis

Spoonbills

Stork

Brolga

Crane

A group of 4 families of wading birds that have similar feeding behaviour. Most species in this group use their long sharp bill to stab prey in shallow water or surrounds. Their bills mostly have small, serrated gripping edges that hold slippery live prey.

Family Ardeidae includes bitterns, herons and egrets.

Bitterns are short-necked, secretive birds that fly with their necks retracted. When disturbed they 'freeze' with their bill pointed skyward, perfectly camouflaged in their reedy wetland environment.

Egrets are also herons and are similar in appearance and habits. Their classification is not clear, but egrets are mostly white and herons are usually brownish. Egrets derive their name from the French word 'aigrette' (pronounced 'egret') which refers to their head plumes used to decorate ladies' headdress in the 19th and 20th centuries. 'Plume hunters' slaughtered huge numbers for the millinery industry, leading to the birds' near extinction.

Both egrets and herons are long-legged waders that feed while wading, often gently stirring the water with one foot to disturb prey, then springing their long necks forward, snatching up fish, crustaceans, insects or other invertebrates. Most egrets and herons nest in colonies and build stick nests high in a tree or shrub. Most herons develop tufts of powder down feathers under the breast and rump plumage that break up as they preen, producing a protective powder that the birds dust through their feathers as a dressing.

Family Threskiornithidae is represented in Australia by three species of ibis and two species of spoonbill.

Ibis have become familiar suburban birds since the early 1970s when large migrations from inland areas occurred. They are easily recognisable with their long sickle-shaped bills used for probing and seizing prey.

Spoonbills are usually seen slowly wading in shallow water sweeping their bill from side to side while stirring up debris and prey from the bottom. Using their specialised spatulate-shaped bills they locate the disturbed prey in the turbid water by touch, filtering it through their sensitive bills.

Family Ciconiidae includes storks, represented in Australia by one species, the Black-necked Stork, or 'Jabiru', a tall, long-legged freshwater forager that spears prey, including fish and snakes, with its powerful, sharp bill.

Family Gruidae includes cranes. Represented in Australia by the elegant Brolga and the similar Sarus Crane.

Brolgas pair for life and are noted for their elaborate, ritualised courtship displays that include graceful dancing, loud trumpeting calls, leaping with outstretched wings, bowing and strutting. Although native Australians have always differentiated the Sarus Crane from the Brolga, the difference was only noted in 1969 by ornithologists and thought to be a recent migrant from Asia. DNA studies have since shown that the endemic subspecies *gilliea* has been present in Australia for several thousand years.

Opposite: Eastern Great Egret in courtship dance.

Australasian Bittern

Botaurus poiciloptilus

650–750mm

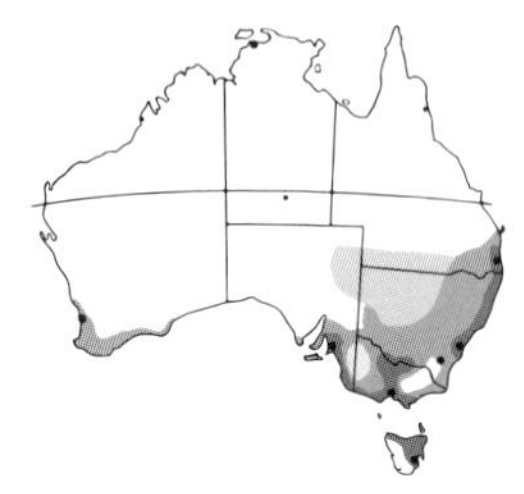

Sexes similar. Secretive, powerfully built and well-camouflaged bittern. Upperparts: Mottled dark brown and black buff. Underparts: Streaked brown and buff. Throat: Whitish. Neck: Brown streak down sides. Bill: Straight, pointed, dark brownish buff. Eyes: Yellow. Face: Pale white stripe around eyes and greenish bare skin patch. Legs: Dull green.
Immature: Paler.

Secretive and rarely seen. Slow movements. Flight is infrequent with deep, slow wingbeats. Part nocturnal.

Habitat: Dense reed beds along edges of swamps, rivers or rice farms.

Feeding: Wading in water in dense vegetation, mostly at night, taking yabbies, eels, fish and small mammals.

Voice: Deep, loud, repetitive booming. Usually at night.

Status: Uncommon, vulnerable.

Breeding: Oct.–Jan. Nest is a flattened platform of water plants over water among reeds / 4–6 eggs.

Werribee, Vic. Deniliquin, NSW.

Left: Stalking prey. Right: When alarmed, birds will freeze with bill pointing skywards.

Australian Little Bittern

Ixobrychus dubius

280–340mm

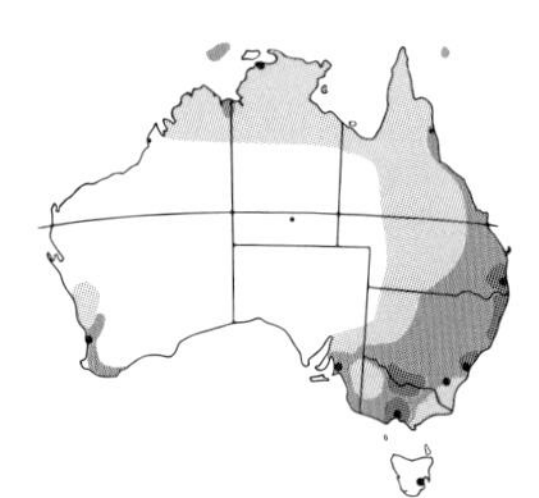

Sexes similar. Smallest Australian heron. Crown, back, tail: Black. Foreneck, underparts: Buff, streaked and spotted brown, with central dark, broken line from throat to belly. Nape, shoulders: Rufous. Wings: Black with large buff patch. Legs: greenish yellow. Bill, lores: Green.
Non-breeding: Bill lores: Yellow. Female: Smaller, duller. Back: Mottled brown. Immature: Browner, heavily streaked.

Secretive and rarely seen. Freezes in an upright pose, bill raised, when disturbed.

Habitat: Freshwater swamps, lakes, rivers with dense reeds, sedges and rush beds.

Feeding: Generally nocturnal. Preys on insects, small fish, frogs. Defends feeding territory.

Voice: Repeated deep croaking 'orrk, orrk, orrk', particularly in the evening or at sunrise and during breeding season.

Status: Uncommon. Threatened. Usually solitary.

Breeding: Oct.–Dec. Loose platform nest in dense rushes / 4–7 eggs.

Deniliquin, NSW. Metroplex Wetlands, Brisbane, Qld.

Left: Male, hunting for prey. Right: Female.

Black Bittern

Ixobrychus flavicollis

550–650mm

Sexes similar. Overall sooty black. Underparts: Yellow stripes from throat to either side of breast, centre neck to belly streaked black and white. Eyes: Pale yellow. Bill: Long, slender, yellow to grey-brown, dark top edge. Legs: Olive-brown. Female: Duller, upperparts browner. Immature: Browner. Upperparts: Brown feathers edged buff.

Roosts in trees by day. Freezes in an upright pose if disturbed.

Habitat: Thickly vegetated edges of ponds, rivers, mangroves, mudflats.

Feeding: Mostly at dusk and dawn. Fish, molluscs, insects.

Voice: Drawn-out cooing. Repetitive booming when breeding.

Status: Uncommon.

Breeding: Sept.–Mar. Sometimes colonially. Stick-platform nest in trees over water / 3–6 eggs.

Centenary Lakes, Cairns, and Daintree River, Qld. Manning Gorge, WA.

Left: Adult. Right: Immature.

Striated Heron

Butorides striata

420–460mm
Wingspan: 600–680mm

Sexes similar. Stocky body and hunched posture. Large head and bill. Glossy black cap and crest. Foreneck: White marks down centre. Eyes: Yellow. Bill: Grey. Legs: Short, dull yellow. **Breeding:** Face, legs flushed orange-red.
2 colour morphs present in Australia. Grey morph: Upperparts: Dark grey-brown. Underparts: Paler to buff. Rufous morph: Variable with soil colour. Common in Onslow area Pilbara, WA. Upperparts: Rufous. Underparts: Pink-brown to buff.
Immature: Grey and white striped underparts and white spotted wings.

Usually solitary. Crouching low, stalks mudflats for prey.

Habitat: Mangroves, intertidal flats, estuaries.

Feeding: Low tide. Uses its heavy, sharp bill to jab prey, taking mudskippers, fish, crabs, crustaceans, molluscs, insects.

Voice: Squawks when alarmed. Squeaking sounds at nest.

Status: Locally common, more common in the north. Sedentary.

Breeding: Feb.–Mar. Scanty stick-platform nest in fork of mangrove branches / 3–4 eggs.

Left: Adult rufous morph. Right: Grey morph.

Nankeen Night-heron

Nycticorax caledonicus

560–680mm
Wingspan: 1100mm

Sexes alike. Slate-black crown and nape with 2–3 slender, long, white nuptial plumes projecting from nape when breeding. Upperparts: Chestnut brown. Underparts: Creamy white. Bill: Black. Eyes: Yellow with white surround. Facial skin: Green or bluish when breeding. Legs: Greenish yellow.
Immature: Black cap. Pale brown, streaked and spotted white.

Mostly nocturnal. Roosts in colonies in trees near water by day or on islands on the ground in treeless areas. Flies with slow, silent wingbeats.

Habitat: Varied. Brackish or fresh, permanent or semi-permanent water. Swamps, lagoons, estuaries with mangroves.

Feeding: Birds leave colony in unison, travelling to wetland feeding grounds. Forage alone from early evening and before dawn for fish and other aquatic prey.

Voice: Guttural croak, often at night.

Status: Common. Nomadic.

Breeding: Sept.–Apr. Stick-platform nest in trees to 25m high / 2–3 eggs.

Left: Breeding adults. Right: Immature.

White-necked Heron

Ardea pacifica

Endemic

910mm
Wingspan: 1600mm

Sexes similar. Breeding: Slate-black back contrasts with white head and neck. Long, thin, iridescent maroon plumes to back and upper breast. Head, neck: White, front faintly spotted. Face: Blue skin. Upper breast, belly, undertail: Dark grey, streaked white. Bill, legs: Black. Eyes: Buff.
Non-breeding: Double-spotting line to foreneck; lacks long, iridescent plumes.
Immature: Similar to non-breeding with heavy black neck spotting.

Stately. Usually solitary. Wades in water or stalks in damp grass. Flies with neck folded. Slow and steady wingbeats and trailing legs. White spots to carpel wing joint prominent in flight.

Habitat: Shallow inland freshwater lakes, wetlands or wet grassy areas. Occasionally coastal mudflats.

Feeding: Insects, crustaceans, marine animals, fish.

Voice: Loud, deep, single or double 'croak' when alarmed.

Status: Moderately common. Nomadic. Appearing in large numbers in good rain. Rarely near coast except in drought.

Breeding: Anytime, depending on availability of food. Platform nest of sticks in trees to 30m high, over or near water / 4–6 eggs.

Left: Breeding. Right: Immature.

Great-billed Heron

Ardea sumatrana

1000–1100mm
Wingspan: 1850–2300mm

Sexes alike. Large, heavily built and overall brownish grey heron. Breeding plumage: Back, breast, nape: Long, silver nuptial plumes. Face: Grey. Throat: White. Eyes: Yellow. Bill: Large, black and yellow on lower base. Non-breeding: Lacks nuptial plumes although some are retained on foreneck.
Immature: Similar, browner, lacks silvery plumes.

Wary and secretive, skulking in mangroves by day.

Habitat: Mangroves, paperbark swamps bordering tidal lakes, estuaries and rivers.

Feeding: Stalks for aquatic animals, fish, crustaceans on mudflats or in shallows, usually at dusk.

Voice: Drawn-out, penetrating, croaking calls, mostly at night.

Status: Uncommon. Sedentary.

Breeding: All months. Untidy platform nest in trees, usually over water / 2 eggs.

Daintree River, Qld.

Left and right: Breeding plumage.

Pied Heron

Egretta picata

450–480mm
Wingspan: 800–900mm

Sexes alike. Small, elegant overall slate-grey heron with contrasting white cheeks, throat and neck. Long black nuchal crest, and dark cap extending below eyes. Back: Long, flowing, blue-black plumes. Neck: Long white plumes around base that hang over blue-black breast. Bill, legs: Yellow. Eyes: Yellow with dark blue facial skin in front and on base of upper bill. Non-breeding: Short crest and plumes.
Immature: Head and neck white. Lacks cap, crest and plumes.

Usually seen in feeding flocks of 30 or more birds. Aggressive towards other species.

Habitat: Salt-or freshwater coastal swamps, billabongs and adjacent grasslands. Rubbish dumps.

Feeding: Communal, taking fish and small invertebrates. During droughts large feeding flocks of several hundred birds form.

Voice: Soft cooing at nest. Harsh croaks.

Status: Common, sedentary. Locally nomadic, following food. Sparse distribution over inland NT, Qld, NSW. Northern birds overwinter in PNG or further north.

Breeding: Sept.–Nov. Colonially, usually in mangroves, small stick-platform nest / 3–4 eggs.

Left: Non-breeding. Right: Breeding.

White-faced Heron

Egretta novaehollandiae

650–690mm
Wingspan: 1000mm

Sexes alike. Most commonly seen heron. Overall pale blue-grey with white face extending to the chin and upper throat and long nuptial blue-grey plumes on back and nape in breeding season. Upper breast, lower neck: Rufous-grey plumes. Legs: Olive-yellow. Bill: Dark grey. Non-breeding: Plumes reduced or absent.
Immature: Paler. Face: Grey. Throat: White. Breast: Pale rufous-grey.

Flies with neck retracted, feet trailing and slow, buoyant, deep wingbeats.

Habitat: Freshwater, brackish or saltwater wetlands, mangroves, dams, tidal mudflats, coastal waters, inland lakes.

Feeding: Waits and watches or stalks prey including fish, eels, frogs, rats, insects.

Voice: Harsh croaking.

Status: Common. Sedentary or nomadic following food supply. Migrating north over winter.

Breeding: Sept.–Nov. In small colonies, stick-platform nest in trees to 20m high / 4–7 eggs.

Left: Adult. Right: Adult in flight – note darker flight feathers.

Eastern Great Egret

Ardea modesta

800–900mm
Wingspan: 950mm

Sexes alike. Largest Australian white heron. Breeding: Overall snowy white with long, erectile, lacy nuptial plumes on back extending past tail. Bill: Black. Lores: Turquoise. Eyes: Red. Legs: Brown.
Non-breeding: Eye, bill, lores: Yellow. Plumes absent.
Immature: Like non-breeding adult.

Usually solitary, mostly seen stalking prey in shallow water.

Habitat: Shallows, intertidal mudflats, wetland shallows, mangroves, reefs, irrigated pastures, sewerage ponds.

Feeding: Strikes and seizes prey including fish, crustaceans, amphibians and insects.

Voice: Loud, low croaking and guttural calls at nest.

Status: Common. Cosmopolitan in temperate climates. Roosts in large flocks, sometimes numbering hundreds of birds.

Breeding: Oct.–Mar. or anytime, depending on availability of food. Colonial. Modest stick-platform nest in tree / 3–6 eggs.

Left: Breeding plumage. Right: Non-breeding.

Intermediate Egret

Ardea intermedia

560–660mm
Wingspan: 900–1050mm

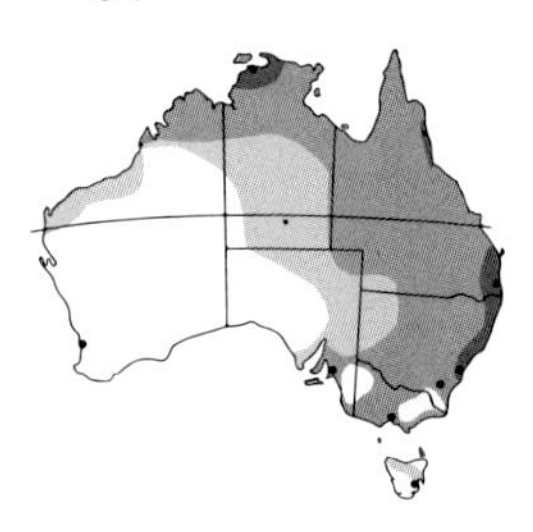

Sexes alike. Overall white with distinctive erectile nuptial plumes that hang like a lacy veil from the breast and back and extend past the tail. Bill, legs: Orange-red. Eyes: Yellow. Facial skin: Blue-green.
Non-breeding: Nuptial plumes absent. Bill, facial skin: Yellow. Legs: Brown.
Immature: Similar to non-breeding adults.

Habitat: Freshwater wetlands, lakes, billabongs, shallow waters and mudflats with emergent plants.

Feeding: Small fish, crustaceans, aquatic animals and insects.

Voice: Usually silent. Loud croaking alarm call. Various guttural sounds in colony at night.

Status: Common.

Breeding: Nov.–Jan. in south. Mar.–May in north. In colonies and with other waterbirds. Saucer-shaped, untidy stick nest in tree forks, 3–10m high / 3–6 eggs.

Left: Breeding pair at the nest. Right: Non-breeding.

Eastern Cattle Egret

Bubulcus coromandus

480–530mm
Wingspan: 900mm

Sexes alike. Overall white egret. Orange-buff spiky nuptial plumes to head, neck, breast and back. Bill: Reddish orange. Eye, eye ring, face skin, legs: Reddish.
Non-breeding: Plumes absent. Overall white, sometimes buff-tinged. Bill, eyes: Yellow. Legs: Olive.
Immature: Similar to non-breeding adult.

Usually seen in flocks, often around cattle.

Habitat: Swamplands, grasslands, farmland with grazing animals.

Feeding: Takes parasites from grazing animals and feeds on insects disturbed by grazing.

Voice: Guttural noises, low croaks.

Status: Common, first arriving in Australia in 1948.

Breeding: All year, depending on food supply. Modest stick-platform nest in trees / 3–6 eggs.

Left: Breeding bird with sticks in courting 'dance'. Right: Non-breeding.

Little Egret

Egretta garzetta

560–650mm
Wingspan: 900mm

Sexes similar. Overall white-plumed bird. Breeding: Nuptial plumes fall from hind crown, lacy plumes on upper breast, wings and mantle. Face: Bare skin flushed red in courtship. Bill: Slender, black with pale yellow lower base. Legs: Black with yellow feet.
Non-breeding: Yellow facial skin. Plumes reduced or absent.
Immature: Lacks plumes. Facial skin: Yellow-green.

Usually solitary, seen feeding in shallow water. Flight is light, buoyant.

Habitat: Tidal mudflats, mangroves, wetlands.

Feeding: Only egret to chase prey, dashes in erratic pursuit, lifting wings to startle aquatic prey including invertebrates and fish or stirs, 'puddles' water to disturb prey.

Voice: Harsh croaks, especially when breeding.

Status: Uncommon in north, rare in south. Nomadic.

Breeding: Oct.–Feb. in south. Mar.–Apr. in north. In colonies, scanty stick nest, in trees or bushes usually over water / 3–5 eggs.

Left and right: Breeding plumage.

Eastern Reef Egret

Egretta sacra

600–650mm
Wingspan: 950mm

Sexes alike. White and grey morphs. Overall white or dark grey. Wispy nuptial plumes on back, breast and nape. Long neck with short, thick legs. Bill: Yellow-grey. Face skin: Greenish-yellow. Eyes: Yellow. Legs: Grey-yellow. Grey morph: Dark grey with white stripe on throat. Plumes: Slate or blue-grey. Bill: Slate brown to grey.
Non-breeding, both morphs: Nuptial plumes absent.
Immature: Similar to non-breeding adult, white or grey morph.

White morph most common in tropical areas; grey morph in temperate areas. Morphs interbreed. Usually solitary, seen stalking prey on reefs or along the tideline.

Habitat: Inshore coral reefs. Intertidal waters, shoreline and rock platforms.

Feeding: Solitary, actively hunts for fish, crustaceans. Raids tern colonies for nestlings.

Voice: Loud croak when alarmed. Guttural calls at nest.

Status: Common.

Breeding: Sept.–Jan. In colonies, nest in low shrubs or rock ledges / 2–5 eggs.

Left: White and dark grey morphs.
Right: Non-breeding white morph.

Below left: Immature grey morph.
Below right: Adult grey morph.

Glossy Ibis

Plegadis falcinellus

490–580mm
Wingspan: 800–950mm

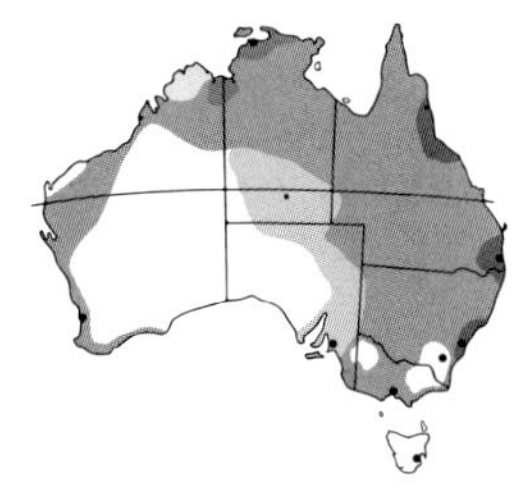

Sexes similar. Australia's smallest ibis. Overall bronze-brown with an iridescent green sheen to wings. Face: Blue-grey facial skin with white border. Eyes, legs: Brown. Bill: Olive-brown, down-curved. Head: Dusky, white streaked.
Breeding: Rich chestnut neck, mantle, underparts. Upperparts: purple-green sheen. Face: Pale bluish.
Female: Smaller. Immature: Duller. Similar to non-breeding with heavy streaked head, neck.

Gregarious, often in flocks of several hundred. Long lines or 'V' formations, short dashing and gliding flight.

Habitat: Mostly inland in well-vegetated swamps, margins of lakes and damp grasslands.

Feeding: Usually in flocks in shallow water on insects, molluscs and crustaceans.

Voice: Nasal croaking. Grunts in breeding season.

Status: Locally abundant. Nomadic. Rare Tas.

Breeding: Sept.–Apr. or in response to good rain. In mixed colonies with other waterbirds, nest placed in low vegetation in water / 4–6 eggs.

Left: Breeding adult. Right: Non-breeding.

Australian White Ibis

Threskiornis molucca

650–750mm
Wingspan: 1100–1250mm

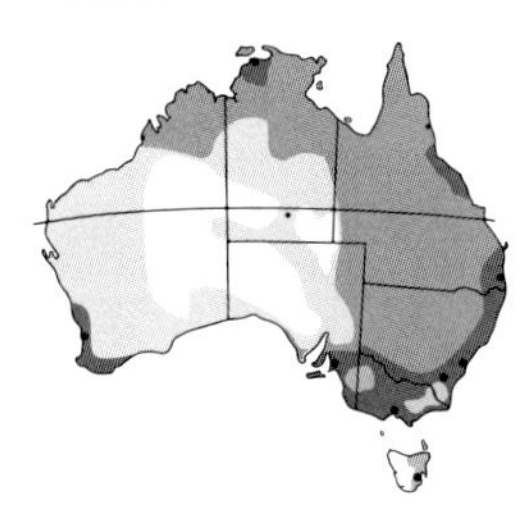

Sexes similar. White body with blackish bare skin on head and upper neck. Wings: White with green-black tipped primaries and lacy plume-like outer secondaries. Eyes: Dark brown. Bill: Long, down-curved, black. Breeding: Bare skin under wings turns scarlet following the line of the wing bones. Nape: Pink bands. Breast: Long, white plumes. Tail: Washed yellow. Legs: Pink.
Female: Smaller. Immature: Similar, but head and neck are black feathered. Lacks wing plumes.

Flies with neck elongated and feet trailing behind tail. Usually seen in lines or 'V' formations, soaring to 300m on thermals.

Habitat: Varied. Freshwater wetlands, swamps, streams, lakes and tidal mudflats. Irrigated pasture, rubbish tips, urban parks.

Feeding: In shallow water for aquatic insects, molluscs, caterpillars, grasshoppers.

Voice: Grunting calls.

Status: Common. Sedentary. Occasionally nomadic depending on water conditions. Young birds disperse, usually northward, up to 1200km.

Breeding: Aug.–Nov. in south. Feb.–May in north. In colonies, stick platform nest in low plants / 2–5 eggs.

Left and right: Breeding plumage.

Straw-necked Ibis

Threskiornis spinicollis

Endemic

680–750mm
Wingspan: 1000–1200mm

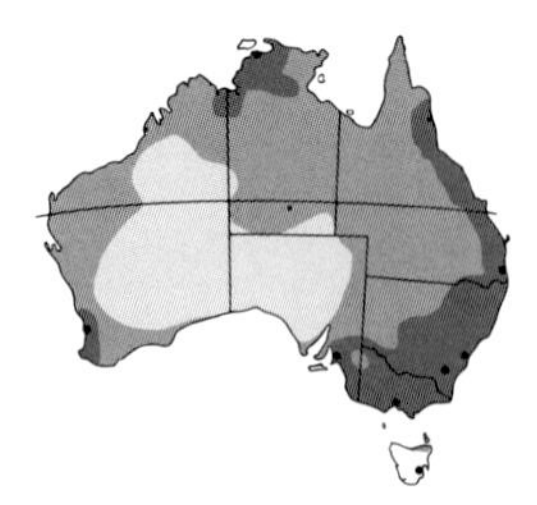

Sexes alike. Australia's most common ibis. Blackish bare skin on head with black, long down-curved bill. Upperparts: Black with iridescent violet-green sheen. Upper neck, belly, rump, tail: Whitish. Breast: Black band and straw-like spiny feathers continuing up to lower neck with bare yellow skin on either side. Underwings: Yellow bare wing patch. Legs: Red.
Immature: Browner. Head mottled. Legs: Black. Lacks neck plumes.

Sometimes seen in pairs, usually in large flocks. Flies in line or 'V' formations, neck extended.

Habitat: Shallows of wetlands, swamps, waterways, pasture, grasslands, parkland, sewerage ponds.

Feeding: Vast quantities of grasshoppers, locusts. Also, frogs, small reptiles, mammals.

Voice: Grunts and croaks.

Status: Common. Abundant. Nomadic, dispersive. Murray–Darling Basin habitat is critical for their breeding and survival.

Breeding: Sept.–Dec. or any time after flooding. In colonies, saucer-shaped nest of reeds and twigs, in swamp plants / 4–5 eggs.

Left: Breeding. Right: Non-breeding.

Royal Spoonbill

Platalea regia

750–800mm
Wingspan: 1400mm

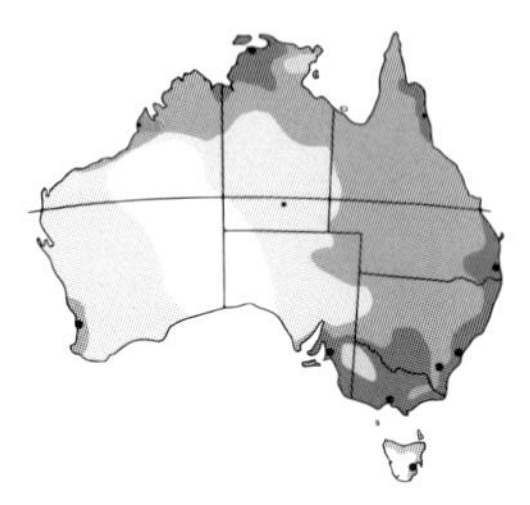

Sexes similar. Overall plumage spotless white with black legs, face and distinctive black spoon-shaped bill. Eyes: Red with yellow patch above. Frons: Red spot. Breeding: White nuptial head crest plumes. Breast: Washed buff.
Non-breeding: Lacks head plumes, white breast.
Immature: Flight feathers tipped black.

Usually seen wading in shallow water, singularly or in pairs. Flight in echelon line (diagonal), steady wingbeats, sometimes soaring.

Habitat: Inland and coastal, larger bodies of water, sewerage ponds, tidal mudflats, wetlands, floodplains.

Feeding: Usually solitary, wading, sweeping bill from side to side for fish, shrimp, crustaceans.

Voice: Low-pitched grunting and croaking.

Status: Common, nomadic.

Breeding: Oct.–May. Loose colonies, nest in trees, usually in water to 9m high / 4 eggs.

Left: Non-breeding. Right: Breeding.

Centre: Mixed group of Royal Spoonbill (centre) and Yellow-billed Spoonbills in breeding plumage.

Yellow-billed Spoonbill

Platalea flavipes

Endemic

800–900mm
Wingspan: 1300mm

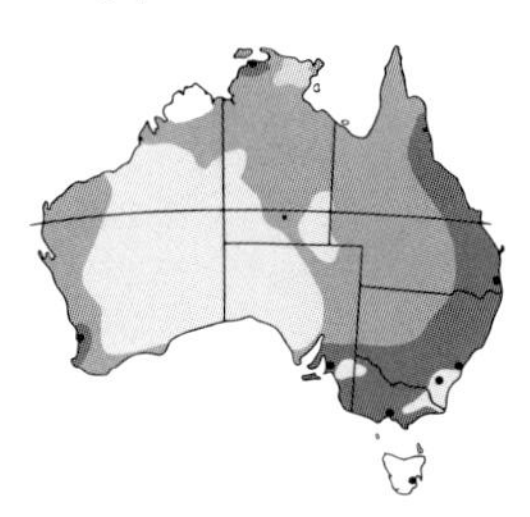

Sexes similar. Distinctive pale yellow or creamy white spoon-shaped bill and greyish yellow facial skin outlined in black. Overall plumage white, often stained. Legs: Greenish yellow. Eyes: Pale pink. Wings: White with black tips to inner flight feathers.
Breeding: Plumage washed pale yellow. Long, spiny, white nuptial plumage on breast. Facial skin: Bluish, outlined in black. Inner wing: Scant, lacy black plumes. Female has shorter bill.
Immature: Lacks black face outline and plumes.

Flies with steady wingbeats and glides, extended neck and trailing legs.

Habitat: Small freshwater swamps and waterways, farm dams, irrigated areas.

Feeding: Usually in muddy waters, wading with slowly sweeping bill for fish, shrimp, crustaceans.

Voice: Usually silent. Grunts, nasal coughs.

Status: Common. Solitary. Nomadic outside breeding season.

Breeding: Sept.–Jan. Singly or in colonies, stick platform nest in trees near water or on ground / 2–4 eggs.

Left: Yellow-billed Spoonbill in breeding plumage. Right: Non-breeding.

Black-necked Stork

Ephippiorhynchus asiaticus

1100–1300mm
Wingspan: 2000mm

Sexes similar. The only stork found in Australia. Tall, stately black and white bird also called 'Jabiru'. Head, neck: Glossy black with iridescent green-purple sheen. Under and upper wings: White with wide black bar. Tail: Black. Underparts, back: White. Bill: Straight, large, black. Eyes: Dark brown. Legs: Long, deep pink.
Female: Yellow eyes. Immature: Overall brown.

Slow, deliberate movements. Flies with neck fully extended. Slow, powerful wingbeats. Soars high on thermals. Long trailing legs.

Habitat: Freshwater lagoons, estuaries, mudflats and wet meadows.

Feeding: Strides in shallow water probing for fish, crabs, frogs, rodents and carrion.

Voice: Clapping and snapping sound made with the bill.

Status: Uncommon. Solitary.

Breeding: Feb.–Jun. Large, untidy stick platform nest in tree in or near water and up to 25m high / 2–4 eggs.

Kakadu NP, NT.

Left: Female feeding. Right: Adult in flight.

Brolga

Grus rubicunda

Endemic

Male: 1050–1250mm
Female: 950–1150mm
Wingspan: to 2400mm

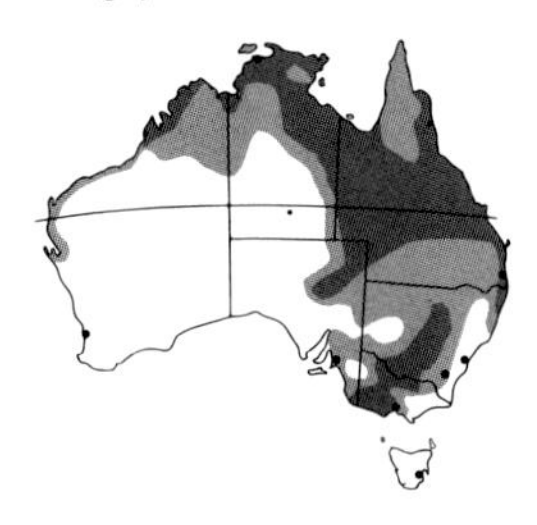

Sexes similar. Large, stately, long-legged crane with overall silver-grey plumage. Head: Bare, pale grey skin on frons, scarlet band across face and bill to nape. Chin: 'Hairy', blackish dewlap bulge. Ear coverts: Grey patch. Wings: Grey with black primaries and a 'bustle' of secondary feathers falling over rump. Bill: Long, straight, grey-green. Eyes: Yellow. Legs: Long, dark grey. Female: Smaller, smaller dewlap. Immature: Pale grey-pink facial skin, small 'bustle'.

Noted for its mutual ritual 'dance' of graceful leaps and bows. Flight is steady, slow, graceful, soaring in thermals. Neck extended, trailing legs.

Habitat: Open wetlands, wet grasslands, vegetated lagoons.

Feeding: In shallows. Mostly sedge tubers insects and molluscs.

Voice: Grating and loud whooping, trumpeting.

Status: Most common in north. Uncommon to rare elsewhere. Nomadic. Concentrated flocks occur at permanent fresh water in the dry season, then disperse to breeding territory in the wet.

Breeding: Sept.–Dec. in south. Feb.–June in north. Grass mound nest in or near shallow water / 2 eggs.

Fogg Dam CR, NT.

Left: Breeding pair at nest. Right: 'Dancing' display.

Sarus Crane

Grus antigone

Male: 1300–1400mm
Female: 1100–1300mm
Wingspan: to 2400mm

Race *gilliae* present in Australia. Sexes similar. Similar to the Brolga, but lacks dewlap under chin. Crown: Grey skin. Face, upperneck: Deep scarlet bare skin. Eyes: Red-brown. Ear coverts: Grey. Bill: Olive-grey. Wings: Grey with paler 'bustle' of secondary feathers falling over rump. Legs: Pink.
Immature: Similar to adults, browner with shorter, darker 'bustle', paler face skin.

Brolga-like 'dancing'. Usually in family groups of 3–5. Strong flight with deep wingbeats, straight neck, long trailing legs. Nesting pairs defend territories of 50–80 hectares.

Habitat: Shallow grass- and sedge-covered marshes, lagoons and adjacent pasture, generally drier habitat than the Brolga.

Feeding: Probes with bill for tubers, emergent shoots, grain, insects. Often in large loose flocks after breeding.

Voice: Trumpeting calls, usually as a duet.

Status: Locally moderately common.

Breeding: Jan.–Mar. Platform of sticks, often in shallow, densely grassed swamps / 2 eggs.

Broomsfield and Hasties Swamps, Qld.

Left: Adult with immature. Right: Adult.

Family Accipitridae, Falconidae

A group of mostly diurnal (day-flying) birds of prey, represented in Australia by two of the world's three families that make up the raptor order Falconiformes.
Raptors have taloned feet that most species use to kill their prey, strong legs to hold their prey, and sharp, hooked bills for cutting and tearing. They vary in size from the large, powerful Wedge-tailed Eagle, able to kill small mammals, to the small Pacific Baza and Black Kite, which have feeble feet and talons and catch insects that they disturb from foliage. Female raptors are larger than males; differences in size between sexes and species helps to lessen competition for the same prey. They build, or modify, substantial existing stick nests.

Family Accipitridae contains the osprey, Pacific Baza, kites, harriers, goshawks, sparrowhawk, buzzard and eagles.
The Eastern Osprey is a cosmopolitan fish-hunting raptor with specialised powerful feet for gripping and carrying large, struggling, slippery fish. The feet have a reversible outer toe and long sharp talons and spines on the soles. Fish are carried headfirst to reduce drag.
The Pacific Baza has weak feet and talons and is specialised to silently patrol the treetops, snatching prey such as stick insects from the outer foliage.
Australia's two species of 'hovering kites' are mostly rodent hunters that hover and silently drop on prey. Its three species of 'soaring kites' soar on long, broad wings with distinctive 'fingertips', searching for fish, small mammals and insects.
Harriers patrol low over open ground, gliding with occasional slow wing flaps, surprising prey as they silently plummet to the ground.
Eagles include some of the world's largest raptors with powerful, long, feather-covered legs; large powerful hooked bills and sharp talons for killing and cutting up prey.
The bird-hunting goshawks and sparrowhawks are specialised to accelerate quickly and to rapidly manoeuvre in flight through wooded areas to attack other birds.

Family Falconidae is represented by six species of falcons, kestrels and the Australian Hobby.
Falcons are specialised for high-speed chases, with long, powerful pointed wings that are swept back during fast manoeuvring. They kill prey by biting its neck vertebrae with a sharp tomial tooth-like projection on their upper mandible.
Kestrels are small, delicate falcons that constantly hover, searching for small prey such as grasshoppers and frogs.
The Australian Hobby is specialised to rapidly manoeuvre in wooded areas, attacking other birds, including birds of an equivalent size.

Opposite: Immature Wedge-tailed Eagle.

Eastern Osprey

Pandion cristatus

500–650mm
Wingspan: 1000–1100mm

Sexes similar. Specialised raptor for fishing. Head, neck, underparts are pale, contrasting with deep glossy brown upperparts. Black band through the eyes and a small bristly crest. Breast: White often streaked brown. Underwings: White, faintly barred black and cream with dark patch on bend of wing. Bill: Black, strongly hooked. Eyes: Pale creamy yellow. Legs: Powerful, light grey.
Female: Larger, more heavily streaked breast. Immature: Darker, heavier streaking, dark eyes.

Usually seen patrolling waters, soaring, hovering, then plunging onto prey, feet first.

Habitat: Coastal shores, mangrove estuaries and major inland rivers, pools and offshore islands.

Feeding: Dives headlong snatching fish, particularly mullet, with its specialised talons.

Voice: Quavering, plaintive whistles and screams.

Status: Cosmopolitan. Usually solitary. Uncommon, sedentary. Occasionally inland rivers.

Breeding: Apr.–Sept. Large bulky nest in a tree or on a cliff face, to 30m / 2–3 eggs.

Left: Note fingered wings in flight.
Right: Adult with prey.

Black-shouldered Kite

Elanus axillaris

Endemic

330–380mm
Wingspan: 950mm

Sexes alike. Elegant, pale grey hawk with a white head, body and tail and black shoulders. Eyes: Red with a black comma patch extending past the eye. Underparts, underwings: White with black tips and black patch on bend of wing. Bill: Black, with small yellow cere. Legs: Yellow. Female: Larger. Immature: Head, back, wings, chest: Streaked golden tan, white tipped, becoming whiter with age. Eyes: Brown.

In singles, pairs or family groups. Flight is swift, shallow wingbeats, soaring high over feeding area.

Habitat: Open woodlands and grasslands, heathlands, saltbush, farmland and parks.

Feeding: Hovers with feet dangling and drops silently onto prey. Feeds almost exclusively on the introduced House Mouse. Also grasshoppers, rats, small reptiles and insects.

Voice: Harsh, wheezing and whistling.

Status: Common, nomadic. Local populations increase in response to mouse plagues.

Breeding: Apr.–Nov. Stick nest in trees 20–25m high / 3–4 eggs.

Left: Immature birds. Right: Adult.

Letter-winged Kite

Elanus scriptus

Endemic

340–380mm
Wingspan: 950mm

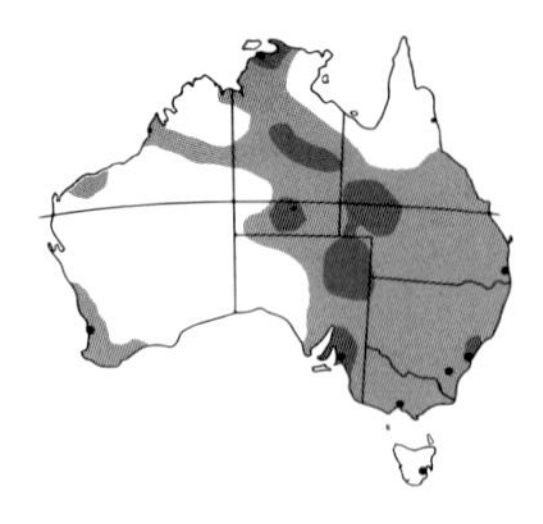

Sexes similar. Large owl-like eyes and distinctive black 'W' across white underwings. Crown: Greyish. Face: White. Eyes: Red with forward, small black patch extending around the eye. Nape, back, rump: Grey. Shoulders: Black patch. Bill: Black. Feet: Cream.
Female: Larger. Immature: Head, back, wings, chest: Streaked golden buff. Eyes: Brown.

Roosts during the day in small groups in leafy trees. Hovers in midair.

Habitat: Fluctuating range. Central inland grasslands and wooded waterways.

Feeding: Nocturnal. Hovers and drops silently onto prey.

Voice: Sharp, penetrating whistled notes.

Status: Eruptive. In good seasons when populations of the native Long–haired Rat are abundant, kites breed continuously, rapidly increase in numbers and disperse to all areas of Australia, except the driest deserts. With drought, rat populations decline as do kites.

Breeding: Oct.–Nov. or in response to abundance of food. Colonies of 2–100 pairs. Bulky stick nest in trees, 1–12m high / 3–6 eggs.

Left: 'W' patterned underwing.
Right: Note owl-like eyes.

Square-tailed Kite

Lophoictinia isura

Endemic

500–560mm
Wingspan: 1500mm

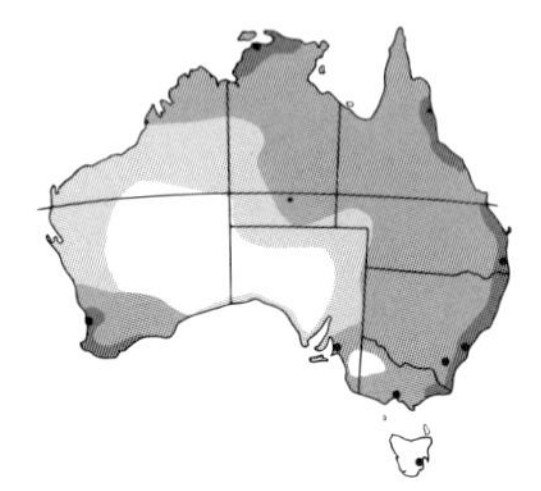

Sexes similar. Overall reddish brown mottled and streaked black, more reddish on the underside. Frons, face, chin: White. Crown: Brown, heavily streaked black. Wings: Grey-brown with black markings. Underwings: Russet coverts, wide light-grey-barred dusky secondaries, deeply fingered wing tips barred black and grey with large cream crescent on wing tips. Tail: Long, square, light grey, tipped black. Eyes: Yellow. Bill: Brownish-yellow tipped black. Legs: Flesh white. Female: Larger. Immature: Similar, finely streaked russet-brown. Eyes: Brown. Upperparts: Mottled black.

Usually solitary, sometimes in pairs in breeding season. Slow, soaring, gliding on upturned wings, seldom flapping, just above the treetops.

Habitat: Open eucalypt forests, woodlands, timbered waterways.

Feeding: Circles low over canopy snatching prey from outer foliage.

Voice: Quavering, yelping, musical rising yelps.

Status: Rare. Sedentary in north. Present in southeast Sept.–Mar. Widely scattered.

Breeding: Aug.–Nov. Bulky stick nest, 12–26m high / 2–3 eggs.

Dry season Kakadu NP, NT.

Left: Adult at nest. Right: Immature.

Black-breasted Buzzard

Hamirostra melanosternon

Endemic

550–600mm
Wingspan: 1500mm

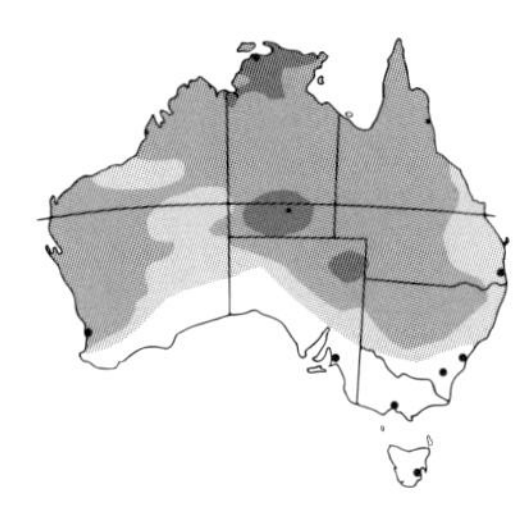

Sexes similar. Identified by its large white wing spots and distinctively short tail. Overall rich rufous and black plumage. Crown: Black with rufous hind crest. Face, breast, nape: Dusky black or pale red-brown. Flanks, lower belly, vent: Rufous-brown. Underwings: Coverts are pale rufous mottled black and white; remainder grey with distinctive white 'windows' at base of flight feathers. Eyes: Light to deep brown. Bill: Pale grey with black tip. Legs: Scaly pale grey. Female: Larger. Immature: More rufous, taking 4 years to reach adult plumage.

Usually solitary. Soars high, glides low scanning for prey. Drops or tosses rocks onto the eggs of ground-nesting birds to suck out the contents.

Habitat: Open forests, grasslands, timbered watercourses in semi-arid areas.

Feeding: Low swoop from perch. Prey mostly taken and eaten on ground. Small animals, birds and birds eggs, insects, carrion.

Voice: Usually silent. Short, sharp, excited yelping at nest.

Status: Uncommon. Sedentary, dispersive. Rare in coastal areas.

Breeding: June–Dec. Large stick platform nest in trees 6–20m high / 2 eggs.

Left: Adult feeding on road-kill kangaroo. Right: Adult in flight.

Pacific Baza

Aviceda subcristata

350–430mm
Wingspan: 1100mm

Sexes similar. Head: Slim, mid-grey with short grey crest and bright yellow eyes. Bill: Black, notched and hooked. Upperparts: Pale slate grey. Breast, nape: Grey. Belly: White, barred rufous-brown. Undertail: White, dark barred with wide terminal band. Vent: Orange-buff. Underwings: White with brown-barred flight feathers, black trailing edge and orange-buff patch. Legs: Short, grey. Female: Paler, brownish wash. Immature: Upperparts: Brown, buff-scalloped. Underparts: Finer barring, buff breast.

Soaring, circling, hovering and plunging display with wings held in a stiff 'V'.

Habitat: Tropical and subtropical, well-watered open forests and adjacent closed forests.

Feeding: Patrols outer canopy. Crashes into foliage to disturb prey. Stick insects preferred when breeding.

Voice: Mellow ascending, descending whistles.

Status: Uncommon. Sedentary.

Breeding: Oct.–Dec. Flimsy nest in leafy trees 15–30m high / 2–3 eggs.

Maiala NP, Qld.

Left: Upperparts. Right: Underparts.

White-bellied Sea-eagle

Haliaeetus leucogaster

Male: 760mm
Female: 840mm
Wingspan: 2200mm

Sexes similar. Large and powerful fish-eating eagle with huge talons. Head, neck, underparts, undertail: White. Back, wings: Dark grey. Underwings: White with black flight feathers. Eyes: Brown. Bill: Large, hooked, grey, dark-tipped. Legs: Cream-white. Female: Larger. Immature: Adult plumage acquired at 3–4 years. First year: Brown with lighter markings, pale head and white panels on underwings.

Patrols skies with slow powerful wingbeats; long, broad, upswept wings. Soars high. Singular, pairs or family groups.

Habitat: Inland, fresh or brackish rivers, coastal shores and offshore islands.

Feeding: Swoops, snatching fish and marine animals from the water surface. Also, waterfowl, nestlings, rabbits are taken. Scavenges on carrion.

Voice: Loud, resonant, metallic 'klanking'. Sustained duets.

Status: Moderately common. Usually sedentary.

Breeding: May–Oct. Huge stick nests in trees 30m or more high, or on the ground on treeless islands / 2 eggs.

Left: Adult on nest with chick. Right: Adult in flight.

Whistling Kite

Haliastur sphenurus

500–550mm
Wingspan: 1500mm

Sexes similar. A mid-sized light brown hawk. Head, neck and underparts are light brown with pale streaking. Upperparts: Mid-brown. Tail: Buff, long, rounded. Wings: Deeply fingered black wing tips. Underwings have pale 'M' shape viewed in flight. Legs, feet: Pale cream. Eyes: Brown. Female: Larger. Immature: More heavily streaked.

Often seen hunting over wetlands, farmlands or low over grass fires. Effortless wheeling flight.

Habitat: Open wooded country near swamps, paddocks with large trees, along rivers and coastal areas.

Feeding: Mostly scavengers. Sometimes live prey. Usually alone, but will congregate at large food source.

Voice: Long, drawn-out descending clear whistle in flight or at nest.

Status: Common. Sedentary. Nomadic during droughts.

Breeding: June–Dec. or opportunistic in good seasons. Bulky stick nest in trees 15m or higher / 2–3 eggs.

Left: Adult at perch. Right: Adult in flight.

Brahminy Kite

Haliastur indus

450–510mm
Wingspan: 1300mm

Sexes alike. Medium-sized hawk with a white head, neck and upper belly. Remainder of plumage is chestnut brown. Wings: Broad, deeply fingered, black wing tips. Eyes: Red-brown. Bill: Pale yellow, strong and hooked. Legs: Cream-yellow.
Immature: Mottled brown.

Glides and sweeps low in slow wheeling flight at up to 50m altitude. Wings held straight out, horizontally. Mostly seen along the shoreline.

Habitat: Mangrove-fringed coastal inlets and mudflats. Rocky shores and beaches.

Feeding: Mostly scavenging on dead fish and marine beachwash. Occasionally snatches small live prey, fish, crabs, animals, insects or steals food from gulls or terns.

Voice: Repeated, quavering, high-pitched 'pee-ah'.

Status: Uncommon. Solitary. Sedentary or locally nomadic.

Breeding: Apr.–Oct. Bulky stick nest in trees up to 18m high / 1–3 eggs.

Left: Adult. Right: Adult in flight.

Black Kite

Milvus migrans

480–500mm
Wingspan: 1200–1500mm

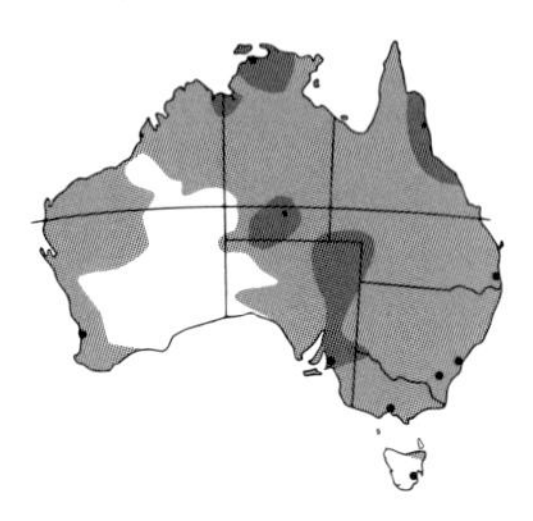

Sexes similar. Medium-sized raptor that can appear almost black from a distance. Plumage is dark brown with light brown markings. Head, neck: Paler brown. Wings: Black '5-fingers' feathers. Tail: Forked, barred darker brown. Underparts: Rufous-brown, dark streaked. Eyes: Brown with dark smudge behind eye. Bill: Black with grey base. Legs: Yellow.
Female: Larger. Immature: Paler, spotted cream.

Congregating in flocks of hundreds. Attracted to bushfires for fleeing small animals. Flight is leisurely, occasional wing flaps, soaring and gliding effortlessly in thermals. Characteristically twists forked tail.

Habitat: Open woodlands, mangrove swamps, tree-lined waterways and homesteads.

Feeding: Scavenges carrion. Feeds on the ground, often around refuse tips. Dives on rodents, reptiles when flushed.

Voice: Plaintive descending quavering 'see-err'.

Status: Abundant to locally common. Greatest numbers during plagues of mice.

Breeding: Sept.–Nov. in south. Mar.–May in north. Stick platform nest to 30m high / 2–3 eggs.

Left: Adult in flight. Right: Adult.

Brown Goshawk

Accipiter fasciatus

Male: 400–430mm
Female: 470–560mm
Wingspan: 1000mm

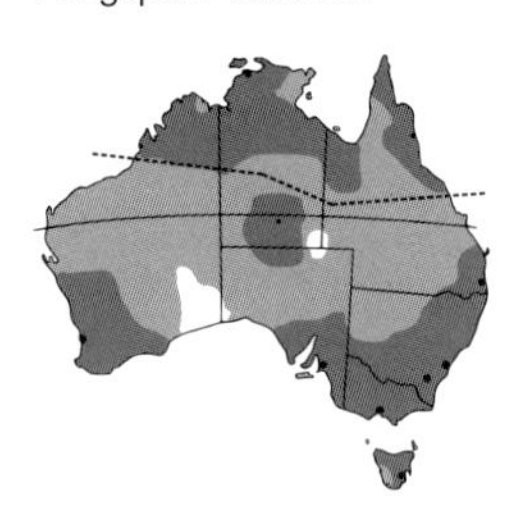

Sexes similar. Bulky with a heavy, hooded 'frown' brow and bright yellow eyes. Upperparts: Slate grey with chestnut collar across upper nape of neck. Underparts: Finely barred cream and pale chestnut. Wings: Slate grey or brownish above and buff and chestnut below. Bill: Blue-grey, dark tip. Legs: Long, yellow with clawed toes.Tail: Long, rounded tip.
Race *didimus,* northern Qld, NT, WA. Smaller, paler.
All females: Larger, small yellow-green cere. Immature: Eyes: Grey-brown. Heavy, dark brown streaking, becoming finer in second year.

Soars high over treetops with wide-spread fingered wing tips and tail. Secretive. Powerful.

Habitat: Dry eucalypt woodlands, open forests, along waterways and farmland.

Feeding: Usually from a concealed perch, pounces on rabbits, small animals, reptiles, small birds, large insects. Prey is consumed back on perch.

Voice: Usually silent. High-pitched shrill chatter, drawn-out descending 'kik-kik-ki-kikik'.

Status: Common. Singularly or pairs. Sedentary in north. In winter immature birds disperse northward.

Breeding: Aug.–Nov. Stick nest in trees 6–30m high / 2–3 eggs.

Left: Adult feeding on a Zebra Finch.
Right: Adult upperparts.

Collared Sparrowhawk

Accipiter cirrocephalus

Male: 280–330mm
Female: 360–390mm
Wingspan: 780mm

Sexes similar. Similar to Brown Goshawk but with more spindly legs with long central toe projecting past claws of other toes and a wide-eye 'stare'. Tail is shorter, square tipped in female and notched in male, and slightly forked when folded. Upperparts: Brownish, slate grey with chestnut-collared nape and grey throat. Underparts, underwing coverts: Narrowly barred rufous and white. Bill: Black. Eyes: Bright yellow.
Female: Larger with grey-cream cere. Immature: Upperparts: Grey-brown. Underparts: Off-white, dark streaked.

Rapid, flickering wingbeats. Soars and glides. Trusting.

Habitat: Open forests and woodlands. Arid interior, tree-lined waterways, parks.

Feeding: Concealed in foliage, darts at high speed to take birds, sometimes larger than themselves, in mid-air. Also, small animals, insects, reptiles.

Voice: Rapid shrill 'ki-ki-ki-ki', staccato chatter.

Status: Common, solitary.

Breeding: July–Dec. Small, rough stick platform in trees 10–30m high / 2–4 eggs.

Left: Female. Right: Female with prey.

Grey Goshawk

Accipiter novaehollandiae

340–540mm
Wingspan: 1100mm

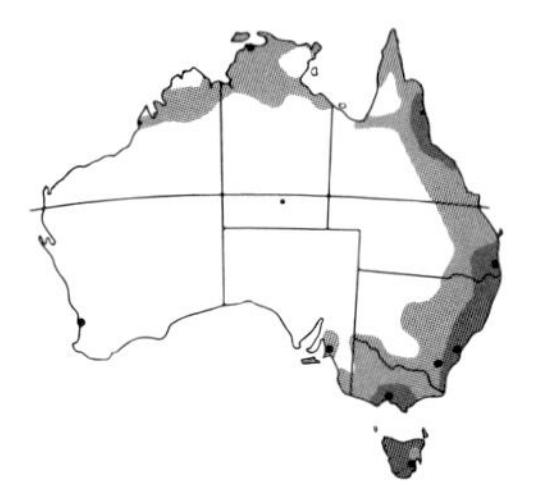

Sexes similar. Overall grey or white with long powerful orange-yellow legs and feet. Bill: Grey with yellow cere. Eyes: Dark red. 2 colour morphs.
White morph: Overall white. Grey morph: Underparts whitish, finely barred grey.
Female: Larger. Immature: White morph: White. Grey morph: Brown wash and wider barring.

Rapid flight, deep wingbeats, interspersed with glides. Soars, glides with upturned wing tips.

Habitat: Coastal, closed tall eucalypt forests, woodlands, rainforests, timber-lined waterways. Grey morph prefers northern, subtropical rainforests. White morph prefers southern tall eucalypt forests.

Feeding: Pursues prey at speed. Striking with powerful toes, small animals, large insects and birds captured mid-air.

Voice: Shrill chattering. Musical, rising, repeated 'swee-swit'.

Status: Uncommon. Solitary. Immature, nomadic.

Breeding: June–Sept. in southeast. Jan.–May in north. Stick nest in trees, 9–30m high / 2–4 eggs.

Lamington NP, Qld. Mt Keira, NSW.

Left: Nesting grey morph with chicks at the nest. Right: Grey morph.

Spotted Harrier

Circus assimilis

Male: 530mm
Female: 600mm
Wingspan: 1450mm

Sexes similar. Large, slender raptor with colourful chestnut underparts, spotted white.
Wings: Long, broad with conspicuous widely fingered black wing tips. Breast, vent, inner wings: Chestnut, spotted white. Face: Chestnut with owl-like 'facial disks' and yellow eyes. Upperparts: Blue-grey barred dark-grey with chestnut shoulder patch. Rump: White. Tail: Boldly banded. Bill: Black.
Female: Larger, browner.
Immature: Mottled and streaked ginger and brown, ginger shoulder patch, fawn rump.

Glides slowly, occasionally flapping. Upturned, widely 'fingered' black wing tips.

Habitat: Open grasslands, inland spinifex plains, saltbush, open woodlands and croplands, windbreaks.

Feeding: Patrols low over vegetation. Dives on prey, taking small mammals, large insects, reptiles, birds.

Voice: Piercing squeaking and rapid chattering.

Status: Moderately common to sparse. Irruptive in response to abundant food.

Breeding: July–Oct. Stick nest 6–20m high in trees / 2–4 eggs.

Left: Adult in flight showing underparts. Right: Adult soaring in flight.

Swamp Harrier

Circus approximans

500–580mm
Wingspan: 1200mm

Sexes similar. Large hawk with a small head and long legs. Upperparts: Dark brown mottled grading to white rump. Tail: Long, grey, faintly barred. Underparts: Buff, streaked rufous grading to off-white belly. Flanks become greyer with age. Eyes: Yellow. Bill: Dusky. Legs: Yellow.
Female: Larger, browner with deeper rufous streaked underparts. Immature: Similar, darker, pale rufous rump.

Glides low over vegetation on long, upswept wings, legs dangling.

Habitat: Wetlands, swamps, tall grass areas, reeds, mangroves and wet pastures.

Feeding: Hunts low, snatching waterfowl or small animals, large insects, frogs, reptiles.

Voice: Usually silent. High-pitched whistles.

Status: Common. Southern birds migrate northward or inland for winter.

Breeding: Sept.–Jan. Stick and grass nest, well hidden in reeds, often over water / 3–6 eggs.

Left: Female in flight. Right: Male soaring in flight.

Red Goshawk

Erythrotriorchis radiatus

Endemic

450–580mm
Wingspan: 1350mm

Sexes similar. Identified by conspicuous barring to underwings and tail. Overall rich reddish brown plumage, strongly mottled black with dark-streaked, whitish grey head and neck.
Tail: Barred grey. Flight feathers: Barred grey-black. Legs: Long, yellow, upperpart covered in rich-rufous feathers. Eyes: Orange. Bill: Black.
Female: Much larger, pale belly. Immature: Brighter rufous head, brown eyes.

Quick, strong wingbeats. Soars with broad '6-fingered' wing tips extended. Secretive.

Habitat: Undisturbed forests, rainforest edges, dense vegetation along waterways.

Feeding: Pursues prey from a concealed perch or scans from high glides Takes mid-sized birds, mammals, reptiles, insects mid-air or from the ground.

Voice: High-pitched chattering, raucous, repetitive 'skeep-skeep-skeep'.

Status: Rare, sparse, vulnerable. Estimates of only 700 breeding pairs remaining. Now restricted to northern Australia. Usually sedentary.

Breeding: Apr.–Nov. in dry season. Bulky stick nest reused each season in trees 10–20m high / 1–2 eggs.

Kakadu NP, NT. Mataranka and Lakefield NP, Qld.

Left: Female at nest. Right: Male.

Little Eagle

Hieraaetus morphnoides

Male: 480mm
Female: 550mm
Wingspan: 1350mm

Sexes similar. Mid-sized, stocky, powerful raptor, particularly harassed by other birds. Overall upperparts mid-brown or dark brown with paler bar across the wings. Head: Black streaked, and a small, black erectile crest. Tail: Square shape when folded, barred below. Legs: Completely feathered to blue-grey talons. Eyes: Brown. Bill: Dark brown.
Pale Morph: Upperparts: Dark brown. Head, neck, underparts: Buff-white finely streaked rufous. Underwings: Pale diagonal bar joining white patch.
Dark morph: Upperparts: Dark brown. Head, underparts: Mid-brown with black streaking. Underwings: Dark with pale patch at end of primaries.
Female: Larger. Immature: Talons: White. Pale morph: Light, bright rufous. Dark morph: Bright, deep rufous.

Soars in tight circles on flat wings with deeply fingered, dark, barred wing tips and fanned, barred, rounded tail.

Habitat: Varied. Coastal and inland open wooded country, grasslands, scrubland.

Feeding: Snatches live prey, mostly rabbits and other small animals, birds, reptiles, insects. Less frequently carrion.

Voice: Rapid, high-pitched 2–3 note whistle.

Status: Common. Sedentary.

Breeding: Aug.–Nov. Large stick nest in trees 10–45m high / 1–2 eggs.

Left and centre: Dark morphs. Right: Immature dark morph in flight. Below: Pale morphs.

Wedge-tailed Eagle

Aquila audax

Male: 900mm
Female:1000mm
Wingspan: 2500mm

Sexes similar. Australia's largest raptor. Distinctive long wedge-shaped tail. Overall blackish brown plumage with dull rufous nape and tawny brown bar across wings. Legs: Densely feathered. Eyes: Hazel brown. Bill: Dull pink to cream with dark tip.
Female: Paler, larger. Immature: Paler, browner and patchy. Adult plumage acquired by 5 years.

Leisurely, powerful wingbeats, soaring, circling, gliding on upturned wings with deeply fingered wing tips. Often seen at 2000m or more riding thermal currents.

Habitat: Variable. Wooded mountain slopes, open country, alpine areas to plains. Less frequently urban areas, beaches.

Feeding: Swoops, snatches prey, mostly rabbits, also wallabies, reptiles, waterfowl, usually consumed on the ground. Scavenges on carrion.

Voice: Repeated double whistle.

Status: Common. Sedentary. Race *fleayi,* Bass Strait Is., Tas., endangered.

Breeding: May–Nov. Large stick-platform nest in a tree or on rock ledge / 1–3 eggs.

Left: Perched adult. Right: Male in soaring flight.

Nankeen Kestrel

Falco cenchroides

Male: 310mm
Female: 350mm
Wingspan: 800mm

Sexes similar. Smallest raptor. Upperparts: Mostly chestnut or mid-rufous with fine streaking. Head: Blue-grey, tinged rufous. Face: White with fine black tear-stripe from front of eyes down side of throat. Tail: Grey, terminating in a black bar and white tips. Bill: Short, notched. Underparts: White, dark-streaked over breast. Wings: Dusky black outer flight feathers. Eyes: Brown, yellow eye ring. Bill: Blue-grey. Legs: Yellow.
Female: Larger, more heavily streaked and more rufous head and tail. Immature: Similar to female, more heavily streaked.

Fast, agile flight or hovers motionless, low to the ground, with rapid wingbeats and its fan-like tail used as a rudder. Typically solitary or in pairs, sometimes congregating in large flocks in good seasons around an ample food supply. Named from its similarity to a buff-coloured cloth once imported from Nanking in China.

Habitat: Open woodland, farmland, urban parks and gardens.

Feeding: Hovers at 10–20m and plunges headfirst at prey, snatching small animals, birds, insects from the ground.

Voice: High-pitched shrill 'ki-ki-ki'.

Status: Common.

Breeding: July–Dec. Nests in tree hollows, building ledges, rock crevices, abandoned nests / 3–7 eggs, usually 4.

Opposite: Male with centipede.

Below left: Male at nest with chicks.
Below right: Female.

Grey Falcon

Falco hypoleucos

Endemic

Male: 340mm
Female: 430mm
Wingspan: 950mm

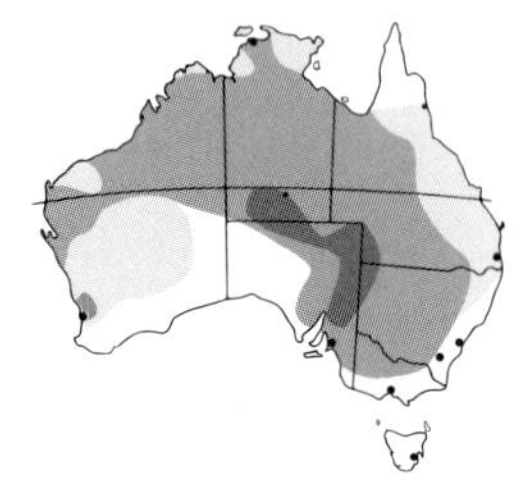

Sexes similar. One of Australia's rarest birds of prey. Overall pale blue-grey with dark primaries. Bill: Grey with black tip and distinctive yellow-orange cere. Eyes: Dark with distinctive yellow-orange eye ring. Face, underparts: Grey-white with fine dark streaks. Underwings: Whitish, dark tipped. Legs: Yellow.
Female: Larger. Immature: Upperparts: Browner, buff feather margins. Underparts: Darker, heavier streaked. Bill, cere: Grey.

Soars and glides on horizontal wings. Circling leisurely or hunts at great speed.

Habitat: Lightly timbered inland plains, farmland, gibber deserts, eucalypt-lined waterways, avoids the driest desert areas.

Feeding: Patrols low over ground or perches motionless. Small mammals, birds, reptiles, insects snatched mostly from the ground.

Voice: Usually silent. Rapidly repeated 'chak' and whining and clucking sounds.

Status: Rare, nomadic.

Breeding: July–Nov. Usually re-lines abandoned hawk or crow nests/ 2–4 eggs, usually 3.

Sturt NP, NSW. Birdsville and Strzelecki Tracks, SA.

Left: Searching for prey. Right: Hunting at great speed.

Brown Falcon

Falco berigora

Male: 450mm
Female: 500mm
Wingspan: 1200mm

Sexes similar. Female larger. Six main colour morphs ranging from overall dusky to sandy with white underparts. Colour forms can be found together in all places except Tas. Tas. birds are consistently brown above and white below. Inland birds are generally light rufous. All forms interbreed. All forms have brown tear-stripe below the eyes and brown patch to side of head. Underwings are pale barred. Eyes: Dark brown. Bill: Blue-grey, black tipped. Cere: Grey-white. Legs: Pale grey, distinctively long, upperparts feathered. Immature: Typically darker than colour of parent. Yellow-buff wash to head and collar, no wing spots and ill-defined tail barring.

Usually the most commonly seen falcon in Australia. Solitary or pairs. Relatively heavy, slow flight. Glides on raised wings. Sometimes hovers.

Habitat: Most habitats, except for dense woodlands. Mostly grassy woodlands, farmland, open forests, urban areas.

Feeding: Scans for prey from high perch or glides, drops silently and snatches small mammals, large insects and small birds.

Voice: Loud, raucous cackles and screeching.

Status: Common.

Breeding: June–Nov. in south. June–Mar. in north. Typically uses abandoned hawk nest or less frequently builds a stick nest or uses tree hollow / 2–5 eggs.

Centre row and below left: Varying adult colour morphs. Below right: Immature.

Black Falcon

Falco subniger

Endemic

Male: 450mm
Female: 550mm
Wingspan: 1100mm

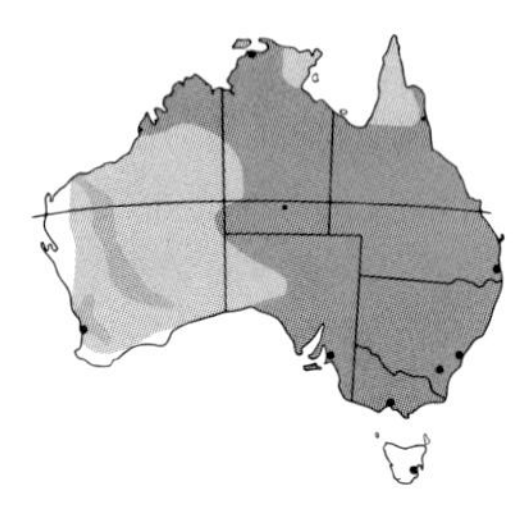

Sexes similar. Largest Australian falcon. Overall dark brown to sooty black with pale upper throat, frons and dark streak below eyes. Eyes: Brown with blue eye ring. Tail: Long, narrow with square tip and notched corners. Wings: Broad, sharply pointed. Legs: Blue-grey, feathered upperparts. Bill, cere: Dark blue-grey.
Female: Larger. Immature: Darker with pale feather margins.

Swift, powerful wingbeats, glides with slightly drooped wings.

Habitat: Sparsely timbered areas, grasslands, wetlands, infrequently in urban areas.

Feeding: Flies at bushes to flush prey or follows human activity, snatching prey flushed by farm equipment. Pirates food from other birds. Kills prey by a bite to the spine with its tomial 'tooth'.

Voice: Harsh chattering. Drawn-out whining calls.

Status: Rare. Nomadic and dispersive, overwinters in more coastal areas.

Breeding: June–Dec. Usurped nests, mostly of hawks, crows or other raptors / 3–4 eggs.

Deniliquin, NSW. Barkly Homestead Roadhouse, NT.

Left: Adult. Right: Immature.

Australian Hobby

Falco longipennis

Male: 300mm
Female: 340–350mm
Wingspan: 900mm

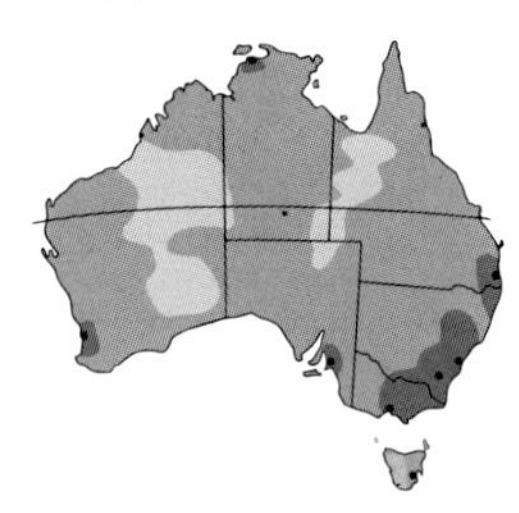

Sexes alike. Similar to the Peregrine Falcon but much smaller. Variable plumage. Birds in arid regions are paler. Cap, face mask: Black. Frons, throat, half-collar: Buff-cream. Eyes: Dark brown with bluish white eye ring. Bill: Light grey, black tipped with dull yellow cere. Upperparts: Blackish grey. Underparts: Rufous-brown with dark streaks. Tail: Square tip with dark barring, tipped white. Wings: Slender, long with dusky flight feathers. Legs: Dull yellow.
Immature: Upperparts: Grey-brown with buff-edged feathers. Underparts: Rufous-brown, indistinct streaks.

Solitary and wary. Flight is high speed, direct, often aerobatic.

Habitat: Open woodland, open forests, margins of lakes, swamps, parks and gardens.

Feeding: Often feeds at dusk. Pursues or dives on prey, mostly birds taken mid-air. Also bats and large insects.

Voice: Whining and shill, harsh chatter or chuckling.

Status: Common. Nomadic. Northern birds more sedentary. Southern birds migrate north, possibly as far as PNG.

Breeding: Sept.–Jan. Nests high in trees, often using abandoned crow or magpie nests / 2–3 eggs.

Tennant Creek, NT.

Left: Nominate race. Right: Immature feeding on an immature swamphen.

Peregrine Falcon

Falco peregrinus

Male: 380mm
Female: 480mm
Wingspan: 1000mm

Sexes similar. Large powerfully built raptor, with a black hood and blue-black upperparts. Bill: Yellow with black tip and yellow cere. Eyes: Dark with yellow eye ring. Throat, breast: Buff-cream with some black spots to lower breast. Underparts: Greyish buff, finely black barred. Tail: Square shape, barred, tipped pale grey. Legs: Yellow.
Female: Much larger. Underparts more russet. Immature: Darker brown and more heavily dark-streaked underparts.

Noted for its speed. Swoops on prey at more than 300km/h. Soars high, stiff, shallow wingbeats and interspersed glides. Presence is marked by noisy panic among other birds.

Habitat: Cliffs, rocky hills, mountains, open timbered waterways, high-rise buildings.

Feeding: Rapid pursuit, snatching mostly small birds and animals with its powerful hooked talons.

Voice: Loud, repeated screams and chattering.

Status: Moderately common. Sedentary.

Breeding: Aug.–Dec. Abandoned nests, recesses in cliff face, tree hollows / 3 eggs.

Left: Immature. Right: Adult with prey.

Family Rallidae, Otididae

Rails

Crakes

Bush-hen

Native-hens

Moorhen

Swamphen

Coot

Bustard

A group of mostly secretive, sedentary, hen-like birds and the Australian Bustard.

Family Rallidae includes rails, crakes, Pale-vented Bush-hen, native-hens, moorhen, coot and swamphen.

There are 16 species represented in Australia, inhabiting wetlands and swampy and damp rainforest areas. Most are difficult to observe, as they generally inhabit dense, reedy vegetation and usually only venture out to feed at first light of the day and twilight. They characteristically have upright tails, which they constantly flick when walking, running or swimming, and long thin toes that allow them to walk on floating aquatic vegetation such as lily pads.

Nests are well hidden in dense vegetation. They are typically bowl-shaped, made from reeds and lined with fine rootlets and grasses. Tall reeds surrounding the nest are often bent over to form a protective hood and provide camouflage from predators above.

Although their flight appears feeble, these species can fly great distances in search of suitable habitat.

Family Otididae is represented in Australia by one species, the Australian Bustard.

This is a large ground-dwelling bird of open grasslands, commonly called the Plains-turkey. One of Australia's largest birds, males are much larger than females at over 1m tall with a wingspan of 2.3m, and an average weight of 6.2kg. Females are considerably smaller than males at 800mm tall, a wingspan of 1.2m and average weight of 3.2kg.

Reluctant to fly, when disturbed they slowly walk away with a 'snooty' appearance, holding their head high with their bill pointed skywards. They may run if alarmed or fly as a last resort and are Australia's heaviest flying bird.

Bustards are noted for their amazingly elaborate courtship displays. A group of males will clear a display arena, which is visited by females for mating. The males, spaced 100m–1000m apart, display to attract the females. Males inflate their huge plume-covered balloon-like throat sacs and release a deep booming vocalisation like a lion's roar every 10–15 seconds, while strutting around dancing from side to side with their tails cocked, heads held high and throat sacs swaying. They finish with their tails fanned over their back while making a booming, croaking noise. They copulate with females that visit their arena.

Extremely beneficial to agriculture, bustards track grasshopper and mouse plagues and gorge themselves on these pests. They have also been observed eating whole cane toads with no ill effects. Once widespread throughout the continent, they are now rarely seen in southeastern Australia and threatened in NSW due to past shooting for food and sport, habitat destruction, and predation by foxes and feral pigs.

Opposite: Australasian Swamphen.

Chestnut Rail

Eulabeornis castaneoventris

430–440mm

Sexes alike. Large thickset rail with a longish pointed tail and thick powerful yellow legs. Blue-grey head, paler under chin. Upperparts: Variable, 2 colour morphs – olive-chestnut in Qld, NT, and olive-grey in the Kimberley region, WA. Underparts: Glossy chestnut. Underwings: Brown. Bill: Long, heavy, greenish. Eyes: Red. Immature: Head: Slate grey. Mantle: Black. Rump: Chestnut brown. Underparts: Pinkish chestnut.

Wary, shy and rarely seen. Flicks tail while moving.

Habitat: Tall, dense mangroves.

Feeding: Struts onto mudflats at low tide to feed mostly on crabs and other crustaceans.

Voice: Loud, screeching, grunting calls.

Status: Uncommon, sedentary.

Breeding: Oct.–Feb. Cupped platform nest in mangrove roots or outer branches 1–3m high / 4–5 eggs.

Middle Arm, NT.

Left: Olive-grey morph. Right: Adult underparts.

Red-necked Crake

Rallina tricolor

260–290mm

Sexes alike. Large crake, distinctively bi-coloured. Head, neck, breast, upper shoulders: Rich chestnut. Upperparts: Dark slate grey. Underparts: Slate grey with faint pale barring. Throat: Paler chestnut. Flanks: Buff-white barring. Flight feathers, undertail: Grey with dull rufous barring. Eyes: Red. Bill: Olive green. Legs: Olive. Immature: Duller. Belly: Barred chestnut.

Usually nocturnal and secretive, flicking tail as it walks. Rarely observed.

Habitat: Along streams in tropical rainforests.

Feeding: Forages along streams at dusk taking frogs, tadpoles, crustaceans, snails, insects.

Voice: Repetitive screeching, grunting.

Status: Uncommon. Mostly sedentary.

Breeding: Nov.–Apr. Nest on ground beside a stump or up to 2m high in pandanus / 3–7 eggs.

Julatten area, Qld.

Opposite: Adult foraging along stream.

Lewin's Rail

Lewinia pectoralis

200–230mm

Sexes similar. Plump rail with a long, pink, dark-tipped bill and extensive barring. Crown, nape: Rich rufous, flecked black. Upperparts: Mottled dark brown and black. Neck, breast: Slate grey. Underparts: Black, finely barred white. Wings: Dusky brown with white barred coverts. Eyes: Brown.
Race *bachipus*, Bass Strait Is. Tas. Generally larger.
Female: Duller, less rufous.
Immature: Duller, faint streaking.

Secretive, shy and elusive. Usually skulks in dense vegetation. Flicks erect, stumpy tail when walking. Flight is awkward. Swims and dives.

Habitat: Thickly vegetated areas adjoining water, reedy swamps, gardens and grasslands.

Feeding: Probes for insects, molluscs, crustaceans, vegetable matter.

Voice: Frog-like croaks and grunts.

Status: Solitary. Sedentary. Uncommon. Race *clelandi*, WA, believed extinct.

Breeding: Aug.–Dec. Saucer-shaped nest, hidden in reeds / 4–6 eggs.

Dharug NP, NSW. Portland, Vic. Heron Is., Qld.

Left: Adult probing for molluscs. Right: Adult calling.

Buff-banded Rail

Gallirallus philippensis

300–320mm

Sexes alike. Elegant colourful rail with a distinctive, long, white eyebrow and chestnut streak through eyes. Crown: Chestnut brown streaked. Nape: Chestnut. Throat, upper breast: Grey. Breast: Orange-chestnut crescent-shaped band. Lower breast, belly, undertail: Black, heavily barred white. Upperparts, wings: Brown, spotted white. Eyes: Red. Bill: Long, straight, flesh brown. Legs: Pink-brown. Cape York, Qld, population smaller, darker with narrower breast band. Immature: Similar, lacks chestnut nape, breast band.

Timid. Usually solitary or in pairs. Runs through cover, flicking its stumpy tail when alarmed.

Habitat: Damp, dense vegetation near swamps, lagoons, watercourses, lawns and gardens.

Feeding: Prefers feeding at dawn and dusk. Insects, small molluscs, seeds, aquatic plants.

Voice: Creaky squeaking, donkey-like repetitive braying.

Status: Common. Some birds migrate northward in winter. Also found on offshore islands. Scattered inland populations.

Breeding: Sept.–Mar. Low nest in dense cover / 5–8 up to 11 eggs.

Opposite: Adult feeding.

Baillon's Crake

Porzana pusilla

160mm

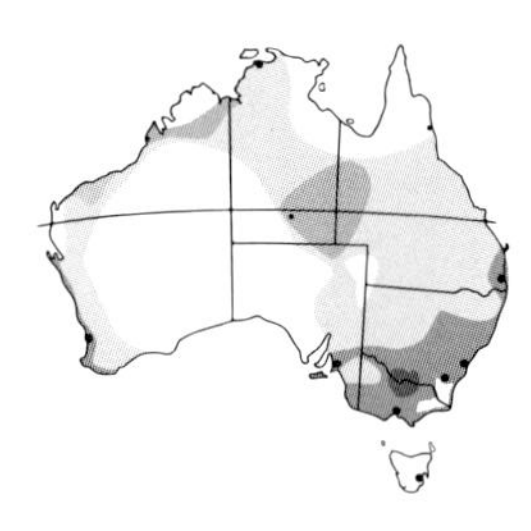

Sexes alike. Australia's smallest crake. Upperparts: Dark cinnamon with large black streaks and small white spots. Breast: Pale blue-grey. Belly, undertail: Barred black and white. Bill: Grey-green. Eyes: Red. Legs: Grey-green. Immature: Paler, indistinct markings.

Secretive and shy. Swims, dives and walks over water plants. Flight is weak, fluttering. Constantly flicks upright tail.

Habitat: Swamps, marshes.

Feeding: Probes in shallow, damp ground for molluscs, seeds, insects and aquatic vegetation.

Voice: Squeaky 'chut' call.

Status: Moderately common. Some birds migrate north to overwinter.

Breeding: Sept.–Feb. Nests among tussocks in shallow water / 4–8 eggs.

Werribee, Vic. Jerrabomberra wetlands, ACT.

Opposite: Foraging among White Snowflake Lily.

Australian Spotted Crake

Porzana fluminea

Endemic

180–200mm

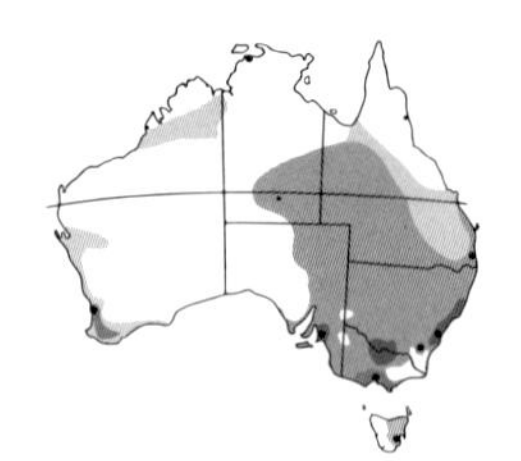

Sexes similar. Largest and most commonly seen Australian crake. Olive-brown upperparts with strong black streaking and small white spots. Face, breast: Dark grey-blue. Belly, flanks: Dark grey barred white. Eyes: Deep red. Bill: Olive with red base to upper mandible. Legs: Olive green.
Female: Paler, less red to base of bill. Immature: Paler with dark and pale brown mottled face and neck.

Usually solitary or in pairs. Flight is laboured and fluttering.

Habitat: Well-vegetated edges of marshes, water channels, swamps, estuaries and saltmarshes.

Feeding: Aquatic plants, molluscs, insects picked from mud.

Voice: Sharp metallic 'crake' and chattering.

Status: Common. Nomadic in inland areas or sedentary near permanent water.

Breeding: Aug.–Feb. Nest in reed beds in shallow water / 3–6 eggs.

Opposite: Adult, stirring water to disturb prey.

Spotless Crake

Porzana tabuensis

180mm

Sexes alike. Plain, dark crake with red eyes and eye ring and reddish legs. Back, wings: Olive-brown. Head, underparts: Dark bluish grey. Undertail coverts: Barred and spotted white. Bill: Black.
Immature: Duller. Throat: Whitish. Eyes: Brown.

Timid. Rarely leaves cover. Weak flight with trailing legs. Seldom flies.

Habitat: Rushes, reeds around swamps and mangroves including offshore islands.

Feeding: Feeds over mud or shallows, taking molluscs, insects, aquatic vegetation.

Voice: Variable. High-pitched squeaking, harsh scolding.

Status: Uncommon. Nomadic.

Breeding: Sept.–Jan. Shallow cup nest, in cover, with 1m long trampled sloping tunnel-like approach to nest in tussock in water / 3–6 eggs.

Fivebough Swamp, Leeton, NSW. Jerrabomberra Wetlands, ACT.

Left: Adult foraging. Right: Undertail coverts.

White-browed Crake

Amaurornis cinerea

160mm

Sexes alike. Tropical crake, walks on lily pads like a jacana. Black crown with a distinctive face pattern, 2 white lines above and below the red eyes. Bill: Olive-yellow, reddish base. Upperparts: Dark brown, mottled paler. Throat, breast, underparts: Pale grey, lighter in centre. Undertail: Buff. Legs: Olive green.
Immature: Similar, paler. Crown: Brown.

Fluttering flight with trailing legs. Runs across waterlily pads.

Habitat: Tropical, coastal, freshwater swamps, lakes, dams.

Feeding: Insects, leeches, worms, slugs, vegetable matter, seeds.

Voice: Chattering. 'Crake' repeated, often in concert.

Status: Locally common. Nomadic or migratory, moving north to PNG in the dry season.

Breeding: Sept.–Feb. Nest in shallows in reeds / 4–8 eggs.

Opposite: Adult foraging among White Snowflake Lily.

Pale-vented Bush-hen

Amaurornis moluccana

260mm

Sexes similar. Secretive, mid-sized waterbird with uniformly dark plumage. Crown, nape, back: Olive-brown. Breast: Slate brown. Belly: Pinkish buff. Undertail: Orange-brown. Eyes: Brown. Bill: Olive-yellow with orange-yellow frontal shield.
Female: Smaller, paler bill.
Immature: Duller. Head: Black.

Shy and mostly nocturnal. Seldom flies. Solitary, pairs or small family groups.

Habitat: Wet grassy edges of rainforests with patches of dense shrubbery. Near swamps, lakes, along creeks. (Associated with lantana thickets on the east coast.)

Feeding: Stalks, wades in shallow water, taking insects, invertebrates, seeds, vegetable matter.

Voice: Shrieking, donkey-like 'nee-you' repeated and single 'tok'. Calls frequently day and night.

Status: Moderately common. Sedentary or partially nomadic.

Breeding: Oct.–April. Saucer-shaped nest with top cover in tall grass / 5–7 eggs.

Opposite: Male (foreground) and female, stalking for prey.

Black-tailed Native-hen

Tribonyx ventralis

Endemic

340–350mm

Sexes alike. Distinctive bill with a bright green frontal shield, orange-red lower mandible and bright yellow eyes. Upperparts: Bronze brown. Throat, breast: Deep blue-grey. Flanks: White spots. Belly: Black. Tail: Black, erect. Legs: Bright pink. Immature: Duller.

Mostly runs or walks with tail cocked. Sometimes flies long distances, in large numbers, to wetland feeding grounds. Usually seen in loose flocks of 5–50 birds.

Habitat: Margins of swamps, lakes, irrigated pasture, rivers and inland claypans.

Feeding: Grass shoots, weeds, herbs, seeds, insects, snails.

Voice: Usually silent. Sharp 'kak' repeated alarm call.

Status: Common. Highly nomadic. Birds can appear unannounced.

Breeding: Any time of year depending on rainfall. Grass cup nest on or near ground, near water / 5–7 eggs.

Left: Adults feeding. Right: Frontal view.

Tasmanian Native-hen

Tribonyx mortierii

Endemic

450–480mm

Sexes alike. Large, heavily built flightless native-hen confined to Tas. Upperparts: Olive-brown. Wings: Darker brown, fine white flecks. Underparts: Blue-grey. Tail: Black. Flank: White patch. Eyes: Bright red. Bill: Yellow. Legs: Dark grey, large feet. Immature: Duller, grey washed. Bill: Grey.

Runs fast when alarmed. Swims strongly. Usually in small groups that include a female and 1 or 2 males.

Habitat: Grassy edges of swampy areas with short Sword Grass and fields with thickets for shelter.

Feeding: Emergent shoots of grass, Sword Grass and insects.

Voice: Harsh scream. Braying 'wee-haw'.

Status: Locally common. Sedentary. Territorial.

Breeding: July–Dec. Nest in cover close to water / 4–9 eggs.

Opposite: Adult feeding chick.

Dusky Moorhen

Gallinula tenebrosa

340–380mm

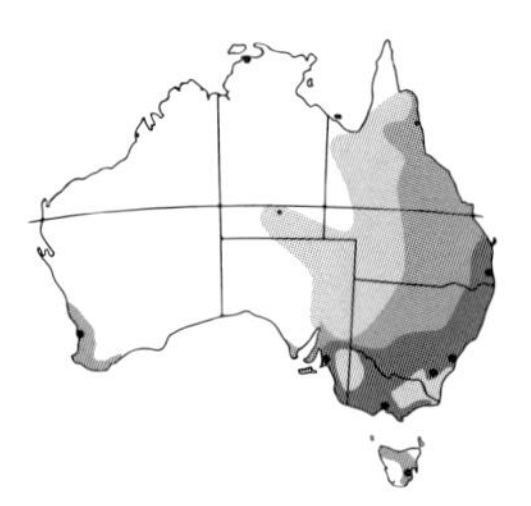

Sexes alike. Drab, dark grey-black plumage contrasts with its distinctive bright red, yellow-tipped bill and large, bright reddish orange frontal shield. Eyes: Brown. Upperparts: Dark brown. Underparts: Dark blue-grey. Undertail: White with broad central black line. Legs: Reddish. Non-breeding: Dull frontal shield, legs are green below the knee. Immature: Duller, white-edged tail.

Runs and swims strongly. Flies with legs trailing. Usually in groups of 2–7 birds.

Habitat: Parkland lakes, permanent swamps, lakes, waterways with grassy banks and reed cover.

Feeding: Vegetable matter, insects, fish, worms, molluscs. Often up-ends when feeding.

Voice: Sharp 'kerk' shrieks.

Status: Common, sedentary, territorial.

Breeding: Aug.–Feb. Nest in reeds in water / 5–8 eggs.

Left: Adult. Right: Immature.

Australasian Swamphen

Porphyrio melanotus

440–480mm

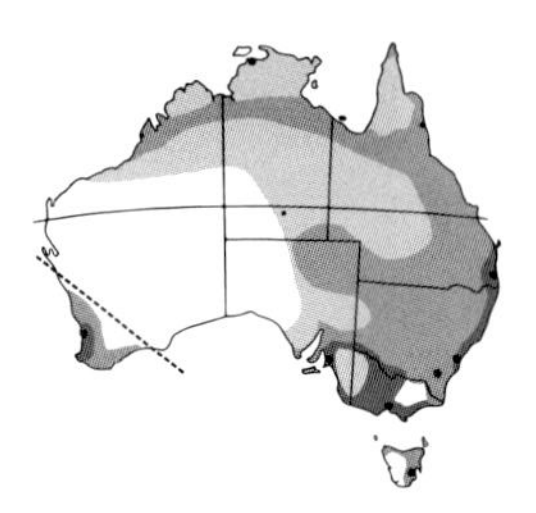

Sexes alike. Common waterhen with a prominent scarlet frontal shield and large, stout, scarlet bill. Head, neck, underparts: Deep purple-blue. Upperparts, wings, uppertail: Dusky black. Undertail: Bright white. Race *bellus*, southwest WA. Sky blue breast, visible in good light. Immature: Duller. Bill: Blackish red.

Typically flicks tail when walking or swimming. Flies to escape danger with legs dangling below the body or trailing behind. Shy, but tame in urban parks.

Habitat: Reedy swamps, river margins, lakes, parkland ponds.

Feeding: Reed shoots, seeds, eggs, small vertebrates, insects, molluscs.

Voice: Usually quiet. Occasional metallic cackling.

Status: Common. Sedentary.

Breeding: Aug.–Feb. Nest in reeds / 3–8 eggs.

Left: Nominate race. Right: Race *bellus*, showing sky-blue breast.

Eurasian Coot

Fulica atra

320–390mm

Sexes alike. Plump, overall slate grey waterbird with a contrasting gleaming white bill and frontal shield. Eyes: Bright red. Legs: Grey, with toes that have wide, flattened lobes.
Immature: Paler. Throat: Whitish. Bill: Greyish.

Strong flight. Mostly aquatic, sleeping on water. Runs across water. Usually in huge flocks on large wetlands.

Habitat: Fresh or brackish water. Wetlands, temporary floodwaters or large ponds.

Feeding: Aquatic plants, stems, seeds, algae. Groups feed at water's edge, in shallows or deep-dive far out.

Voice: Abrupt 'krek' and shrill calls.

Status: Common, sedentary or nomadic. Often in large groups or winter flocks.

Breeding: Aug.–Feb. Nest of sticks and reeds, in shallows or on islands / 6–7 eggs.

Left: Adult feeding. Right: Adult.

Australian Bustard

Ardeotis australis

Male: 1100–1200mm
Wingspan: 2300mm
Female: 800mm
Wingspan: 2000mm

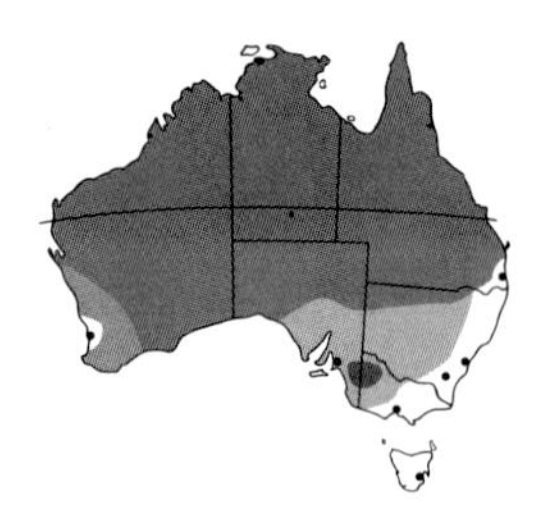

Sexes similar. Tall, stately bird with an erect posture. Crown: Black. Wings, tail: Chestnut, finely barred black. Wing coverts: Black and white patterned. Face, neck, belly, breast: White, with black band to breast. Eyes: Yellow, with white eyebrow. Bill: Whitish. Legs: Pale yellow. Female: Smaller, with brown crown and no breast band. Immature: Similar to female. Heavily barred neck.

Usually solitary or in pairs. Freezes when disturbed, then slowly walks away with head held high. Strong, slow flight with neck and legs extended. Complex courtship display.

Habitat: Wooded, grassy plains, dry woodlands, mulga, spinifex, saltbush scrub.

Feeding: Omnivorous. Insects, grass seeds, fruits. Mice and grasshoppers in large quantities.

Voice: Deep booming, rising and falling.

Status: Uncommon. Loss of habitat. Extinct in settled areas.

Breeding: Jan.–Mar. in north. Sept.–Nov. in south, or anytime in suitable conditions. Nest in grass / 1–2 eggs.

Bruce Hwy, Rockhampton, Qld.

Left: Male. Right: Male in display.

Family Scolopacidae, Rostratulidae, Haematopodidae, Recurvirostridae

Snipes

Godwits

Curlews

Whimbrel

Sandpipers

Tattlers

Greenshank

Redshank

Turnstone

Dowitcher

Knots

Sanderling

Stints

Ruff

Phalarope

Oystercatchers

Avocet

Stilts

A grouping of 4 families dominated by migratory wading birds.

Family Scolopacidae includes snipes, godwits, curlews, sandpipers, phalarope, tattlers, greenshank, redshank, turnstone, dowitcher, knot, stints, Ruff, Sanderling and Whimbrel.
A group of mostly non-breeding migratory waders, also called 'shorebirds', they are noted for their annual journey of many thousands of kilometres. They are commonly found on beaches, rocky coastal shores, tidal mudflats or wetlands over summer. Each species of wader is a specialised feeder with a different shape and length of bill to exploit a specific food niche. Some have long, slender bills for probing mudflats or shorelines, others have short blunt bills for picking up surface prey, while those that feed while wading have varying-length legs, influencing the food niche they utilise.
Many waders breed in Siberia or the Arctic Circle in the northern summer and spend the southern summer in Australia, usually returning to the same beach. Some waders occasionally overwinter in Australia, or arrive early and can be seen in their breeding plumage. The migration flights are amazing feats of endurance. A Bar-tailed Godwit tracked by satellite travelled 11,700 km in 8 days, flying non-stop from its breeding grounds in Iceland. The Red-necked Stint completes a 23,000km round trip each year from its breeding grounds in eastern Siberia and western Alaska to summer in Australia.
There are three species of *Gallinago* snipe – Latham's, Pin-tailed and Swinhoe's. They are hard to separate in the field and rely on close examination of tail patterns and number of tail feathers for accurate identification.

Family Rostratulidae contains the Australian Painted Snipe. This beautifully patterned rail-like wader is Australia's sole representative. It was considered a subspecies of the Greater Painted Snipe of Africa and Asia, but recent genetic analysis has shown that it is a separate species that has been isolated from the 'true snipes' for approximately 19 million years. The breeding behaviour is unusual in that it has reversed parental roles. The female is more colourful and larger. The male builds the nest while the female mates and lays eggs in several nests, leaving each male to look after incubation duties and raising the young.

Family Haematopodidae includes oystercatchers.
It is represented by two species in Australia. They feed at sandy beaches, probing with their distinctive red bills.

Family Recurvirostridae includes stilts and the avocet.
The Red-necked Avocet feeds on aquatic insects and crustaceans, sweeping its distinctive upcurved bill from side to side while wading in shallow water. Unlike other shorebirds, it can swim readily and sometimes up-ends to feed like a duck.

Outstanding wader-watching locations:
Cairns Esplanade, Qld. Kooragang Is., NSW. Werribee, Vic. Orielton Lagoon, Hobart, Tas. Price Saltfields, Yorke Peninsula, SA. Roebuck Bay and Broome Bird Observatory, WA. Buffalo Creek, NT.

Opposite: Red-necked Avocets feeding.

Latham's Snipe

Gallinago hardwickii

270–290mm

Sexes alike. Australia's largest snipe, cryptically plumed for superb camouflage. High, rounded head and extremely long, straight, olive-brown bill. Eyes: Large, brown with dark eye stripe. Crown: 2 broad, brown stripes. Upperparts, shoulders: Cryptic brown, black and buff. Wings: Dark brown flight feathers. Underparts: Cream, white and flecked black on upper throat and upper breast extending to flanks. Tail: Rufous band and tipped white (16–18 feathers). Legs: Olive-grey. Immature: Similar.

Quail-like, erratic flight when flushed. Wary. Solitary or small groups.

Habitat: Wet grasslands, swamps, mangroves, irrigated areas.

Feeding: Walks briskly probing for worms, aquatic larvae and crustaceans, often at dawn and dusk.

Voice: In flight, loud, rasping 'crek'.

Status: Locally common non-breeding migrant.
Arrive: Aug. Depart: Apr.

Breeding: Northern islands of Japan and East Asia.

Jerringot Reserve, Geelong, Vic.

Left: Adult cryptic plumage. Right: Underparts.

Pin-tailed Snipe

Gallinago stenura

250–270mm

Sexes, immature: Similar. Smallest of the 3 snipes. Difficult to distinguish in the field from Swinhoe's Snipe. The Pin-tailed Snipe has a stumpy, square tail with 'pin-like' outer tail feathers. Upperparts: Dark brown cryptic-patterned dark rufous. Breast: Streaked buff. Underparts: White. Tail: Short; rufous with black band and white tips (24–28 feathers). Short legs.

Usually only seen when flushed. Low, flat, heavy, zigzagging flight.

Habitat: Edges of freshwater wetlands, swamps, lakes with emergent vegetation and dryer claypans.

Feeding: Picks food from surface, including insects, small aquatic life, earthworms, some plant matter.

Voice: Nasal, metallic 'tchet' when flushed.

Status: Regular but rare non-breeding migrant.
Arrive: Aug. Depart: Mar.
Some juvenile birds overwinter in northern Australia.

Breeding: Siberia.

Broome Bird Observatory, WA.

Left: Adult cryptic plumage. Right: Back view showing tail.

Swinhoe's Snipe

Gallinago megala

240–260mm

Sexes alike. Cryptic, mottled black, brown and buff upperparts. Underparts: Brown and grey with brown flecks to breast and throat. Crown: Dark-brown stripes. Eyes: Dark brown. Bill: Long, straight, olive-brown. Tail: Black with ochre band and white tipped (20–24 feathers). Legs: Olive-grey.
Immature: Similar.

When flushed from daytime cover flight is erratic and quail-like.

Habitat: Wet grasslands, swamps, and billabongs on coastal plains.

Feeding: Feeds at night probing mud and soft ground for caterpillars, worms, crustaceans and vegetable matter.

Voice: Rasping 'shrek' in flight.

Status: Moderately common non-breeding migrant.
Arrive: Nov. Depart: Mar.

Breeding: Northeast Asia.

Knuckey Lagoons CP, NT.
Broome Bird Observatory, WA.

Opposite: Adult Feeding.

Black-tailed Godwit

Limosa limosa

360–430mm

Sexes similar. Long, straight, dark-tipped bill with pink base. Head, neck, underparts: Buff-grey. Upperparts: Uniformly grey-brown. Rump: White. Tail: Black. Wings: Broad white bar and dusky flight feathers. Eyes: Dark brown with white eyebrow. Breeding plumage: Male: Head, mantle, upper breast: Russet. Eyebrow: Buff. Tail: White with black sub-terminal bar. Female: Head, neck: Buff. Immature: Similar but browner.

Distinctive acrobatic flight. Solitary or in small flocks.

Habitat: Coastal tidal estuaries, fresh or brackish, shallow muddy lagoons and swamps.

Feeding: Plunges bill deep into mud for worms, molluscs, crustaceans, insects.

Voice: Variable sharp 'witta-wit'.

Status: Moderately common non-breeding migrant. Arrive: Aug. Depart: May.

Breeding: Northern Eurasia.

Broome Bird Observatory, WA. Cairns Esplanade, Qld. Kooragang Is., NSW.

Left and right: Non-breeding.

Bar-tailed Godwit

Limosa lapponica

Male: 380–420mm
Female: 420–450mm

Sexes similar. Large, stocky wader with mottled grey-brown upperparts. Lower back, rump: White with brown scalloping. Bill: Long, pink, dark-tipped, slightly upturned. Eyes: Brown with white eyebrow. Underparts: Cream-white with grey-brown washed breast. Legs: Olive-grey. Breeding plumage: Male: Rich rufous-brown head, neck, mantle. Upperparts: Richly mottled grey, brown, black. Underparts: Overall rufous-brown.
Female: Larger, duller. Immature: Similar, browner upperparts with buff-edged feathers.

Distinctive flight in large close flocks, rising, falling, banking together. Steady, even wingbeats. Flocks can number tens of thousands.

Habitat: Coastal sandy beaches, tidal mudflats, estuaries.

Feeding: In shallows, probing, sweeping for invertebrates, crustaceans. Often feeds at low tide on moonlit nights. Female ventures furthest out.

Voice: Sharp 'witta-wit', 'kak' alarm call.

Status: Common non-breeding migrant, more common in the north. Arrive: Aug. Depart: May.

Breeding: Northeastern Siberia.

Opposite: Non-breeding, with breeding bird (centre).

Little Curlew

Numenius minutus

300–330mm

Sexes similar. Tiny curlew with a shortish, slightly down-curved, grey bill and mottled dark brown and buff upperparts. Crown: Dark brown. Eyes: Large, dark with buff eyebrow. Throat, breast: Streaked chestnut brown. Underparts: Whitish fawn. Legs: Grey-brown.
Female: Slightly larger.
Immature: Similar to adult.

Usually wary, sometimes tame in urban areas. Freezes when disturbed. Runs swiftly. Flight is buoyant.

Habitat: Open, dry grasslands, including airfields, urban lawns.

Feeding: Probes cracks in soil for insects, caterpillars, seeds.

Voice: Chattering, rising sharp 'ti-ti-ti' when alarmed.

Status: Non-breeding migrant, moderately common. Arrive: Sept. Depart: May. More common in north. Some birds move south to avoid monsoonal rains.

Breeding: Siberia and Mongolia.

Opposite: Non-breeding.

Eastern Curlew

Numenius madagascariensis

Critically Endangered
580–620mm

Sexes alike. Large, stocky wader with a distinctive, very long, down-curved black bill with a pink base. Overall plumage: Brown and buff streaked. Legs: Long, grey. Eyes: Brown with a white eye ring.
Female: Larger (world's largest wader in body size and bill length). Immature: Similar, shorter bill.

Wary and extremely shy. Quick to take flight if disturbed. Usually seen singularly or in pairs. Slow, deliberate wingbeats.

Habitat: Sandy beaches, intertidal mudflats, estuaries and mangrove swamps.

Feeding: Probes mostly for small crabs, also worms, crustaceans, molluscs.

Voice: Distinctive. Mournful but melodious 'cur-lew'.

Status: Moderately common, non-breeding migrant. An estimated 75% of the world's population winter in Australia. Critically endangered due to loss of habit along the migration staging sites between China and Korea. Arrive: Sept. Depart: May. Some birds overwinter, but migrate northward.

Breeding: Siberia and Manchuria.

Left: Adult feeding. Right: Adult in flight.

Whimbrel

Numenius phaeopus

400–430mm

Sexes alike. Mid-sized curlew with distinctive dark streaks along the crown and a long, slightly down-curved brown bill with pink lower base. The bill is shorter and less curved than the Eastern Curlew. Upperparts: Streaked mid-brown. Underparts: White, barred brown on flanks. Rump: White. Eyes: Dark brown with whitish eyebrows. Legs: Olive-grey.
Immature: Similar.

Wary. Singularly or in small flocks. Flight is light and fast with white rump and streaked tail an aid to field identification.

Habitat: Estuaries, mudflats, mangroves, exposed reefs, sewerage ponds, meadows and fields.

Feeding: In small groups, probing for worms, molluscs, crustaceans and insects. Fruit and seed also taken.

Voice: Musical high-pitched 'ti-ti-ti'.

Status: Moderately common non-breeding migrant. Arrive: Aug. Depart: Apr. Rare in southern part of range.

Breeding: Arctic Circle. Tundra of Alaska and Siberia.

Left: Adult. Right: In flight.

Terek Sandpiper

Xenus cinereus

230–260mm

Sexes alike. Pale wader with a very long, gently upward-curving black bill with an orange base. Frons, crown: Grey-brown, high and steep. Eyes: Brown with pale eyebrow. Upperparts, sides of neck: Grey-brown, centre mark to feathers. Underparts: White. Wings: White trailing edge, noticeable in flight, giving a flickering effect. Legs: Long, orange with partly webbed feet. Breeding plumage: Darker with 2 black lines along back. Feathers have dark centre patch. Immature: Like breeding adults, darker.

Bobs head, dashes quickly, probes and runs while feeding.

Habitat: Tidal mudflats, brackish swamps, coastal mangroves, sheltered estuaries, sandbars.

Feeding: Crouches while foraging for worms, molluscs, crustaceans and insects.

Voice: Musical, piping trills. Calls in flight.

Status: Moderately common non-breeding migrant. Most common on the northern and eastern coasts of Australia. Arrive: Aug. Depart: Apr.

Breeding: Northern Russia and Finland.

Opposite: Non-breeding adult feeding.

Common Sandpiper

Actitis hypoleucos

200–210mm

Sexes alike. Small sandpiper with a relatively long body, short greyish yellow legs and a medium-length fine, straight, brown bill. Upperparts: Dark olive-brown with black fine streaking, barring. Eyes: Brown, whitish eye ring, eyebrow and dark stripe through eye. Underparts: White with grey-brown breast. Tail: Brown, white outer feathers.
Breeding plumage: Richer, heavy streaking to breast. Immature: Similar to non-breeding adult.

Often seen in flight, low, close to water, fast flicking wingbeats and alternating glides showing white trailing edge and distinct wing bars. Constant nervous movement and tail bobbing.

Habitat: Steep edges of mudbanks, shallows, pools, farm dams, inlets with mangroves, rarely open shoreline.

Feeding: Mostly insects. Snatched from rock crevices, cracks in soil or from leaves. Actively chases prey.

Voice: Piping swallow-like 'twee-wee-wee'.

Status: Regular but uncommon non-breeding migrant.
Arrive: July–Aug. Depart: May.

Breeding: Northern Eurasia.

Left and right: Non-breeding.

Grey-tailed Tattler

Tringa brevipes

260–270mm

Sexes alike. Mid-sized, mostly grey wader with a long, straight, grey bill. Dark brown eyes with white eyebrow above and dark brown eye stripe. Breast: Grey. Underparts: White. Legs, feet: Yellow-olive.
Breeding plumage: Richer, finely barred brown underparts.
Immature: Similar to non-breeding. Pale spotted upperparts.

Most commonly seen wader. Singularly or in flocks of 50 or more. Constantly bobs head and flicks rear body up and down. Flight is low with clipped wingbeats. On landing, spreads wings before folding.

Habitat: Coastal intertidal rockpools, shallows, beaches, mangrove mudflats.

Feeding: Worms, molluscs, crustaceans and particularly crabs before return migration.

Voice: Flute-like variations of 'troo-eet'.

Status: Non-breeding migrant.
Arrive: Sept. Depart: Apr. Adults arrive first. First-year birds about 4 weeks latter. More common in the north.

Breeding: Siberia.

Left: Remnant breeding plumage.
Right: Non-breeding.

Wandering Tattler

Tringa incana

280–290mm

Sexes immature: Alike. Very similar to Grey-tailed Tattler, but slightly larger and darker with thicker bill and eyebrow that does not extend past the eye.
Upperparts: Plain dark grey.
Throat: White. Underparts: Light grey to white, sometimes with traces of breeding plumage.
Bill: Dark grey with nasal groove extending two-thirds along bill. Eyes: Brown. Legs, feet: Greenish yellow.
Breeding plumage: Darker.
Underparts: Heavily barred except for white central belly.

Fast graceful flight. Bobs head and body. Runs around rockpools as it feeds.

Habitat: Intertidal rockpools, coral reefs, less frequently mudflats, sandbars.

Feeding: Crabs, molluscs, crustaceans.

Voice: Differs from Grey-tailed Tattler. Rapid, ascending, descending trill, usually in flight.

Status: Regular but uncommon non-breeding migrant.
Arrive: Sept. Depart: Mar.

Breeding: Siberia to Alaska and British Columbia.

Left: Breeding plumage. Right: Non-breeding.

Common Greenshank

Tringa nebularia

320–350mm

Sexes alike. Large, pale wader with a medium long, grey-green bill slightly upturned and darker at base. Head: Pale grey, finely streaked. Eyes: Brown with whitish eye ring. Upperparts: Mid-grey, feathers finely edged white and notched black. Back, rump: White, seen in flight. Underparts: White, with streaking to sides of breast. Legs: Grey-green to yellowish.
Breeding plumage: Richer, chestnut-grey-black. Throat, breast, sides: Heavy streaking.
Immature: Similar to non-breeding. Darker upperparts and heavier streaking to throat and breast.

Wary, active, excitable. Solitary, pair or less often in small groups.

Habitat: Edges of fresh, brackish or saline lakes, pools and inland wetlands and billabongs.

Feeding: Briskly wading in shallows or in mud, snatching worms, crustaceans, molluscs.

Voice: Staccato whistling. Shrill 'kew-kew-kow'.

Status: Common non-breeding migrant. Arrive: Sept. Depart: Apr.

Breeding: Northern Eurasia.

Left: Non-breeding. Right: Breeding plumage.

Marsh Sandpiper

Tringa stagnatilis

210–220mm

Sexes alike. Resembles the Common Greenshank, but smaller, more slender, with distinctive and very long, yellowish green legs and a fine long, straight, black bill. Crown, nape: Finely streaked pale grey-brown. Eyes: Brown with whitish surround. Frons, throat: White. Upperparts: Pale grey-brown feathers with paler scalloped edges. Lowerback, rump: White, wedge-shaped. Tail: White barred grey-brown. Underparts: White, with pale streaking to sides of breast. Underwings: White. Legs: Olive green.
Breeding plumage: Upperparts: Rich cinnamon-buff, heavily streaked breast and sides.
Immature: Like non-breeding.

Singularly or in small groups. Nervous, fast dashing in half circles when feeding. Bobs head. Flight is fast, clipped wingbeats, trailing feet.

Habitat: Shallow edges of fresh or saline marshes, mostly inland. Occasionally beaches, mudflats.

Feeding: Forages in shallows for aquatic insects, crustaceans, molluscs. Sometimes swims and pecks water surface.

Voice: Soft 'tee-oo' or sharp alarm 'yip'.

Status: Moderately common non-breeding migrant. Some overwinter. Arrive: Aug. Depart: May.

Breeding: Northern Eurasia.

Left and right: Non-breeding.

Common Redshank

Tringa totanus

270–290mm

Sexes alike. Medium-sized wader with long red legs and a fine straight bill with an orange-red base. Upperparts: Dark grey-brown, heavily streaked. Rump: White, wedge-shaped rump. Tail: Barred. Underparts: White with grey-brown streaking.
Non-breeding: Paler, duller, less streaked.
Immature: Warmer brown wash.

Wary. Often in mixed flocks. Strong, erratic flight.

Habitat: Tidal mudflats, sandbars, mangroves, coastal wetlands and salt marshes.

Feeding: Forages day or night for worms, molluscs, crustaceans and insects.

Voice: High-pitched yelping. 'Theehu-hu' alarm call.

Status: Rare, non-breeding summer migrant. More common in northwest Australia.

Breeding: Eurasia.

Roebuck Bay, WA.

Left: Non-breeding. Right: Remnant breeding plumage.

Wood Sandpiper

Tringa glareola

200–230mm

Sexes alike. Small, slender and solitary wader with a straight, mid-length, black bill. Head: Dark, streaked white. Eyes: Dark, ringed white below a long white eyebrow. Chin: White. Breast, throat: Heavily streaked. Upperparts: Dark brown, boldly flecked white. Rump: White. Underparts: White. Tail: Finely barred black.
Non-breeding: Duller, greyer, smaller eyebrow.
Immature: Similar to non-breeding; warmer browns with buff flecks. Chin, throat, breast: Washed brown, lightly streaked.

Flight is high, fast, zigzagging.

Habitat: Shallow freshwater swamps, inland waterholes, often fringed with River Red Gums.

Feeding: Mostly aquatic insects. Rapidly probes in mud for worms, fish, spiders, beetles, vegetable matter.

Voice: High-pitched whistling, rapid 'chi-chi-chip'.

Status: Regular but uncommon non-breeding migrant. Arrive: Aug.–Sept. Depart: Apr.

Breeding: Scandinavia to Siberia and Manchuria.

Left and right: Non-breeding.

Ruddy Turnstone

Arenaria interpres

220–240mm

Sexes similar. Distinctively marked, mid-sized wader with tortoiseshell-like brown, black and chestnut mottled upperparts. Head: Mostly white with a black-streaked crown and a black face pattern. Underparts: White. Breast: Black and white. Legs: Short, bright orange. Bill: Black, short, slightly upturned.
Non-breeding: Greyer with less or no chestnut colouring. Duller, grey brow. Head: More streaked, greyer.
Female: Duller. Immature: Similar to non-breeding adult; browner.

Constant running while feeding. Confiding; it allows close observation.

Habitat: Rocky shores, coral cays, tidal reefs, beaches with shingle and seaweed.

Feeding: Named for its habit of turning beach debris over to expose prey, including insects, worms, small crustaceans. May break open mussels and barnacles using its strong bill.

Voice: Grunting, rattling, ringing 'kee-oo'.

Status: Moderately common non-breeding migrant; most common in the north. Arrive: Sept. Depart: May. Small groups sometimes overwinter.

Breeding: Siberia and Alaska.

Left: Breeding. Right: Non-breeding.

Asian Dowitcher

Limnodromus semipalmatus

320–350mm

Sexes alike. Large wader with a distinctive long neck, long dark legs and long, straight, snipe-like black bill. Head: Dark cap, and a bold white eyebrow. Upperparts: Mottled grey-brown, flecked white. Breast: White, mottled light brown. Underparts: White with dusky flecks under tail.
Breeding plumage: Russet head and underparts. Faint, whitish eyebrow. Shoulders, back: Russet-brown with dusky feather centres.
Immature: Similar to non-breeding, but darker with brown washed breast.

Flight is swift and steady with bill pointed downward.

Habitat: Tidal sandflats and mudflats, sometimes inland.

Feeding: Walks jerkily, rapidly plunging bill into mud to slowly extract prey, often working in small groups.

Voice: Usually silent.

Status: Rare non-breeding migrant. Vagrant down east coast. First recorded in 1971, increasing in numbers.
Arrive: Sept. Depart: Apr.

Breeding: Siberia, Mongolia to Manchuria.

Roebuck Bay, WA.

Left: Breeding plumage. Right: Non-breeding.

Great Knot

Calidris tenuirostris

Critically Endangered
280–300mm

Sexes similar. Largest sandpiper. Thickset. Similar to Red Knot but has longer bill and more distinct mottling. Crown: Distinct, dark streaking. Upperparts: Grey with light brown tipped feathers. Underparts: White, sometimes with brown spotting to breast and flanks. Eyes: Dark with indistinct white eyebrow. Bill: Dusky, stout, straight. Legs: Olive-grey. Breeding plumage: Richer coloured and strongly patterned black and white breast. Female: Slightly larger. Immature: Similar to non-breeding; darker and brownish washed breast.

Strong, direct flight in flocks.

Habitat: Intertidal, sheltered coastal sandflats or mudflats and often inland lakes.

Feeding: Often in mixed flocks, thrusting bill, probing for insects, molluscs, worms, crustaceans.

Voice: Usually silent. Double-noted soft whistle in flight, taken up by flock and becoming continuous.

Status: Locally abundant non-breeding migrant. Arrive: Sept. Dep.: May. Some may overwinter.

Breeding: Northern Siberia.

Kooragang Is., NSW.

Left: Breeding plumage. Right: Non-breeding.

Red Knot

Calidris canutus

Critically Endangered
250mm

Sexes similar. Plump wader with shortish, straight, black bill. Upperparts: Mid-grey with paler feather tips. Rump: White with irregular dark-brown crescents. Eyes: Brown with white eyebrow. Underparts: White with pale-grey spotting, barring on breast, flanks.
Breeding plumage: Russet, mottled black upperparts. Chestnut below.

Female: Duller. Immature: Similar non-breeding, browner.

Form huge flocks containing thousands of birds that rise rapidly together and land together. Steady wingbeats, swift flight.

Habitat: Tidal sandflats and mudflats.

Feeding: Head down, moving together in compact flocks at low tide. Methodical, deep probing for invertebrates, insects, molluscs, crustaceans.

Voice: Usually silent. Low, throaty 'knut-knut' when feeding. Whistled alarm call. Piercing shrill call in flight.

Status: Common non-breeding migrant. Arrive: Aug.–Sept. Depart: May.

Breeding: Northeastern Siberia to Alaska.

Left: Coming into breeding plumage. Right: Non-breeding.

Sanderling

Calidris alba

180–200mm

Sexes alike. Tiny, pale grey wader with bright white underparts contrasting with black legs and black, short bill. Breast: Light grey wash. Upperparts: Pale pearly grey. Crown: Fine black streaks. Face: White. Wings: Blackish brown with dark grey shoulder, broad white wing bar and trailing edge seen in flight. Eyes: Brown.
Breeding plumage: Upperparts, throat, breast: Richly patterned chestnut, black and white. Overwintering birds occasionally take on breeding plumage. Immature: Similar to non-breeding but stronger blackish markings. Breast: Pale buff.

In small flocks, running together along edge of breaking waves. Take off in unison. Flight is swift, direct, low with shrill calls.

Habitat: Sandy ocean beaches.

Feeding: Head down in small groups probing the sand as the waves recede for small invertebrates.

Voice: Soft twittering when feeding and shrill 'twick-twick'.

Status: Moderately common non-breeding migrant.
Arrive: Sept. Depart: Apr.

Breeding: Inside Arctic Circle.

Roebuck Bay, WA.

Left: Remnant breeding plumage. Right: Non-breeding.

Red-necked Stint

Calidris ruficollis

150mm

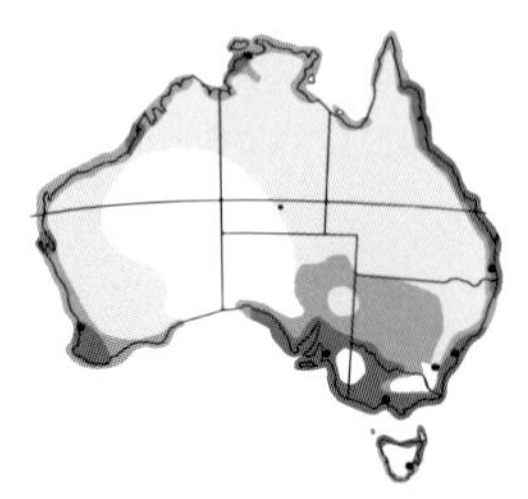

Sexes alike. Tiny grey-brown mottled sandpiper with white wing bars viewed in flight. Crown: Dark streaked. Underparts: White, grey wash to sides of breast. Rump, tail: White with dark central streak. Eyes: Brown with white eyebrow and indistinct dark eye stripe. Bill: Black, short, tapering with a small 'knob' tip. Legs: Black.
Breeding plumage: Face, throat, neck, upper breast: Pink-chestnut, becoming brighter and extending over back.
Immature: Similar to non-breeding but browner with warm buff feather margins.

Compact flocks, wheeling, gliding. Runs while feeding. Mostly found in coastal areas.

Habitat: Mudflats, beaches and swamp margins.

Feeding: Rapidly probes for worms, crustaceans, insects.

Voice: High-pitched, abrupt 'chit'.

Status: Common non-breeding migrant. Arrive: Aug. Depart: Apr. Many birds overwinter.

Breeding: Eastern Siberia to Alaska.

Left: Remnant breeding plumage. Right: Non-breeding.

Long-toed Stint

Calidris subminuta

100–160mm

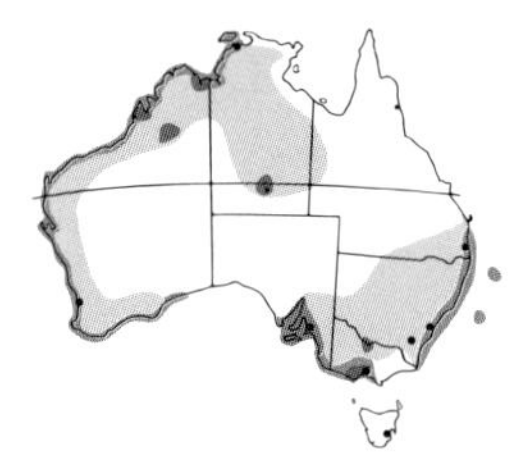

Sexes alike. Tiny wader, similar to the Red-necked Stint but taller and slimmer. Bold white wing bar viewed in flight. Upperparts: Mottled, grey-brown with dark feather centres. Crown: Blackish streaked. Eyes: Dark brown, long white eyebrow. Shoulders: Dark spotted. Breast: Grey washed with dusky streaks, paler towards centre. Underparts: White. Rump, tail: Black, white sides. Bill: Blackish, short. Legs: Long, greenish yellow with long centre toe, trailing in flight.
Breeding plumage: Upperparts darker, chestnut mottled.
Immature: Similar to breeding adult but darker with long buff eyebrow.

When flushed, flies snipe-like with rapid wingbeats, twisting and calling.

Habitat: Coastal and inland freshwater open swamps, muddy shallows of drying ponds or floodplains.

Feeding: Slowly pecks surface, sometimes probes, in between darting runs.

Voice: Sharp, rapid chirruping and trilling when disturbed.

Status: Uncommon non-breeding migrant. Arrive: Aug. Depart: Apr. Sparse distribution. Some birds overwinter. Vagrant in Tas.

Breeding: Siberia.

Swamps around Perth, WA.

Left: Breeding plumage. Right: Non-breeding.

Pectoral Sandpiper

Calidris melanotos

200–230mm

Sexes alike. Similar to the Sharp-tailed Sandpiper. Upperparts: Mottled brown with dusky feather centres and faint white wing bar in flight. Crown: Dark to warm-brown. Eyes: Dark with pale eyebrow. Tail: Outer feathers part white, black tipped. Breast: Grey-buff, heavily streaked with a defined edge joining white belly. Underparts: White. Underwings: Mostly white.
Breeding plumage: Richer colour with a russet wash.
Immature: Similar to non-breeding, but more russet.

When flushed, may freeze or fly zigzag, before dropping to cover. Upright stance.

Habitat: Fresh or saline swamps and streams.

Feeding: Methodical, head down, probing for small crustaceans, worms, insects and aquatic vegetation.

Voice: Shrill, piping 'krrrt' when flushed.

Status: Uncommon non-breeding migrant. Arrive: Sept., mostly to the east coast then disperses sparsely across the inland. Depart: Mar.– Apr.

Breeding: Siberia, Alaska, Canada.

Left: Breeding plumage. Right: Non-breeding.

Sharp-tailed Sandpiper

Calidris acuminata

190–220mm

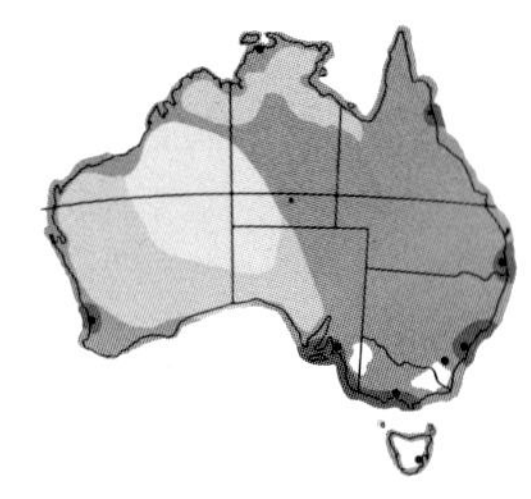

Sexes alike. Distinctive, fine pointed tail and plumpish belly. Upperparts: Mottled dark brown and buff. Long, fine, white wing bar seen in flight. Crown: Rufous and black streaked. Breast: Grey tinted with irregular buff-grey streaks. Underparts: White. Tail: Outer feathers part white, tipped black. Eyes: Dark brown. Bill: Dark grey, short and straight. Legs: Dark olive.
Breeding plumage: Chestnut crown with rich rufous chest and upperparts, densely overlayed with black chevrons.
Immature: Similar to adult breeding plumage, plainer.

Groups fly swiftly in unison.

Habitat: Fresh or brackish water. Mudflats, mangroves, sandflats, inland swamps, lagoons.

Feeding: Often in large flocks. Head bent, probing in shallows for aquatic insects, worms and aquatic vegetation.

Voice: Shrill, piping notes. 'Pleep' call in flight.

Status: Common, abundant non-breeding migrant. Arrive: Aug. Depart: Apr. Some birds overwinter.

Breeding: Siberia.

Left and right: Non-breeding plumage.

Curlew Sandpiper

Calidris ferruginea

Critically Endangered
200–210mm

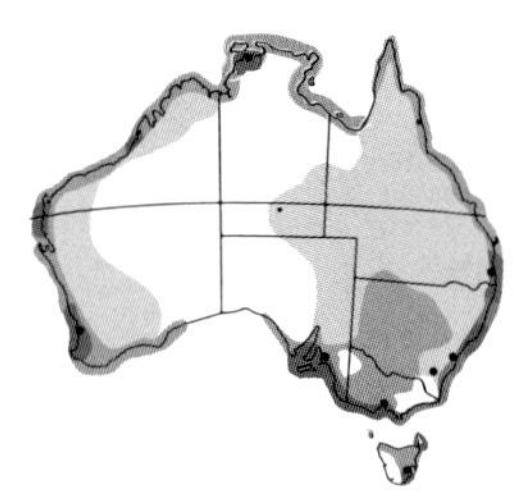

Sexes alike. Small wader with a long, gently down-curved black bill and black legs. Upperparts: Grey-brown with white wing bar in flight. Rump: White with black-tipped, white tail. Underparts: White. Eyes: Brown with white eyebrow. Legs: Black.
Breeding plumage: Head, body: Rich, reddish chestnut with fine barred sides of breast, flanks and richly patterned, buff, chestnut and black back.
Immature: Similar, browner with buff-tinted breast.

Congregates in large flocks, sometimes containing thousands of birds. Wheeling, chasing flight in compact flocks. Appears hunched when wading in deep water or walking.

Habitat: Mostly coastal estuarine and tidal mudflats and sandflats.

Feeding: Walks, wades, pecking and probing for insects, worms, crustaceans.

Voice: Twittering, chirping.

Status: Common non-breeding migrant. Arrive: Sept. Depart: Apr. Some birds overwinter. Sparse inland.

Breeding: Northern Siberia.

Left: Non-breeding. Right: Breeding plumage.

Broad-billed Sandpiper

Limicola falcinellus

170–190mm

Sexes alike. Long, brown-olive bill with a down-curved tip and broad base. Crown: Streaked black with double white eyebrow. Upperparts: Grey feathers with pale edges and dark centres. Underparts: White with grey streaked breast. Wings: Dusky with white wing bar and black trailing edge in flight. Rump, tail: White with dark central line and dark-tipped tail.
Breeding plumage: Rich rufous wash and black feather centres.
Immature: Paler with buff washed breast.

Solitary or occasionally in small flocks with other waders. More common in the north.

Habitat: Intertidal mudflats, sandbanks, wetlands, sewerage ponds and reefs.

Feeding: Often in deep water, head submerged, actively jabbing for worms, crustaceans, molluscs, insects.

Voice: Soft trilling and scolding chattering.

Status: Uncommon non-breeding migrant. Arrive: July. Depart: Apr.–May.

Breeding: Northern Siberia.

Cairns Esplanade, Qld. Broome Bird Observatory, WA.

Left: Non-breeding. Right: Remnant breeding plumage.

Ruff (Reeve)

Philomachus pugnax

Male: 290–300mm
Female: 240–270mm

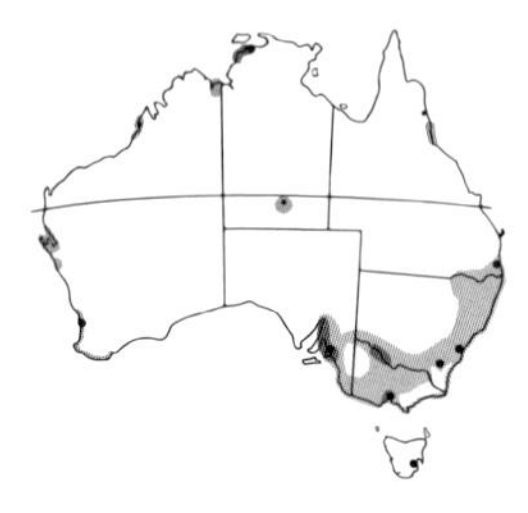

The male is called a Ruff, the significantly smaller female is called a Reeve. Non-breeding sexes are otherwise similar. Upperparts: Mid-grey-brown with feathers edged buff. Crown: Dusky streaked. Neck, frons: White, flecked. Underparts: Dull white with brown wash on breast. Wings: Dusky brown flight feathers with faint white wing bar visible in flight. Eyes: Dark brown. Legs: Yellow-green to grey-brown. Bill: Short, brown with yellow base, small 'knob' tip. Breeding plumage: Male: Erectile neck 'ruff' ear tufts, yellow facial 'lumps'.
Female: Darker, heavily patterned, russet-washed back. Immature: Upperparts: Dark buff-edged feathers. Face, neck, underparts: Buff.

Habitat: Edge of freshwater lagoons and estuarine mudflats.

Feeding: Probes for larvae, invertebrates and crustaceans.

Voice: Low grunts. Mostly silent.

Status: Uncommon non-breeding migrant. Arrive: Sept. Depart: Apr.

Breeding: Eurasia.

Saltfields, Yorke Peninsula, SA.

Right: Non-breeding Ruff (background), smaller female (Reeve) with some remnant breeding plumage.

Red-necked Phalarope

Phalaropus lobatus

170–200mm

Sexes similar. Non-breeding: Pale grey upperparts. Frons, crown: White grading to dusky. Face: White with dusky grey stripe through eyes to ear coverts. Neck: White sides. Wings: Dark grey coverts, white edging and 2 white wing stripes. Underparts: White, grey wash on side of breasts and flecked flanks. Bill: Black. Legs: Dark blue-grey.
Breeding plumage: Female brighter than male. Upperparts: Sooty grey, buff edged. Neck, breast: Broad, bright rufous stripe. Chin, throat: White. Underparts: Dusky grey lower breast, remainder white. Immature: Similar to non-breeding, but darker.

Mostly aquatic. Swims high in water.

Habitat: Freshwater and saline coastal marshes and bays.

Feeding: Swims or short flight chasing insects. Stirs shallow water by spinning, then plucking food from surface or lunging. Mostly insects, also molluscs, crustaceans, worms, vegetation.

Voice: Soft 'chick' in flight.

Status: Irregular migrant. Regular in northwest WA, but many sight records in southeast. Arrive: Sept. Depart: Apr.

Breeding: Northern Europe, Asia.

Port Headland, WA.

Left: Breeding female. Right: Non-breeding male.

Australian Painted Snipe

Rostratula australis

Endemic

240–300mm

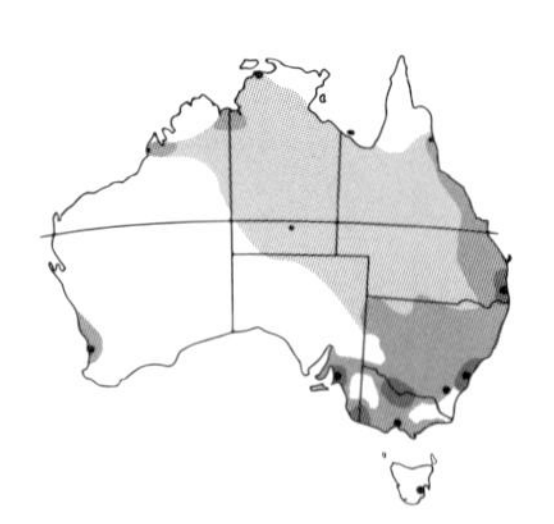

Female is larger, brighter and more colourful. Bill: Long, pinkish and slightly down-curved. Head, nape, throat: Deep chestnut with white stripe on crown and white comma-shaped eye patch. Upperparts: Patterned metallic green, black, grey and buff with a 'V' line from shoulders over upper back.
Male, immature: Smaller, duller, greyer with buff comma-shaped eye patch and crown stripe. Buff wing spots.

Rarely seen, secretive, freezes when approached. Flight is slow, irregular wingbeats, trailing legs. Walks bobbing body up–down. Seen mostly at dawn and dusk.

Habitat: Shallow, fresh, brackish wetlands with muddy margins and some dense cover.

Feeding: Bobs head while probing soft mud for insects, invertebrates and plant material.

Voice: Usually silent. Soft booming in display. Buzzing alarm call.

Status: Vulnerable.

Breeding: Oct.–Dec. in south. Mar.–May in north. Male constructs nest, raises young. Scrape in mud nest / 4 eggs.

Deniliquin, NSW. Merin Merin GR, Clunes, Vic.

Left: Male. Right: Female.

Australian Pied Oystercatcher

Haematopus longirostris

480–520mm

Sexes alike. Large black and white wader with bright red legs, bill and eye ring. Black head, breast and upperparts are sharply defined by the white underparts. White wing bars seen in flight. Underwings: White, broadly edged black. Rump: White. Tail: White, broadly tipped black.
Female has a longer bill.
Immature: Similar to adults. Brown eye. Dark-tipped bill and brownish plumage.

Flocks form in autumn and winter.

Habitat: Sandy beaches, sandbars, mudflats, estuaries.

Feeding: Probes for molluscs and crustaceans. Uses bill for prising, stabbing and hammering.

Voice: Loud, piping 'kleep-kleep' call.

Status: Common. Sedentary.

Breeding: Aug.–Jan. Scrape in sand nest / 2–3 eggs.

Left: Teaching young to forage. Right: Adult in flight.

Sooty Oystercatcher

Haematopus fuliginosus

Endemic

480–520mm

Sexes similar. Only entirely black-plumed shorebird. Bright red eyes, eye ring and bill. Legs are pink.
Northern birds have larger eye rings and bill.
Female: Larger, thinner bill.
Immature: Similar to adults, duller, browner, with buff edging on wing coverts. Eye ring, bill: Orange. Legs: Grey.

Usually singular or in pairs.

Habitat: Rocky coasts, coral reefs and estuaries. Beaches less frequently.

Feeding: Crustaceans, sea worms and molluscs. Uses chisel-like bill to prise open shells.

Voice: Loud piping and high-pitched 'kleep-kleep' call.

Status: Moderately common. Sedentary.

Breeding: June–Jan. Nest is lined with shells, seaweed, in a depression in sand or rock crevice, frequently on offshore islands / 2–4 eggs.

Left: Adult. Right: Adult in flight.

Red-necked Avocet

Recurvirostra novaehollandiae

Endemic

400–450mm

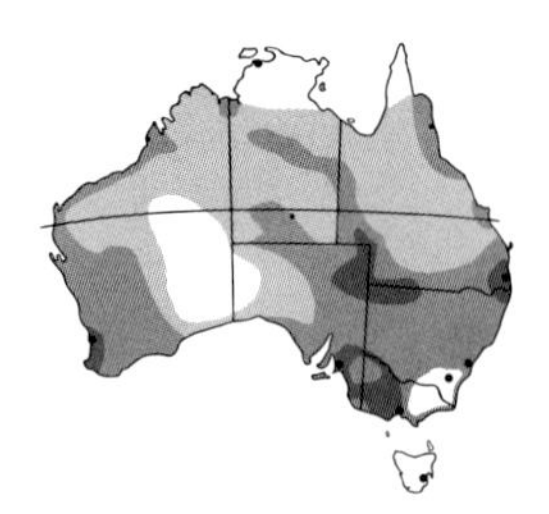

Sexes alike. Distinctively patterned wader with an upturned, long and slender, black bill. Head, neck: Bright chestnut. Eye ring: White. Body: White with 2 black streaks on back. Wings: White with 2 broad black bars. Legs: Blue-grey, feet are partially webbed.
Immature: Similar to adults.

Usually seen in compact flocks of 100 or more birds, roosting, flying or feeding together. Fluttering flight with quick, shallow wingbeats. Trailing legs.

Habitat: Brackish estuaries, tidal flats, marshes, large shallows, fresh- or saltwater wetlands.

Feeding: Swimming or wading in shallow water, sweeping bill back and forth, catching tiny aquatic invertebrates with each sweep, or pecking in mud for insects, worms, crustaceans, molluscs, seeds.

Voice: Trumpeted 'too-toot'. Wheezing, barking.

Status: Common. Highly nomadic.

Breeding: Inland after rain or flooding, on exposed islands. Scrape in ground nest / 4 eggs.

Left: Adult; in flight the legs extend beyond the tail. Right: Adult hunting in the shallows.

White-headed Stilt

Himantopus leucocephalus

360–390mm

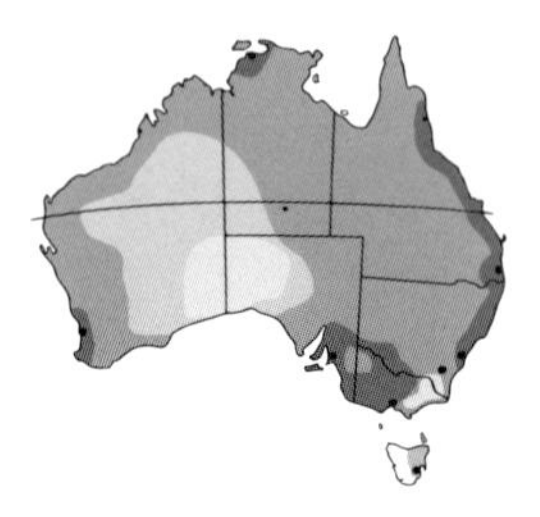

Sexes alike. Distinctive large, black and white wader with very long orange-red legs and a straight black bill. Hind neck, upperparts: Black divided by broad white collar and red eyes. Body: White.
Immature: Similar. Head, hind neck: Light grey.

Usually seen in small family groups. Flight is laboured with long trailing legs.

Habitat: Shallow edges of saline or freshwater ponds, estuaries, swamps.

Feeding: Slowly stalking in shallows for minute organisms. Seizes prey from on or near surface, mainly aquatic insects but also molluscs and crustaceans.

Voice: High-pitched barking.

Status: Common. Nomadic, depending on conditions and available food.

Breeding: Aug.–Dec. Nests in loose colonies, depression in mud or sand or a small vegetation mound, close to water / 4 eggs.

Left: Adult. Right: Adult calling in flight.

Banded Stilt

Cladorhynchus leucocephalus

Endemic

380–410mm

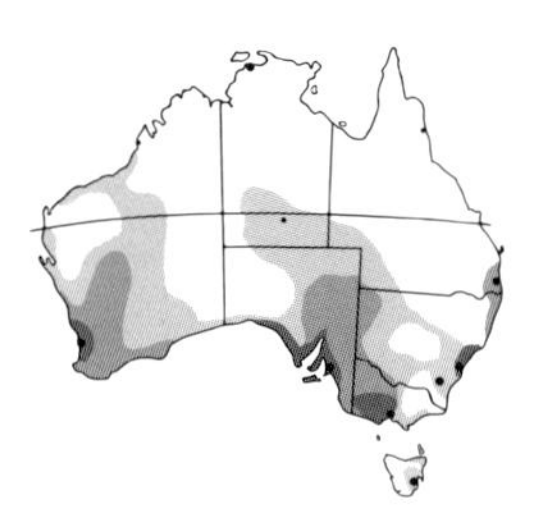

Sexes alike. Plump stilt with a broad, reddish brown breast band and long pink legs. Head, body: White. Wings: Brown-black with white trailing edge panel. Bill: Long, fine, black. Eyes: Brown. Non-breeding: Breast band: Mottled.
Immature: Lacks breast band.

In flocks at all times. Synchronised fluttering or fast flight. Shallow wingbeats with legs trailing. Extremely secretive when breeding.

Habitat: Shallow, open, salt lakes, estuaries, inlets.

Feeding: Swimming, wading and probing almost exclusively for brine shrimps and crustaceans.

Voice: Yelping notes and wheezing calls.

Status: Uncommon. Highly nomadic. In small groups or large flocks numbering thousands.

Breeding: Arid inland, after rain or flooding, on exposed islands. Scrape in ground nest / 3–4 eggs.

Dry Creek Salt Works, SA.

Left, right: Stalking for prey. Below: Flock in synchronised flight.

9

Family Charadriidae, Jacanidae, Burhinidae, Glareolidae, Pedionomidae, Turnicidae

Plovers

Dotterels

Lapwings

Jacana

Stone-curlews

Pratincoles

Plains-wanderer

Button-quails

A group containing additional members of the wader group of birds

Family Charadriidae also includes the two Australian species of lapwing.
Previously called plovers, they have now been placed in the cosmopolitan sub-family Vanellinae. These ground-dwelling birds are related to waders and named for their characteristic quick, then hesitant wingbeat flight. The Masked Lapwing is also called the Spur-winged Plover because their wings are armed with yellow spurs at the carpel joint that are used when defending their nest in the breeding season.

Family Jacanidae contains the jacana.
It is represented in Australia by the Comb-crested Jacana. Birds of the wetlands, they have long, spurred wings, long legs and amazingly long toes that allow them to walk across aquatic vegetation. When disturbed, both adults and chicks can dive and remain motionless underwater with only their nostrils and bill above water. Adult birds may also carry eggs and chicks under their wings as a defence against predation or rapid changes in water level. 'Jacana' is Brazilian in origin; however, the name Lotusbird is well established in Australia.

Family Burhinidae contains stone-curlews.
It is represented by two species of semi-nocturnal, ground-dwelling, plover-like waders. Although awkward-looking birds, they can run quickly and their flight is direct and fast. The young walk within hours of hatching, and eggs or chicks are carried under the adult's wing if danger threatens. Young birds are taught to feed by adults dropping food in front of them. If disturbed, the Bush Stone-curlew will freeze motionless, relying on its cryptic plumage for camouflage, while the Beach Stone-curlew runs from danger, only reluctantly taking flight.

Family Glareolidae includes pratincoles.
It is represented in Australia by two species. Pratincoles are aberrant, wader-like birds with short legs and long wings. Elegant in flight, they weave and dive like swallows when catching insects in the air. They are also graceful on the ground as they run and dart after insects like typical waders. Pratincoles pair for life and share incubation duties; the changeover is announced by pebble-throwing or calling.

Family Pedionomidae contains the endemic Plains-wanderer.
It is the sole family member. Critically endangered, its origins trace back 60 million years. Totally unique, it has no close relatives in the world. Although it looks and acts like a button-quail, it is not related. Females are larger and brighter than males and are dominant in courtship. They leave the care of the young to the male, usually unaided. The young walk from the nest upon hatching.

Family Turnicidae includes Button-quails.
It is represented by seven species in Australia, five of which are endemic. Button-quails are small and quail-like, but are unrelated to the native *Coturnix* spp. of 'true quail'. They are thought to be an ancient group and have recently been placed in the wader order Charadriiformes. The young run on hatching and fly within one to two weeks. The endangered Buff-breasted Button-quail, the largest of the species, is the only Australian bird never photographed.

Opposite: Immature Comb-crested Jacana resting on a lily pad with one leg tucked up under its body and its long toes partially revealed.

Pacific Golden Plover

Pluvialis fulva

250mm

Sexes alike. Mid-sized plover with an upright stance. Frons, face: Pale buff with whitish eyebrow and dark ear patch. Crown, hind neck, upperparts, breast: Brown with golden edged feathers. Belly, undertail: White. Bill: Black. Legs: Dark grey. Breeding plumage: Crown, hind neck, upperparts: Gold-buff speckled dark. Face: Black with broad, white stripe extending from brow. Breast: Mottled golden brown, grey. Chin, throat, underparts: Black.
Immature: Similar to non-breeding, more buff.

Swift, graceful flight in synchronised groups.

Habitat: Rocky coastal areas, beaches, sandflats, estuaries, marshes. Less frequently inland.

Feeding: At low tide picks up molluscs, crustaceans, insects and vegetable matter.

Voice: Piping 'too-weet' alarm call.

Status: Moderately common. Arrive: Aug. Depart: Apr. Some birds overwinter.

Breeding: West Alaska to Siberia.

Werribee, Vic.

Left: Breeding. Right: Non-breeding adult.

Grey Plover

Pluvialis squatarola

290mm

Sexes similar. Crown, hind neck, upperparts: Mottled grey-brown with pale, notched-pattern feather edges. Face: Whitish with faint pale grey eyebrow. Bill: Black, bulky. Underparts: Pale grey, buff flecked. Tail: White, strongly barred black. Underwings: Mostly white with black 'armpit'.
Breeding plumage: Crown, hind neck, upperparts: Strongly patterned black and silvery grey. Frons through to lower breast: Long white stripe. Face, chin, throat, belly: Black. Vent: White. Birds in moult phase look patchy.
Immature: Similar to non-breeding but darker.

Timid and wary.

Habitat: Seashore, coastal grassy mudflats and marshes.

Feeding: Runs, stops and pecks, feeding on crustaceans, invertebrates.

Voice: Plaintive: 'pee-o-wee'.

Status: Regular non-breeding migrant. Arrive: Aug. Depart: Apr. Locally common, mostly in southern WA. Some young birds may overwinter.

Breeding: Arctic.

Buffalo Creek, NT. Cairns Esplanade, Qld.

Left: With remnant breeding plumage. Right: Non-breeding.

Little Ringed Plover

Charadrius dubius

140–160mm

Sexes similar. Non-breeding: Head, upperparts: Grey-brown with pale scalloped feather margins. Bill: Black. Breast band: Grey-brown and incomplete. Collar, underparts: White.
Breeding plumage: Distinctive facial pattern and black breast band under white collar. Yellow eye ring and distinctive white eyebrow. Crown, upperparts, back, wings: Grey-brown with a white nape collar. Underparts: White. Bill: Black. Legs: Orange-pink.
Immature: Browner with 'scalloped' pale feather margins. Indistinct face mask.

Usually solitary or in pairs, but migrate in small groups. Wary. Active, with erratic movements. Runs and freezes. Wades. Flies with pointed backswept wings.

Habitat: Muddy edges of lakes, swamps, tidal areas.

Feeding: Forages in mud for insects and worms.

Voice: Descending, clear 'pee-ooo'.

Status: Rare but regular visitor, more common in the north. Arrive: Sept. Depart: Mar.

Breeding: Europe to western Asia.

Leanyer TP, Darwin, NT.

Left: Breeding plumage. Right: Non-breeding.

Red-capped Plover

Charadrius ruficapillus

Endemic

140–160mm

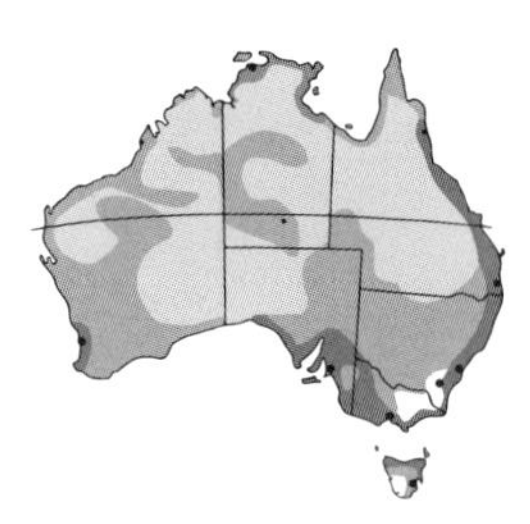

Small, plump, active wader. In breeding plumage the male has a bright reddish brown crown and nape. Face: White with frons finely edged black and fine black line through eyes. Upperparts: Grey-brown, narrow, white wing bar and dusky flight feathers. Underparts: White with short, reddish brown and black stripes extending from the shoulder. Bill, legs: Black. Eyes: Brown. Female: Head, breast: Duller with indistinct black markings. Non-breeding plumage: Plain grey upperparts. Underparts: White.
Immature: Mottled upperparts, paler.

Usually in pairs or small groups. Scurries across flats, stopping suddenly if disturbed, flies short distance, runs again.

Habitat: Sandflats and mudflats along coast and tidal inlets and inland salt lakes.

Feeding: Insects and small crustaceans.

Voice: Trilling and churring.

Status: Moderately common. Sedentary or nomadic.

Breeding: Sept.–Jan. Scrape in ground nest / 2–3 eggs.

Left: Female on eggs. Right: Breeding male with chick under wing on nest.

Double-banded Plover

Charadrius bicinctus

180–190mm

Sexes similar. Non-breeding: Plain grey-brown upperparts with white frons, lores and eyebrow stripe extending to nape. Breast: White, brown bar at base of neck and faint, often broken brown bars on lower breast. Underparts: White. Bill: Black. Breeding plumage: Male: Crown, nape: Dark grey-brown. Face: White with black eye mask. Breast: White with black bar on base of neck and chestnut bar on lower breast. Underparts: White. Female: Duller, paler, browner. Immature: Similar to adult, indistinct breast bars.

Habitat: Tidal mudflats, beaches, inland swamps, edges of freshwater or brackish lakes.

Feeding: Probes for worms and insects.

Voice: Piping whistle. Musical trill. Rapid 'chip-chip'.

Status: Common non-breeding migrant. Arrive: Feb. Depart: Oct.

Breeding: New Zealand.

Botany Bay, NSW. Werribee, Vic. Bruny Is., Tas.

Left: Breeding male. Right: Non-breeding plumage.

Lesser Sand Plover

Charadrius mongolus

190–210mm

Sexes alike. Mid-sized plover with a short, stout black bill and large dark eyes. Non-breeding: Crown, upperparts, partial breast band: Grey-brown. Face: Grey with white stripe through eye, grey lores. Underparts: White. Wings: Pale wing bar in flight. Underwing coverts: White. Legs: Greenish grey.
Breeding plumage: Bright chestnut cap, nape and broad breast band. Lores, ear coverts, frons: Black. Underparts: White. Legs: Greenish.
Males brighter. Immature: Similar to non-breeding, lightly scalloped upperparts.

Often in association with the similar Double-banded and Greater Sand Plovers.

Habitat: Coastal estuaries, tidal mudflats, mangroves, sandy beaches.

Feeding: Wet sand and mud, short runs and dips, stealthy approach to worm holes, tugging worms, also crabs, molluscs.

Voice: Short 'derrit-drit' when disturbed and short trilling.

Status: Non-breeding migrant. Arrive: Sept. Depart: Apr. More common in the north. Some birds may overwinter.

Breeding: Siberia.

Left: Breeding plumage. Right: Non-breeding.

Greater Sand Plover

Charadrius leschenaultii

210–230mm

Sexes alike. Mid-sized plover with a stout black bill and dark eyes. Non-breeding: Upperparts, shoulder: Pale grey-brown. Breast: Grey-brown breast band, sometimes with a broken centre. Frons, throat, underparts: White. Lores, ear coverts: Grey-brown. Bill: Black, thick. Eyes: Dark. Legs: Olive-grey.
Breeding: Male: Crown, nape, breast band: Rusty brown. Face: Black mask enclosing white frons. Female: Duller. Immature: Similar to non-breeding, paler.

Usually in flocks or solitary. When disturbed, rises then drops back to the same position.

Habitat: Coastal areas, tidal areas and mudflats.

Feeding: Small crustaceans, molluscs.

Voice: Trilling long and short notes.

Status: Uncommon non-breeding migrant. Arrive: Oct. Depart: Mar.

Breeding: Northern Asia.

Broome, WA. Buffalo Creek, NT.

Left: Breeding plumage. Right: Non-breeding.

Oriental Plover

Charadrius veredus

230–250mm

Sexes similar. Large, elegant plover. Non-breeding: Long greenish brown to yellow-orange legs. Upperparts: Grey-brown with pale rufous wash. Eyebrow, chin, throat: Light buff. Breast: Grey-brown band. Underparts: Whitish buff. Eyes: Dark brown. Bill: Black with paler base. Breeding plumage: Upperparts: Dark brown. Face, chin, neck: White with a fine, dark brown line behind eye. Breast: Broad, red-brown band with black lower edge. Underparts: White. Immature: Similar to non-breeding, more buff.

Extremely wary. Swift, erratic and zigzagging flight. Bobs head, often rests on one leg.

Habitat: Dry inland plains, claypans, airfields, seashores.

Feeding: Often forages at night for insects, crustaceans.

Voice: Sharp. Piping 'clink'. Melodious trilling in flight.

Status: Regular non-breeding migrant. Most common on northern coast and inland. Arrive: Sept. Depart: Mar.

Breeding: Mongolia, Manchuria.

Lee Point, Darwin, NT.

Left and centre: Non-breeding plumage. Right: Breeding.

Inland Dotterel

Charadrius australis

Endemic

200–210mm

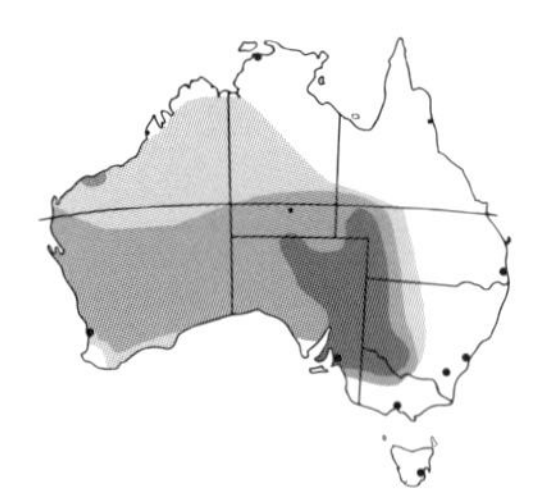

Unique wader well adapted to its arid inland habitat. Breeding plumage: Bold black stripe across the crown, below eyes and looped around the nape and shoulders to form a bold 'Y' pattern on the breast extending to the belly. Frons, face, throat: Sandy buff. Crown, upperparts: Yellow-buff streaked. Underwings: Pale chestnut, dusky flight feathers. Eyes: Dark brown. Bill: Short, slender, dull yellow, tipped black. Legs: Dull grey-yellow.
Non-breeding: Similar, paler, black bands to head, nape, breast less distinct.
Sexes similar. Immature: Similar to non-breeding, paler, indistinct black banding.

Well camouflaged, crouches motionless when disturbed, flies as last resort. Small flocks move in unison. Flight is low, swift, with even wingbeats.

Habitat: Gibber plains, semi-arid regions.

Feeding: Mostly nocturnal, eats insects, beetles, ants, grasshoppers, seeds. Browses on saltbush leaves for moisture.

Voice: Usually silent. Contact 'kuick' call. Alarm deep 'kroot'.

Status: Moderately common.

Breeding: Apr.–Oct. Deep scrape in open ground, loosely lined nest / 3 eggs.

Left and right: Breeding plumage.

Black-fronted Dotterel

Elseyornis melanops

160–180mm

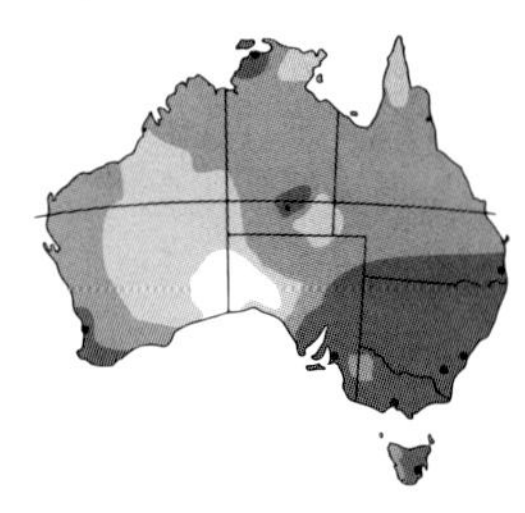

Sexes alike. Small plover with a bold, black face mask and V-shaped breast band extending around the nape. Bright red eye ring and bright red, black-tipped bill. Eyebrow: White, extending across nape. Chin, throat, underparts: White. Crown: Grey-brown, dark streaked. Upperparts: Grey-brown, pale streaked. Wings: Chestnut shoulder patch, black flight feathers, pale coverts. Legs: Short, dull pink.
Immature: Similar, paler, dull pink eye ring in pale grey band.

Buoyant flight with erratic flapping wingbeats.

Habitat: Dry riverbeds, edges of lagoons, swamps, dams.

Feeding: Running and pecking for crustaceans, aquatic insects, seeds, rarely wades, feeding alone or in pairs.

Voice: High-pitched metallic 'tink', 'tizzing' in alarm.

Status: Common. Sedentary, or nomadic in drought conditions.

Breeding: Aug.–Jan. or after inland rain. Solitary, scrape in ground nest or among pebbles / 2–3 eggs.

Opposite: Adult feeding.

Hooded Plover

Thinornis rubricollis

Endemic

190–210mm

Sexes alike. All-black head and throat with red eye ring and stout, red bill, tipped black. Hind neck: White crescent stripe. Mantle: Black crescent stripe splits into short breast band. Underparts: White. Upperparts: Pearly grey. Rump, tail: Black, broadly edged white.
Immature: Upperparts pale brown, indistinct grey collar.

Usually in pairs or small groups. Flies quickly and settles again. Runs, squats.

Habitat: Grassy sand dunes, sandy beaches, estuaries, coastal lakes. In southwest WA, along shores of inland salt lakes.

Feeding: Along beaches at high tide as waves retreat, running quickly, bobbing for insects, small crustaceans, invertebrates.

Voice: Barking like 'kep-kep' and short piping notes.

Status: Uncommon, rare. Heavily impacted by human activity.

Breeding: Aug.–Jan. Nest is a depression lined with pebbles and seaweed, on ocean beaches above high-tide mark / 2–3 eggs.

Bruny Is. and Maria Is., Tas.

Opposite: Adult pair feeding.

Red-kneed Dotterel

Erythrogonys cinctus

Endemic

180mm

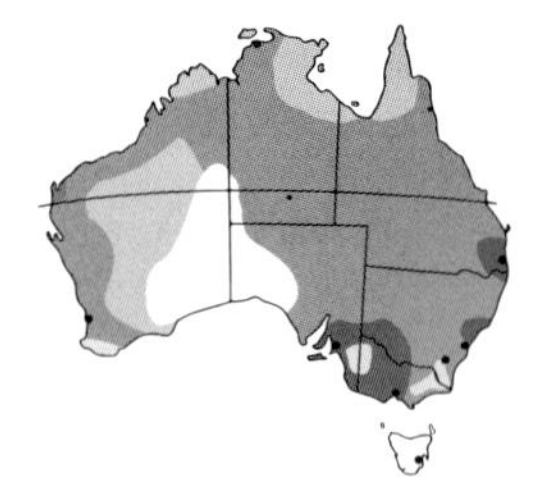

Sexes alike. Small, plump wader. Black crown, nape, face and wide breast band. Chin, throat: Pure white. Upperparts: Rich bronze brown. Underparts: White, with flanks grading from black to chestnut. Wings: Broad white edge. Underwings: White, tipped and marked black. Tail: White sides and tip. Bill: Red, tipped black. Legs: Red to 'knee', grey below.
Immature: Upperparts: Mid-brown. Underparts: White with faint breast band.

Upright stance, bobs head. Feet trail in flight. Gregarious, usually in groups of 50 or more.

Habitat: Edges of shallow freshwater or brackish swamps, wetlands, claypans, floodwaters with vegetation.

Feeding: Wades in shallows or probes mud along shoreline for aquatic insects. Sometimes swims.

Voice: Loud trilling in flight.

Status: Common. Nomadic in dry periods.

Breeding: Sept.–Dec. Nest is scrape in ground or saucer of twigs, under a tussock and near water / 2–4 eggs.

Fivebough Swamp, Leeton, NSW. Kelly's Swamp, ACT.

Left: Adult in flight. Right: Adult feeding.

Banded Lapwing

Vanellus tricolor

Endemic

250mm

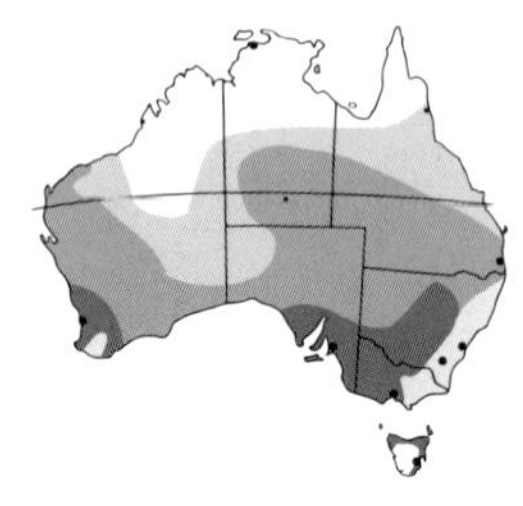

Sexes similar. Distinctive red wattle above a yellow bill. Crown, sides of neck, breast band: Black. Eyes, eye ring: Yellow with a bright white stripe through eyes extending to the nape. Chin, throat: White. Underparts: White. Tail: White, terminating in a broad black band tipped white. Upperparts: Grey-brown. Wings: White diagonal bar, black flight feathers seen in flight. Legs: Pink to grey.
Female: Small wattles and wing spurs. Immature: Similar to adult, browner.

Usually in pairs. Aggressively defends nest.

Habitat: Dry, open, short-grass or stony country, sparse open acacia or eucalypt woodlands, mudflats, farms, often far from cover.

Feeding: Forages on the ground for seeds, vegetable matter, invertebrates.

Voice: High-pitched, harsh 'kew-kew' staccato call. Often at night.

Status: Moderately common. Nomadic following inland rains.

Breeding: June–Nov. Small colonies, scrape in ground nest / 3–5 eggs.

Avalon Airfield, near Werribee, Vic.

Left: Adult. Right: Inflight.

Masked Lapwing

Vanellus miles

330–380mm

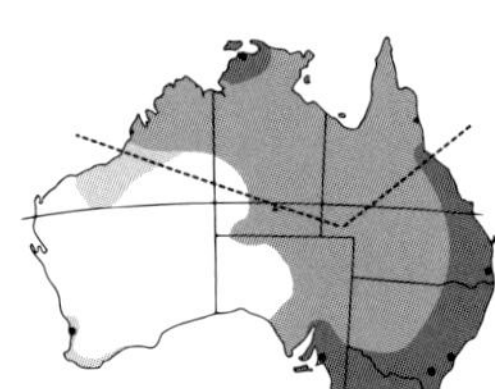

Sexes similar. Distinctive yellow wattles. Crown, nape: Black. Upperparts: Brown. Underparts: White. Wings: Yellow, black-tipped spurs on 'shoulder'. Rump: White. Tail: White with broad black terminal band tipped white. Underwings: White with black flight feathers. Legs: Red. Northern birds, race *miles* Masked Plover have larger wattles extending behind the eyes and white necks. Southern birds, race *novaehollandiae*, Spur-winged Plover are larger with a black hind neck, sides of breast and smaller wattles. Interbreeding occurs between subspecies. Female, all: Larger, brighter. Immature: Similar, mottled crown, upperparts darker, buff-edged feathers. Spurs, wattles small or absent.

Usually wary, ground dwelling, in pairs. Quick wingbeats.

Habitat: Open, short-grassed areas near swamps, lagoons, mudflats. Urban parks, gardens.

Feeding: Ground feeding, taking worms, insects.

Voice: Often calls at night. Staccato chatter. Penetrating 'keek-keek' repeated alarm call.

Status: Common, sedentary, wanders locally.

Breeding: July–Dec. in south. Nov.–June in north. Scrape in ground nest / 3–4 eggs.

Left: Nominate race, Masked Plover at nest. Right: Race *novaehollandiae*, Spur-winged Plover.

Comb-crested Jacana (Lotusbird)

Irediparra gallinacea

200–240mm

Sexes similar. Fleshy comb on frons that changes from red to yellow when excited. Crown, back of neck, breast band, tail and flight feathers: Black. Upperparts: Dark olive-brown with iridescent highlights. Face, throat, foreneck: White, edged yellow-buff. Underparts: White. Underwings: Black. Bill: Pink, black tipped. Eyes: Yellow with dark stripe in front. Legs: Long, olive green, extremely long toes. Immature: Lacks breast band. Small, pink comb. Upperparts: Cinnamon.

Agile, runs or walks over floating vegetation with head bobbing. Freezes when disturbed. Swims submerged. Swift, low flight with long trailing legs.

Habitat: Permanent, deep freshwater swamps, billabongs.

Feeding: Aquatic plants, seeds, insects mostly taken from floating vegetation.

Voice: Trumpet-like alarm. Soft twittering, piping, often in flight.

Status: Common. Sedentary. Locally abundant in the north. Less common in the south.

Breeding: Sept.–May. Raft of vegetation nest / 3–4 eggs.

Fogg Dam CR, NT. Lake Argyle, WA.

Left: Walking over floating vegetation searching for prey. Right: At nest.

Beach Stone-curlew

Esacus magnirostris

530–580mm

Sexes alike. Large, thickset, cryptically plumaged wader with a large head and massive black uptilted bill with a yellow base. Legs: Stout, yellow, with thick 'knees'. Underparts: White. Immature: Duller.

Wary and secretive. Shelters in mangroves during the day. May run fast, but generally moves slowly with head down. Flight is fast with stiff wingbeats. Singles, pairs or small groups.

Habitat: Reefs, beaches and mudflats and offshore islands in the north.

Feeding: Mostly at dawn and dusk, particularly for crabs, hammered open and often washed before eating. Also other crustaceans and insects.

Voice: High-pitched repeated wailing.

Status: Uncommon. Sedentary.

Breeding: Oct.–Feb. Scrape in sand nest / 1–2 eggs.

Yule Point, Cairns, Qld.

Opposite: Small group feeding at beach.

Bush Stone-curlew

Burhinus grallarius

Endemic

500–580mm

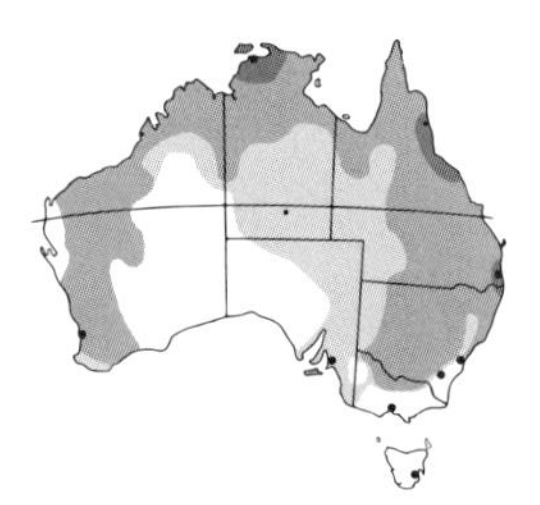

Sexes alike. Tall, slim, cryptically plumaged wader with long, olive-brown legs with thick 'knees'. Upperparts: Grey-brown streaked with black and rufous. Whitish band across wings. Eyes: Large, yellow with long white eyebrow and dark brown streak that extends down side of neck. Frons: White patch. Underparts: White, streaked black and brown.
Northern birds are more rufous. Immature: Paler

Mostly nocturnal, sheltering during the day, squatting in pairs or loose flocks. May freeze or lie flat if disturbed. Flight is fast, direct, with long legs dangling. Flies reluctantly and rarely during the day. Noted for its remarkable courtship dance.

Habitat: Open woodland, mallee, forest edges, inland waterways.

Feeding: Ground feeding on insects, crustaceans, lizards and small mammals.

Voice: Usually at night, eerie high-pitched, spine-tingling, wailing 'wee-loo' repeated.

Status: Sedentary. More common in north. Vulnerable in Vic.

Breeding: July–Jan. Nest is a scrape in ground / 2 eggs.

Cairns Cemetery, Mt Molloy, and Julatten area, Qld.

Left: Long legs dangling in flight. Right: Adult pair.

Australian Pratincole

Stiltia isabella

Endemic

200–230mm

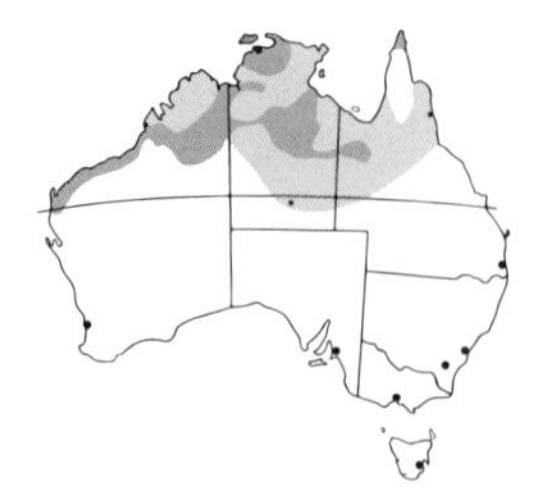

Sexes alike. Graceful and slender bodied. Breeding plumage: Buff brown, rufous-tinged shoulders and duller lower back. Bill: Red, black tip. Eyes: Dark brown. Wings: Dusky. Chin, throat: Pale buff, darker on breast. Underparts: White with brown band across upper belly, flanks.
Non-breeding: Bill: Brownish red, black tip. Breast band reduced to flanks. Belly: White.
Immature: Similar to non-breeding. Bill: Dark grey. Upperparts: Browner, mottled.

Swallow-like flight. Rising, falling, often zigzagging in flocks.

Habitat: Dry, gravelly, open coastal plains, gibber plains, claypans, treeless plains.

Feeding: Hunts insects on the wing like swallows or dashes bobbing across ground, feeding like a plover.

Voice: Sweet, rising–falling whistles in flight. Persistent calling in summer storms.

Status: Common to uncommon. Migrating Feb.– Apr. High flying to far north or offshore islands.

Breeding: Aug.–Jan. In loose colonies, 2 eggs laid on ground.

Mamukala Wetlands, Kakadu NP, and Knuckey Lagoons CP, Darwin, NT.

Left: Breeding plumage. Right: Non-breeding.

Oriental Pratincole

Glareola maldivarum

230mm

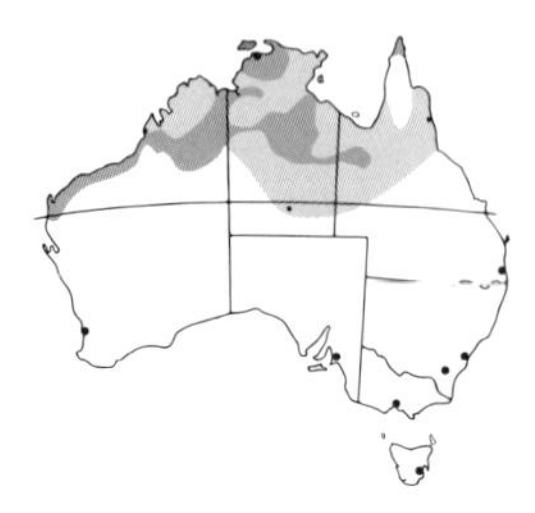

Sexes alike. Long, pointed, black wings and black, forked, white-tipped tail. Upperparts, breast: Dull olive-brown. Throat: Buff with thin streaked-black edging. Underparts: White. Bill: Black, red at gape. Underwings: Chestnut coverts.
Breeding: Similar to non-breeding but brighter, throat light buff, with sharp black edging. Gape of bill: Bright red.
Immature: Similar to non-breeding but paler. Lacks black throat edging.

Swallow-like flight, appearing in large flocks. Rest together on ground.

Habitat: Swamp edges and claypans.

Feeding: Ground feeding for grasshoppers, termites, crickets or hawks for insects.

Voice: Noisy 'chick-chick' call. In flight, 'too-wheet, too-wheet'.

Status: Regular non-breeding migrant. Locally abundant to uncommon. Arrive: Oct. Huge flocks arrive in the northwest and disperse mostly inland. Depart: Apr.

Breeding: Pakistan, India, Southeast Asia, Japan, Philippines.

East Point Reserve, NT, in Oct.

Left: Breeding. Right: Non-breeding.

Plains-wanderer

Pedionomus torquatus

Endemic Critically Endangered

Male: 150mm
Female: 170mm

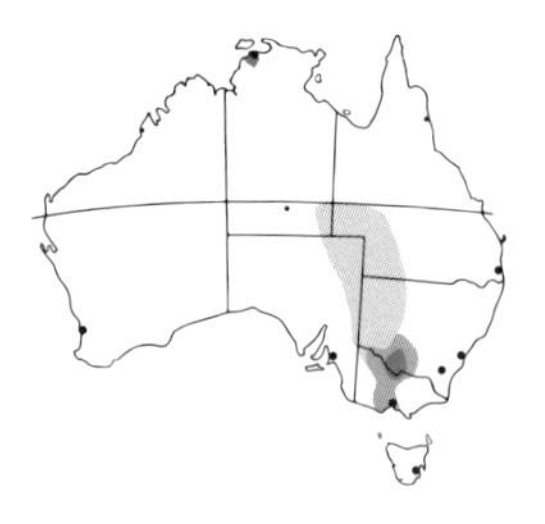

Small and highly unique, quail-like grassland bird, with origins tracing back 60 million years. Upperparts: Brown, mottled. White wing bar displayed in flight. Cheeks: Buff patch around eye. Eyes, bill: Pale yellow. Throat: White with black spots to sides. Neck: Spotted, marked, dark brown collar. Belly: Off-white. Legs: Cream.
Female: Larger than male. Spotted black and white neck. Breast: Crescent-shaped rufous band. Lower belly: Buff, scalloped brown.

Squats, freezes when alarmed. Runs. Whirring short flight with legs trailing. Usually solitary or groups of 3–5 birds.

Habitat: Sparse native grasslands and flat sparsely grazed plains.

Feeding: Nocturnal. Seeds and insects.

Voice: Sad, repetitive 'coo'.

Status: Rare, endangered. Sedentary in suitable habitat.

Breeding: June–Jan. Deep scrape in ground, grass-lined nest / 2–4 eggs.

Deniliquin area, NSW.

Left: Male. Right: Female.

Red-backed Button-quail

Turnix maculosus

120–130mm

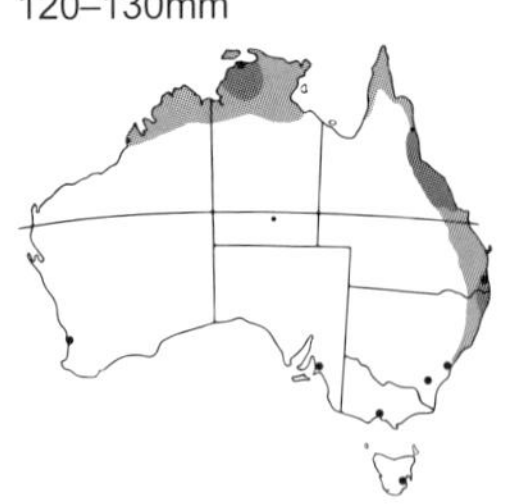

Small and stocky, cryptically plumed, ground-dwelling bird. Densely mottled upperparts, barred grey-rufous with cream streaks and dull rufous collar. Shoulders: Rich rufous, black flecked. Bill: Yellow. Eyes: White. Throat, side of face: Buff. Breast: Tawny. Underparts: Buff with black scalloping on flanks.
Female: Similar, larger, brighter, brighter yellow bill and larger, richer, rufous collar. Immature: Similar, brown eyes, darker head.

Rarely seen, shy and elusive. Freezes when disturbed. Rarely flies. In pairs or small coveys.

Habitat: Grasslands, open and savannah woodlands with grassy ground layer, pastures and crops.

Feeding: Forages on ground for seeds and insects.

Voice: Female: Loud 'oom' in courtship. Male: Silent.

Status: Vulnerable in NSW. Locally common.

Breeding: Oct.–July. Depression nest under grass cover / 4 eggs.

Holmes Jungle–Micket Creek area, NT. North Pine Dam, Petrie, Qld.

Left: Male. Right: Female.

Black-breasted Button-quail

Turnix melanogaster

Endemic

Male: 150–160mm
Female: 170–180mm

Sexes similar. Rare, large, plump, pale-eyed button-quail, endemic to coastal eastern Australia. Female: Larger with distinctive black hood, face, throat and white eyes. Crown: Thin rufous stripe with spotted white eyebrow. Breast, flanks: Black, heavily scalloped white with plain black centre. Belly, undertail: Dark grey. Upperparts: Chestnut, mottled black, grey, white.
Male: Similar to female but smaller and paler. Immature: Similar to male, duller.

Usually in pairs or small coveys. Rarely seen. If disturbed, will freeze or run. Flight is reluctant and clumsy.

Habitat: Coastal and near-coastal regions, clearings at the edge of rainforests, vine scrub.

Feeding: Fossicks in leaf litter for insects and seeds.

Voice: Female: Low, repeated booming. Male: Clucking sound.

Status: Rare. Vulnerable.

Breeding: Sept.–Mar. Nest is a shallow depression at base of shrub or grass clump / 3–4 eggs.

Inskip Point, Qld.

Left: Male. Right: Female.

Chestnut-backed Button-quail

Turnix castanotus

Endemic

Male: 160mm
Female: 180mm

Sexes similar. Small, cryptically coloured, ground-dwelling native grassland bird. Pale grey finely speckled whitish face and breast.
Female: Larger. Upperparts: Rich, dark rufous, lightly barred, mottled black and white. Crown: Finely striped rufous and grey. Rump: Rufous. Underparts: Grey, spotted white breast. Belly, undertail: Plain pale grey. Bill: Greyish cream. Eyes, legs: Yellow. Immature: Similar, darker. Eyes: Brown.

Usually in coveys of 6–20. Runs with neck extended.

Habitat: Sandstone ridges with spinifex, dry savannah woodlands.

Feeding: Nocturnal. Grass seeds and some insects.

Voice: Female: Low repeated mournful 'oom' contact call.

Status: Moderately common, sedentary.

Breeding: Dec.–May. Domed or partially domed nest in shallow depression at base of shrub or grass clump / 4 eggs.

Umbrawarra Gorge NP, NT.

Opposite: Female.

Red-chested Button-quail

Turnix pyrrhothorax

Endemic

Male: 130mm
Female: 150mm

Similar to Little Button-quail, but darker upperparts. Male is paler, whiter throat and belly. Upperparts: Mid-grey-brown mottled, barred black and cream. Eyes: Cream.
Female: Similar. Larger, richer colours. Underparts: Orange-rufous. Immature: Similar. Eyes: Dark.

Found in singles, pairs or coveys. Noisy chatter when flushed.

Habitat: Grasslands, open grassy woodlands, rainforest margins and spinifex.

Feeding: Mainly nocturnal. Seeds and insects.

Voice: Female: Repeated rapid 'oom' in breeding season and gurgling sounds.

Status: Common.

Breeding: Oct.–Mar. Shallow depression nest at base of grass tuft, sometimes domed / 4 eggs.

Deniliquin area, NSW.

Left: Female. Right: Male.

Little Button-quail

Turnix velox

Endemic

Male: 130–140mm
Female: 140–150mm

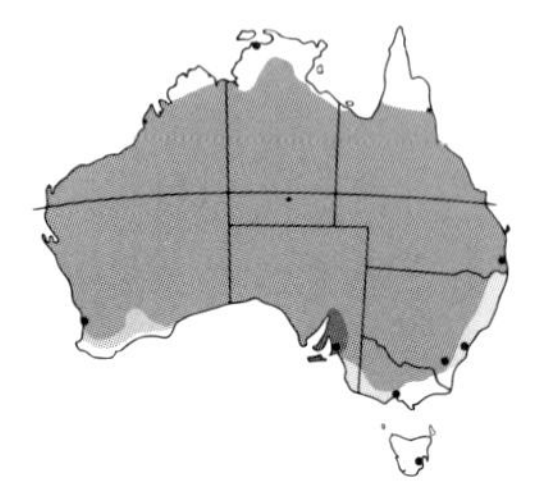

Sexes similar. Female: Chestnut streaked white upperparts. Crown: Pale cinnamon, mottled rufous. Mantle: Plain cinnamon. Rump: Faintly mottled cinnamon-buff. Breast, flanks: Buff. Undertail: Whitish. Eyes: Pale yellow. Legs, feet: Pink. Male is similar to female, but smaller with darker upperparts and dark-streaked crown. Immature: Similar, duller. Eyes: brown.

Solitary or small coveys. Often seen in flight with whirring wings and visible, distinctive white flanks.

Habitat: Open woodlands, arid inland, reaching coast erratically.

Feeding: Mainly nocturnal. Native grass seeds, some insects.

Voice: Female: Repeated, mournful 'oop' in breeding season, often at night.

Status: Common. Highly nomadic, flying long distances for food.

Breeding: All months, rain dependent. Shallow depression nest at base of grass tuft, usually partly domed / 3–5 eggs.

Deniliquin area, NSW.

Left: Female. Right: Male.

Painted Button-quail

Turnix varius

170–200mm

Sexes similar. Small ground-dwelling bird with a pale greyish face, pale eyebrow, red eyes. Crown, nape: Black-grey streaks and white spots. Upperparts: Rufous, heavily barred, mottled black, pale grey-white streaks. Breast, flanks: Olive-grey 'painted' with cream spots edged black. Belly, undertail: Plain cream.
Female: Larger, brighter rufous upperparts, finer breast spots. Head: Greyer. Immature: Duller. Eyes: Brown.

Pairs or family groups. Quickly runs, head up. Whirring, fast-weaving, explosive flight when flushed.

Habitat: Temperate and eastern tropical forests with deep leaf litter. Lightly forested areas, open woodlands, mulga, brigalow and mallee.

Feeding: Forages in leaf litter for insects and seeds by spinning about on alternate legs, leaving tell-tale circles. Active at dusk, night, early morning.

Voice: Female: Booming courtship call. Male: Silent.

Status: Moderately common. Sedentary or nomadic. Race *scintillans*, WA, vulnerable.

Breeding: Aug.–Mar. Depression in ground, nest is sometimes partially domed / 4 eggs.

You Yangs RP, Vic.

Left: Female. Right: Male.

Buff-breasted Button-quail

Turnix olivii

Endemic Endangered

Male: 180mm
Female: 200mm

The largest button-quail and the only Australian bird never to have been photographed. Similar to the Painted Button-quail but larger and paler. Female: Upperparts: Pale buff-cinnamon, faintly mottled black and rufous with pale streaks. Crown: Rufous streak. Eyes: Yellow. Bill: Stout, greyish cream. Breast, flanks: Olive-cream, faintly spotted. Belly, undertail: White. Male: Similar to female, paler, black-spotted neck. Immature: Similar. Eyes: Dark.

One of Australia's least known birds. Rarely seen due to small size and excellent camouflage. Birds run and crouch and rarely fly.

Habitat: From the Coen area south to Chillagoe, Qld. Open tropical wooded grasslands, heaths, and fringes of swamps.

Feeding: Seeds and insects.

Voice: Female: Low, repeated, mournful 'oom' contact call.

Status: Rare, seasonally nomadic. Endangered.

Breeding: Dec.–May. Domed or partially domed nest in shallow depression at base of shrub or grass clump / 3–4 eggs.

Mt Molloy, Julatten area, Qld.

Opposite: Typical, open tropical wooded grassland. Buff-breasted Button-quail habitat.

10 Family Laridae

Skuas

Jaegers

Noddies

Terns

Gulls

Family Laridae is the gull family, a large group of familiar, mostly coastal birds.

Skuas are the powerful scavengers and predatory 'pirates of the sea'. Aggressive birds, they are noted for chasing other birds, forcing them to disgorge, then eating the stolen food.

They gorge on carrion such as whales and penguins by using their strong, hooked upper bill to tear flesh. They will steal eggs from nesting seabirds and follow ships, squabbling over refuse. On land, skuas often swallow stones to aid digestion.

Jaegers are similar to, but smaller than, skuas. Their flight is distinctive, swift, agile and falcon-like. They are also predatory pirates that can force much larger birds, such as shearwaters and gannets, to disgorge their food. Most often they harass and steal food from smaller birds such as terns and gulls, but they also hunt for their own food, including fish, crustaceans and carrion.

Gulls are familiar, mostly coastal birds that are often referred to as 'seagulls', although they only rarely venture out to sea. Adult gulls found in Australia are overall white bodied with black-tipped wings, grey backs and bright red legs and bills. However, the immature gulls are mottled brown and have a different eye, bill and leg colour, making identification of them difficult.

The well-known Silver Gull takes two years to attain full adult plumage, while larger species such as the Pacific Gull take four years. Since the 1950s the Silver Gull population has exploded due to the availability of food scraps at parks and garbage tips. They have established huge breeding colonies on offshore islands, leaving little room for other species of breeding seabirds.

Terns are similar to gulls but generally have a more slender build and finer bills. Their flight is distinctively graceful. They are long lived, with records of members of some species living for 25–30 years. Like gulls, they usually breed in large colonies. The Black-naped Tern nests on the Great Barrier Reef. Those birds that nest on the northern end of the reef time their laying with summer moon phases when small surface fish are readily available. Migrating terns may travel vast distances. In a tracking study in 2013, Arctic Terns left their breeding ground in the Netherlands for their southern summer grounds on the north-eastern Antarctic coast. The average distance for the return trip was 90,000 km; one bird recorded 91,000 km. This is the longest known migration by any animal.

Noddies and noddy terns are similar to terns but are mostly tropical and oceanic. They are named for their head-nodding courtship displays. They usually breed in large colonies of 100,000 or more on offshore islands.

Opposite: Silver Gull.

South Polar Skua

Stercorarius maccormicki

510–660mm
Wingspan: 1500mm

Sexes alike. Large, stocky bird. Outstretched wings have a white 'flash'. Bill: Hooked, black. Eyes: Brown. Legs: Black. 2 colour morphs and intermediates. Light morph: Upperparts dark brown; pale buff head; pale buff underparts. Dark morph: Overall dark brown streaked, mottled rufous-buff, pale base of bill. Immature: Similar.

Follows ships. Agile in flight. Aggressive and predatory.

Habitat: Open oceans, occasionally inshore waters.
Feeding: Fish and krill. Harasses food from seabirds. Trawler scraps. Scavenging. When nesting takes rats, rabbits, birds, carrion, eggs. Often swallows stones to help grind food.
Voice: Usually silent in Australian waters. Gull-like shrieks.
Status: Regular but uncommon winter visitor.
Breeding: Antarctic.

Left and right: Light morph.

Brown Skua

Stercorarius antarcticus

510–660mm
Wingspan: 1500mm

Sexes alike. Larger, stockier, more aggressive than the South Polar Skua. Overall dark brown plumage streaked and mottled rufous and buff. Outstretched wings have a white 'flash'. Bill: Short, strong, hooked, black. Legs: Black. Eyes: Dark brown. Moult phase: Mottled and worn. Immature: Smaller, mottled.

Agile, powerful flapping and gliding flight. Aggressive, predatory. Follows ships.

Habitat: Open oceans. Occasionally inshore waters.
Feeding: Similar to South Polar Skua (above).
Voice: Usually silent in Australian waters. Sharp screeches in aerial attack.
Status: Moderately common. Regular visitor to southern seas.
Breeding: Dec.–Mar. Subantarctic islands, including Macquarie Is. Nest is a depression on ground / 2 eggs.

Left: Adult chorus. Right: Immature.

Pomarine Jaeger

Stercorarius pomarinus

480–540mm
Wingspan: 1700–1900mm

Sexes alike. Similar to Arctic Jaeger but heavier and longer with larger white wing 'flash' and shorter, twisted, spoon-shaped streamers; short in non-breeding birds. Breeding plumage to head, neck: Buff-yellow, absent in non-breeding birds. Legs: Blue-black. Light morph: Black cap. Dark morph: Less common, dusky brown, often with black cap. Head, neck: Rufous buff. Immature: Browner. Short streamers. Barred, mottled brown and white.

Solitary or flocks of 40–50. Constant deep wingbeats, infrequent gliding.
Habitat: Oceanic. Occasionally large bays.
Feeding: Floats, dives, taking fish, crustaceans, molluscs, birds. Scavenger. Harasses gulls. Follows trawlers.
Voice: Harsh 'whtch-you'.
Status: Moderately common Sept.–Apr.
Breeding: Arctic Circle.

Off Cape Otway, Vic.

Left: Intermediate morph.
Right: Pale morph.

Arctic Jaeger

Stercorarius parasiticus

460–500mm
Wingspan: 1250–1400mm

Sexes alike. Distinctive, long, pointed, central tail feathers when breeding. White wing flashes. Bill: Dark grey, hooked tip. Eyes: Brown. Legs: Black. Dark morph: Uniformly dusky brown, often with a paler nape and white patch near wing tip. Pale morph: Creamy white belly, rump. Crown: Dark cap to below eyes. Breast: Sides greyish brown. Upperparts: Grey-brown. Non-breeding: Brown barred neck, short tail feathers. Immature: Lacks pointed tail feathers. Bill: Grey, tipped black. Indistinctly mottled, barred. Follows ships. Pelagic. Only come ashore to breed.
Habitat: Oceanic. Occasionally large bays and estuaries.
Feeding: Harasses seabirds. Scavenger. Fish, carrion.
Voice: Gull-like shrieking. High, nasal squealing.
Status: Common. Oct.–Apr. More common in east.
Breeding: Arctic tundra.

Opposite: Dark morph (left) and pale morph.

Long-tailed Jaeger

Stercorarius longicaudus

500–550mm
Wingspan: 1000–1150mm

Sexes similar. Small, slender jaeger. Breeding: Blackish cap, grey upperparts separated by yellow-tinted white collar. Long tail plumes up to 200mm long. Wings: Grey, dark trailing edge. Non-breeding: Duller. Throat, flanks, undertail coverts: Dark barred. Uppertail coverts: Pale barred. Streamers shorter. Bill: Grey, black tipped.
Immature: Dull, pale barred underparts.
(Dark morph is rare.)

Flies higher than other jaegers. High tern-like flight. Swims buoyantly. Ignores ships.

Habitat: Oceanic. Offshore waters, entrances to bays.

Feeding: Flies low, hovers and dips, snatches prey. Pirates food from smaller gulls and terns.

Voice: Usually silent at sea.

Status: Regular but uncommon migrant. Oct.–May.

Breeding: Arctic.

Left: Breeding adult. Right: Immature.

Common Noddy

Anous stolidus

380–450mm
Wingspan: 850mm

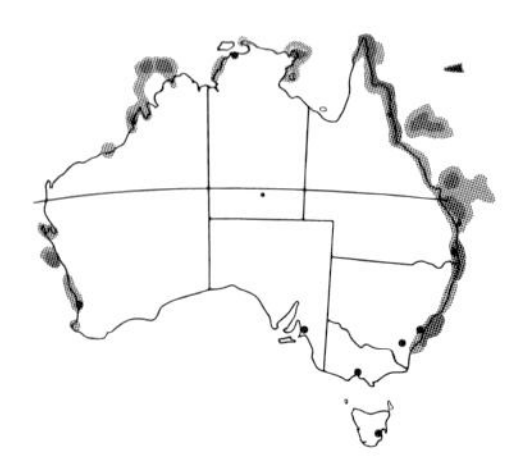

Sexes similar. Largest and brownest noddy. Overall brown plumage with a whitish frons and crown with defined fine black edge from bill to eyes. Underwings: Pale with dark edges. Bill: Slender, black. Tail: Long, pointed, notched. Eyes: Dark brown. Legs: Black.
Immature: Similar to adults but lacks cap. Wings: Lightly mottled.

Courtship includes bowing and nodding and male feeding female a freshly caught fish.

Habitat: Offshore islands, coral cays, tropical seas.

Feeding: Captures prey from the surface, mostly squid, small fish. Also pandanus fruit and insects.

Voice: Grunting, guttural sounds.

Status: Abundant in all oceans.

Breeding: Sept.–Nov. in south. Mar.–June in north. In dense colonies, nest of rough twigs, in grasses, seaweed, in a low tea-tree or on ground / 1 egg.

Left: Adult. Right: Adult in flight.

Black Noddy

Anous minutus

360–400mm
Wingspan: 650mm

Sexes similar. Overall plumage sooty black with silvery white cap sharply defined and black lores. Tail: Short, wide fork. Bill: Slender, long, black.
Immature: Similar to adults, browner. Frons: White.

Fast, agile flight, dipping, skimming along wave face.

Habitat: Great Barrier Reef, Coral Sea. Islands and coral cays.

Feeding: Surface-snatching fish, aquatic animals from inshore surf. From early morning and returning to colonies before dusk.

Voice: Harsh 'kir', rattling 'chor'.

Status: Common.

Breeding: Sept.–Dec. Nest of grass, roots, guano in shady tree / 1 egg.

Heron Is, Qld.

Left: At nest. Right: Flight.

Lesser Noddy

Anous tenuirostris

310–330mm
Wingspan: 600mm

Sexes similar. Overall sooty brown except for light grey frons and crown, grading to dark grey nape. Lores: Pale grey. Bill, legs: Black. Eyes: Dark-brown.
(The endemic Australian Lesser Noddy, race *melanops*, is vulnerable. Houtman Abrolhos Is. breeding grounds.)
Immature: Similar, lighter crown.

Habitat: Coastal seas around Abrolhos Is., WA. Reefs and lagoons with dense mangroves.

Feeding: Forages around reefs, lagoons and out to sea. Skims, hovers low over water, taking prey from the surface. Each morning flies in regular lines to fish, returning in the afternoon.

Voice: Rattling alarm. Purring.

Status: Locally common. Large flocks in breeding season, small flocks at other times.

Breeding: Aug.–Dec. Abrolhos Is., WA. In colonies. Seaweed nest in mangroves / 1 egg.

Left: Adult with nesting material. Right: Underparts.

White Tern

Gygis alba

280–330mm
Wingspan: 760mm

Sexes alike. Mid-sized overall white tern with large black eyes and eye ring. Bill: Long, black tapering to fine point. Immature: Washed grey-brown upperparts. Variable patch behind eye. Nape, mantle, upperwings: Light brownish speckles.

Buoyant, fluttering, effortless flight, or travels fast with deep wingbeats.

Habitat: Oceanic. Tropical islands including Norfolk, Lord Howe.

Feeding: Hovers, flutters, dips, picking up prey.

Voice: Soft, resonating buzz. Status: Locally common.

Breeding: Oct.–Mar. Nests in colonies on sheltered rock ledges, crevices or in Norfolk Island Pines. Egg placed in depression or on a branch / 1 egg.

Left: Feeding chick. Right: In flight.

Grey Ternlet

Procelsterna cerulea

250–300mm
Wingspan: 500–600mm

Sexes alike. Also called Grey Noddy. Small, grey noddy. Head, underparts, underwings: Pale grey. Upperparts: Blue-grey. Bill, eyes, eye ring, legs: Black. Immature: Similar with brownish wash and browner crown.

Graceful, fluttering flight, sometimes settles on water.

Habitat: Tropical, subtropical oceanic islands and surrounding waters.

Feeding: In compact flocks, dipping and fluttering over water and plucking or pattering surface with web feet. Small crustaceans, fish, squid.

Voice: Purring 'kirrr-kirrr'.

Status: Vagrant to locally common. Common Lord Howe and Norfolk Is. Casual visitor east coast, often beach-washed after storms.

Breeding: Sept.–Feb. Lord Howe and Norfolk Is. Scanty matted grass nest on rock crevice or ledge / 1 egg.

Opposite: Adult group at nesting ledge.

Bridled Tern

Onychoprion anaethetus

350–410mm
Wingspan: 760mm

Sexes alike. Similar to the Sooty Tern but with distinctive long white eyebrow and frons. Crown: Black cap. Upperparts: Dark grey-brown. Underparts: Whitish grey. Tail: Forked with streamers. Immature: Dark sooty brown, scalloped buff upperparts.

Habitat: Open seas, offshore islands, rarely inshore waters.

Feeding: In flocks, often nocturnally, plunge-dive or pick small fish from surface.

Voice: Staccato yapping. Scolding notes.

Status: Common, usually in flocks, sedentary in the north. Uncommon in the south.

Breeding: Sept.–Jan. in enormous colonies. Nest is a scrape in ground in sheltered location on offshore islands / 1 egg.

Left: Adult. Right: Adult in flight.

Sooty Tern

Onychoprion fuscata

380–460mm
Wingspan: 900mm

Sexes alike. Black crown, upperparts, wings and deeply forked tail with long streamers. Frons: Broad white patch. Underparts, underwing coverts: White. Eyes: Brown. Legs: Black. Non-breeding: Finely mottled white crown and upperparts. Immature: Speckled, sooty-brown upperparts and grey breast.

Pelagic outside breeding season. Buoyant, graceful, soaring flight.

Habitat: Oceans, islands of tropical and subtropical waters.

Feeding: Mostly at dusk or night. Snatches small fish, crustaceans from just below the surface.

Voice: Noisy, persistent screaming at the nest, barking and growling.

Status: Common.

Breeding: Sept.–Dec. in huge colonies. Offshore islands including Lord Howe Is. Scrape nest in sandy ground / 1 egg.

Left: Breeding pair. Right: Underparts.

Fairy Tern

Sternula nereis

220–270mm
Wingspan: 500mm

Sexes alike. Mostly white with a black cap extending in front of eyes. Bright orange-yellow bill. Wings, tail: Pale grey. Legs: Orange.
Non-breeding: Bill: Dull yellow, black tip. Crown: Streaked white.
Immature: Similar to non-breeding, mottled crown.

Often travel in large flocks of several thousand birds.

Habitat: Inshore near sand beaches, coastal wetlands.

Feeding: Plunging from 5m for fish. Scavenging along waterline.

Voice: High-pitched scolding.

Status: Moderately common.

Breeding: Nov.–Feb. Colonial. From Broome, WA, south to Botany Bay, NSW. Nest is a scrape, often in clumps of Pigface or Beach Daisy.

Peron Peninsula, WA.

Left, right: Adult.

Little Tern

Sternula albifrons

Endangered

210–240mm
Wingspan: 500mm

Sexes alike. Olive or yellow black-tipped bill and black line through eyes to bill. Overall white with black crown, nape, lores. Upperparts, tail: Pale grey. Wings: Pale grey, black outer feathers. Underparts: White. Legs: Yellow.
Non-breeding: Similar. Bill, legs: Black. Crown: Receding, pale grey or flecked black-white.
Immature: Similar. Darker, flecked brown upperparts.
Rapid fluttering flight. Sociable.

Habitat: Beaches, coral reefs, estuaries, harbour entrances.

Feeding: Plunge-dives for small surface fish.

Voice: Rasping, high-pitched squeaks and chattering.

Status: Uncommon.

Breeding: Apr.–July in north. Aug.–Jan. in southeast. In small colonies on beach just above high tide line, scrape / 2 eggs.

Left: Breeding. Right: Non-breeding.

Gull-billed Tern

Gelochelidon nilotica

340–420mm
Wingspan: 760–860mm

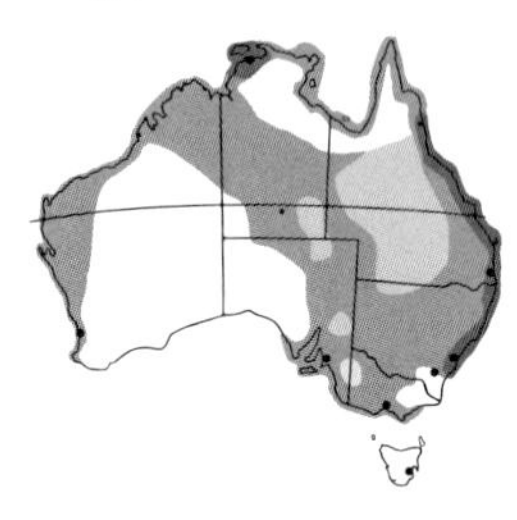

Sexes alike. Mostly an inland bird, rarely ventures out to sea. Breeding: Black crown to below eyes. Bill, legs: Black. Upperparts: Pale grey. Underparts: White. Wings: White, dark tipped, trailing edge. Tail: Short, forked.
Non-breeding: Head: White, small black patch before eye; larger patch behind extending over ear coverts.
Immature: Streaked crown and mottled upperparts.

Habitat: Extending far inland, more coastal after breeding.

Feeding: Aerial dipping. Hawks for insects. Does not dive.

Voice: Usually silent. Harsh 'ka-huk, ka-huk' when breeding.

Status: Resident east coast. Summer migrant to WA Aug.–Apr. Vagrant Tas. (Smaller Asian race *affinis* summer migrant to northeastern Australia.)

Breeding: Sept.–May. in good seasons in colonies on islands in inland lakes / 2–4 eggs.

Left: Breeding. Right: Non-breeding.

Caspian Tern

Hydroprogne caspia

580–540mm
Wingspan: 1400mm

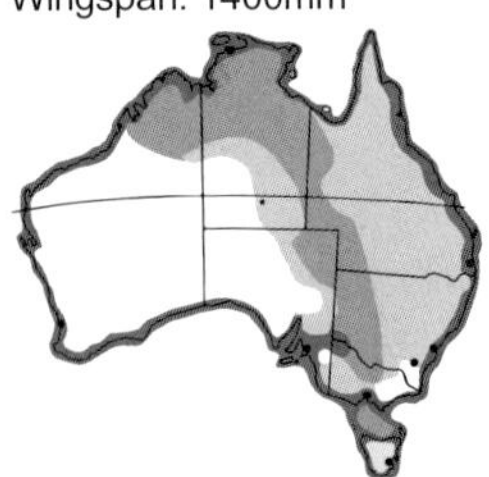

Sexes alike. World's largest tern. Massive black-tipped scarlet bill. Short crest. Crown, frons: Black to below eyes. Back, wings, tail: Pale grey. Remainder of bird is white. Legs: Blackish. Eyes: Brown. Tail: Slight fork.
Non-breeding: Frons: White. Crown: Streaked white.
Immature: Similar to non-breeding.
Aggressive, unsociable. Steady wingbeats in flight.

Habitat: Coastal. Patrols surf, shallow estuaries, rivers, lakes.

Feeding: Hovers 10–20m, plunges headlong for prey.

Voice: Deep, harsh screams.
Status: Moderately common. Sedentary or nomadic.

Breeding: Oct.–Feb. in south, most months in north. Singularly or in small colonies. Deep scrape, usually unlined / 1–3 eggs.

Left: Breeding plumage. Right: Non-breeding.

White-winged Black Tern

Chlidonias leucopterus

220–240mm
Wingspan: 600mm

Sexes alike. Breeding: Black with white tail and upperwing coverts. Bill, legs: Red. Eyes: Brown. Non-breeding: Crown: Black streaked grey. Frons, upperparts, underparts, underwings: Grey. Bill, legs: Black. Immature: Similar to non-breeding with greyer shoulders, dusky mantle and primaries.

Usually in flocks, buoyant flight, frequently dipping to surface.

Habitat: Swamps, wetlands, estuaries, coastal lagoons, bays and harbours.

Feeding: Wheeling, hovering, skimming water surface for insects.

Voice: Harsh screech.

Status: Uncommon but regular non-breeding migrant. Some overwinter. Vagrant Tas.

Breeding: Eastern Eurasia.

Left: Non-breeding. Right: Breeding plumage.

Whiskered Tern

Chlidonias hybrida

260mm
Wingspan: 700mm

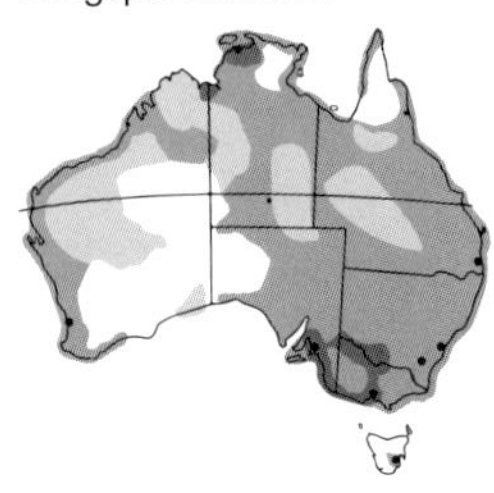

Sexes alike. Largest marsh tern, with bright red legs and pointed bill. Head: Large black cap to below eyes. White cheeks and sides of neck. Upperparts: Pale grey. Tail: Grey, slightly forked. Underparts: Dark grey, white under tail. Underwings: White. Non-breeding: Bill, legs: Dark red to black. Crown: Streaked black and white. Nape: Black. Underparts: Light grey. Immature: Similar to non-breeding, brownish shoulder.

Often in flocks up to 100. Flying lower into the wind than other terns. Circles, hovers, swoops.

Habitat: Inland wetlands, fresh or brackish lakes, irrigated pasture, occasionally coastal, offshore.

Feeding: Hawks for insects or plucks fish and small aquatic prey from water surface.

Voice: Harsh, piping shrieks.

Status: Common. Nomadic.

Migratory, moving north in winter.

Breeding: Oct.–Dec. Large colonies, nest of aquatic plant matter in shallow water / 3 eggs.

Left: Non-breeding. Right: Breeding plumage.

Roseate Tern

Sterna dougallii

350–380mm
Wingspan: 750mm

Sexes alike. Distinctive, deeply forked tail. Breeding: Long, white streamers. Underparts: White with pink wash. Crown: Black to below eyes. Underwings: Dark leading edge to wing tips. Bill: Variable, black with red base or red, tipped black. Eyes: Dark brown. Legs: Red. Upperparts: Pearl grey. Underwings: Black wedge on outer leading edge and white trailing edge to primaries. Non-breeding: Lacks pink wash. Mottled crown, shorter streamers. Legs, bill: Black. Immature: Similar to non-breeding with dark grey shoulder bar and black outer primaries.

Habitat: Marine, coastal, coral reefs, islands.

Feeding: Plummets vertically for small fish.

Voice: Sharp, rattling alarm.

Status: Common to uncommon.

Breeding: Sept.–May. In colonies on offshore islands, cays, Great Barrier Reef and off mid-coast WA. Scrape in sand or coral / 1–3 eggs.

Lady Elliott Is., Qld.

Left: Adult. Right: Immature.

White-fronted Tern

Sterna striata

400–420mm
Wingspan: 800mm

Sexes alike. Long, slender, black bill. Frons: White. Crown: White grading to speckled to black, slightly flatterned. Nape: Black. Upperparts: Pale grey with deeply forked tail. Underparts, underwings: White with fine black edge to primaries. Eyes: Brown. Legs: Red-brown. Breeding plumage: Long tail streamers. Crown, frons: Black with white line between bill and frons. Immature: Broad black shoulder bar, barred wing coverts. Crown: Mottled.

Graceful flight. Follows trawlers.

Habitat: Temperate oceanic and coastal waters.

Feeding: Hovers, swoops from 6–10m to pick small fish.

Voice: High-pitched whistling.

Status: Moderately common winter migrant from NZ, often young birds. In flocks May–Nov.

Breeding: NZ. Since 1979 some breeding on Bass Strait islands.

Left: Adult at nest site. Right: Breeding adult in flight.

Black-naped Tern

Sterna sumatrana

300–320mm
Wingspan: 600mm

Sexes alike. All year overall white with pale grey upperparts and a black crescent extending from around eyes to broad band across nape. Wings: Dark lower edge to primaries. Bill, legs: Black. Eyes: Brown. Tail: Deeply forked with long streamers. Immature: Crown: Mottled brown. Back, wings: Mottled grey, brown. Shorter tail.

Habitat: Coral cays, coastal and inshore lagoons.
Feeding: In flocks. Skims, hovers over shallow water, snatching prey from surface.
Voice: Harsh scolding notes. Rattling alarm call.
Status: Locally common. Sedentary.
Breeding: Sept.–Jan. In colonies, Great Barrier Reef. Scrape in sand / 2 eggs.

Left: Adults. Right: Adult in flight.

Common Tern

Sterna hirundo

310–370mm
Wingspan: 800mm

Sexes alike. Deeply forked tail, dark brown eyes.
Breeding: Overall light grey, white rump, black crown. Bill, legs: Black.
Non-breeding: Bill: Black. Frons: White. Crown: Receding, black-white streaked to all black. Upperparts: Pale grey with black shoulder bar. Tail: Dark grey. Underparts: White.
Immature: Similar to non-breeding, paler, greyer.

Adults attack intruders at nest including humans or dogs.

Habitat: Saline or brackish estuaries, beaches, sandbars, bays, harbours, less frequently wetlands. Returns to the same location each year.
Feeding: Mostly small fish from surface or shallow-plunging from 2–3m. Also insects.
Voice: Brisk 'kik-kik-kik'.
Status: Moderately common visitors Oct.–Apr. Many immature birds overwinter.
Breeding: temperate Europe, Asia and Nth America.

Opposite: Non-breeding plumage.
Right: Breeding plumage.

Arctic Tern

Sterna paradisaea

320–370mm
Wingspan: 800mm

Sexes alike. Similar to the Common Tern but squatter with shorter bill and legs.
Non-breeding: Bill: Black. Frons: White grading to black crown, hind neck and nape. Upperparts: Mid-grey, dusky stripes on outermost primaries. Rump: White. Tail: White, deeply forked. Underparts: White. Legs: Black.
Breeding: Bill, legs: Bright red. Crown, frons: Black.
Immature: Similar to non-breeding, dusky upperparts.

Buoyant and graceful flight.

Habitat: Oceanic.

Feeding: In cold waters. Circling, dipping and plunging for fish, plankton, crustaceans. During migration flies continuously without feeding.
Voice: Grating 'kee-kee' or high pitched 'kee-yaah'.
Status: Regular in small numbers. Migrates from Arctic Circle to Antarctic in summer, passing through Australian waters in Sept.–Apr. Immature birds may overwinter or stay for several seasons.
Breeding: Arctic and subarctic.

Left: Non-breeding. Right: Breeding.

Lesser Crested Tern

Thalasseus bengalensis

380–400mm
Wingspan: 900mm

Sexes alike. Bright orange bill. Crown, frons: Black separated by fine white line, white spots on frons. Upperparts: Grey. Underparts, neck: White. Non-breeding: Face, frons: White with receding crown streaked white. Bill: Duller.
Immature: Upperparts: Mottled brownish and sooty primaries, outer tail.

Habitat: Tropical coastal waters, offshore islands.

Feeding: Over shallow water or deep-sea plunge-diving for small fish.

Voice: Loud calls and rasping.

Status: Common. Nomadic.

Breeding: May–Nov. Large colonies, scrape in ground nest / 1 egg.

Left: Non-breeding. Right: Breeding colony in Qld.

Crested Tern

Thalasseus bergii

440–480mm
Wingspan: 950–1000mm

Sexes alike. Crown, frons: Black to below eyes and separated from bill by white line. Shaggy crest. Upperparts: Mid-grey. Underparts, neck: White. Bill: Light yellow. Legs: Black. Non-breeding: Duller. Receding black crown, streaked white.
Immature: Dusky. Bill: Greenish yellow. Crown: Mottled.

Swift, powerful flight and paired mating display flight.

Habitat: Inshore waters and estuaries. Often roosts on jetties.

Feeding: Plunge-dives for fish from 10–15m high.

Voice: Loud 'cawing', rasping.

Status: Common.

Breeding: Sept.– Dec. in south. Mar.–June in north. Offshore islands, large noisy colonies, scrape on ground / 1 egg.

Left: Non-breeding. Right: Breeding.

Pacific Gull

Larus pacificus

Endemic

600–650mm
Wingspan: 1300–1500mm

Sexes alike. Australia's largest gull. Large, red-tipped, yellow bill. Head, neck, underparts: White. Back, wings, tail: Black, white trailing edge on wings and white sub-terminal band on tail. Eyes: White. Birds in SA, WA, have red eyes. Birds take 4 years to reach adult plumage.
Immature: Overall brown. Bill, Eyes: Black. Bill: Cream with brown tip. Frons: Whitish. Legs, eyes: Grey. Sub-adults: As adults with white parts mottled light-brown.

Habitat: Coastal waters, offshore islands, swamps, rubbish tips.

Feeding: Fish, squid, crabs, small birds, eggs. Drops hard-shelled prey from height to break open. Scavenges for scraps.

Voice: Mournful 'kow, kow, kow'.

Status: Moderately common.

Breeding: Sept.–Dec. Offshore islands, elevated location, grass nest on ground / 2 eggs.

Port Phillip Bay, Vic.

Left: Sub-adult (foreground) and immature. Right: Adult.

Kelp Gull

Larus dominicanus

400–450mm
Wingspan: 1250–1400mm

Sexes similar. Similar to the Pacific Gull but smaller, with red spot to tip of lower bill. Overall white with black back. Wings: Black with small, white 'window' in wing tips, white tip primaries, white trailing edge. Eyes: White, red rimmed. Legs, feet: Greenish yellow with black claws. Full adult plumage takes about 5 years.
Female: Smaller. Immature: Dark brown, glossy black bill, dark eye, grey-brown legs. Year 2: Mottled white shoulder and tail. Bill: Greyer. Feet: Bluish white.

Pairs or small flocks. Deep, slow wingbeats alternating with glides.

Habitat: Coastal bays, beaches, reefs, offshore islands.

Feeding: Similar to Pacific Gull, but more aggressive.

Voice: Crying, yelping, laughing repetitive notes, usually loud.

Status: Uncommon.

Breeding: Sept.–Dec. Bulky grass nest on ground / 2–3 eggs.

Bruny Is., Tas.

Left and right: Adult.

Silver Gull

Chroicocephalus novaehollandiae

400–450mm
Wingspan: 940mm

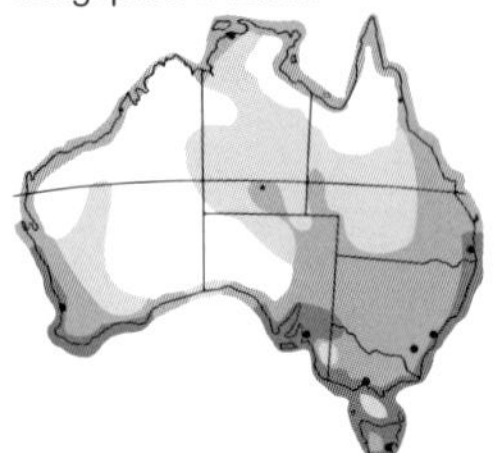

Sexes alike. Common seagull, overall white with grey back. Bill, legs: Bright red. Wings: Tipped black and white. Eyes: White with red eye ring.
Immature: Upperparts: Pale grey, mottled brown and buff. Legs, bill: Grey. Sub-adult: Brownish red bill and legs.

Often in large flocks.

Habitat: Coastal and inland lakes, particularly around settled areas near water, rubbish tips.

Feeding: Small fish, aquatic life, scavenges for scraps.

Voice: Raucous squeaks and cackles.

Status: Common to abundant. Sedentary and dispersive.

Breeding: Aug.–Nov. In large colonies on offshore islands. Scrape in sand nest, lined with seaweed / 1–3 eggs.

Left: Immature. Right: Adult.

11

Family Columbidae

Pigeons

Doves

Family Columbidae is a cosmopolitan group of pigeons and doves well represented in Australia.

The term 'dove' generally refers to the smaller members of the group, and 'pigeon' to larger members.

Pigeons and doves can be divided into two groups: the ground-feeding seed-eaters and the generally beautifully coloured, arboreal, fruit-eating birds.

Generally, ground-feeding, seed-eating pigeons and doves from dry inland and more open habitats have more subdued colours and patterns, which provide protection from predators such as hawks and falcons above, by providing camouflage with their grassy surroundings.

Some species, such as the White-quilled Rock-Pigeon, show colour variations within subspecies that reflect the local soil colouration.

Generally fruit-eating doves, also called fruit pigeons, are birds from rainforest and hot humid habitats. Although many have colourful plumage, they are still well camouflaged when feeding in the rainforest canopy on native fruits.

Most pigeons and doves need to drink regularly, especially ground-feeding species, because of their dry seed diet. Most pigeons drink by dipping their bills into water and sucking it up, taking less time to drink and allowing them to still watch for danger. (Most other bird species scoop water in their bill and tilt their head back to drink.)

Both parents produce a highly nutritious 'crop milk', which they feed to the young. Crop milk is secreted from the lining inside the bill and regurgitated and has higher levels of protein and fat than human or cow milk.

Typically, nests are a loose platform of fine interlaced twigs placed in a tree or shrub. Some inland species nest on the ground.

Many native pigeon species were decimated by the clearing of vast tracts of their habitat and large-scale shooting for 'sport' and eating in the mid-19th century. Some species, such as the Topknot Pigeon, have adapted to changing conditions and have begun re-establishing their numbers dramatically since the 1950s through protection from hunting and the spread of Camphor Laurel trees on their previously cleared habitat, the berries of which are now part of their staple diet.

Some ground-feeding pigeons, such as the Common Bronzewing, have probably benefited from clearing for grazing. Others, such as the Crested Pigeon, have re-established their populations by expanding their range. Until recently a bird of the arid and semi-arid zones of inland Australia, this species, after a large inland drought, has successfully colonised coastal areas across the country and become a common urban bird in most towns and cities.

Still, many species are threatened or endangered by land clearing, logging and predation by cats and foxes.

Opposite: Crested Pigeon in courting display.

White-headed Pigeon

Columba leucomela

Endemic

380–410mm

Sexes similar. Large pigeon with a distinctive white head, neck and breast. Bill: Red tipped, pale yellowish. Eye ring: Red. Back, wings, tail: Black with purple-green sheen on back and rump. Lower breast, belly: Grey. Female: Duller. Immature: Duller, greyer crown and nape. Bill: Brownish, white tipped.

Wary and secretive. Usually seen in pairs or small groups of 12–15 birds.

Habitat: Lower canopy of coastal rainforests, remnant scrub and suburban parks and gardens.

Feeding: Arboreal, feeding on fruit and seeds, especially native laurels and the introduced Camphor Laurel tree (weed).

Voice: Loud 'whoo' repeated 3 times. Contact call a low 'oom'.

Status: Moderately common. Nomadic, follows ripening rainforest fruits.

Breeding: July–Mar. Typical scanty twig nest in dense foliage or vines / 1 egg.

Lamington NP and Bunya Mountains NP, Qld.

Left: Adults roosting. Right: Adult male.

Brown Cuckoo-dove

Macropygia phasianella

390–450mm

Sexes similar. Distinctively long, tapering tail. Upperparts: Dark brown with metallic sheen on neck. Eyes: Blue with fine red eye ring and white line under eye. Underparts: Pale brown. Female: Darker throat and chest and chestnut crown. Immature: Duller, more heavily barred and scalloped.

Strong, graceful, low flight. Usually in groups of up to 10.

Habitat: Coastal rainforests, forest margins and wooded areas adjacent to streams.

Feeding: Mostly arboreal, foraging in low branches for fruit, berries and seeds.

Voice: Sharp, 3-note 'wook-a-wook', a commonly heard sound of the rainforest.

Status: Locally common in rainforests or regrowth areas. Populations decimated by shooting and clearing of rainforests.

Breeding: All year, mostly Sept.–Jan. in the south, earlier in the north. Minimal stick nest, up to 6m high / 1–2 eggs.

Lamington NP, Qld. Morton NP, NSW.

Left: Female. Right: Male tail detail.

Pacific Emerald Dove

Chalcophaps longirostris

230–270mm

A small, plump dove with overall mauve-brown plumage, with bright iridescent green wings, white shoulder patch and 2 white bars across back. Rump: Dark brown-black. Eyes: Dark. Bill: Bright red.
Female: Duller with grey shoulder patch. Immature: Chestnut with brown barring on head and body and some green on wings.

Solitary or in pairs, mostly walking under cover. When disturbed, will often walk rather than fly away. Flight is usually low, direct, fast with regular wingbeats.

Habitat: Coastal rainforests and wet sclerophyll forests, heathlands, mangroves and gardens.

Feeding: Ground feeder. Often seen picking through rainforest litter for fallen seeds and fruit. Occasionally feeds on figs and fruit pecked from trees.

Voice: Monotonous, low-pitched, penetrating 'coo-coo-coo'.

Status: Solitary usually. Common in suitable habitat.

Breeding: Any time of year, mostly Jan.–Apr. in NT. Sept.–Dec. in the east. Scant nest of interwoven twigs / 2 eggs.

Daintree NP, Qld. Howard Springs NP, Darwin, NT.

Left: Female on nest. Right: Male showing white bars to back.

Common Bronzewing

Phaps chalcoptera

Endemic

280–360mm

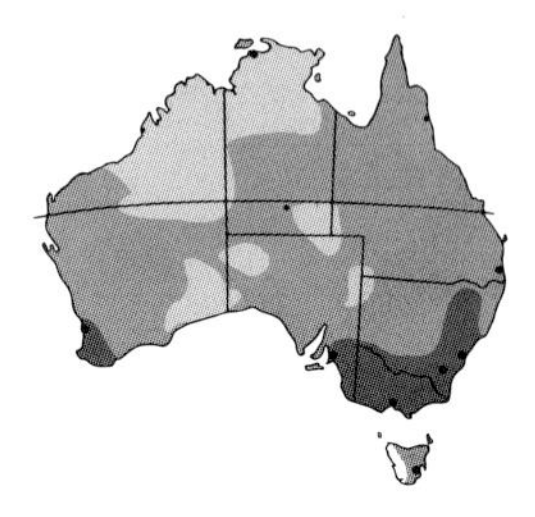

Sexes similar. Medium-sized, plump pigeon with purple-brown upperparts, with buff barring and distinctive metallic spots of blue, green and red on wings and white line under eyes. Frons: Cream. Breast: Grey-buff. Female: Duller with grey frons and more pronounced stripe under eyes. Immature: Duller.

Shy and wary, flying close to the ground when disturbed. Usually solitary or in pairs. Large groups occur around water at dusk.

Habitat: Wet and dry forests, woodlands, mallee, heaths, farmland and gardens. Avoids dense rainforests.

Feeding: Forages on ground for fallen seeds, grass, weed seeds and spilt grain. Drinks frequently.

Voice: Deep, resonant 'oom' repeated continuously with 3 second intervals in breeding season.

Status: Most widespread native pigeon, common in suitable habitat.

Breeding: Variable, mostly Aug.–Dec. Typical twig nest in tree or bush, 1–10m high / 2 eggs.

Left: Male. Right: Female.

Brush Bronzewing

Phaps elegans

Endemic

250–230mm

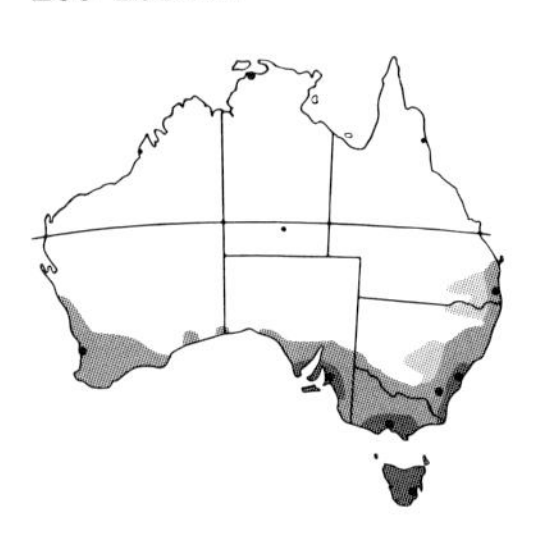

Sexes similar. Similar to the Common Bronzewing but stockier with chestnut patch on chin and chestnut line running through eye. Crown: Chestnut-buff. Wings: Dark olive-brown with bronze iridescent blue-green bars. Back, shoulders: Chestnut. Breast: Blue-grey.
Female: Grey frons. Immature: Duller, brownish. Lacks chestnut throat patch and collar.

Habitat: Dense coastal woodlands with tea-tree and banksia scrub, coastal dunes, dense wet forests with dense undergrowth.

Feeding: Entirely ground feeding on small seeds, particularly acacia seeds. Drinking at dawn or dusk.

Voice: Low, deep resonant 'hoot' call, often continued for several minutes. Higher pitched than Common Bronzewing.

Status: Populations in some areas decimated by loss of habitat and predation by foxes and cats.

Breeding: All year, mostly Oct.–Jan. Nest is slightly cupped, on or near ground / 2 eggs.

Waychinicup NP, WA.

Left: Male. Right: Female.

Crested Pigeon

Ocyphaps lophotes

Endemic

310–360mm

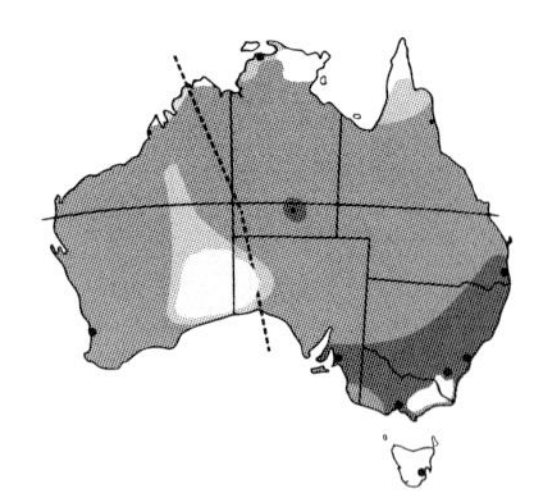

Sexes alike. Stocky with a distinctive, thin, upright black crest. Mostly overall grey-brown plumage grading to pinkish underparts. Wings: Brownish with prominent black bars and metallic green and purple patches. Eye ring: Pinkish red. Birds in western SA, WA, have narrower wing bars.
Immature: Duller.

Usually in small groups of 5–6. Takes flight with vigorous whistling wingbeats alternating with long glides on stiff, outstretched wings. Flicks its long tail when alighting.

Habitat: Avoids heavily timbered country. Prefers lightly wooded grasslands near water. Farmland, parks and gardens.

Feeding: Ground feeding on native grass, fallen acacia and eucalypt seeds. Exotic weed seeds, spilt grain in settled areas.

Voice: Loud 'coo' or rising 'coo-oo'.

Status: Common. Sedentary.

Breeding: All year, mostly Sept.–Mar. Nest is a frail platform of twigs in tree up to 5m high / 2 eggs.

Opposite: Breeding pair.

Flock Bronzewing

Phaps histrionica

Endemic

280–310mm

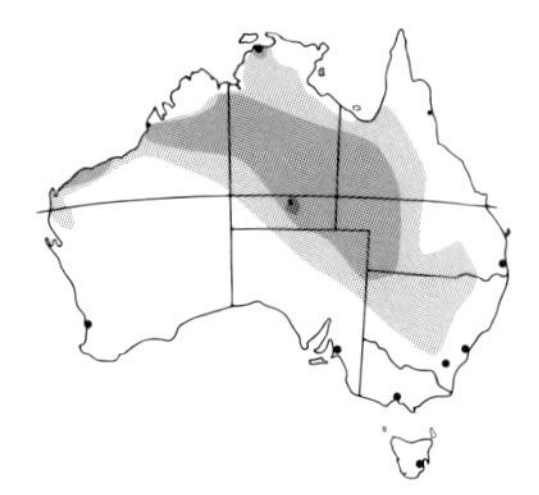

Sexes similar. Large, nomadic sandy-coloured pigeon, stockier than the Common Bronzewing. Upperparts: Sandy copper plumage. Head: Black with white frons, ear markings and throat bib. Underparts: Grey. Female, immature: Head patterned pale sandy brown instead of black.

Gregarious, highly nomadic, often in flocks of hundreds.

Habitat: Treeless grassy plains and saltbush plains, mulga, never far from water.

Feeding: Ground feeding on soft native grass seeds, native herbs.

Voice: Usually silent. Occasional soft 'coo-coo' when breeding.

Status: Uncommon in south of range due to intensive sheep grazing on inland plains and loss of native grasses. Locally common in north.

Breeding: Spring in the south. Following the wet season in the north. Typical nest is a scrape on the ground under grass tussock / 2 eggs.

Barkly Tablelands, NT.

Left: Male (foreground) and female.
Right: Male in flight.

Spinifex Pigeon

Geophaps plumifera

Endemic

200–240mm

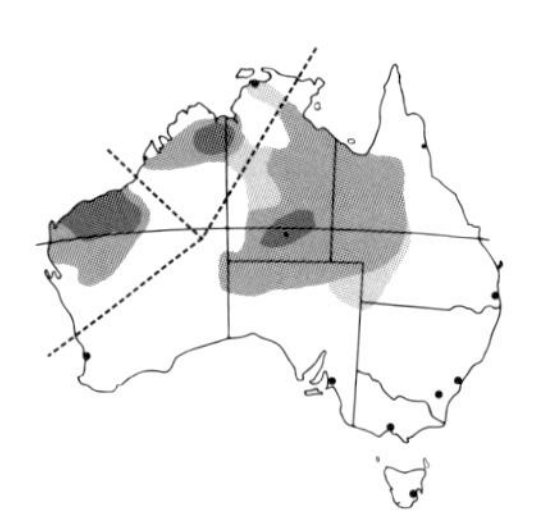

Sexes alike, all races. Nominate race: Northwest NT, WA. Conspicuous, erect crest and large, red eye skin patch. Upperparts: Reddish brown with black barring on wings and side of neck. Throat: Black. Breast: Reddish brown with dark bar. Belly: White.
Central inland birds, race *leucogaster*, similar to nominate race, but belly is patchy white. Mid-WA birds, race *ferruginea*. Breast: Rufous-brown. Belly: Rufous brown with dark bar. Immature: Lacks black and white markings to face and throat.

Habitat: Arid, stony, spinifex grasslands and rocky hills near water supply.

Feeding: Ground feeders foraging for grass and spinifex seed in small groups up to 15.

Voice: Low-pitched, repeated 'woo-coo-up-coo-up' and deep guttural 'coo-r-r-r'.

Status: Common.

Breeding: All year, dependent on rain for food supply. Nest is a scrape in the ground, sheltered in spinifex clump / 2 eggs.

Ormiston Gorge, NT.

Left and centre: Nominate race.
Right: Race *ferruginea*.

White-quilled Rock-Pigeon

Petrophassa albipennis

Endemic

280–300mm

Sexes alike. Plump, ground-dwelling pigeon with a long rounded tail and short wings. Overall brown with white scalloping and white wing patch viewed in flight. Throat: Black with white spots.
NT, race *boothi*, is smaller with almost no white on wing.
Immature: Less white spotting on throat.

Usually seen in pairs or small flocks. Freezes when disturbed.

Habitat: Sandstone escarpments with spinifex and broken rocks.

Feeding: Foraging among rocks for grass, legume seeds.

Voice: Loud 'coo-carook'.

Status: Moderately common. Sedentary.

Breeding: Mostly Mar.–Nov. Nest of sticks and spinifex stalks in a rock crevice or on a shaded ledge 1–6m high / 2 eggs.

Mitchell Falls Walk, WA.
Keep River NP, NT.

Opposite: Race *boothi*, NT.

Chestnut-quilled Rock-Pigeon

Petrophassa rufipennis

Endemic

280–310mm

Sexes alike. Similar to the White-quilled Rock-Pigeon, but with bright chestnut patches across outer wings when viewed in flight. Head, neck: Dark brown with grey scalloping. Throat: Off-white. Wing patch: Chestnut viewed in flight.
Immature: Similar, duller.

Squats and freezes when disturbed, merging into rocks.

Habitat: Rocky sandstone hills and gorges with patches of trees along the escarpments of northwest NT.

Feeding: Ground feeding on acacia seeds, grass and legume seeds.

Voice: Repetitive, loud 'coo-carook'.

Status: Locally common in its restricted range.

Breeding: Mar.–Sept. Stick platform nest in a rock crevice or on ledge, lined with spinifex / 2 eggs.

Kakadu NP, NT.

Left: Bright chestnut wing patches revealed in flight. Right: Adult upperparts.

Partridge Pigeon

Geophaps smithii

Endemic

250–280mm

Sexes alike. Medium-sized with overall olive-brown plumage and distinctive red or yellow facial markings. Throat: White. Bill: Grey-black. Underparts: Pinkish brown with small black-barred central patch on breast. Belly, flanks: White, V-shape.
Nominate race: Red facial skin.
Kimberley area, WA, race *blaauwi*, has yellow-ochre facial skin.
Immature: Duller with upperparts flecked fine chestnut, grey eye rings and brown eyes.

Rarely flies. Freezes when disturbed, or 'explodes' into a vertical flight; birds disperse in all directions.

Habitat: Tropical eucalypt woodlands and savannah woodlands near water.

Feeding: Ground feeding. Prefers seeds of legumes such as *Swainsona* spp., acacia, herbs and native grasses.

Voice: Low, long rolling 'coo'.

Status: Vulnerable. Sedentary.

Breeding: All year, mostly Mar.–Oct. Nest is a scrape on the ground hidden in dense grassy area / 2 eggs.

Kakadu NP, NT. Mitchell Falls Walk, WA.

Left: Nominate race. Right: Race *blaauwi*.

Squatter Pigeon

Geophaps scripta

Endemic

260–320mm

Sexes alike. Mostly grey-brown with distinctive black and white stripes on the face and throat, and orange-red or blue-grey bare eye skin. Throat: White, Breast: Blue-grey with deep white V-shape below. Bill: Grey-black. Wings: Brown with pale edges that give a mottled effect.
Nominate race: Blue-grey facial skin.
Northern Qld, Race *peninsulae* has orange-red facial skin.
Immature: Duller with upperparts flecked fine chestnut, grey eye rings and brown eyes.

Groups or small flocks. Rarely flies. Freezes if disturbed or runs erratically with neck extended. May burst into flight with rapid, clapping wingbeats, showing the broad dark-tipped outer tail.

Habitat: Dry savannah woodlands or grassy plains, near water.

Feeding: Ground feeding, taking seeds and invertebrates.

Voice: Low 'coo-cwoop' repeated.

Status: Sedentary or partly nomadic. Locally common in north. Vulnerable in south.

Breeding: All year after rain, mostly May–June. Nest is a grass-lined scrape on the ground / 2 eggs.

Carnarvon Gorge NP, Mt. Molloy and Julatten area, Qld.

Left: Nominate race. Right: Race *peninsulae*.

Diamond Dove

Geopelia cuneata

Endemic

190–240mm

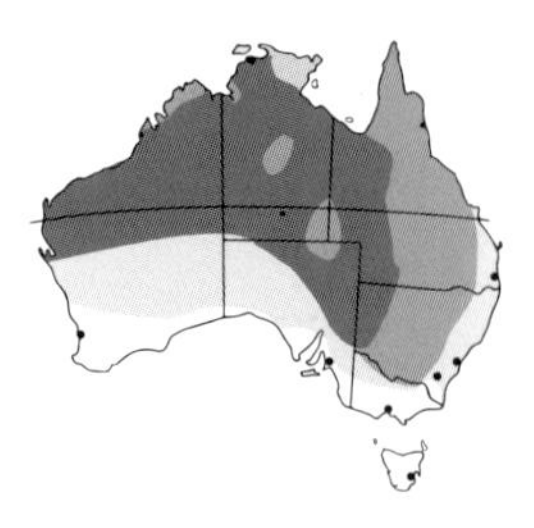

Sexes similar. Smallest Australian dove. Mostly blue-grey with fine white spots on the grey-brown wings. Eyes: Red with thick red eye rings. Head, throat, chest: Blue-grey. Belly: White. Female: Duller with browner foreparts. Immature: Grey-black stripes to upper parts.

Gregarious. Congregates in large flocks around waterholes. Swift, direct undulating flight.

Habitat: Widespread. Prefers dry grassy woodlands, especially in inland areas, in arid and semi-arid areas, spinifex and dry mulga, with nearby water.

Feeding: Ground feeder on small seeds, some insects, including ants.

Voice: High-pitched melancholy 4-note 'coo' repeated.

Status: Locally common. Nomadic, following fresh seed after rains.

Breeding: Any time of year following rain, mostly Sept.–Jan. Interwoven flimsy grass nest in low shub or scrubby tree / 2 eggs.

Opposite: Adult male.

Peaceful Dove

Geopelia placida

200–240mm

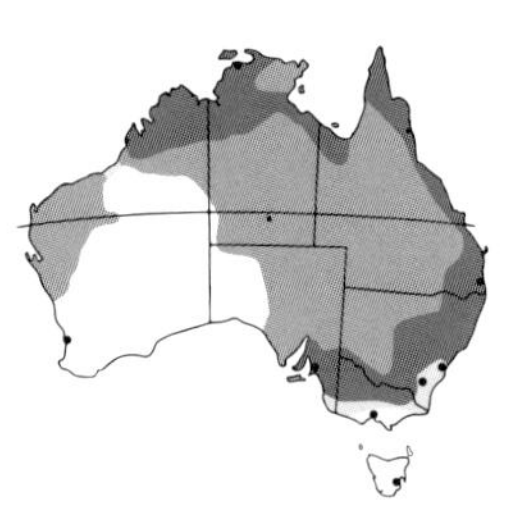

Sexes alike. Similar to the Bar-shouldered Dove but smaller. Tiny pink-grey and grey-brown dove, with prominent fine barring on the upper breast. Upperparts: Grey-brown, barred with black streaks on neck, rump, wings and back.
Immature: Duller.

Undulating flight reveals their rufous-coloured underwing coverts and grey flight feathers.

Habitat: Open forests, scrubland, especially with acacias, open, well-grassed woodlands, all with water nearby. Parks, gardens and farmland. Offshore islands.

Feeding: Ground feeder, prefers small seeds of native grasses and small shrubs, occasionally insects.

Voice: A high-pitched, 3-noted 'doodle-a-do'.

Status: Common in watered areas throughout their range.

Breeding: Any month. Mostly Oct.–Jan. Small, frail platform nest of twigs and roots, 1–12m high in leafy shrub or tree / 2 eggs.

Left: Pair drinking. Right: Female at nest with young.

Bar-shouldered Dove

Geopelia humeralis

270–300mm

Sexes alike. Long-tailed dove with boldly black scalloped, bright rufous coppery nape and mantle. Upperparts: Dark grey-brown with distinctive black barring to upper wing coverts and pronounced red-bronze to hind neck. Face, head: Blue-grey with black flecks to crown. Eye rings: Blue-grey, turning red-brown when breeding. Chest: Blue-grey grading to cream underparts. Tail: Long white tipped.
Immature: Duller.
Swift, level, direct flight with long, loud whistle, revealing chestnut patch on flight feathers and cinnamon wing lining. Usually in pairs. Flocks gather where food is concentrated.

Habitat: Widespread near water in woodlands with grassy understorey, coastal and semi-arid shrub. Mangroves near creeks and pandanus thickets in the north. Well-treed gardens.

Feeding: Feeds in short grass near shelter, taking grass seeds, spilt seed, bulbs of some sedges.

Voice: High-pitched, cheery 4-note 'cuckoo-cuck-oop'.

Status: Common. Sedentary or locally nomadic.

Breeding: Any time of year, mostly Sept.–June in south, Feb.–Apr. in north. Twigs and grass nest, usually placed in a shrub or tree fork, to 6m high / 2 eggs.

Left: Tail fanned in courtship display. Right: Note the copper-coloured nape feathers.

Wonga Pigeon

Leucosarcia melanoleuca

Endemic

380–450mm

Sexes alike. Large, plump pigeon with a small head, long tail, short broad wings and grey upperparts with a distinctive white 'V' to upper breast. Lower breast, undertail: Boldly marked white with black crescents and streaks. Lores: Black. Frons, chin: White. Eyes: Red-brown. Bill, legs, feet: Deep pink to red.
Immature: Brownish.
Solitary. Shy. Freezes when disturbed. Takes off with 'explosive' clattering wingbeats.

Habitat: Dense coastal eucalypt forests with acacia understorey, rainforests, adjacent highlands.

Feeding: Ground feeding, foraging for fallen fruits, eucalypt and acacia seeds.

Voice: High-pitched, loud, monotonous whistled 'wonk'.

Status: Common in some areas, but populations decimated by wide-scale shooting and habitat loss. Sedentary.

Breeding: All year, mostly Oct.–Jan. Dish-shaped nest of sticks, twigs and grass, often high up in a tall tree / 2 eggs.

Royal NP, NSW. Lamington NP, Qld.

Left: Dust bathing to maintain feathers in good condition. Dust smothers parasites and absorbs excess oils. Right: White 'V' to upper breast.

Torresian Imperial-Pigeon

Ducula spilorrhoa

380–440mm

Large, mostly tree-dwelling white pigeon with black primaries and tail. Undertail coverts: Scaly black and white.
Immature: Duller.

Birds that reside on offshore islands travel in large flocks to feeding areas on coastal mainland each morning, returning in evening. Strong flight.

Habitat: Coastal rainforests, mangroves, offshore islands.

Feeding: Mostly arboreal feeding, primarily on rainforest fruits.

Voice: Loud, deep 'woop-woooo'.

Status: Birds from Top End to east Qld are breeding migrants from PNG. Arrive in large numbers from Aug. Depart in Feb.–Apr. Birds from southern range are sedentary, locally nomadic. Birds form breeding colonies on offshore islands. Locally common to moderately common in north, uncommon further south due to habitat loss and shooting.

Breeding: Sept.–Jan. Mostly on offshore islands. Substantial stick nest, low in mangroves / 1 egg.

Daintree Village and Iron Range NP, Qld.

Left and right: Adult birds.

Banded Fruit-dove

Ptilinopus cinctus

Endemic

380–440mm

Sexes alike. Striking black and white appearance. Head, neck, upper breast: White with a distinctive black breast band. Rump: Grey. Uppertail: Black, terminating in a broad grey band. Underparts: Mid-grey.

Shy, usually well concealed in foliage.

Habitat: Rainforest with Anbinik trees (a relic rainforest tree species that dominates the monsoon rainforests), forested gorges and plateaus in Arnhem Land, NT. Never far from water.

Feeding: Quietly feeds on rainforest fruits, especially figs.

Voice: Deep, low repeated 'coo'.

Status: Uncommon.

Breeding: May–Nov. (Dry season.) Flimsy nest of twigs, 2–5 m high / 1 egg.

Kakadu NP, NT.

Opposite: Adult bird.

Wompoo Fruit-dove

Ptilinopus magnificus

350–450mm

Sexes alike. Largest fruit-dove. Striking bright green upperparts and light blue-grey head and neck. Breast: Deep purple. Belly, undertail, wing bar: Deep yellow. Birds in northern and far north Qld are small and brighter. Immature: Duller with light purple, blotched green breast.

Despite size and bright colours, difficult to see in dense rainforest canopy. Best located by their call and falling fruit where feeding above.

Habitat: Dense tropical and subtropical rainforests.

Feeding: Arboreal, takes small fruits. Rarely comes to ground.

Voice: Deep loud 'wallock-a-woo' and soft 'wom-poo'.

Status: Locally common in the north. Once common south to Illawarra region, NSW, now rare due to loss of rainforest habitat.

Breeding: June–Jan. Flimsy stick platform nest, 5–20m high at end of palm frond or acacia branch / 1 egg.

Lamington NP, and Iron Range NP, Qld.

Left: Adult feeding on White Cedar fruits. Right: Adult feeding on Bangalow Palm fruits.

Superb Fruit-dove

Ptilinopus superbus

220–240mm

Small, spectacularly colourful, upper-canopy-dwelling pigeon. Purple crown with broad orange-red hind neck. Upperparts: Green, spotted black. Throat, breast: Grey-flecked purple with wide blue-black breast band. Belly: White with green barred flanks.
Female: Blue crown patch and green breast. Lacks black band and orange-red on neck. Immature: Similar to female without blue crown patch.

Arboreal and well camouflaged in foliage. Best located by sound of falling fruit when feeding.

Habitat: Open forests with native fruits, rainforests and adjacent mangroves.

Feeding: Rainforest fruits. Native laurels important source of food in Qld.

Voice: Low, resonant repeated 'oom', rising to loud 'woops'.

Status: Partially nomadic. Moderately common in north. Uncommon south of Richmond River, NSW due to destruction of habitat.

Breeding: June–Feb. Typical nest, up to 10 m high / 1 egg.

Iron Range NP, Qld.

Left: Male. Right: Female.

Rose-crowned Fruit-dove

Ptilinopus regina

220–240mm

Sexes similar. Small, colourful, foliage-dwelling pigeon named for its rose pink crown. Head, breast: Grey marked with green. Upperparts: Bright green.
Tail: Yellow tip. Lower breast: Magenta. Orange yellow belly and undertail.
Top End, NT, WA, race *ewingii.* Paler crown. Underparts more yellow.
Female: Duller. Immature: Mostly green above. Lacks crown patch.

Habitat: Dense rainforests, mangroves, paperbark swamps and wet sclerophyll forests from Derby, WA, across the north to Bellinger River, NSW.

Feeding: Arboreal, best detected by falling fruit when feeding.

Voice: Loud cooing, accelerating in speed and decreasing in pitch, and 2-syllable 'coo', with explosive single 'coo'.

Status: Moderately common in north of their range, nomadic following ripening rainforest fruits. Vulnerable in NSW.

Breeding: Variable, mostly Sept.–Jan. Flimsy twig platform nest in rainforest vine or tree to 30m high / 1 egg.

Fogg Dam CR, botanic gardens, and East Point Reserve, Darwin, NT. Green Is., Qld.

Left: Nominate race. Right: Race *ewingii.*

Topknot Pigeon

Lopholaimus antarcticus

Endemic

400–450mm

Sexes similar. Large, pale grey pigeon with a distinctive double grey crest sometimes curling forward and partly down over bill, and a rusty-coloured crest swept over the crown to nape. Upperparts: Dark grey. Wings: Dark grey, rounded with black flight feathers. Underparts: Pale grey. Tail: Black, pale grey central band. Bill, eyes: Red.

Female: Smaller, paler crest. Immature: Similar to female.

Often seen in fast-moving flocks.

Habitat: Coastal rainforests and adjacent woodlands.

Feeding: Completely arboreal. Flocks follow seasonal ripening of rainforest and palm fruits, including Camphor Laurel fruits in absence of native fruits. Relies on rain water and dew for moisture.

Voice: Usually silent. Single, low, throaty 'coo' when feeding.

Status: Nomadic. Moderately common in north. Uncommon in south due to habitat loss. Once flocks of several thousand birds provided food for early settlers. Today flocks number 50–200.

Breeding: June–Dec. Platform stick nest up to 30m high / 1 egg.

Lamington NP, Qld. Royal NP, Morton NP, NSW.

Left: Adult. Right: Head detail.

Barbary Dove

Streptopelia risoria

Introduced

270–300mm

Overall creamy buff with a black half-collar fringed white. Head, chin: Paler. Flight feathers: Brown with pale edges. Belly: Paler, sometimes pinkish. Legs, feet: Deep pink. Eyes: Dark red. Immature: Lacks half-collar. Eyes: Yellowish.
An aviary escapee that will establish in urban areas. Established in Adelaide.

Feeding: Scraps, seeds.

Voice: Repeated 'koo-kooroo', high-pitched laugh notes.

Status: Pest species that is a threat to endemic species.

Breeding: All year / 2 eggs.

Report sightings: National Animal Pest Alert 1800 084 881.

Left: Barbary Dove.

Laughing Dove

Streptopelia senegalensis

Introduced

250–270mm

Brown with blue-grey patch on wings. Head, neck, breast: Mauve-pink with black-spotted chest. Lower belly: White.

Introduced from Africa in 1860s to Perth zoo. Now present throughout Perth and suburbs, also Kalgoorlie and Esperance, WA. Colonises bushland.

Feeding: Scraps, animal feed, grain.

Voice: Cooing, laughing notes.

Status: Abundant, sedentary. Established in southwestern WA.

Breeding: All year / 2 eggs.

Centre right: Laughing Dove.

Spotted (Turtle) Dove

Streptopelia chinensis

Introduced

300–330mm

Brown upperparts, black and white collar. Underparts: Fawn. Habitat urban areas, scrubby thickets, grain-growing areas.

Introduced from China. Now present in coastal eastern Australian urban and grain-growing areas.

Feeding: Scraps, animal feed, spilt grain.

Voice: Musical, persistent cooing.

Status: Abundant, expanding in settled areas. Replaces the native Bar-shouldered and Peaceful Doves.

Breeding: All year / 1–2 eggs.

Below left: Spotted (Turtle) Dove.

Rock Dove (Feral Pigeon)

Columba livia

Introduced

310–340mm

Extremely variable, but generally blue-grey with glossy purple-green sheen.

Introduced from Europe in early 1800s, now widespread across urban areas and crop margins.

Feeding: Scraps and spilt grain.

Voice: 'Coo' or 'rackety-coo'

Status: Abundant.

Breeding: Year-round / 2 eggs.

Below right: Rock Dove (Feral Pigeon).

Family Cacatuidae, Psittacidae

Cockatoos

Galah

Corellas

Cockatiel

Lorikeets

Parrots

Rosellas

Ringneck

Budgerigar

A group containing 2 families of usually colourful birds with similar feeding habits: cockatoos and parrots. This group has evolved to exploit almost every habitat in Australia.

Family Cacatuidae includes cockatoos, corellas and the Galah and Cockatiel.
Cockatoos exploit the woody fruits of many native plants by extracting seed with their large specialised bill, or feeding on the ground on fallen seed, or digging for corms or roots. All cockatoos nest in tree hollows and require large old-growth trees that can accommodate hollows of sufficient size. Some species are threatened through loss of old-growth trees.
Corellas are small ground-feeding cockatoos and include the Long-billed Corella, with its long, curved upper mandible, used for digging up roots and bulbs for food.
Although the Cockatiel looks like a small slender parrot, it is actually the smallest member of the cockatoo family.

Family Psittacidae includes the parrots, lorikeets, Fig-parrot, rosellas, ringneck and Budgerigar.
This is a diverse and colourful group consisting of approximately 40 species.
Lorikeets are arboreal feeders that have brush-like tongues for extracting nectar and pollen from flowering plants.
The Budgerigar is noted for the enormous green and gold flocks that form when food is plentiful, often containing tens of thousands of birds. When drinking, the flock members take turns, with smaller groups continuously taking to the air and swirling about to confuse predators.
The tiny Fig-parrots are Australia's smallest parrots. Even smaller than the Budgerigar, they are found in rainforests where they feed on soft fruit with a preference for figs. Rosellas and ringnecks are mostly quiet ground feeders, feeding on grass seeds and fallen seeds. They also forage in eucalypts for insects, seeds, fruits, nectar and larvae.
The remaining species, loosely referred to as parrots, are usually ground foragers, except for the Swift Parrot, which has a lorikeet-like brush-tipped tongue for extracting nectar and pollen from honey-flora and a lorikeet-like flight. Parrots belonging to the genus *Neophema* are commonly known as grass parrots.
The Night Parrot is a bird of the remote inland spinifex regions that eats spinifex seeds and nests in clumps of spinifex. Until recently believed extinct, with no known sightings since 1912, it was 'rediscovered' in 2013 in western Queensland and the first live photographs were taken. In 2017 it was also discovered in Western Australia.
Most parrot species rely on hollows in old-growth trees for nesting. Those that have different nesting habits include the 'anthill parrots' of northern Australia. These parrots excavate a tunnel into a termite mound leading to a nesting chamber. The Rock Parrot nests in rock cavities and the Ground and Night Parrots construct a grass nest on or near the ground, well hidden in dense cover.
Many parrots will nest in garden nest boxes with dimensions selected to suit the chosen species and the nest box placed at the correct height above the ground.

Opposite: Yellow-tailed Black Cockatoos. Female above, male below.

Palm Cockatoo

Probosciger aterrimus

Size: 560mm

Sexes similar. Australia's largest cockatoo. Overall black plumage with a spectacular crest and crimson bare-skin cheek patches that change to scarlet when alarmed or excited. Bill: Massive, elongated, grey-black with the larger upper mandible partly crossing over the smaller lower mandible, enabling it to crack hard nuts and break large sticks. Eyes: Dark-brown.
Female: Smaller red facial skin patch and bill. Immature: Smaller, short crest, white-tipped bill and pale yellow scalloping to underparts.

Flight is slow with deep wingbeats. Often in flocks. Unique territorial display using a large stick to drum on dead tree limbs, creating a loud noise.

Habitat: Rainforests and adjacent savannah woodlands.

Feeding: Powerful bill for cracking large seeds, nuts and rainforest fruits. Arboreal and ground feeders.

Voice: Piercing, metallic whistles, wails or screeches in flight.

Status: Uncommon.

Breeding: Aug.–Feb. Large tree-hollow nest, usually in a eucalypt / 1 egg.

Iron Range NP, Qld.

Left: Male. Right: At nest hollow.

Red-tailed Black Cockatoo

Calyptorhynchus banksii

Endemic

Size: 550–600mm

Overall black with rounded crest and broad, red band on lower undertail. Bill: Large, grey-black. Eyes: Dark grey. Several races with differing habitat, food preferences, size, bill size and colour of female tail bands.
Female: Brown-black with small yellow spots to head, neck and wings. Underparts: Lightly barred yellow-orange. Undertail: Barred orange-red or yellow in southwest WA. Bill: Paler. Immature: Similar to female.

Active, noisy, gregarious. Usually seen in large flocks, also pairs. Slow and buoyant flight. Deeply fingered wings.

Habitat: Dense forests, open woodlands, arid inland, WA wheat belt.

Feeding: Northern birds take seeds of grasses, trees and sometimes mangroves. Southwestern birds prefer upper canopy and woody seeds of Marri. Inland birds mostly ground feeders on grass and weed seeds, proteas, casuarinas. Isolated southeast birds prefer seeds of Brown Stringybark.

Voice: Loud, grating 'rur-rak'.

Status: Nomadic. Uncommon. Most common in the north. Endangered in the southeast.

Breeding: Mar.–Apr. and July–Oct. in south. May–Sept. and Apr.–July in north. Large tree hollow nest / 1 egg.

Kakadu NP, NT.

Left: Male. Right: Female.

Glossy Black Cockatoo

Calyptorhynchus lathami

Endemic

Size: 480–500mm

Smallest black cockatoo. Plumage is overall brownish black with a distinctive broad red to orange-red tail band. Crest: Small, brownish black. Bill: Grey, bulbous.
Female: Yellow markings to head and side of neck. Less red on tail band with dark bars. Immature: Similar to female, darker head and wings spotted with yellow.

Usually seen in groups of 2–3, sometimes up to 10. Entirely arboreal.

Habitat: Coastal forests and open woodland, usually where sheoaks grow.

Feeding: Quietly and almost exclusively on sheoak seeds. They have preferred 'feeding trees'. Wood-boring insects also taken. Drink in late afternoon.

Voice: Prolonged, soft 'tarr-red, tarr-red'.

Status: Generally uncommon. Nomadic, dependent on ample supply of sheoak seed. Race *halmaturinus* isolated, Kangaroo Is., endangered.

Breeding: Mar.–Aug. Nest in large tree hollow, very high, prefers eucalypts / 1 egg.

Left: Adult female and male. Right: Female.

Yellow-tailed Black Cockatoo

Calyptorhynchus funereus

Endemic

560–650mm

Overall brownish black, faintly yellow-scalloped with yellow cheek patch, yellow tail band, short crest and powerful bill. Eyes: Dark brown with pink eye ring. Upper bill: Dark grey. Southern birds more prominently scalloped.
Female: Brighter, larger yellow cheek patch. Eye rings: Grey. Upper bill: Whitish. Immature: Small, dull yellow cheek patch.

Usually seen in flocks of 10–20. Flight is slow, buoyant, with full wingbeats.

Habitat: Coastal eucalypt forests, banksia heathland, adjacent ranges and pine plantations.

Feeding: In small to large noisy flocks. Seeds of forest trees including sheoaks, eucalypts, banksias. Extracts wood-boring grubs.

Voice: Loud, wailing 'whee-la' contact call. Harsh screeching.

Status: Locally nomadic. Locally common. Moves from high country to lowlands or coast in winter.

Breeding: Mar.–Apr. in north. July–Jan. in south. Nest in large tree hollow / 2 eggs.

Royal NP, NSW (winter). Mt Wellington and Bruny Is., Tas.

Left: Female. Right: Male eating the seeds of Old Man Banksia.

Carnaby's Black Cockatoo

Calyptorhynchus latirostris

Endemic Endangered

540–560mm

Overall dusky black with off-white feather edges creating a scalloped pattern. Tail: Black with white panels. Cheeks: White patch. Eyes: Dark brown, pink eye ring. Bill: Grey-black.
Female: Pale grey bill. Eye rings: Grey. Immature: Similar to adult, duller, smaller white areas. Grey eye rings.
Roving flocks form in autumn and winter.

Habitat: Dry inland eucalypt woodlands and mallee scrub. Parks and gardens after breeding completed.

Feeding: Predominantly seeds, preferring banksia, hakea, grevillea seeds; some flowers and nectar. Often ground feeding on fallen nuts.

Voice: Drawn-out, wailing 'whee-la', often in flight. Harsh screeching.

Status: Endangered.

Breeding: July–Sept. Large old-growth tree hollow, at least 100 years old. Favours Salmon Gum and Wandoo trees / 1–2 eggs.

Kings Park and Porongurup NP, WA.

Left: Male feeding in Red Swamp Banksia. Right: Female.

Baudin's Black Cockatoo

Calyptorhynchus baudinii

Endemic

520–570mm

Similar to Carnaby's Black Cockatoo but has longer and narrower deeply curved black bill. Overall black-brown plumage with fine, pale whitish scalloping. Cheeks: White patch. Tail: Broad, white band towards tip. Eye rings: Pink.
Female: Pale bill, grey eye rings. Larger white cheek patch. Immature: Similar to female.

Gregarious, often form large flocks in non-breeding season.

Habitat: Moist, forested areas dominated by Marri.

Feeding: Bill specialised to extract seed from their diet staple, Marri. Wood-boring insect larvae extracted from trees, also ground feeder on fallen seed or herbs. Feeds quietly.

Voice: Drawn-out 'whee-la' whistle. Harsh screeching.

Status: Vulnerable.

Breeding: Aug.–Nov. Nest in large hollow, favours Karri trees / 2 eggs.

Cape Naturaliste and Porongurup NP, WA.

Opposite: Male (foreground) and female background.

Gang-gang Cockatoo

Callocephalon fimbriatum

Endemic

320–360mm

Small and stocky with a distinctive scarlet head and crest. Overall slate grey plumage with feathers edged in light grey, giving a scalloped effect. Short tail and long broad wings. Female: Grey head with lemon-barred, light grey plumage. Underparts, cheeks: Pinkish barred. Immature: Similar to female, but males have a red-barred crown and short crest.

Gregarious, mostly arboreal. Flight is heavy, wheeling, swooping with deep wingbeats.

Habitat: Cooler, wetter coastal to high-mountain ranges in heavily wooded areas. Parks and gardens.

Feeding: Feeds quietly. Prefers seeds of eucalypts, acacia and native Cypress Pine.

Voice: Distinctive 'creaky-door' raspy screech. Soft 'growling' while feeding.

Status: Common. Sedentary. Vulnerable NSW. Winter migration to lower altitudes. Immature birds are nomadic. Introduced Kangaroo Is.

Breeding: Oct.–Jan. High nest in tree hollow / 2–3 eggs.

Australian National Botanic Gardens, ACT. Barren Ground NR, NSW.

Left: Male with female drinking. Right: Female taking seeds from Coast Wattle.

Major Mitchell's Cockatoo

Cacatua leadbeateri

Endemic

390mm

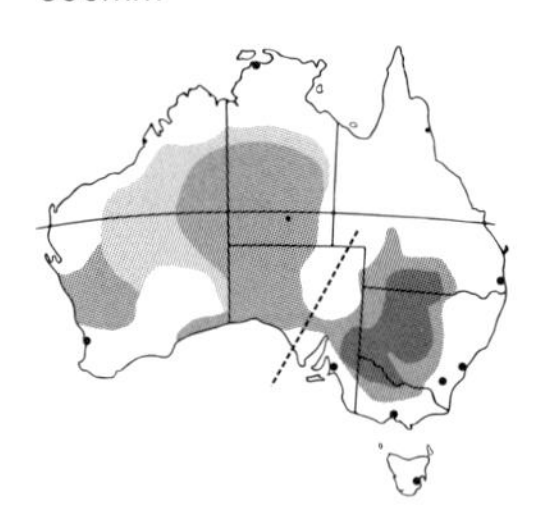

Sexes similar. Spectacular pink and white cockatoo with a distinctive, forward-curved crest with broad white tip above scarlet and yellow bands. Crown, upperparts, tail: White. Head, breast: Pink. Eye ring: White. Underwings: Deep pink with white edges.
Race *mollis,* Pink Cockatoo WA, NT, part SA. Lacks yellow bands in crest.
Female: Less scarlet in crest. Eyes: Red. Immature: Similar to female, duller.

Flight is low with rapid wingbeats, alternating with brief glides. Non-breeding birds form wandering flocks. Bonded pairs remain resident in their large territory.

Habitat: Arid and semi-arid scrublands, lightly timbered grasslands, casuarina stands, mallee and cypress pines.

Feeding: Ground or arboreal feeder, seeds and fruits, nuts, roots and insect larvae.

Voice: Harsh alarm screeching. 2-note quavering screech contact call. Commonly heard in flight.

Status: Generally in decline due to loss of habitat, lack of suitable tree hollows and illegal trapping.

Breeding: May–Dec. Nests in tree hollow 5–20m high / 2–4 eggs.

Wyperfeld NP and Hattah–Kulkyne NP, Vic.

Left: Pair, male (foreground) and female. Right: Feeding on emu-bush fruits.

Galah

Eolophus roseicapilla

Endemic

350mm

Sexes alike. Deep pink face, neck and underparts. Crest, crown: Pale pink. Upperparts: Pale grey. Eyes: Dark brown. Eye ring: Reddish. Bill: Bone coloured. Nominate race WA to SA.
Race *albiceps*, southeast Australia. Crest, crown: Whitish. Eye ring: Grey.
Race *kuhli*, tropical north Qld, NT, WA. Smaller crest, overall paler. Eye ring: Whitish. Hybrids occur where ranges overlap.
Female: Red or light-brown eyes. Immature: Similar, duller. Eyes: Pale brown, whitish eye ring.

Usually seen in flocks of about 30 birds. Flight is swift, wheeling.

Habitat: Woodlands with cypress pines, casuarinas and eucalypts.

Feeding: Usually ground feeders, taking grass seeds, herbs and roots, new leaf growth, saltbush, blossoms, insects and larvae. Spilt grain and cereal crops.

Voice: High-pitched screech, repeated.

Status: Common. Sedentary. Immature nomadic. Vulnerable NSW, Vic.

Breeding: Feb.–July in north. July–Dec. in south. Nests in tree hollows or nest boxes / 2–4 eggs.

Left: Female showing bright pink underparts. Right: Male at nest hollow.

Long-billed Corella

Cacatua tenuirostris

Endemic

380–400mm

Sexes similar. Distinguished by its extremely long, whitish, deeply curved upper mandible. Overall white plumage with a faint yellow tint to underwings and tail. Eye ring: Large, pale blue-grey. Face, breast: Strongly marked orange-red.
Female: Paler with shorter upper bill. Immature: Less red to face and smaller bill.

Flight is swift with shallow wingbeats.

Habitat: Woodlands, open forests and adjacent grasslands. Urban parkland.

Feeding: Ground feeding, corms, seeds, roots and bulbs. Uses long upper mandible for digging up roots and bulbs. Feeds in flocks with 'sentinel' posted to warn of danger.

Voice: Harsh alarm screech. 2-syllable 'yodel' contact call, constant in flight.

Status: Secure, but dependent on large trees with hollows for nesting.

Breeding: Aug.–Dec. Nests in tree hollow in large eucalypt / 2–4 eggs.

Deniliquin area, NSW. Halls Gap area, Vic.

Left: Male. Right: Female feeding on everlasting daisy seeds.

Western Corella

Cacatua pastinator

Endemic

400–450mm

Sexes alike. Similar to the Long-billed Corella but with a smaller bill and larger crest. Overall white plumage with rich yellow underwings and undertail. Lores: Orange-red. Eye ring: Large, blue-grey.
Immature: Similar, shorter bill. Overall washed pale yellow.

Gregarious, in noisy flocks. Strong flight with shallow wingbeats. Often flying high.

Habitat: Wheat belt and remnant eucalypt woodlands dominated by Wandoo, Marri or Jarrah, near water. Grassy woodlands near watercourses.

Feeding: Forages on the ground. Grass seeds. Digs for corms, roots of introduced grasses.

Voice: Shrill alarm shrieks. Chuckling.

Status: Vulnerable.

Breeding: Aug.–Oct. Nests in tree hollow / 1–4 eggs.

Dalwallinu in summer and Perup NR, WA.

Opposite: Perched group of adult birds.

Little Corella

Cacatua sanguinea

350–400mm

Sexes alike. Small white cockatoo with stiff, white, erectile crown feathers. White plumage with yellow wash under wings and tail. Salmon pink between eyes and bill. Eye ring: Blue-grey. Bill: Whitish, short.
Immature: Smaller, similar, shorter bill, pale eye ring.

In breeding season, usually seen in pairs or small flocks; larger flocks form after breeding. Flight is strong with rapid flapping wingbeats.

Habitat: Variable. Grasslands, open woodlands, semi-arid areas, monsoon woodlands, farms, some urban areas. Roosts near water.

Feeding: Mostly ground feeding on grass seeds, other seeds. Digs for corms, legumes and insect larvae.

Voice: Varied shrieks and loud whistles.

Status: Nomadic. Locally common. Often in flocks of several thousand.

Breeding: Opportunistic, commonly June–Oct. Nest in tree hollow / 2–3 eggs.

Left: Little Corella feeding on seedpods of Mount Morgan Wattle. Right: Underwings washed yellow.

Sulphur-crested Cockatoo

Cacatua galerita

480–550mm

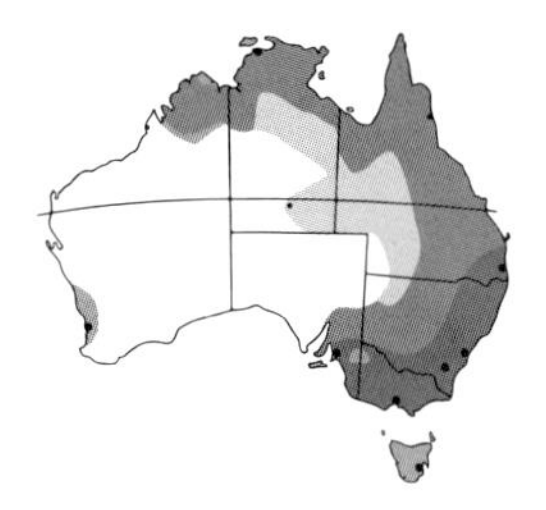

Sexes alike. Familiar, noisy white cockatoo with a distinctive forward-curving bright yellow crest, raised when alarmed. Plumage overall white, washed yellow under wings and tail. Eyes: Dark brown with fleshy, white eye ring. Powerful, slate-black bill.
Female: Red-brown eyes.

Gathers in large flocks outside the breeding season. Noisy and conspicuous. Relatively tame.

Habitat: Most types of timbered country. Forests, woodlands, cultivated land, parks and gardens.

Feeding: Usually in small groups, while a 'sentinel' keeps watch from a nearby perch. Often foraging on the ground. Grass seeds, herbaceous plants, grains, fruits, berries, weed seeds, insects, larvae are taken.

Voice: Harsh, raucous screech.

Status: Locally abundant to common.

Breeding: Aug.–Jan. in south. May–Sept. in north. Nest high in a tree hollow, prefers eucalypts near water / 2 eggs.

Left: Small flock. Right: Exiting the nest hollow.

Cockatiel

Nymphicus hollandicus

Endemic

290–320mm

Smallest member of the cockatoo family. Mostly grey with a dainty, pointed crest and long, tapered tail. Ear patch: Rich orange. Crest, frons, face: Bright lemon. Wing: White patch.
Female, immature: Duller colours. Face: Pale lemon wash. Crest: Grey. Tail: Grey, barred faint lemon.

Usually seen in small flocks. Flight is direct and swift.

Habitat: Drier areas in open or lightly timbered country, usually near water.

Feeding: Forages mostly on the ground in pairs or flocks. Favours seeds of native grasses, herbs, acacia, also berries and fruit.

Voice: Warbling, loud 'queel-queel'.

Status: Seasonally migratory in the south, arriving in spring – leaving in late summer. Nomadic in the north.

Breeding: Aug.–Dec. in south, Apr.–Aug. in north. Tree hollow nest near water / 4–6 eggs.

Left: Male at nest hollow. Right: Female landing.

Scaly-breasted Lorikeet

Trichoglossus chlorolepidotus

Endemic

230–240mm

Sexes alike. Only all-green-headed lorikeet. Overall emerald green plumage. Breast: Green feathers edged with yellow giving a scale-like effect. Underwings: Spectacular orange-red. Bill, eyes: Red.
Immature: Duller. Shorter tail, brown bill and eyes.

Associates with other lorikeets. Noisy and conspicuous. Swift and direct flight.

Habitat: Open timbered areas with flowering trees and shrubs. Parks and gardens.

Feeding: Arboreal feeder on nectar and pollen. Also insects, fruits and seeds.

Voice: More shrill than the Rainbow Lorikeet. Constant chatter when feeding.

Status: Common, more common in the south. Nomadic, following flowering honey-flora.

Breeding: Aug.–Jan. in south, earlier in north. Nest in deep, high tree hollow / 2 eggs.

Left: Pair leaving nest hollow. Right: Taking nectar from Black Wattle showing patch of underwing colour.

Rainbow Lorikeet

Trichoglossus moluccanus

300mm

Sexes similar. Brightly coloured arboreal feeder with a bright blue head, bright green upperparts with a lime green half-collar, and bright green tail. Breast: Red-yellow. Belly: Blue. Female: Smaller with shorter bill. Immature: Similar to female, duller.

Usually in pairs or flocks. Flight is fast, straight or wheeling. Although brilliantly plumed, well camouflaged while feeding but easily located by noisy chatter and falling blossoms.

Habitat: Open forests, heathlands, rainforests, parks and gardens.

Feeding: Arboreal feeders with a brush-tipped tongue for mopping up nectar and pollen from eucalypt blossoms and other honey-flora. Also insects, fruits and seeds.

Voice: Metallic screeching in flight. High-pitched chatter when feeding.

Status: Abundant to common. Vagrant in Tas. Introduced to Perth.

Breeding: Aug.–Jan. Tree hollow nest near water, or nest box / 2 eggs.

Left: Feeding in 'Peaches and Cream' Grevillea. Right: Breeding pair at nest hollow.

Red-collared Lorikeet

Trichoglossus rubritorquis

300mm

Sexes similar. Similar to the Rainbow Lorikeet but with an orange-red collar over the nape. Previously thought a subspecies of the Rainbow Lorikeet.

Female: Paler belly.

Habitat: Open eucalypt forests, coastal heathlands. Parks and gardens.

Feeding: Arboreal feeder on nectar and pollen, mainly from eucalypts. Also insects, fruits and seeds.

Voice: Metallic screeching in flight. High-pitched chatter when feeding.

Status: Abundant to common. Nomadic, following the availability of flowering trees and shrubs.

Breeding: Sept.–Oct. through to Feb.–Mar., corresponding with the wet season and the abundance of food. Nest in a tree hollow, near water / 2 eggs.

Left: Feeding in Darwin Woollybutt. Right: Male showing underparts.

Varied Lorikeet

Psitteuteles versicolor

Endemic

190–230mm

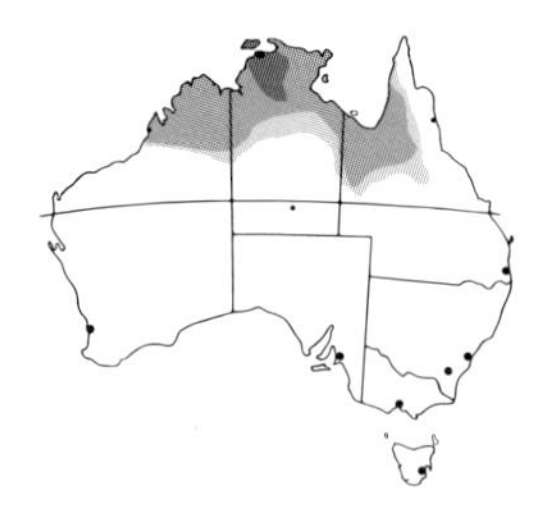

Small, tropical, softly coloured lorikeet. Overall pale green with finely streaked yellow-green plumage. Eye ring: White. Crown, frons: Bright red. Breast: Streaked pale pink. Underwings: Yellowish-green.
Female: Duller. Crown: less green. Immature: Similar to female, duller. Bill: Brown.

Usually in large, noisy flocks. Fast, direct flight with an audible whirring as they pass overhead.

Habitat: Dense monsoonal woodlands and melaleuca swamps. Parks and gardens.

Feeding: Arboreal foraging in blossoms for nectar and pollen. Also insects, fruits and seeds. Aggressive towards other species when feeding.

Voice: High-pitched screech. Constant chatter when feeding.

Status: Common. Nomadic, following flowering trees.

Breeding: Opportunistic. Usually Apr.–Aug. Hollow tree limb nest near water / 2 eggs.

Mt Isa, Qld. Kakadu NP, NT.

Left: Male feeding in Red Bloodwood. Right: Female, showing olive-green crown.

Musk Lorikeet

Glossopsitta concinna

Endemic

200–230mm

Sexes similar. Medium-sized, sturdy lorikeet with mostly green plumage and a blue crown. Frons: Red band, extending behind the eyes. Mantle: Bronze-brown. Breast: Yellow on sides, green below.
Female: Crown is less blue.
Immature: Dark bill.

Usually associates with other lorikeets. Flight is swift, direct, with whirring wings.

Habitat: Open forests with flowering eucalypts and banksias, farmland, urban parks and gardens.

Feeding: Arboreal feeder, preferring the upper canopy. Main diet is nectar, pollen, also lerp and scale.

Voice: Metallic shrieks in flight, constant chatter when feeding.

Status: Common, but in decline. Nomadic, following flowering trees.

Breeding: Opportunistic. Commonly Aug.–Jan. Tree hollow nest / 2 eggs.

Chiltern–Mt Pilot NP and You Yangs RP, Vic.

Left: Adult feeding in Lemon-scented Gum. Right: Feeding on Red-flowered Yellow Gum.

Little Lorikeet

Glossopsitta pusilla

Endemic

160–180mm

Sexes alike. Small lorikeet with bright green plumage. Face: Red. Bill: Black. Eyes: Yellow. Mantle: Bronze-brown. Underwings: Yellow.
Immature: Less red on face with brown bill and eyes.

Usually high in the canopy. Flight is swift, direct and high, with rapid whirring wingbeats and constant calling.

Habitat: Open forests, woodlands with tall trees, usually along watercourses. Also mountain eucalypt forests, parks and gardens.

Feeding: Arboreal, foraging in the upper canopy in small flocks, mostly for pollen and nectar and some fruit.

Voice: Noisy chatter when feeding. High-pitched screech in flight.

Status: Sporadically distributed. Generally nomadic.

Breeding: July–Jan. Tree hollow nest near water / 3–5 eggs.

Chiltern–Mt Pilot NP and You Yangs RP, Vic.

Left: Leaving the nest hollow. Right: Adult underparts.

Purple-crowned Lorikeet

Glossopsitta porphyrocephala

Endemic

170–185mm

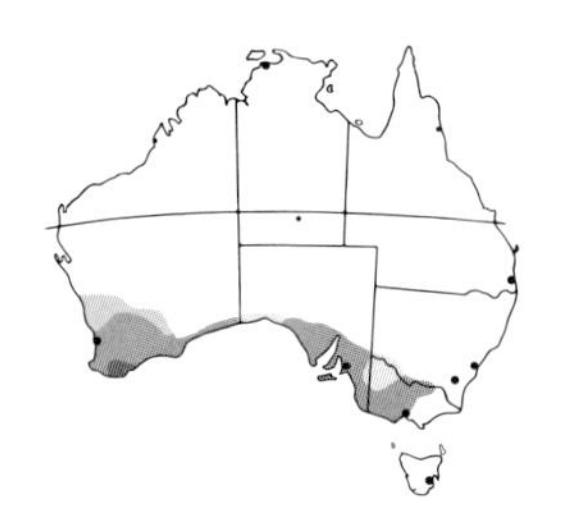

Sexes alike. Tiny, well-camouflaged and more often heard than seen lorikeet, with a deep purple crown. Frons, ear coverts: Orange-red. Upperparts: Bright green. Underparts: Greenish yellow. Belly: Pale blue. Underwings: Crimson. Bill: Small, black.
Immature: Lacks purple crown, duller.

Usually found in small groups, often with little and musk lorikeets. Fast straight flight.

Habitat: Prefers open forests with large-flowered eucalypts; also mallee, dry forests and coastal parks and gardens. Only endemic lorikeet in southwestern WA.

Feeding: Arboreal foraging for nectar and pollen. Fruit, berries, insects and larvae also taken.

Voice: Loud 'tsit-tsit-tsit'. Twittering when feeding.

Status: Common. Less common NSW. Nomadic, following flowering events, sometimes wandering further inland.

Breeding: Aug.–Dec. Tree hollow nest often near water / 3–4 eggs.

Left: Feeding in banksia tree. Right: Underparts.

Double-eyed Fig-parrot

Cyclopsitta diophthalma

130–150mm

Smallest Australian parrot. Three isolated populations. Overall bright green. Underwing: Wide bands of pale yellow. Tail: extremely short.
Immature: Duller, less red patches.
Race *marshalli*, Marshall's Fig-parrot, far north Cape York. Face: Larger red frons and cheek patches, edged finely with blue around eyes. Female: Blue frons, pale cheek patches.
Race *macleayana*, Macleay's Fig-parrot Cairns region, north Qld. Face: Red frons. Red cheek patch edged blue. Female: Red frons, paler blue around eyes and pale cheek patches.
Race *coxeni*, Coxen's Fig-parrot, NSW and Qld border region. Palest face markings. Frons: Blue. Smallest red cheek patches edged blue.

Often found with Musk and Little Lorikeets. Flight is high and rapid.

Habitat: Lowland tropical rainforests with fruit-bearing trees.

Feeding: Silently forages high in the canopy, in small flocks or pairs. Seed eaters, taking the kernels of figs or other rainforest fruit. Also nectar, insect larvae.

Voice: High-pitched 'tseet'.

Status: Uncommon. Race *coxeni* is critically endangered due mainly to land clearing. To help preserve them, plant *Ficus* spp. as well as *Syzygium*, *Elaeocarpus*, *Litsea* spp. and Silky Oak.

Breeding: Aug.–Oct. Nests in tree hollows / 2 eggs.

Race *macleayana,* Cairns Esplanade and Centenary Lakes, Qld. Race *marshalli,* Iron Range NP, Qld.

Top left: Race *marshalli*, male. Top right: Race *marshalli*, female.

Centre left: Race *macleayana*, male, feeding on native fig. Centre right: Race *macleayana*, female.

Eclectus Parrot

Eclectus roratus

420–480mm

Unusually, the female is the most brilliantly coloured. The male is overall bright emerald green with scarlet underwings and side of belly. Bill: Large, orange. Female is overall scarlet with blue belly, blue eye rings and black bill. Immature: Similar to adults. Bill: Brown.

Usually seen in pairs or small flocks. Flight is rapid with slow, shallow wingbeats.

Habitat: Forest canopy of tropical rainforests.

Feeding: Berries, fruits, seed, blossoms and nectar taken from the canopy.

Voice: Strident raucous cries echo through rainforest. Male call resembles 'quork'. Female has a harsh screeching whistle. Also a variety of bell-like sounds and chuckling.

Status: Uncommon. Sedentary.

Breeding: July–Feb. High up in a large tree hollow / 2 eggs.

Iron Range NP, Qld.

Left: Male. Right: Female at nest hollow.

Red-cheeked Parrot

Geoffroyus geoffroyi

220–250mm

Stocky, short-tailed bright green parrot with a rose-red face and bluish violet crown. Bill: Red above, grey-brown lower bill. Eyes: Pale yellow. Underwings: Blue.
Female: Brownish head and grey bill. Immature: Green head. Yellow bill.

Usually seen in pairs or family groups. Swift, direct flight.

Habitat: Dense tropical rainforests, usually along watercourses. Occasionally adjacent woodlands.

Feeding: Arboreal, noisy foragers for seeds, nuts, blossoms, nectar and fruit, preferring figs.

Voice: Loud, metallic chattering and screeching.

Status: Sedentary. Uncommon. Also in PNG and adjacent islands.

Breeding: Aug.–Dec. Excavates nest cavity in a high rotting tree limb / 3 eggs.

Iron Range NP, Qld.

Left: Female on left and male on right. Right: Male.

Australian King Parrot

Alisterus scapularis

Endemic

400–450mm

Only endemic parrot with a red head. Upper bill: Red, tipped black. Eyes: Bright yellow. Back, wings: Dark green. Breast, belly: Red. Lower back, narrow nape band: Deep blue. Tail: Blue-black.
Female: Mostly green head, breast and upperparts. Belly, underparts: Red. Bill: Brownish red. Immature: Brownish bill, chest and throat.

Usually in pairs or family groups. Distinctive, rapid, undulating flight.

Habitat: Heavily timbered mountain forests and rainforests, parks, gardens. Migrates to coastal plains in winter. Returns to highlands to breed.

Feeding: Usually in outer foliage, taking native fruit, nuts, nectar and blossom. Sometimes ground feeding on fallen fruit and seed. Orchard pest where native fruit is removed.

Voice: Male has 3–4 musical notes. Shrill flight call.

Status: Common.

Breeding: Oct.–Jan. Nests in deep hollow in a tall tree / 5 eggs.

Left: Male taking the fruit of Coast Beard-heath. Right: Female.

Red-winged Parrot

Aprosmictus erythropterus

320mm

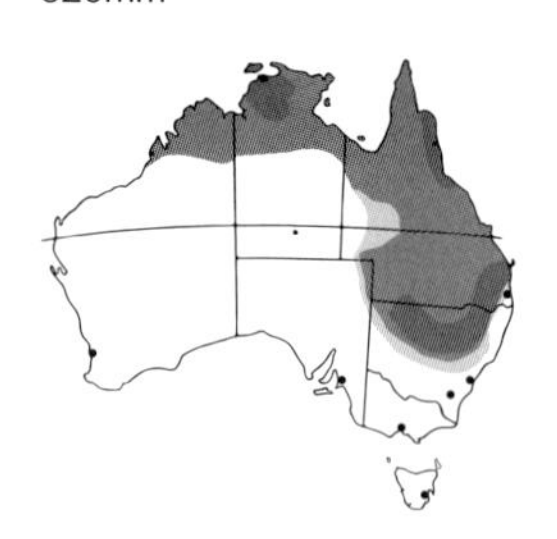

Bright-green with vivid, large, scarlet shoulder patch. Rump: Blue, usually hidden. Tail: Green, edged and tipped yellow. Eyes, bill: Orange.
Female and immature: Yellowish-green plumage and smaller wing patch. Rump, lower back: Pale-blue. Eyes: Dark.

Arboreal, in pairs or small groups, usually only coming to ground to drink. Flight is erratic with long full strokes, pausing on each stroke.

Habitat: Variable. Subtropical, arid inland, eucalyptus woodland along watercourses, acacia scrub and mangroves.

Feeding: Arboreal, taking seeds, nectar, pollen, flowers, fruits including mistletoe. Also insects and their larvae.

Voice: Soft chatter when feeding. Shrill, metallic 'crillik-crillik' call.

Status: Conspicuous and common.

Breeding: May–Feb. in north. Aug.– Feb. in south. Deep hollow in a tall tree, usually near water / 4–6 eggs.

Left: Male (foreground) and female feeding on the fruit of Sandpaper Fig.

Superb Parrot

Polytelis swainsonii

Endemic

400mm

Graceful, slender, mostly bright green parrot with a long, dark-green tail. Frons, crown, cheeks, neck: Bright yellow with broad, bright red throat band. Bill: Reddish brown. Eyes: Red. Female: Pale blue-green face, tinged pink. Immature: Similar to female, brown eyes.

Flight is swift and effortless.

Habitat: Inland slopes of Great Dividing Range, Murray–Murrumbidgee valley, adjacent plains and floodplains. Much of its preferred habitat is listed as endangered.

Feeding: Forages on a wide variety of plants from ground level to tree canopy. Lerp taken in winter. Never far from water.

Voice: Deep, throaty warbles and twitter.

Status: Vulnerable due to habitat loss and illegal trapping.

Breeding: Sept.–Dec. Nests in high tree hollow / 4–6 eggs.

Deniliquin area, NSW.

Left: Male feeding on native grass seeds. Right: Female.

Regent Parrot

Polytelis anthopeplus

Endemic

400–420mm

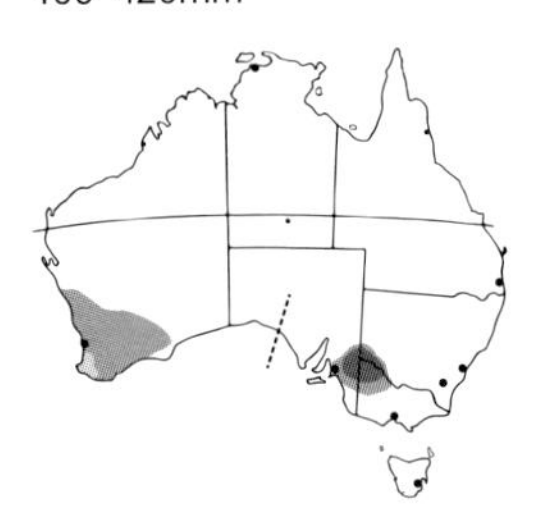

Two isolated races separated by the Nullarbor Plain.
Western nominate race: Duller in comparison. Head, underparts, rump: Olive-yellow.
Eastern race *monarchoides*: Bright yellow head, underparts and rump: Back: Olive-yellow. Wings: Yellow with lower broad, red band. Tail: Long, tapering, dark blue-grey. Bill, eyes: Orange.
Female, immature: Duller olive green, pinkish wing patch.

Flight is swift and graceful, sometimes erratic.

Habitat: Eastern race: Semi-arid interior, mostly associated with the Murray River system, River Red Gum and mallee woodlands. Western race: Open forests and woodlands with Salmon Gums.

Feeding: Ground foraging, usually in pairs or small groups, on seeds of native grasses, also fruits, buds, flowers, insect larvae, psyllids, lerp. Drinks before feeding and before roosting.

Voice: Rolling warble or harsh 'carrak-carrak' in flight.

Status: Vulnerable. More common in WA. Seasonal migration, moving inland after good rain.

Breeding: Aug.–Jan. Hollow branch / 3–5 eggs.

Along Murray River. Hattah–Kulkyne NP, Vic. Dryandra SF, WA.

Opposite: Race *monarchoides*, male, extracting seeds from acacia seedpods.

Princess Parrot

Polytelis alexandrae

Endemic

400–450mm

Sexes similar. Mid-sized parrot with an extremely long, slim, blue-green tail. Pale olive-green plumage with pale blue-grey head and nape. Chin, throat, upper breast: Pale rose-pink. Shoulders: Lime-green patches. Rump: Violet-blue. Tail: Long, tapered olive-green. Thighs: Green, rose pink. Eyes: Bright orange-red. Bill: Orange-pink. Female, immature: Duller, shorter tail.

Usually found in pairs or small flocks. Flight is undulating with irregular wingbeats, with long, streaming tail.

Habitat: Arid sand dunes, scrubland, mulga and trees along watercourses.

Feeding: Small flocks foraging mostly on the ground for seeds. Blossoms and nectar also taken.

Voice: Usually silent. Rolling, strident notes in flight. Twittering.

Status: Highly nomadic. Rare. Vulnerable.

Breeding: Sept.–Jan. Nests in tree hollow, usually a eucalypt, near water / 3–6 eggs.

Canning Stock Route, WA.

Left: Male taking flight. Right: Male.

Green Rosella

Platycercus caledonicus

Endemic

290–360mm

Sexes similar. Only parrot species confined to Tas. Head, upperparts: Bright yellow. Frons: Red band. Cheeks, throat, shoulder patch, underwings: Blue. Upperparts: Dark green with black mottling. Tail: Green and blue. Undertail coverts: Lightly mottled red.
Female: Similar, duller, shorter tail. Immature: Mostly green. Head, underparts: Yellow-green.

Active, noisy. Small foraging flocks form after breeding.

Habitat: Confined to Tas. and offshore islands. Wooded areas and parks and gardens on fringes of towns.

Feeding: Usually quietly feeds early morning and late afternoon. Favours fruit and seeds and nectar from eucalypt blossom. Occasionally raids orchards.

Voice: Loud, piercing flight notes. Softer whistled notes.

Status: Common.

Breeding: Sept.–Jan. Nests in high tree hollow / 4–5 eggs.

Left: Upperparts. Right: Male feeding on flower buds.

Crimson Rosella

Platycercus elegans

Endemic

350–380mm

Sexes similar. Several colour forms across its range.
Nominate race: Eastern and southeastern Australia. Mostly bright crimson with bright blue cheek patches. Back: Black feathers with crimson margins. Underwing: Blue. Tail: Blue. Eyes: Brown. Bill: Bone.
Female: Duller. Immature: Mostly dark green with blue cheek patches. Crown, throat, breast vent: Red.
Race *nigrescens*, northern Qld. Smaller, darker.
Race *flaveolus*, Yellow Rosella, inland rivers, primarily Murray and Murrumbidgee Rivers. Prefers River Red Gums for nesting. Mostly yellow with red frontal band and blue cheek patches. Sexes similar. Immature: Olive-green back and rump and yellow-green underparts.
Races *subadelaidae* and *fleurieuensis*, collectively called Adelaide Rosella, Mount Lofty Ranges and adjacent region including suburban Adelaide. Head, neck, underparts: Orange-yellow. Upperparts: Feathers mostly black but broadly edged yellow. Immature: Mostly dark green with red markings.
Race *subadelaidae* found in far north SA is overall more yellow and paler than race *fleurieuensis*.

Habitat: Coastal and mountain forests and nearby parks, gardens and farmland. Favours tall eucalypt forests.

Feeding: Often ground feeding, foraging for fallen seed, preferably eucalypt and grass seeds. Nectar, fruit and scale insects also taken.

Voice: Shrill piping whistles.

Status: Common.

Breeding: Sept.–Jan. Nests in hollow, preferably high in tall eucalypt or nest box / 4–8 eggs.

Yellow Rosella common along Murray River, SA.

Centre left: Immature, feeding in Rock Correa by extracting nectar from base of snipped-off flower. Centre right: Male at nest hollow in Brittle Gum.

Below left: Race *flaveolus*, Yellow Rosella. Below right: Race *fleurieuensis* Adelaide Rosella.

Eastern Rosella

Platycercus eximius

Endemic

350–380mm

Sexes similar. Vividly coloured parrot with a bright red head and breast with prominent white cheek patches. Lower breast: Yellow. Belly: Lime green. Undertail coverts: Red. Back, wings: Greenish yellow with black 'scale-like' pattern. Female: Patchy red-yellow nape. Immature: Duller.

Despite its vivid colours it is surprisingly well camouflaged when among foliage or on the ground feeding.

Habitat: Open eucalypt woodlands and grasslands. Urban parks, gardens and farmland.

Feeding: Mostly a quiet ground feeder on grass and fallen seeds. Also forages, frequently in eucalypts for insects, seeds, fruits and nectar.

Voice: Musical 'pee-pity, pee-pity' and metallic 'kwink-kwink-kwink' in flight.

Status: Common. Sedentary.

Breeding: Aug.–Jan. Nests in a hollow, usually a eucalypt or nest box / 4–8 eggs.

Left: Female and male bathing. Right: Male.

Pale-headed Rosella

Platycercus adscitus

Endemic

280–340mm

Sexes alike. Head, nape: White with a yellow tint. Cheek: White patch. Lower breast, belly: Blue. Undertail coverts: Red. Back, wings: Yellow with black 'scale-like' pattern. Underwings: Blue. Birds north of Cairns have violet cheek patch.

Interbreeds with the Eastern Rosella. Hybrids occur where their ranges overlap. Usually silent when feeding, noisy chattering at other times.

Habitat: Lightly timbered woodlands, grasslands, parks, gardens and farmland.

Feeding: Similar to Eastern Rosella.

Voice: Mostly a quiet ground feeder.

Status: Sedentary. Common, particularly along watercourses.

Breeding: Feb.–June in north. Sept.–Dec. in south. Nests in tree hollow / 4–8 eggs.

Canungra, Qld.

Left: Female and male at birdbath. Right: Male extracting the seeds from acacia seedpods.

Northern Rosella

Platycercus venustus

Endemic

290–320mm

Sexes alike. Only rosella found in far northwestern Australia. Predominantly pale yellow with a distinctive black cap and conspicuous white cheek patches that are partly blue in west of range. Back: Yellow with black 'scale-like' pattern. Underparts: Yellow with faint black scalloping. Undertail coverts: Scarlet. Tail: Blue and green, tipped with white. Immature: Duller. Head, breast: Flecked red.

Usually found alone or in pairs. Occasionally in small groups.

Habitat: Savannah woodlands and littoral forests.

Feeding: Usually a quiet ground forager on grass seeds. Fruit, nuts and insects are also taken.

Voice: 2-note metallic call.

Status: Uncommon.

Breeding: June–Aug. Nests in tree hollow / 4–5 eggs.

Koolan Is. and Kununurra, WA. Caranbirini CR and Gunlom Falls, Kakadu NP, NT.

Left: Side view showing black cap. Right: Underparts showing black scale pattern.

Western Rosella

Platycercus icterotis

Endemic

250–300mm

Smallest and quietest rosella. Head, upperparts: Bright red. Cheeks: Distinctive yellow patches. Back, wings: Dark green with black 'scale-like' pattern. Shoulder, underwings: Deep blue. Rump: Green. Tail: Green with blue margins. Female: Similar. Head, breast: Mottled red and green. Frons: Red. Cheek patches: Duller. Belly: Duller red and green. Immature: Lacks cheek patches and red underparts.

Usually in pairs or small groups.

Habitat: Open eucalypt forests, sparse woodlands, cultivated land and orchards.

Feeding: Usually ground foraging on native grass seed. Also acacia seed, other seeds, nuts, fruit, blossom and insects and their larvae.

Voice: Soft musical whistles. High-pitched ringing 'tring-see, tring-see'.

Status: Moderately common. Mostly sedentary.

Breeding: Aug.–Dec. Tree hollow / 4–5 eggs.

Wungong Gorge Walk, Bungendore Park and Dryandra SF, WA.

Left: Female at nest hollow. Right: Male.

Australian Ringneck

Barnardius zonarius

Endemic

320–440mm

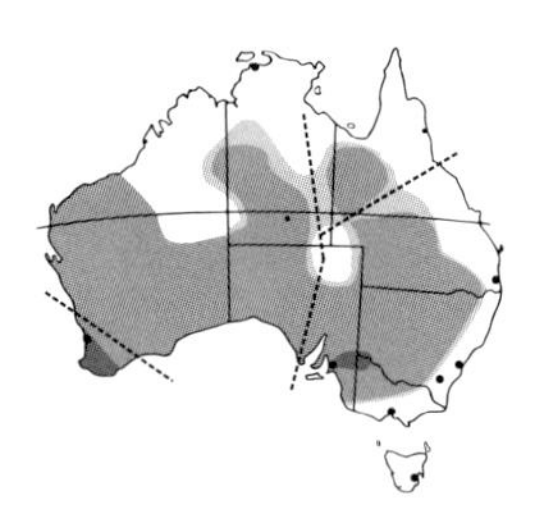

Large parrot with size and colour variations across its range. Nominate, Port Lincoln Parrot. NT, SA, WA, excluding southwest WA. Bright green with black head and dark blue cheeks with a narrow yellow collar. Belly: Yellow. Tail: Green, tipped with blue. Sexes similar. Female, duller, browner cap.
Race *barnardi*, Mallee Ringneck. Qld, NSW, Vic., to SA. Deep blue back and mantle with green head. Frons: Red band. Lower breast: Orange-yellow. Collar: Yellow. Favours tall wet eucalypt forest.
Race *macgillivrayi*, Cloncurry Parrot, northern inland Qld. Lacks red frontal band.
Race *semitorquatus*, Twenty-eight Parrot, southwestern WA. Larger. Belly: Green. Frontal band: Red. Female: Dull head. Underwing: Light bar. Underparts: Pale yellow.

Habitat: Mostly dry eucalypt woodlands and mallee, near water.

Feeding: Forages for seeds of grasses, fallen acacias, cypress pines. Also nectar and pollen, preferably from eucalypt blossoms. Insects and their larvae also taken.

Voice: *B. zonarius*: High-pitched 'kwink'.
Race *barnardi*: High-pitched ringing 'kling'.
Race *macgillivrayi*: Ringing 'kling'.
Race *semitorquatus*: Double-noted 'twenty-eight' call.

Status: Common.

Breeding: Nominate race and race *semitorquatus*: Aug.–Feb. Interbreeding occurs where their ranges overlap.
Race *barnardi*: Aug.–Jan.
Race *macgillivrayi*: Feb.–June.
Nests in tree hollows / 4–5 eggs.

Centre left: Nominate race, Port Lincoln Parrot, lacks red frontal band. Centre right: Race *semitorquatus*, Twenty-eight Parrot.

Below left: Race *macgillivrayi*, Cloncurry Parrot. Below right: Race *barnardi*, Mallee Ringneck.

Greater Blue Bonnet

Northiella haematogaster

Endemic

280–380mm

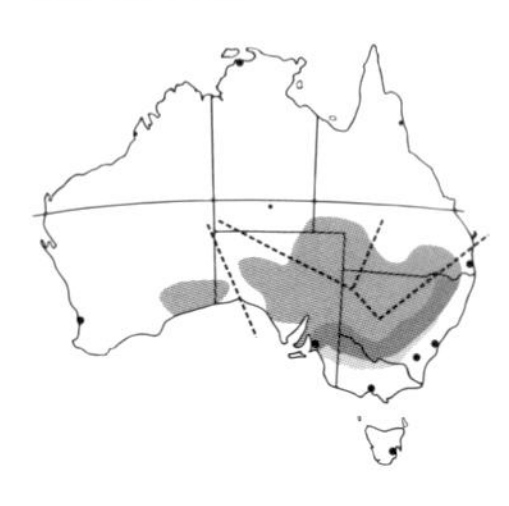

Sexes alike, all races. Distinctive blue frons and face. Variations in size and colour across its range.
Three races. Nominate race: Eastern and southern NSW, Vic. SA: Mostly brown with a blue face, red belly and yellow vent. Upperwing coverts: Olive-yellow.

Race *haematorrhous*, southern Qld, NSW. Similar to nominate with red undertail coverts.

Race *pallescens*, Lake Eyre Basin region, northern SA, Qld. Similar to nominate, but paler with red undertail coverts.

Habitat: Open and semi-arid woodlands.

Feeding: Ground foraging for seeds, mostly acacia and saltbush.

Voice: Harsh 'chuck', penetrating whistle.

Status: Common.

Breeding: Aug.–Dec. Nests in deep tree hollow 7m high or more / 4–7 eggs.

Wyperfield NP, Vic. Port Augusta area, SA.

Left: Nominate race.

Naretha Blue Bonnet

Northiella narethae

Endemic

280–380mm

Sexes similiar. Similar to Greater Blue Bonnet but smaller, with turquoise frons and red undertail coverts.

Habitat: Vegetated fringe of Nullabor Plain, WA, SA. (See map for Greater Blue Bonnet.)

A molecular study in 2015 led to reinstatement as a separate species.

Eyre Bird Observatory, WA.

Above Right: Naretha Blue Bonnet.

Red-capped Parrot

Purpureicephalus spurius

Endemic

280–390mm

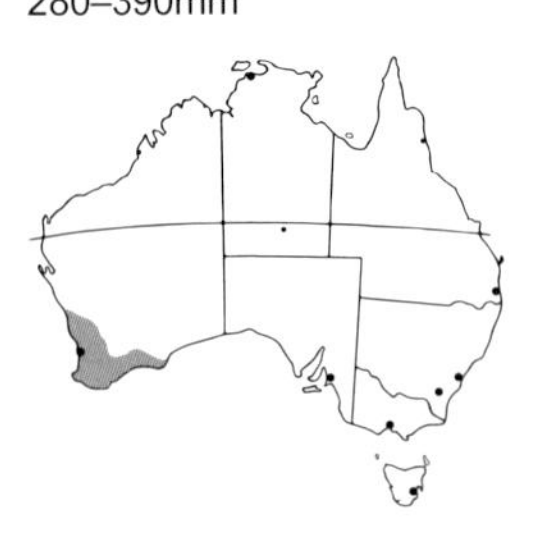

Sexes similar. Deep crimson crown and frons. Cheeks and rump: Bright yellow-green. Breast, belly: Purple. Lower flanks and thighs: Red. Upperparts, tail: Dark green. Under wings: Blue. Eyes: Dark brown.
Female: Duller with red crown flecked green. Immature: Mostly green with light cheek patches.

Undulating rosella-like flight.

Habitat: Marri and Jarrah forests and woodlands. Parks and gardens. Sometimes orchards.

Feeding: Feeds quietly, almost exclusively in Marri trees, foraging mostly for seeds, also nectar and some fruit.

Voice: Metallic shriek in flight.

Status: Sedentary. Usually common.

Breeding: Aug.–Dec. High tree hollow / 4–6 eggs.

Bungendore SF, WA.

Left: Immature feeding on manuka fruits. Right: Male.

Swift Parrot

Lathamus discolor

Endemic

Critically Endangered
250mm

Sexes alike. Slender-tailed, lorikeet-like parrot, mostly bright green with bright bluish purple crown. Frons, throat: Bright red, bordered with narrow yellow band. Underwing coverts, undertail coverts: Distinctive bright red. Tail: Long, pointed and lorikeet-like.
Immature: Duller.

Often seen with other lorikeets.

Habitat: Dry sclerophyll eucalypt forests and woodlands. Parks and gardens. Relies on old-growth Tasmanian Blue Gums, for food, shelter and nesting sites.

Feeding: Brush-like tongue similar to lorikeets. Mostly arboreal, foraging primarily for nectar, but also seeds, fruit, psyllids and lerp.

Voice: Chattering when feeding. Metallic 'clink-clink' flight call.

Status: Uncommon. Endangered. Migrates from Tas. to the mainland after breeding.

Breeding: Breeds exclusively in Tas. Sept.–Dec. coinciding with flowering of Tasmanian Blue Gums. Nest is high in a tree hollow, usually eucalypt / 3–5 eggs.

Bruny Is. and Tas. parks, particularly in the south in spring and summer.

Opposite: Feeding in Swamp Mahogany.

Red-rumped Parrot

Psephotus haematonotus

Endemic

240–270mm

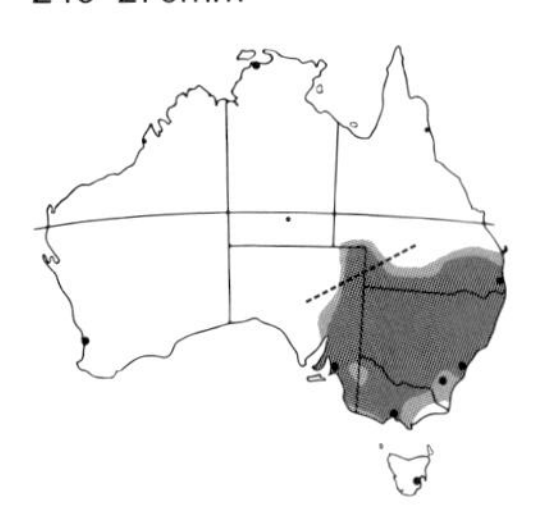

Mid-sized, slender-bodied parrot, with an emerald-green head and breast and a bright red rump. Upperparts: Blue-green, with bright yellow shoulder patch and blue-green wings. Tail: Long, tapered, blue and green, tipped white. Underwings: Blue. Race *caeruleus*, Lake Eyre Basin region. Paler. Female: Dull olive-green. Rump: Green. Shoulder patch: Pale blue. Immature: Duller.

Mostly seen in small flocks or pairs, foraging on the ground. Flutters or scurries from one patch of shade to another. Flies quickly into trees when disturbed.

Habitat: Lightly timbered grasslands, mallee, urban parks, gardens and farmland.

Feeding: Native grass seeds, leaves, flower buds and fruits.

Voice: Musical warble. Shrill whistle.

Status: Moderately common. Congregates in flocks in winter. Sedentary.

Breeding: Aug.–Dec. Nests in a hollow limb, usually a eucalypt, often near water, or nest box / 3–7 eggs.

Opposite: Male (foreground) and female.

Mulga Parrot

Psephotus varius

Endemic

270–320mm

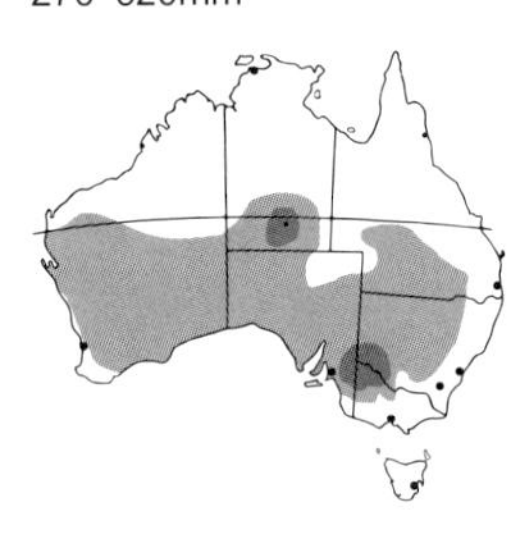

Mostly bright bluish-green with yellow frons, red patch on crown and yellow shoulder patch. Belly: Yellow marked with red. Underwings: Blue. Tail: Green and blue, tipped white. Female: Overall olive-green with dull red neck patch. Frons: Orange-yellow. Breast: Brownish green. Belly: Green. Immature: Duller.

Usually seen on ground, either solitary, in pairs or small flocks.

Habitat: Lightly timbered mallee and mulga woods.

Feeding: Mostly quietly on the ground in the early morning or late afternoon, taking grass seeds, herbaceous plants and berries.

Voice: Mellow 'jeep-jeep-jeep' whistle. Mostly silent.

Status: Sedentary. Wanders locally in search of food. Moderately common, but in decline due to habitat loss.

Breeding: July–Dec., or after good rain. Tree hollow nest / 4–6 eggs.

Opposite: Male (on left) and female.

Golden-shouldered Parrot

Psephotus chrysopterygius

Endemic Endangered

230–280mm

Rare, slender, upright parrot. Mostly blue with yellow shoulder patch. Crown: Black. Frons: Yellow. Back, shoulders: Green-brown. Lower belly, vent: Pinkish red. Female: Mostly light yellow-green with bronze crown and nape. Belly to vent: Faint blue, mottled faint red-orange.

Habitat: Well-grassed tropical savannah woodlands with termite mounds.

Feeding: Ground foraging, preferring native grass seed and herbaceous plants.

Voice: Soft whistled 'peep'.

Status: Rare and endangered due to loss of habitat and large-scale illegal trapping. Once found over most of Cape York, now restricted to 2 small areas.

Breeding: Sept.–Apr. Excavates a 40cm long, north–south tunnel in termite mound with end nesting chamber / 3–5 eggs.

Rare at Staaten River NP, Qld.

Opposite: Male (foreground) and female.

Hooded Parrot

Psephotus dissimilis

Endemic

260–280mm

Similar to the closely related Golden-shouldered Parrot, it is differentiated by geographic location, a black crown that extends to below the eyes and a larger yellow shoulder patch. Undertail coverts: Bright scarlet-orange mottled yellow.
Female: Dull olive, lacks black crown. Frons, face: Pale blue. Undertail coverts: Pale pinkish orange, mottled yellow.
Immature: Similar to female with yellowish bill. Males have darker head.

Habitat: Open, dry, eucalypt forests and grasslands with termite mounds.

Feeding: Native grass seeds and herbaceous plants. Similar to Golden-shouldered Parrot.

Voice: Sharp 'chissic-chissic'.

Status: Uncommon. Sedentary. Wanders locally.

Breeding: Apr.–July. Excavated nest in termite mound / 3–6 eggs.

Around Pine Creek and Katherine, NT.

Left: Male. Right: Female.

Budgerigar

Melopsittacus undulatus

Endemic

180mm

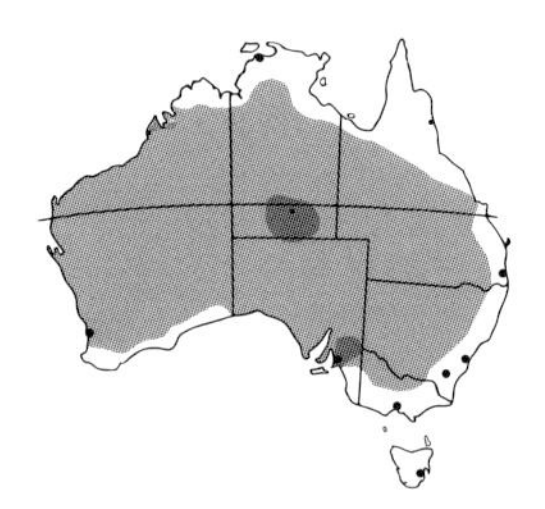

Sexes similar. Small, long-tailed, mostly bright green and yellow parrot with upperparts barred black and yellow. Face: Yellow. Cheek: Feathers tipped violet-blue. Eyes: White or yellow. Cere: Dark blue.
Female: Light blue to white cere, turning brown when breeding.
Immature: Duller, brown eyes.

Widespread, often seen in large, densely packed flocks. In the morning often congregate at water troughs and tanks.

Habitat: Widespread. Arid and semi-arid woodlands, scrubland, mallee, mulga and spinifex deserts, near water.

Feeding: Ground foragers, usually in flocks of up to 100 birds, feeding almost entirely on seeds of native grasses and herbs, saltbush. Dry seed diet provides little moisture and birds need to be near water to drink each day.

Voice: Animated sharp chatter when feeding. Warbled 'chirrip' flight call.

Status: Common. Highly nomadic.

Breeding: Aug.–Jan. in south. June–Sept. in north. Prolific breeding when conditions are favourable. Nests in a hollow in a tree, log or timber post.

Left: Adult male. Right: Male feeding female in the nest hollow.

Bourke's Parrot

Neopsephotus bourkii

Endemic

220–230mm

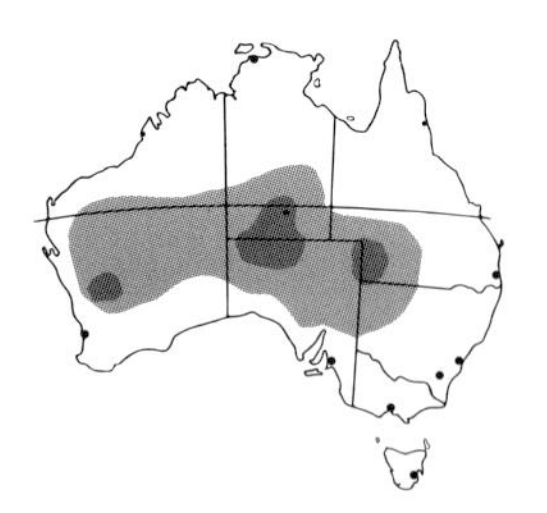

Small, mostly pinkish brown parrot with blue frons and white around eyes and face. Underparts: Salmon-pink belly and pink-scalloped patterned breast. Flanks, rump, undertail: Pale blue. Wings: Yellow-white scalloping and blue patch on wing bend.
Female: Similar. Less or no blue on frons. Immature: Similar to female, less pink on breast.

Semi-nocturnal, active after dusk and before dawn, sometimes at night. Drinks at dawn and dusk. Flutters to cover if disturbed

Habitat: Mulga and acacia scrub in arid and semi-arid areas, near water.

Feeding: Small groups of 4–6. Ground foraging for grass seeds, legume seeds and herbs.

Voice: Soft 'chu-wee' in flight. Twittering when feeding.

Status: Common to uncommon. Sedentary and nomadic. Population declining in areas with rabbit infestations and overgrazing.

Breeding: Aug.–Dec. or after good rain. Nests in small tree hollow, usually acacia or casuarina tree / 3–6 eggs.

Nallan Station, WA. Eulo Bore, Qld. Kunoth Well, NT.

Opposite: Courting pair, female (on left) and male.

Blue-winged Parrot

Neophema chrysostoma

Endemic

210–240mm

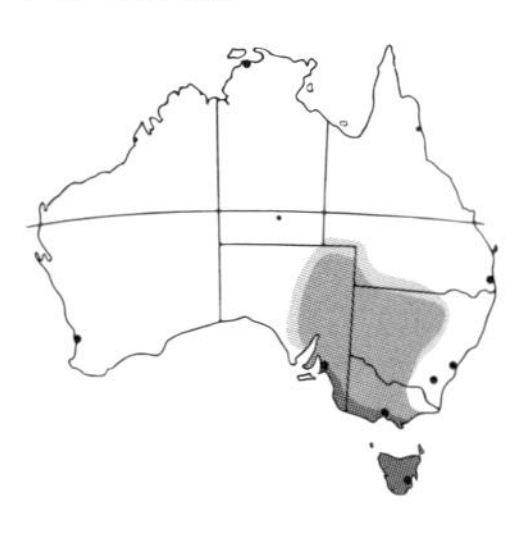

Sexes similar. Slender grass parrot with olive green upperparts, yellow face mask and dark blue frons. Wings: Green with large blue patch. Breast: Green. Belly: Yellow, occasionally orange-yellow. Female: Duller, greener. Immature: Indistinct face mask and frons.

Gregarious. Extremely well camouflaged. When disturbed, they fly off as one seeking cover. Birds that breed in Tas. have a defined seasonal migration route to and from Tas. to mainland, leaving in Mar., returning in Oct. Some birds overwinter in Tas.

Habitat: Varied, open habitats including saltbush plains, mallee, woodlands and coastal dunes.

Feeding: In groups, foraging on ground on grass and herbaceous plant seeds.

Voice: Soft twittering and tinkling.

Status: Common.

Breeding: Oct.–Jan. in Tas., southeastern coastal SA and southern Vic. Nests in hollows in trees, stumps / 4–6 eggs.

In autumn, birds gather at Woolnorth, Tas., before crossing to the mainland. Werribee TP, Vic.

Left: Female (on left) and male. Right: Male.

Elegant Parrot

Neophema elegans

Endemic

220–280mm

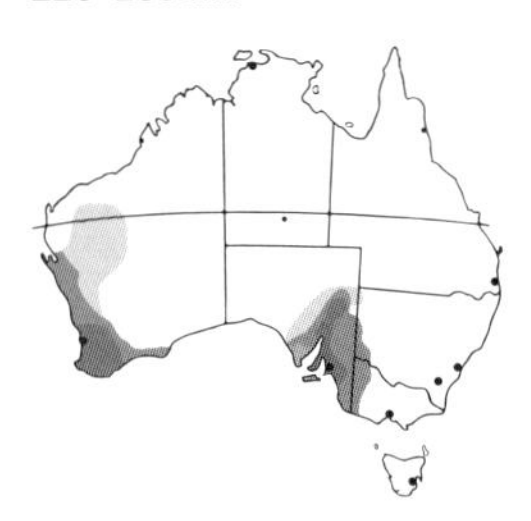

Sexes similar. Similar to the Blue-winged Parrot but distinguished by brighter yellow markings and less blue on wings. Yellow-olive above. Frontal band: Blue, extending past the eyes. Female: Duller. Immature: Similar to female, lacks frontal band.

Typical grass parrot, usually seen in flocks of 20 or more. Freezes when disturbed, only seeking cover at the last minute.

Habitat: Open country, mallee, semi-arid acacia scrubland, saltbush plains, lightly timbered grasslands, coastal sand dunes.

Feeding: Forages on ground for seeds. Eastern birds frequently seen with Blue-winged Parrots.

Voice: Sharp 'zit-zit' in flight. Soft twittering when feeding.

Status: 2 separate populations separated by the Nullarbor Plain. Common generally. Partially nomadic in the west.

Breeding: Aug.–Nov. Nests in hollow tree or stump / 4–6 eggs.

Stirling Ranges and Dryandra SF, WA. Port Gawler CP, and Price Saltfields, Yorke Peninsula, SA. (Permission required for entry.)

Left: Male. Right: Immature.

Rock Parrot

Neophema petrophila

Endemic

220–240mm

Sexes similar. Stocky grass parrot, endemic to SA and confined to coastal habitats. Upperparts: Dull olive. Underparts: Olive-yellow. Face: Pale blue with dark blue frontal band. Wings: Light blue patch. Tail: Blue and olive, tipped with yellow. Eyes: Grey. Female, immature: Less blue on face.

Often seen flying high and swiftly between offshore islands and mainland, circling overhead before landing.

Habitat: Exposed coastal areas including sand dunes, rocky offshore islands, saltmarsh and heath areas, usually within 10km of coastline.

Feeding: Ground foraging in pairs or small flocks for seeds and fruits of salt-adapted plants.

Voice: Repeated 'tsit-tsit'.

Status: Uncommon, usually sedentary.

Breeding: Aug.–Dec. on offshore islands. Nests in a depression under rock ledge or crevice / 4–5 eggs.

Port Gawler CP, SA. Bremer Bay and Rottnest Is., WA. Two Peoples Bay and Mount Manypeaks IBA, WA.

Opposite: Male.

Orange-bellied Parrot

Neophema chrysogaster

Endemic Endangered

220–250mm

Sexes similar. Small grass parrot with bright green upperparts and pale green face with blue and yellow frontal band. Throat, breast, upper belly: Green to yellow. Lower belly, vent: Bright orange. Wing borders, underwings: Blue. Tail: Green and blue, tipped with yellow. Eyes: Brown. Female: Slightly paler. Immature: Similar, indistinct blue-yellow frons.

Habitat: Prefers coastal areas. Open damp grasslands, tidal flats, sedge swamps, coastal dunes. In the breeding season favours forests with Smithton Peppermints.

Feeding: Ground forager on seeds, herbaceous plants and berries.

Voice: Distinctive, metallic 'buzzing' alarm call. Soft warbling when feeding.

Status: Rare and endangered due to loss of habitat.

Breeding: Nov.–Dec. Mainland birds migrate to Tas. to breed, nesting in tree hollow near button grass plains / 4–5 eggs.

Melaleuca, Tas. (mid Oct.–Mar.)

Left: Female. Right: Male showing bright orange vent.

Turquoise Parrot

Neophema pulchella

Endemic

200–220mm

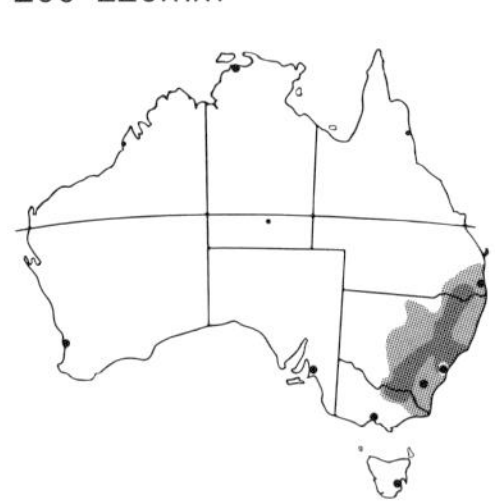

Small, lightly built grass parrot with bright green crown and upperparts and turquoise face and throat. Shoulders: Turquoise blue grading to deep blue with upper chestnut-red patch. Underparts: Bright yellow. Underwings, wing edges: Deep blue. Tail: Green tipped with yellow. Eyes: Brown. Female: Duller, lacks chestnut-red patch. Lores: White. Throat, breast: Green. Immature: Similar to females.

Habitat: Open forests, timbered grasslands on mountain slopes.

Feeding: Ground foraging in pairs or small flocks for grass and herbaceous plant seeds. Also nectar, scale insects.

Voice: Penetrating whistle in flight. High-pitched chatter when feeding.

Status: Uncommon. Nomadic. Threatened Vic. Vulnerable NSW. Once the most common parrot in western Sydney.

Breeding: Aug.–Dec. Nest in tree hollow / 4–5 eggs.

Girraween NP, Qld. Back Yamma SF, NSW. Chiltern–Mt Pilot NP, Vic.

Left: Male feeding in native grassland. Right: Female.

Scarlet-chested Parrot

Neophema splendida

Endemic

190–220mm

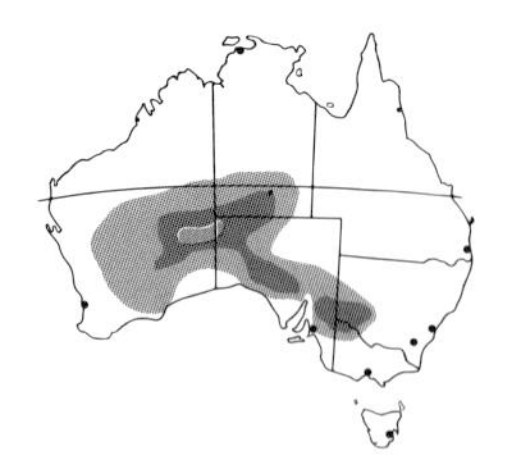

Small, brightly coloured, quiet, unobtrusive grass parrot with bright green upperparts and bright blue face, throat and cheek. Breast, belly: Brilliant scarlet. Underparts: Bright yellow. Underwings, wing edges: Pale blue. Tail: Green and blue, tipped with yellow. Eyes: Brown. Female, immature: Similar, paler blue face. Breast: greenish-yellow.

Habitat: Spinifex sand plains, acacia scrubland, mallee, mulga scrub, open woodlands.

Feeding: Ground foraging for grass and herbaceous plant seeds. Thought to obtain moisture from succulent plants when water is not available.

Voice: Soft twittering.

Status: Uncommon to rare. Sporadically disperses in inland areas. Nomadic following food availability.

Breeding: Aug.–Jan., or following good rain. Tree hollow nest in acacia or small eucalypt / 3–5 eggs.

Great Victoria Desert, SA–WA. Gluepot NR, SA.

Left: Male. Right: Female feeding young in nest hollow.

Eastern Ground Parrot

Pezoporus wallicus

Endemic

300mm

Sexes alike. Mostly bright green with black and yellow streaks and long, narrow, barred tail. Frontal band: Small orange-red band.
Immature: Heavier black breast marks. Lacks red frontal band. Eyes: Brown.

Nocturnal. Shy, elusive and rarely seen. Prefers to run when flushed. Flight is straight and direct over short distances.

Habitat: Damp coastal heathlands and grasslands.

Feeding: Feeds at night on grass and herb seeds.

Voice: High-pitched ascending scale 'tee-tee-tee-ea', heard early morning and at dusk.

Status: Vulnerable NSW due to loss of habitat and frequent fires.

Breeding: Sept.–Dec. Well-hidden grassy nest on the ground / 3–4 eggs.

Strathan area, Tas. Barren Grounds NR, and Green Cape, Ben Boyd NP, NSW.

Left: Adult. Right: Immature.

Western Ground Parrot

Pezoporus flaviventris

Endemic Critically Endangered

300mm

Similar to the Eastern Ground Parrot but with brighter yellow underparts and more indistinct black barring to belly and undertail coverts.

Nocturnal. Shy, elusive and rarely seen. Prefers to run when flushed. Flies in zigzag pattern with rapid wingbeats and short gliding phase.

Recent genetic studies suggest that the Western Ground Parrot split from the Eastern Ground Parrot about 2 million years ago. They are one of the world's rarest birds, with an estimated 140 birds remaining, due mainly to land clearing, predation by foxes and cats, and inappropriate fire regimes.

Habitat: Older, dense and unburnt coastal heaths with a diverse range of low-growing mature shrubs and sedges.

Feeding: Seeds of native sedges, especially Telegraph Rush, a tufted, grass-like perennial sedge. Also seeds of bottlebrush, banksia and grevillea.

Voice: Variable high-pitched call at dusk and dawn answered by adjacent birds.

Status: Critically endangered.

Nesting: Little is known about breeding behaviour. Thought to nest in spring, with nest a lined depression on the ground. Female probably lays 3–4 eggs.

Opposite: Adult feeding.

Night Parrot

Pezoporus occidentalis

Endemic Endangered

220–250mm

Sexes alike. Believed extinct until 2013, when it was rediscovered in western Queensland. In 2017 it was also discovered in Western Australia. Mostly bright green streaked with yellow and black. Belly: Bright yellow. Underwings: Prominent white to pale yellow bar visible in flight. Tail: Narrow, short, brown with yellow and green barring. Highly nomadic and elusive. This is a nocturnal mainly ground-dwelling bird.

Habitat: Arid and semi-arid areas, including sand hills, floodplains, claypans, edges of salt lakes, watercourses, waterholes with low, dense vegetation including spinifex.

Feeding: Ground forager for seeds, including seeds of spinifex, and green plant material.

Voice: Low 2-note whistle.

Status: Endangered. Extremely rare.

Breeding: Opportunistic, but usually in July–Aug. Nest tunnelled into spinifex clump / 2–4 eggs.

Opposite: Night parrot on Bush Heritage's Pullen Pullen Reserve, Qld.

Family Centropodidae, Cuculidae

Coucal

Koel

Cuckoos

Bronze-cuckoos

A group containing the Pheasant Coucal and the cuckoos.

Family Centropodidae contains approximately 30 species worldwide. The Pheasant Coucal is Australia's sole representative.
Previously regarded as a member of the cuckoo family, DNA testing and other research has now shown it to be only distantly related. Unlike other cuckoo species, Pheasant Coucals are ground dwellers. They build their own stick platform nest, lined with grass and leaves and well hidden in dense grassy vegetation or in a pandanus thicket. Birds form pairs for life. The male usually incubates the eggs and both parents feed the young.

Family Cuculidae contains approximately 79 species.
Cuckoos are represented in Australia by 12 species. Mostly graceful, arboreal birds, they parasitise the nests of over 100 species of Australian birds.
Australian cuckoos are 'brood parasites'. They lay an egg in a host bird's nest that contains eggs similar in appearance to their own. The cuckoo usually disposes of the host bird's egg after depositing its own egg. Alternatively, the cuckoo egg hatches faster than the host bird's egg and the newly hatched cuckoo instinctively ejects any other eggs or nestlings from the nest, thus monopolising the food supply.
The exception is the Channel-billed Cuckoo, the world's largest cuckoo. The chick does not eject the host's eggs or nestlings, but as the Channel-billed Cuckoo chick grows fastest and monopolises the food supplied by the foster parents, the host's young rarely survive. Parasitising the nests of larger birds such as magpies, currawongs and ravens, the male Channel-billed Cuckoo harasses potential host birds, causing them to leave their nest, at which the female drops in to lay her eggs. The Channel-billed Cuckoo and Pacific Koel are the only Australian cuckoos that often lay more than one egg in the host nest.
Most Australian cuckoos are summer migrants, including the Oriental Cuckoo, a non-breeding migrant that travels from Eurasia each summer. The Channel-billed Cuckoo, the world's largest cuckoo, is a breeding migrant that winters in New Guinea.
Two races of the Pacific Koel migrate to summer in Australia from two different regions. One race is from New Guinea and summers in northern Australia. The other, larger race is from Indonesia and the Philippines and visits the eastern states of Australia. These birds mainly parasitise the nests of larger honeyeaters, including wattlebirds and Blue-faced Honeyeaters.
Small cuckoo species eat insects and especially hairy caterpillars, while the larger cuckoos feed on fruits, especially figs, and mice, lizards and nestlings.

Opposite: Immature Channel-billed Cuckoo being fed by a Currawong host parent.

Pheasant Coucal

Centropus phasianinus

500–700mm

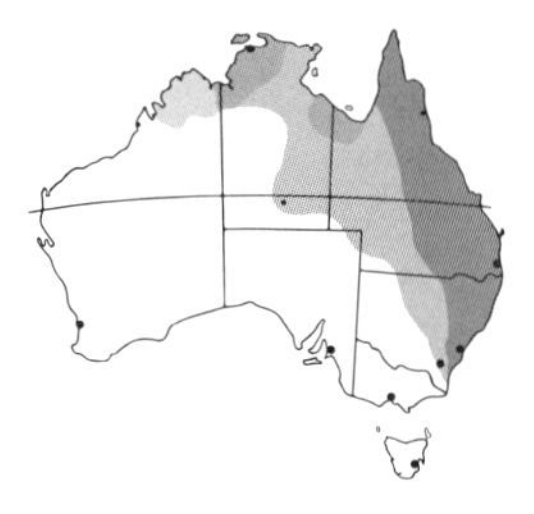

Sexes similar. Pheasant-like, ground-dwelling cuckoo with a long, broad, barred tail and short rounded wings.
Breeding: Black bill, head, neck and underparts. Upperparts: Rich reddish brown with brown and cream barring on wings. Tail: Black, barred orange. Eyes: Red. Non-breeding: Head, upperparts: Cinnamon brown, streaked white. Underparts: Straw brown. Eyes: Pale.
Female: Larger. Immature: Paler, mottled, similar to non-breeding with orange speckled head, neck, upperparts.

Non-parasitic. Slow, weak flight and glides. Runs rather than flies. Clumsy, flapping leaps.

Habitat: Dense understorey. Damp, grassy woodlands, canefields. Parks and gardens.

Feeding: Small reptiles, frogs, mice, large insects, eggs, chicks.

Voice: Series of booming, falling–rising notes, 'oop-oop-oop' repeated constantly.

Status: Common, sedentary.

Breeding: Aug.–Feb. Sticks and grass platform hidden in grassy tussocks, often Sword Grass, or up to 2m high in pandanus thickets / 3–5 eggs.

Left: Breeding. Right: Non-breeding.

Pacific Koel

Eudynamys orientalis

390–450mm

Also called the Eastern Koel. Large, parasitic cuckoo identified by its overall glossy black plumage tinged with blue and green and distinctive long tail. Eyes: Bright red.
Female: Upperparts: Brown with white spotting. Head: Black. Tail: Dark brown with white bars. Underparts: Cream with thin, brown bars. Immature: Crown: Cream. Upperparts: Rufous, barred with large white spots.

Singular or in pairs.

Habitat: Tall forests. Rainforest fringes, riverside scrub, eucalypt forests, parks and gardens.

Feeding: Arboreal feeder, native figs, fruits, berries.

Voice: Known for its loud, ascending, monotonous call of 'koo-el', from which it gets its name. Female quieter.

Status: Common breeding migrant. Arriving late Sept. Departing in Mar.

Breeding: Nov.–Jan. Parasitises nest, frequently of large honeyeaters such as Red Wattlebird, friarbirds, also figbirds / 1 egg.

Left: Male feeding on Bangalow Palm fruits. Centre: Male. Right: Female.

Channel-billed Cuckoo

Scythrops novaehollandiae

570–700mm

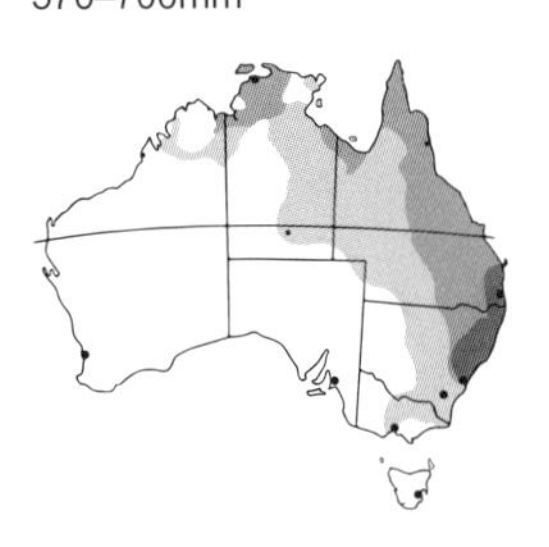

Sexes alike. Largest cuckoo that occurs in Australia and is unmistakable with its massive, pale grey-blue bill. Upperparts: Grey with darker scalloped pattern on back and wings. Tail: Long, grey, tipped white. Undertail: Black and white barring. Eyes, eye ring: Red. Immature: Buff head, brownish upperparts. Lacks red eye ring.

Shy. Mostly roosts and feeds in the tops of tall trees. Spreads wings and tail when feeding.

Habitat: Rainforests, woodlands, forests with tall trees, along watercourses.

Feeding: Mostly native figs, also other native fruits, occasionally insects and bird eggs.

Voice: Loud, raucous 'quork' constantly repeated, dropping in pitch.

Status: Moderately common. Arriving from Indonesia and New Guinea in Sept.; leaving by Mar.

Breeding: Oct.–Dec. Parasitises nests of large bird species such as Australian Magpie or Pied Currawong / 1–2 eggs.

Opposite: Adult showing its massive bill.

Horsfield's Bronze-cuckoo

Chalcites basalis

150–170mm

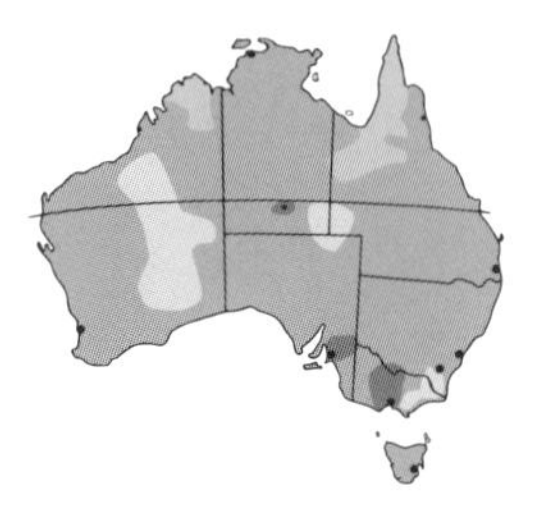

Sexes similar. Distinctive long, dark eye stripe and whitish eyebrow stripe. Upperparts: Scalloped, olive-brown with copper sheen. Throat: Streaked dark brown. Underparts: Cream with dull brown incomplete breast, flank bars. Uppertail: Edged with orange-brown. Undertails: Grey, black barred white to sides and tip. Eyes: Red to dark brown.
Female: Less distinct barring to underparts. Immature: Duller, faint or no breast bars.

Male calls day and night in breeding season.

Habitat: Mostly lightly wooded areas.

Feeding: Mostly insects, larvae and hairy caterpillars.

Voice: High-pitched, mournful, descending whistle.

Status: Common. Northern birds are sedentary. Southern birds move northward in autumn.

Breeding: July–Feb. in south. Most of the year in the north. Parasitises mostly domed nests, or cup-shaped nests / 1 egg.

Left: Male showing iridescent upperparts. Right: Note incomplete barring.

Shining Bronze-cuckoo

Chalcites lucidus

150–170mm

Sexes similar. Race *plagosus* present in Australia. Nominate race is an occasional east coast migrant. Coppery bronze head and mantle with white flecks on frons. Back, wings: Iridescent bronze-green. Face: White, mottled copper-bronze. Underparts: White with copper-bronze barring.
Female: Crown duller. Immature: Duller, underparts white, barring on sides.

Solitary.

Habitat: Coastal forests, rainforests, parks and gardens.

Feeding: Hairy caterpillars, moths, beetles, insects and their larvae.

Voice: Repeated, short, rapid, high-pitched whistles.

Status: Common. Seasonal migration northward for winter.

Breeding: Aug.–Jan. Usually parasitises domed nests, mostly gerygones and Yellow-rumped Thornbills / 1–16 eggs a season.

Left and right: Race *plagosus*.

Little Bronze-cuckoo

Chalcites minutillus

140–150mm

Sexes similar. Small and inconspicuous. Metallic green-bronze upperparts and white with copper-bronze barred underparts. Red eyes, eye ring, with white eyebrow.
Female: Brown eyes and white eye ring. Immature: Similar to female, barring only on flanks.
Race *russatus*, Gould's Bronze Cuckoo: Rich rufous-washed upperparts, neck and tail. Eye ring: Red. Underparts: Dark bronze barring.
Female: Rusty breast and tail. Eye ring: Pale grey or tan. Immature: Similar to female, barring only on flanks.

Gould's and Little Bronze Cuckoos were once considered separate species. However, they hybridise where their ranges overlap. Gould's range from Cape York Peninsula to Rockhampton area, Qld.

Habitat: Monsoon forests, mangroves, stringybark forests, acacia thickets.

Feeding: Usually feeds in the upper canopy. Insects, larvae, caterpillars.

Voice: High-pitched whistled trilling, quickly repeated.

Status: Uncommon.

Breeding: Oct.–Jan. Parasitises domed nest, often of gerygones / 1 egg.

Middle Arm, Darwin, NT. Cairns Botanical Gardens, Qld.

Left: Nominate male. Right: Nominate female.
Below: Race *russatus,* Gould's Bronze Cuckoo. Male (left) and female (right).

Pallid Cuckoo

Cacomantis pallidus

Endemic

280–320mm

Sexes similar. Large, slender cuckoo with mid-grey upperparts. Whitish grey underparts. Undertail: Dusky black with white barring. Bill: Brown, slightly down-curved. Eyes: Dark with yellow eye ring and ill-defined brown eye stripe. Nape: White spot.
Occasionally, rufous morphs with upperparts barred reddish-brown and black.
Female: Browner. Immature: Upperparts: Heavily streaked dark brown and white. Underparts: Paler.

Solitary. Raises and lowers tail on landing.

Habitat: Open timbered areas.

Feeding: Perch and pounce. Large insects, grasshoppers, prefers hairy caterpillars.

Voice: Distinctive, repeated, loud piping. Rising 8 or more notes, mournful whistle, often at night.

Status: Moderately common. Migratory. Concentrated in southeastern Qld, NSW, Vic., Tas., in the breeding season, then dispersing Australia-wide at other times.

Breeding: Sept.–Jan. Parasitises nests, commonly of Yellow-faced, Singing and White-plumed Honeyeaters.

Left and right: Adults showing distinctive tail barring.

Black-eared Cuckoo

Chalcites osculans

190–210mm

Sexes alike. Largest bronze-cookoo. Distinctive black eye stripe, whitish eyebrow stripe and throat. Upperparts: Dull grey-brown with a dull metallic sheen. Underparts: Cream fawn. Uppertail: Grey-brown, white tipped. Undertail, cream with brownish bars.
Immature: Paler.
A usually, quiet shy bird, retreats through the undergrowth when disturbed.

Habitat: Dryer areas, woodlands, mallee.

Feeding: Perch and pounce. Insects, larvae, prefers hairy caterpillars, mostly taken from the ground.

Voice: A plaintive, descending soft whistle. Repeated 'feww, feww'.

Status: Uncommon. Migratory, moving into the subcoastal areas of southeast and southwest Australia for the summer.

Breeding: Mar.–July inland. June–Oct. in west. Aug.–Dec. in east. Parasitises mostly domed nests. Usually Speckled Warbler nests in the east and Redthroat nests in central and western Australia / 1 egg.

Left: Adult with hairy caterpillar. Right: Adult showing upperparts.

Fan-tailed Cuckoo

Cacomantis flabelliformis

250–270mm

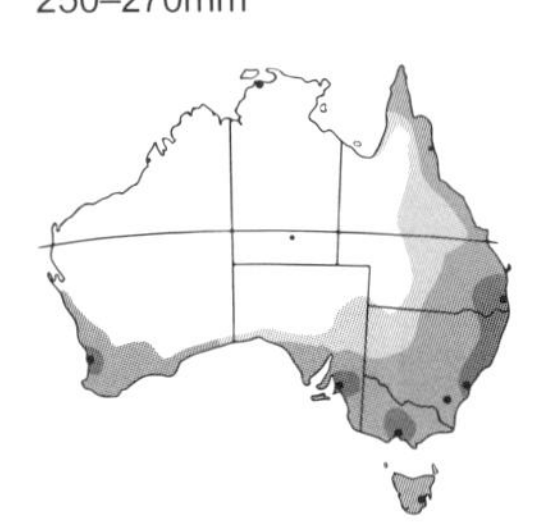

Sexes similar. Slender cuckoo with slate-grey upperparts and bright yellow eye ring around dark eyes. Underparts: Light rufous with paler belly. Undertail: Black with white barring.
Uppertail: White tipped.
Female: Paler rufous underparts.
Immature: Eye ring: Greenish yellow. Upperparts: Rich brown. Underparts: Grey-brown finely barred, russet wash.

Characteristically fans and tilts tail on landing. Solitary. Often sings throughout the year.

Habitat: Widespread. Rainforests, forests, woodlands, heaths, parks and gardens.

Feeding: Perch and pounce. Insects, hairy caterpillars, moths. Taken in mid-air or from the ground and consumed after returning to the perch.

Voice: Rapid, descending, mournful trill, repeated. Sings throughout the year, including winter.

Status: Common. Tas. birds migrate to the mainland in the non-breeding season.

Breeding: Aug.–Dec. Parasitises thornbill nests, particularly of Brown Thornbills, also flycatchers, fairy-wrens and scrubwrens / 1 egg.

Left and right: Adult male showing its distinguishing bright yellow eye ring.

Chestnut-breasted Cuckoo

Cacomantis castaneiventris

230–250mm

Sexes similar. Similar to Fan-tailed Cuckoo, but smaller and more richly coloured. Upperparts: Deep grey, metallic sheen. Underparts: Cinnamon. Undertail: Broadly banded. Female: Paler underparts. Often heard but rarely seen.

Habitat: Dense tropical rainforests, monsoon forests and wet gullies.

Feeding: Inconspicuously forages in dense canopy or sometimes watches from a low perch for insects, larvae; prefers caterpillars.

Voice: Descending, 3-whistled call repeated and trill 'preeer-preeer-preeer'.

Status: Uncommon, sedentary.

Breeding: Oct.–Dec. Parasitises domed nests, frequently of Tropical Scrubwren and Lovely Fairy-wren / 1 egg.

Iron Range NP, Qld (Sept.–Mar.).

Left: Male. Centre: Female. Right: Adult showing upperparts.

Brush Cuckoo

Cacomantis variolosus

200–240mm

Sexes similar. Distinctive square-shaped tail with white tip and barring underneath. Upperparts: Grey-brown, darker and mottled on wings. Underparts: Pale, grey, buff wash.
Race *dumetorum*, northern Qld, NT, WA: Smaller.
Female is similar to male but paler underparts with pale grey barring on chest. The uncommon barred morph has darkly barred underparts and streaked upperparts. Immature: Upperparts: Strongly barred dark brown. Underparts: Barred white and brown.

Solitary. Usually high in the upper canopy.

Habitat: Wooded areas including rainforests and mangroves. Occasionally parks and gardens.

Feeding: Insects and larvae, prefers hairy caterpillars.

Voice: Slow, melancholy descending 6–8 notes.

Status: Moderately common. Sedentary in the north. Southern birds migrate northward to overwinter.

Breeding: Oct.–Feb. in south. Most of year in north. Parasitises nests, commonly nests of robins, fantails, wrens and woodswallows / 2–4 eggs deposited in different nests.

Left: Male. Right: Female.

Oriental Cuckoo

Cuculus optatus

300–320mm

Sexes similar. Secretive cuckoo. Grey with rufous-washed upperparts, yellow eye ring and yellow base of bill. Upper breast: Grey. Underparts: White strongly barred dark brown. Tail: Dark grey, white barring. Undertail: Distinctive white tips seen in flight. Legs, feet: Orange-yellow. Occasionally female rufous morphs with upperparts barred reddish brown and black, and underparts strongly barred black. Immature: Dark red-brown upperparts with black barring.

Solitary or in pairs. Falcon-like, swift and graceful flight. Raises and lowers tail on landing.

Habitat: Forests, woodlands, in dense foliage.

Feeding: Forages on the ground or in foliage for large insects and their larvae.

Voice: Repeated 'kak-ak' (similar to Dollarbird).

Status: Uncommon, rare summer non-breeding migrant. Nov.–Apr.

Breeding: Northern Eurasia.

Left: Rufous morph female. Centre and right: Adult.

14

Family Strigidae, Tytonidae, Podargidae, Caprimulgidae, Aegothelidae

Hawk-owls

Masked Owls

Frogmouths

Owlet-nightjar

Nightjars

A grouping of five families that includes true owls and owl-like birds.

Family Strigidae contains hawk-owls.
Hawk-owls are true owls and include the 'eagle-sized' Powerful Owl, Australia's largest owl. Typical of hawk-owls, it has large, yellow forward-looking eyes that provide precise binocular night vision; asymmetrical ears that give precise directional hearing; and soft plumage allowing silent flight. It is able to detect the smallest of prey from a considerable distance. The nest is a deep hollow lined with a bed of wood chips in a living or dead tree.

Family Tytonidae contains masked owls.
Masked owls are also true owls and similar to hawk-owls but have dark eyes surrounded by a heart-shaped facial disk. The facial disk captures and concentrates sound so the bird can calculate the exact direction and distance of prey, allowing efficient hunting in complete darkness. They have large wings and soft specialised plumage that allows them to fly quietly and slowly as they seek prey.

Family Podargidae contains frogmouths.
Often confused with owls, frogmouths are nocturnal birds of prey related to nightjars. They lack the powerful feet and curved talons of owls. Named for their wide frog-like gape used to catch insects, frogmouths roost on a tree branch by day and are well camouflaged, their cryptic plumage and stance making them appear like a dead, broken branch. Their nest is a flimsy platform of criss-crossed twigs, usually on a horizontal fork in a eucalypt.

Family Caprimulgidae contains nightjars.
Nightjars hunt aerially for insects at night. They nest on the ground, laying their eggs directly on the leaf litter. During the day they roost alone, well camouflaged in leaf litter on the forest floor.

Family Aegothelidae contains the Australian Owlet-nightjar.
Australia's sole representative of this family, the Australian Owlet-nightjar is a small nocturnal bird that roosts and nests in small tree hollows during the day and hunts insects at night. The nest is lined with fresh eucalypt or acacia leaves, which are replenished with fresh leaves as needed. Despite being one of the most widespread nocturnal birds, it is rarely seen. It often nests or roosts in nest boxes.

Opposite: Tawny Frogmouth. Its cryptic plumage resembles the bark of stringybark eucalypts, a favoured roosting tree.

Powerful Owl

Ninox strenua

Endemic

600–650mm
Wingspan: 1000–1350mm

Sexes similar. Australia's largest owl. Upperparts: Dark brown with white and pale brown barring. Eyes: Orange-yellow. Face: Grey-brown. Tail: Rounded. Underparts: Whitish with dark greyish V-shaped markings. Feet: Extremely large, yellow-orange, sharp talons. Female: Smaller. Immature: Paler, dark eye patches. Underparts, crown: White, lightly streaked.

Breeding pairs have a permanent bond and live together.

Habitat: Varied, with tall trees, including woodlands, forests and rainforests, most commonly roosting along creeks. Sometimes parks and gardens.

Feeding: Nocturnal. Swoops prey up in its talons, mostly tree-dwelling mammals such as possums, flying foxes. Also large birds, rats, rabbits.

Voice: Loud, slow 'whooo-hooo'.

Status: Moderately common, sedentary. Vulnerable in Qld, NSW. Endangered Vic.

Breeding: May–Oct. Large deep tree hollow in eucalypt 12–40m high / 2 eggs.

Left: Adult with head turned 180 degrees. Right: Adult with captured Brush-tailed Possum prey from previous night.

Southern Boobook

Ninox boobook

300–360mm
Wingspan: 700–850mm

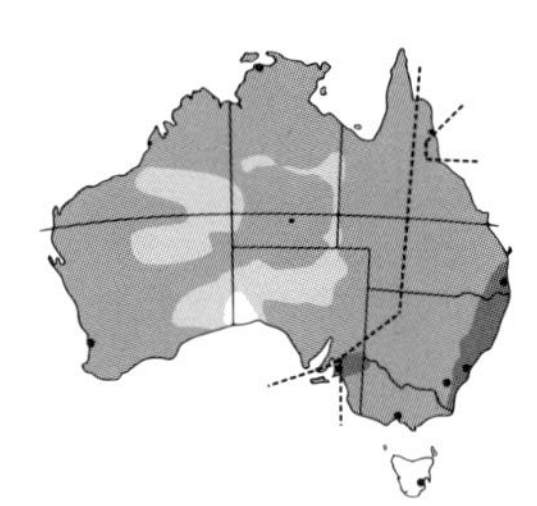

Sexes alike. Distinctive dark brown facial disk around yellow eyes. Upperparts: Dark brown, white spots on back, wings. Underparts: Cream with heavy reddish brown, mottled streaks. Bill: Light bluish grey. Colour and size variations across range. Nominate race *boobook*, eastern north Qld to Adelaide, SA: Largest.
Race *lurida*, Red Boobook, northeast Qld rainforests: Smaller, darker.
Race *ocellata*, inland Qld, NT, SA, WA and interior: Paler, streaked underparts.
Race *halmaturina*, Kangaroo Is. and adjacent mainland.
Female: All are larger and deeper coloured. Immature: Paler, buffy white streaked underparts. Dark eye patches.

Habitat: Varied. Wooded areas, eucalypt forests, woodlands and desert scrub.

Feeding: Nocturnal. Perch and pounce or prey seized mid-air. Prey includes small mammals, small birds, moths, spiders and insects.

Voice: Familiar, monotonous, soft-toned 'mo-poke'.

Status: Common. Some Tas. birds overwinter on mainland.

Breeding: Aug.–Jan. Nest in a tree hollow, 3–5m high / 2–5 eggs.

Left: Nominate race. Centre: Race *lurida*, Red Boobook. Right: Race *ocellata*.

Morepork

Ninox novaeseelandiae

290mm

Similar to the Southern Boobook but smaller, much darker and has richer brown plumage. Upperparts are more reddish, darker and covered in large white spots with scattered white spots on wings. Underparts are whiter and extensively spotted. It is distinguished by its bright, yellow eyes.

Some birds disperse to southern Vic. in autumn and winter.

Habitat: Forested areas.

Feeding: Nocturnal. Perch and pounce or prey seized mid-air. Prey includes small mammals, small birds, moths, spiders and insects.

Voice: Drawn-out 'mo-poke'.

Status: Common.

Breeding: Aug.–Jan. Nest in a tree hollow, 3–5m high / 2–5 eggs.

Left and right: Adult Morepork showing distinguishing, much brighter eyes than Southern Boobook.

Barking Owl

Ninox connivens

350–450mm
Wingspan: 850–1200mm

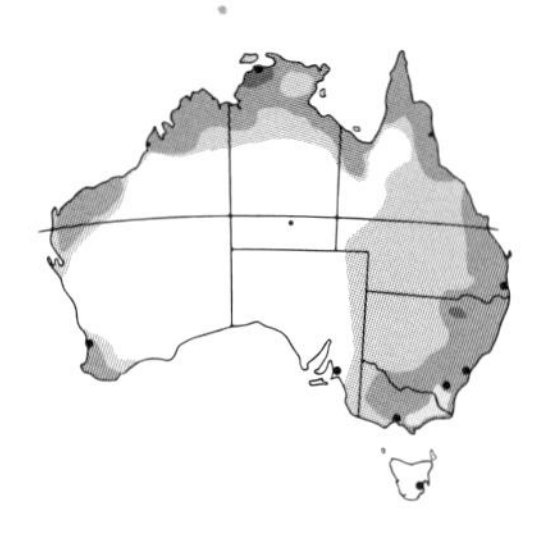

Sexes similar. Female smaller. Distinctive large, bright yellow eyes and 'woof-woof' barking call. No facial disk. Upperparts: Grey-brown, with wings and lower back spotted white. Underparts: White with vertical brown streaking. Feet: Large, yellow.
Race *peninsularis*, Northern Qld, NT, WA: Smaller. Underparts streaked rich, rufous brown. Immature: Similar, less streaked, whitish neck.

Territorial pairs form at 2 years.

Habitat: Forests, woodlands, eucalypt savannahs, farmland with remnant bushland.

Feeding: Hunts hawk-like. Relies mostly on sight for hunting, usually in more open areas. Insects, small mammals, birds, reptiles, also rodents and rabbits.

Voice: Call-and-answer duets. Rapid, repetitive barking 'woof-woof'. High-pitched quavering, loud, human-like scream.

Status: Moderately common. Sedentary. Vulnerable NSW. Endangered Vic.

Breeding: July–Oct. Nest in tree hollow 3–10m high / 2–3 eggs.

Gunlom Falls, Kakadu NP, NT.

Left: Nominate race. Right: Race *peninsularis*.

Sooty Owl

Tyto tenebricosa

Male: 370–430mm
Female: 450–500mm
Wingspan: 850–1150mm

Sexes similar. Large black eyes. Dark grey upperparts spotted white. Flat, heart-shaped pale grey face disk, bordered black. Underparts: Pale grey, spotted white, irregular black bars. Tail: Short. Legs: Feathered with large, powerful feet, black talons. Female: Larger. Immature: Paler.

Roosts in deep shade, hollows or caves. Sedentary.

Range: Southern Qld to Vic.

Habitat: Densely wooded gullies of rainforests, tall wet forests.

Feeding: Nocturnal, hunting for mammals, birds, rodents, insects.

Voice: Female: Descending, long, whistling scream. Male: Short scream. Harsh screeches.

Status: Moderately common.

Breeding: Opportunistic. High nest in deep tree hollow, usually a eucalypt, sometimes cave or crevice / 1–2 eggs.

Royal NP, NSW.

Centre left: Sooty Owl.

Lesser Sooty Owl

Tyto multipunctata

Endemic

300–380mm
Wingspan: 800–950mm

In most respects, similar to Sooty Owl but smaller, paler and more heavily white spotted. Face: Whitish grey disk, dark around eyes and bordered black. Underparts: Light grey with grey-black V-shaped markings. Legs: Whitish.

Northern Range Qld. See Sooty Owl map.

Habitat: Tall, tropical mountain rainforests and adjacent eucalypt forests.

Voice: Similar, but softer.

Julatten area Qld.

Centre right: Lesser Sooty Owl.

Rufous Owl

Ninox rufa

450–550mm
Wingspan: 1000–1200mm

Sexes similar. Australia's only tropical owl. Large, robust and hawk-like. Reddish brown upperparts with fine, pale brown bars, long tail and indistinct blackish brown facial disk. Eyes: Greenish yellow. Bill: Light grey. Underparts: Buff with close brown bars. Legs: Feathered. Race *meesi*, far north Qld: Tends to be smaller and paler.
Race *queenslandica*, central Qld. Larger and darker.
Female: All are smaller, darker. Immature: White, barred rufous.

Rarely seen. Roosts singularly or in pairs. Powerful flight, hawks, swoops and glides.

Habitat: Monsoon forests, rainforests, rainforest fringes, dense forests and woodlands.

Feeding: Takes birds, small arboreal mammals, insects snatched in aerial chase.

Voice: Mournful 'woo-hoo'.

Status: Uncommon. Sedentary.

Breeding: June–Sept. Tree hollow, 8–40m high /1–2 eggs.

Julatten area, Qld. Darwin Botanic Gardens, NT.

Left: Nominate race, NT, WA. Right: Race *queenslandica*, Qld.

Eastern Barn Owl

Tyto delicatula

300–400mm
Wingspan: 900–1300mm

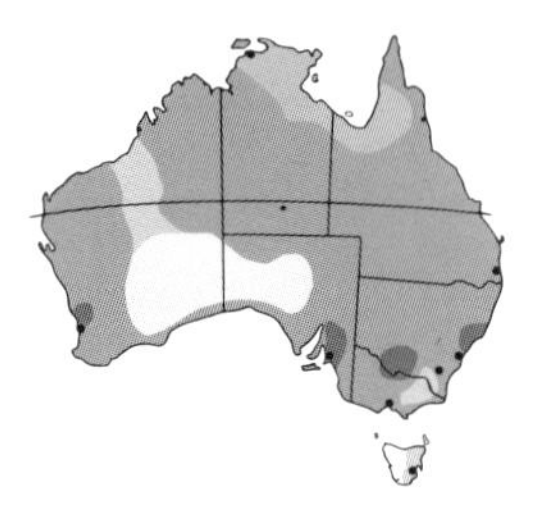

Sexes similar. Slim and pale with an upright posture and a distinctive, heart-shaped facial disk. Pearly grey upperparts with black spots tipped white. Underparts: White with scattered dark spots. Wings: Beige markings. Facial disk: White with brown edging. Eyes: Small, black.
Female: Larger. Immature: Similar.

Solitary or in pairs. Crouches and spreads wings when threatened.

Habitat: Most habitats, but prefers lightly wooded areas.

Feeding: Nocturnal. In flight, mostly uses acute hearing to snatch small mammals, insects, small reptiles, birds, mostly from the ground. Common House Mouse is favoured.

Voice: Drawn-out rasping screech.

Status: Cosmopolitan. Common. Highly nomadic.

Breeding: Opportunistic. Usually nests in a deep tree hollow 15–25m high. Occasionally in caves or buildings / 2–6 eggs.

Left: Adult. Right: Adult with rodent prey.

Australian Masked Owl

Tyto novaehollandiae

350–400mm
Wingspan: 900–1300mm

Sexes similar. Similar to the Eastern Barn Owl but larger, more heavily built and plumage is usually darker. Large feet and talons, larger than the Eastern Barn Owl. There are four races; each race has several morphs.
Upperparts: Blackish brown with grey and white speckles.
Back, wings: Beige barring.
Underparts: Beige, with black spots to flanks and breast. Bill: Whitish. Facial disk: White or beige or reddish brown with deeper brown around the eyes.
Feet: Powerful feathered legs and large talons.
Female: All are significantly larger than male. Immature: Similar to adults, paler.
Nominate race found across most of Australia.
Race *kimberli*, across tropical northern Australia: Smallest and palest with white underparts and pale grey upperparts.
Race *castanops*, Tasmanian Masked Owl: Largest race and endemic to Tasmania. The female is darker and is the world's largest *Tyto* owl. Occasionally intermediate in colour.
Race *melvillensis*, confined to Melville Is.: Small and endangered.

Habitat: Coastal forests, woodlands with tall trees. Roosting in forested gullies, tree hollows, occasionally caves.

Feeding: Nocturnal. Preferring edges of open forests. Scanning from a low branch and pouncing on rodents, rabbits, other small mammals. Also tree-dwelling mammals, birds and insects.

Voice: Drawn-out rasping, hissing screech.

Status: Uncommon to rare. Vulnerable. Locally endangered.

Breeding: Mar.–May, Oct.–Nov. Nests in deep tree hollow, usually a eucalypt, or sometimes on a cave ledge / 2–3 eggs.

Barrington Tops NP and Murramarang NP, NSW. Cobourg Peninsula, Garig Gunak Barlu NP, NT. Mt Wellington and Truganini Reserve, Tas.

Centre left and right: Nominate race. Below, left and right: Race *castanops*, Tasmanian Masked Owl.

Eastern Grass Owl

Tyto longimembris

320–380mm

Wingspan: 1000–1200mm

Mid-sized owl with distinguishing very long, mostly bare legs that hang below or behind in flight. Small eyes with dark 'tear' patch. Upperparts: Dark brown washed orange-yellow, spotted beige. Wings: Barred dark brown and beige with whitish spots. Facial disk: White, heart-shaped, brown edging to top. Underparts: White, lightly spotted brown. Undertail: White with dark bars. Female: Similar, larger. Facial disk, underparts: Washed orange-yellow and dark spotted. Immature: Similar, darker.

Ground-dwelling. Nocturnal.

Habitat: Tall grasslands, swampy heathlands, grassy plains, floodplain sedges, crop fields.

Feeding: Gliding, hovering a few metres above ground for rodents, small mammals, insects.

Voice: Usually quiet. Harsh screeches.

Status: Uncommon. Vulnerable NSW. Usually resident in the northeast. Western populations nomadic and fluctuate markedly with food (rodent) supply.

Breeding: Opportunistic, Mar.–June usually. Trampled platform nest in dense grass / 3–8 eggs.

Ingham and Julatten area, Qld. Kooragang Is. NSW. Holmes Jungle NP, NT.

Left: Female at nest on ground. Right: Adult in flight.

Tawny Frogmouth

Podargus strigoides

Endemic

380–550m

Sexes similar. Often confused with owls, but are more closely related to nightjars. Well camouflaged with overall cryptic plumage. Eyes: Yellow with indistict pale whitish eyebrow. Olive-grey bill with a wide gape, bristled tufts and long dark malar line.
Nominate race, east coast Qld, NSW, Vic, Tas, southeast SA: Mottled grey and silver grey or brown. Crown, back, wings: Black streaked. Underparts: Paler grey, black streaked. Race *brachypterus,* inland NT, Qld, NSW, SA, south of Broome WA. Smaller, reddish brown. Race *phalaenoides,* tropical north: Half the size of nominate race. Paler grey mottled upperparts, plainer underparts. Females are sometimes rufous. Female: All races, paler with a slighter bill. Female plumage can vary from grey to rufous. (Males always grey.) Immature: Similar.

Territorial. Well camouflaged. Roosts on a branch during the day. When disturbed, freezes, with bill pointing upwards imitating a broken limb.

Habitat: All habitats, except arid areas without trees. Parks and gardens.

Feeding: Nocturnal, hunting after dusk and before dawn, dropping from perch onto prey including grasshoppers, spiders, other insects, mammals, birds, frogs.

Voice: Low, resonant bursts, 'oom-oom-oom' repeated monotonously.

Status: Common. Sedentary.

Breeding: Aug.–Dec. Loose stick platform nest in tree fork 3–25m high / 2–3 eggs.

Centre, left and centre: Nominate race. Centre right: Race *brachypterus.*

Below left: Race *brachypterus.* Below centre: Race *phalaenoides.* Below right: Race *phalaenoides,* rufous female.

Papuan Frogmouth

Podargus papuensis

500–600mm

Sexes similar. Largest (lengthwise) frogmouth with a large head and solid, grey bill with bristled tufts. Upperparts: Mottled dark reddish brown or dark grey, blotched white along the wings. Underparts: Fawn, mottled and streaked mostly white and also blackish. Eyes: Red. Tail: Long, barred. Male is heavier, darker. Female: Smaller, more rufous, less mottled.

Nocturnal. Singular, occasionally in pairs. Freezes when disturbed.

Habitat: Rainforest margins, mangroves, eucalypt and paperbark forests along watercourses, swamp areas.

Feeding: Large insects, also small reptiles, frogs, rodents, small birds. Prey is scooped up in wide gape.

Voice: Drumming, low, booming 'oom'. Snaps bill.

Status: Moderately common. Sedentary.

Breeding: Aug.–Jan. Loose stick platform nest in tree fork, 6–20m high / 1–2 eggs.

Daintree River, Julatten area and Lamington NP, Qld.

Left: Papuan Frogmouth family with more rufous and subtly patterned female (on left), chick (centre) and darker, more boldly patterned male (on right).

Right: Male with female obscured.

Marbled Frogmouth

Podargus ocellatus

370–410mm

Sexes similar. Smallest, slimmest frogmouth. Red-brown or grey upperparts finely streaked, mottled black and white, with blackish crown and dark grey, heavier mottled wings. Frons: Pale. Underparts: Paler. Bill: Grey with prominent, barred-brown and cream plumed bristles. Eyes: Orange-red with whitish eyebrow. Tail: Long, graduated.
Race *marmoratus*, far north Qld: Smaller with longer bristles, longer tail and barred wings. Female: Plainer, browner. Immature: Similar, brown eyes, brown eye stripe.

Solitary.

Habitat: Far northern range: Rainforests, gullies along watercourses with ferns and Bangalow Palms. Southerly range: Moist lowlands, tall dense vine forests.

Feeding: Nocturnal. Hunts from a perch, mostly for large insects.

Voice: Repeated 'koo-loo' and descending 'cackling'. Bill snapping.

Status: Uncommon. Sedentary.

Breeding: Oct.–Aug. Twig and moss nest on a horizontal branch or on top of epiphytes 3–15m high / 1–2 eggs.

Iron Range NP, Mt Glorious and Lamington NP, Qld.

Left and right: Adult showing distinguishing orange eyes.

Australian Owlet-nightjar

Aegotheles cristatus

210–240mm

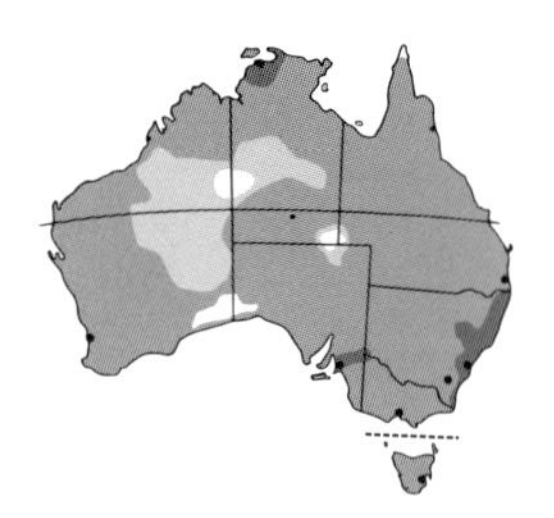

Sexes similar. Smallest of Australia's nocturnal birds. Plumage colour varies regionally. Upperparts: Pale grey to dark grey to reddish grey, finely barred black. Head: 2 black stripes to nape. Eyes: Dark brown, large, non-reflective. Bill: Small, wide and fringed with whiskers. Underparts: Pale, faintly speckled and barred grey.
Four colour morphs. Coastal southeastern and southern WA are dark grey. Northern and arid inland birds are reddish grey. Race *tasmanicus* is confined to Tas.
Female: All are larger. Immature: Similar, markings blurred.

Solitary. Several roosting hollows.

Habitat: Wooded areas and scrubland with suitable tree hollows.

Feeding: Nocturnal. Usually large insects, spiders. Perch and pounce or caught in mid-air.

Voice: High-pitched churring.

Status: Common. Territorial. Race *tasmanicus* vulnerable.

Breeding: Aug.–Dec. Nest in a tree hollow, cliff crevice, or frequently nest boxes. 1–5m high / 2–5 eggs.

Left: Adult. Right: Female at nest hollow.

Spotted Nightjar

Eurostopodus argus

290–310mm

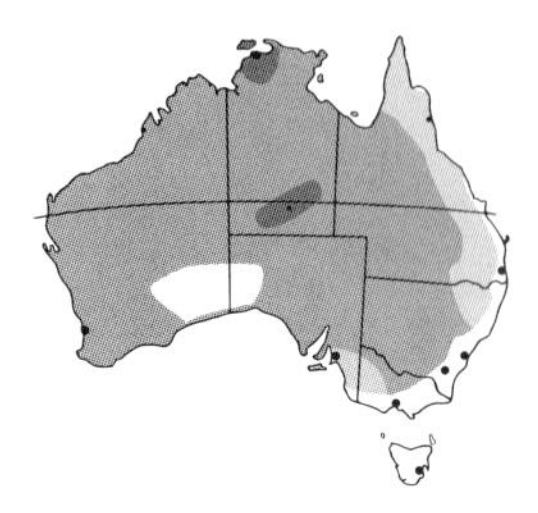

Sexes alike. Larger and more colourful nightjar, perfectly camouflaged among the red or brown soils, rocks, twigs, sticks and leaves of its preferred drier environments. Similar to the White-throated Nightjar, but smaller. Throat: White. Wings: Large white spots.
Immature: Browner, paler.

Usually occurs singularly. Often small groups in migration.

Habitat: West of the Great Dividing Range in drier open, dry woodlands, forests, mallee; stony rises with leaf litter, sparse inland vegetation.

Feeding: Usually nocturnal, flying close to ground for moths, beetles and other insects.

Voice: Musical gobbling call 'caw-caw-caw-gobble-gobble-gobble'.

Status: Moderately common.

Breeding: Sept.–Jan. in south. Dec.–Mar. in north. Egg laid on leaf litter / 1 egg.

Opposite: Adult well camouflaged for typically drier inland environments.

White-throated Nightjar

Eurostopodus mystacalis

320–370mm

Sexes alike. Often heard but not often seen. Upperparts: Mottled, streaked black, brown, beige and grey. Throat: Black with white side patches. Underparts: Dark brown with fine grey barring to breast grading to tan on undertail coverts. Bill: Black. Eyes: Brown, highly reflective.

The largest and darkest Australian Nightjar.

Habitat: Forests, woodlands with ample leaf litter and stony areas.

Feeding: Usually nocturnal, hunting close to ground for moths and other insects.

Voice: Rising series of low 'whook-whook-whook', accelerating to 'laughter'. Also a low crooning.

Status: Moderately common. Rarely seen. Southern birds migrate north in winter to Qld or PNG.

Breeding: Sept.–Nov. Egg laid in leaf litter / 1 egg.

Bunyip SP, Vic.

Opposite: Adult, exceptionally well camouflaged, resting on the forest floor during the day.

Large-tailed Nightjar

Caprimulgus macrurus

270–290mm

Sexes alike. Similar to the White-throated Nightjar but with distinct white patches on wings, best seen in flight. Throat: White band. Tail: Long, pointed with broad white outer tail feathers. Eyes: Brown. Bill: Short, wide, fringed by bristles.
Immature: Duller.

Buoyant and erratic flight.

Habitat: Edges of rainforests dense woodlands.

Feeding: Most active at dusk, flying close to ground for moths and other flying insects.

Voice: Axe-like monotonous 'chop' territorial call, continuously repeated.

Status: Moderately common. Sedentary.

Breeding: Aug.–Jan. Eggs are laid in leaf litter / 2 eggs.

Boonooroo–Tuan area, Qld. East Point, Darwin, NT.

Opposite: Adult nightjar roosting during the day on the ground in leaf litter, relying on its cryptic plumage for camouflage.

15 Family Apodidae, Alcedinidae, Halcyonidae, Meropidae, Coraciidae

Swifts

Swiftlet

Kingfishers

Rainbow Bee-eater

Dollarbird

A varied group containing birds from five different families.

Family Apodidae includes swifts and the endemic Australian Swiftlet. Most species are regular or occasional visitors from the northern hemisphere. The small Australian Swiftlet is the only species that breeds in Australia.
Swifts are aerial specialists, feeding on insects and drinking while flying. Australia's largest swift, the White-throated Needletail, has been observed catching aerial insects at heights of over 1km. Swifts' legs are poorly developed and weak, making walking difficult.
Swifts build cup-shaped nests high up on the walls of dark caves, using plant fibre cemented to the cave wall with saliva. Some species, including the Australian Swiftlet, are echolocating species that emit sharp clicking sounds for orientation inside the usually dark breeding cave. The Australian Swiftlet is also the only bird in the world with a sibling incubation strategy. Two eggs are laid 27 days apart, so that the first nestling has feathers when the second egg is laid and then incubates the second egg. The first nestling fledges the day before the second egg hatches. This strategy allows the parents to raise two young in the often-short breeding season.

Family Alcedinidae contains the plunge-fishing river kingfishers, which include the Azure Kingfisher and Australia's smallest kingfisher, the Little Kingfisher. These birds have a permanent waterfront territory and nest in tunnels up to 1m long, drilled into stream banks.

Family Halcyonidae contains the 'tree kingfisher' group. This family includes two kookaburra species and seven widespread dry-land-feeding kingfisher species. Tree kingfishers nest in tree hollows or excavate hollows in arboreal termite mounds.
The Laughing Kookaburra, the largest of the kingfishers, occupies the same family territory all year round. Their iconic chorus of morning laughter is part of a ritual to advertise their territory and warn off other kookaburra groups. They nest in large, flat-floored tree hollows that have direct access to the entrance so the young can excrete over the edge. Male and female kookaburras form permanent bonds and other members of the group help the breeding pair in nesting duties.
Other members of this group either nest in hollows or excavate nesting chambers, often in termite mounds. Some, such as the Torresian Kingfisher, will choose either option. In 2015 the Torresian Kingfisher, previously a sub-species of the Collared Kingfisher, was recognised as a full species.

Family Meropidae contains the colourful Rainbow Bee-eater. This is the only Australian member of this family. Like other bee-eaters, the Rainbow Bee-eater removes the stings of bees and wasps before eating the insects or feeding them to their young. They are communal birds living in groups of 20–30. When nesting, they excavate a tunnel up to a metre long with a nest chamber at the end, usually in a riverbank.

Family Coraciidae contains the Dollarbird. This is the sole representative of the 'roller' family found in Australia, named for their aerial rolling courtship display flight. The Dollarbird is named for the round silver-blue wing patches seen in flight, said to resemble the Spanish silver dollar coin.

Opposite: Sacred Kingfisher at nest hollow.

House Swift

Apus affinis

130–140mm

Sexes alike. Stocky build. Overall blue-black plumage. Rump: White. Throat: White. Wings: Paler below. Tail: Short with shallow fork when closed, square when fanned.

Fluttering flight; spending most of their life in the air, they catch insect prey and drink on the wing.

Habitat: Over open areas.

Feeding: Small, flying insects.

Voice: Shrill twittering.

Status: Uncommon summer vagrant, but perhaps becoming a more regular visitor.

Breeding: North Africa to Southeast Asia and Japan.

Left: White rump visible in flight. Right: Adult.

Australian Swiftlet

Aerodramus terraereginae

Endemic

110–115mm

Sexes similar. Only swiftlet that breeds in Australia. Dark brown upperparts with faint greyish-white band on rump. Wings: Long, black sheen. Tail: Black, shallow notch. Underparts: Brownish grey. Eyes: Dark brown.
Immature: Paler, duller.

In bad weather, often seen circling high.

Habitat: High over cliffs, forested ranges, grasslands, offshore islands.

Feeding: Small flying insects.

Voice: High-pitched 'cheep'. Echolocating, emitting clicking notes in caves.

Status: Locally common. Nesting sites located by flocks circling above.

Breeding: In colonies of hundreds, in caves in tropical coastal Qld. Sept.–Feb. Small cup-shaped nest of grasses, twigs cemented to wall with saliva / 2 eggs.

Left: Exiting dark nesting cave using echolocating. Right: Detail of nesting colony.

White-throated Needletail

Hirundapus caudacutus

200–220mm

Sexes alike. Large swift with a distinctive, long-curved 500mm wingspan and short, square tail with 'needle-like' protruding feathers. Upperparts: Sooty brown with greenish sheen. Head: Dark olive. Frons, throat, lower flanks, undertail coverts, vent: Whitish. Undertail: Greenish black.
Immature: Duller markings.

Usually in flocks. Fast and high flight, has been observed catching insects at altitudes of over 1km. Roosts in trees.

Habitat: Aerial, mostly east of the Great Dividing Range over wooded areas, farmland, water.

Feeding: Aerial insectivores at very high altitudes or less frequently low to the ground.

Voice: High-pitched chatter.

Status: Non-breeding summer migrant. Arrive: Oct. Depart: Mar.–Apr.

Breeding: from Siberia to Japan.

Left: Small flock. Right: Flight.

Pacific Swift

Apus pacificus

180–210mm

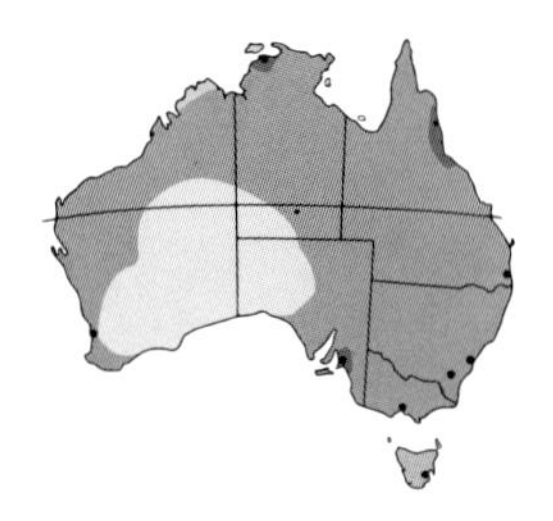

Sexes alike. Also called Fork-tailed Swift. Distinctive, long, deeply forked tail and long, narrow, scythe-shaped, finely pointed wings. Body: Sooty black. Throat, chin, rump: White.

Often in flocks ranging from 100–1000. In Australia, flocks have been recorded reaching tens of thousands while feeding on insect swarms.

Habitat: Aerial, often following low-pressure systems, mostly inland plains, less frequently coastal and urban areas.

Feeding: Aerial insectivores over water or open ground. At high altitudes or as low as 1m.

Voice: Shrill chattering.

Status: Common. Non-breeding summer migrant. Arrive: Oct. Depart: Mar.–Apr.

Breeding: Eastern Asia.

Left: Adult, showing white rump patch. Centre: Tail fully spread. Right: Tail forked when closed.

Azure Kingfisher

Ceyx azureus

170–190mm

Sexes alike. Rich azure blue upperparts with white throat and neck plumes. Bill: Long, black. Eyes: Dark, with small orange spot before eyes. Underparts: Orange-tan, blue to sides of breast. Legs, feet: Orange with 3 toes, 2 pointed forward.
Race *ruficollaris*, tropical north: Smaller, intense blue flanks.
Race *diemenensis*, endemic to Tas.: Largest and deepest hued. Smaller bill.
Immature: Duller. Whitish legs and feet.

Usually seen hunting from a low branch overhanging water.

Habitat: Coastal watercourses, mangroves, wetlands.

Feeding: Dives for fish and aquatic life. Usually returning to perch to batter prey before swallowing it whole.

Voice: A shrill flight whistle.

Status: Moderately common. Race *diemenensis* endangered.

Breeding: Sept.–Jan. in south. Oct.–Apr. in north. Excavates tunnel and nest chamber in riverbank / 4–7 eggs.

Daintree River, Qld. Royal NP, NSW. Croajingolong NP, Vic. Middle Arm, Darwin, and Yellow Waters cruise, Kakadu NP, NT.

Left: Nominate race. Right: Race *ruficollaris*.

Little Kingfisher

Ceyx pusillus

120–130mm

Sexes alike. Rich, glossy blue upperparts. Underparts: White with blue on sides of breast. Eyes: Dark brown with white spot in front. Neck: Broad white plumes. Legs, feet: Black with 3 toes, 2 pointed forward.
Nominate race, occasional visitor to the tip of Cape York.
Race *halli*, coastal Qld: Larger, deeper blue with least blue to sides of breast.
Race *ramsayi*, Top End, NT: Large blue patches on sides of breast, sometimes forming a complete breast band.
Immature: Duller, pink legs.

Wary. Usually seen on a low perch, motionless, scanning the water. Deep-diving after prey. Sometimes hovers.

Habitat: Mangrove-lined estuaries and rainforests along watercourses.

Feeding: Fish, other aquatic life.

Voice: Shrill, high-pitched whistle in flight.

Status: Uncommon. Solitary. Resident.

Breeding: Oct.–Mar. Drills tunnel 10–15cm long with a nest chamber at the end, in the side of embankment or between mangrove roots, in epiphyte or decaying stump / 4–5 eggs.

Daintree River and Centenary Lakes Botanic Gardens, Cairns, Qld. Yellow Waters cruise, Kakadu NP, NT.

Left: Race *halli*. Right: Race *ramsayi*.

Buff-breasted Paradise-Kingfisher

Tanysiptera sylvia

290–320mm

Sexes similar. Spectacular kingfisher with iridescent royal blue crown and wings and a white patch on the centre of the back. Bright orange-yellow underparts and distinctive long tail streamers that protude 180mm beyond rest of tail.
Eyes: Dark with black eye stripe. Rump: White. Bill, legs: Orange-red.
Female: Duller, protruding tail feathers shorter. Immature: Dark bill, no protruding tail feathers, fine grey scalloping to breast.

Although brightly coloured it is difficult to see in rainforest canopy.

Habitat: Lowland coastal rainforests.

Feeding: Hunts prey from ground or low branch, taking insects, small reptiles, invertebrates.

Voice: Series of 'chop-chop' notes, repeated.

Status: Uncommon. Breeding migrant from New Guinea. Appears early Nov.–Apr.

Breeding: Nov.–Jan. Excavates nest in termite mound, with a nesting chamber at end of a 150mm long tunnel / 3–4 eggs.

Iron Range NP and Cairns area, Qld.

Left: Male showing underparts. Right: Male.

Laughing Kookaburra

Dacelo novaeguineae

Endemic

400–450mm

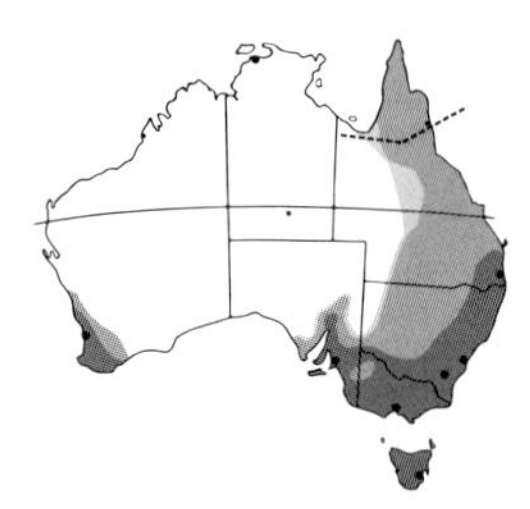

Sexes alike. Whitish head with mottled brown on crown and wide, pronounced brown eye stripe. Underparts: Whitish. Upperparts: Brown with mottled blue wing patches and variable blue rump. Tail: Rufous brown, barred black, white outer feathers and tip. Bill: Very large, upperpart black, lower bill beige-grey. Race *minor*, far north Cape York, Qld: Smaller.
Immature: Shorter tail, black bill.

A familiar bird well adapted to urban environments. May become tame.

Habitat: Areas with large eucalypts with nest hollows. Forest edges, clearings, woodlands, parks and gardens.

Feeding: Perches, scanning for prey, pouncing on insects, crustaceans, small mammals, reptiles or birds.

Voice: Rollicking laughing chorus advertises territory.

Status: Common. Resides in family groups. Introduced to southern WA and Tas.

Breeding: Sept.–Jan. In tree hollows or sometimes in suitable nest boxes / 1–4 eggs.

Left: Immature. Right: Adult.

Blue-winged Kookaburra

Dacelo leachii

400–480mm

Sexes similar. Distinctive large bill, pale eyes and mostly blue wings. Head: Off-white with brown streaks and brown eye stripe. Bill: Dark bluish grey above, pale pinkish below.
Eyes: White. Upperparts: Brown with blue shoulders, rump and mottled blue on wings. Tail: Deep blue, white tip. Breast, underparts: Finely scalloped buff. Race *occidentalis,* far western WA: Larger, whiter, less streaked head. Breast, underparts: Finely scalloped rufous-buff.
Female: All have rufous tail barred black with a pale tip.
Immature: Similar to female, paler head, brown underparts.

More retiring than the Laughing Kookaburra.

Habitat: Eucalypt woodlands along watercourses, paperbark swamps, savannah woodlands, farmland, parks, gardens.

Feeding: Sits motionlessly scanning for prey. Insects, reptiles, frogs, aquatic life, spiders, snakes or small birds.

Voice: Harsh, loud howls with trills and twittering.

Status: Common. Sedentary. Locally nomadic.

Breeding: Sept.–Jan. In tree hollow to 25m high / 2–5 eggs.

Kakadu NP, NT. Broome Bird Observatory, WA.

Opposite: Nominate race male (left) and female (right).

Yellow-billed Kingfisher

Syma torotoro

180–210mm

Sexes similar. Distinctive, rusty orange head, neck and bill. Nape: Black patch. Eyes: Black with black partial eye ring in front of eye. Throat: White. Back, wings: Dull green. Rump, tail: Blue. Eyes: Dark with black eye stripe. Underparts: Light orange-grey.
Female: Black crown and paler underparts. Immature: Similar, neck and breast flecked black. Bill mostly black.

Usually solitary. Both sexes have crests. Male crest is used in breeding display.

Habitat: Fringes of rainforests, monsoon forests and mangroves, preferably adjoining eucalypt woodlands.

Feeding: Usually hunts from a low exposed branch, large insects, small reptiles, earthworms.

Voice: Mournful, descending trill and whistles.

Status: Uncommon.

Breeding: Nov.–Jan. Excavated nest chamber in arboreal termite mound / 3–4 eggs.

Iron Range NP, Qld.

Left: Female. Right: Male.

Forest Kingfisher

Todiramphus macleayii

190–220mm

Sexes similar. Violet-blue upperparts with turquoise blue back and rump. Nape, throat: White collar. Underparts: White, sometimes buff flanks. Eyes: Dark brown with a white patch in front and a broad black eye stripe. Frons: 2 large white spots. Bill: Long, straight, mostly black with lower pink stripe. Underwings: White patch seen in flight.
Race *incinctus*, eastern Australia: Smaller white wing patches. Greener back.
Female: All have incomplete neck collar. Immature: Duller, darker crown, buff scalloped underparts.

Excavates the nest hollow in an arboreal termite mound by flying straight at it from several metres, striking it with great force.

Habitat: Savannah woodlands, open coastal forests, mangroves, wooded swampy areas.
Feeding: Hunts from low, exposed branch or ground, aquatic life, insects, small reptiles, worms.
Voice: Rapid, harsh, trilling and whistles.
Status: Moderately common. Solitary or pairs. Southern birds migrate north for winter.
Breeding: Aug.–Dec. In a termite mound / 3–6 eggs.

Kakadu NP, NT.

Left: Female. Right: Male.

Torresian Kingfisher

Todiramphus sordidus

230–270mm

Sexes similar. Colour variations across range, becoming browner from east to west. Dull, blue-green upperparts with a broad white collar and white spot before eyes. Eyes: Dark with black eye stripe. Wings: Blue, green. Underparts: White. Tail: Upperpart blue. Bill: Large, heavy.
East coast: Greenest.
Tropical north, Qld, NT, WA: Brown-green. Black eye stripe and dark crown.
Pilbara region, WA: Brownest, smallest.
Female: Duller. Immature: Duller, flecked plumage.

Flight usually swift, low, direct.

Habitat: Mangroves, tidal flats, estuaries, nearby parks, gardens.
Feeding: Picks prey from tidal flats and shallow water, favours crabs. Insects captured mid-air. Also lizards, occasionally birds.
Voice: 2-note call 'kik-kik', repeated rapidly several times.
Status: Common. Vulnerable NSW. Mostly sedentary in north. Southern birds migrate north and overwinter on offshore islands, New Guinea.
Breeding: Sept.–Mar. Usually in hollow in mangrove tree or hollow drilled into termite nest close to mangroves 4–15m high / 2–4 eggs.

East Point and Middle Arm, Darwin, NT. Daintree River and Cairns Esplanade, Qld.

Left: Male from east coast. Right: Male from the tropical north.

Red-backed Kingfisher

Todiramphus pyrrhopygius

Endemic

200–240mm

Sexes similar. Small kingfisher with a distinctive bright rufous rump and green and white streaked crown. Eyes: Brown, with pronounced brown eye stripe extending around nape. Wings, tail: Blue-green. Nape: White collar extending to white underparts. Bill: Large, black, pale below.
Female: Duller, paler, less streaking to head. Immature: Duller, white flecked shoulders, scalloped nape collar and sides of breast.

Quiet and unobtrusive.

Habitat: Mallee, mulga, savannahs, woodlands. Often far from water, obtaining enough moisture from insects for survival. Avoids forests.
Feeding: Hunts insects, frogs, reptiles and small mammals from an open perch.
Voice: Mournful descending whistles.
Status: Common. Highly nomadic. Southern birds move to northern Australia for winter.
Breeding: Aug.–Feb. In tree hollow, or tunnels into a vertical creek bank / 3–6 eggs.

Hattah-Kulkyne NP, Vic. Strzelecki Track, SA.

Left: Female feeding on a reptile. Right: Male.

Sacred Kingfisher

Todiramphus sanctus

190–230mm

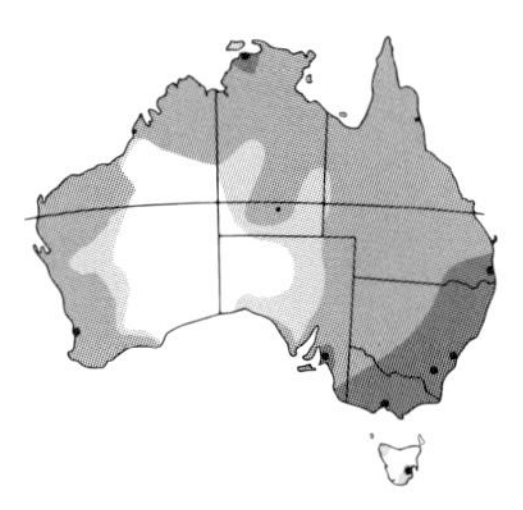

Sexes similar. Mid-sized bird with a turquoise green crown and back and turquoise blue wings, rump and tail. Eyes: Beige spots before dark eyes and black eye stripe. Nape: Broad, white collar. Underparts: Light beige to white. Female: Whiter. Upperparts: Greener. Immature: Duller, brownish scalloping to underparts and nape.

Solitary. Usually hunts from a low, exposed branch, returning to same perch to eat.

Habitat: Mangroves and all open wooded areas except arid inland areas.

Feeding: Rarely eats fish, preferring terrestrial prey. Small reptiles, insects, beetles, crustaceans.

Voice: Repeated 'kee-kee-kee'.

Status: Common. Many birds move north for winter.

Breeding: Sept.–Mar. Excavated nest chamber in arboreal termite mound, steep riverbank or hollow limb / 3–6 eggs, often 2 broods in a season.

Left: Male. Right: Immature.

Rainbow Bee-eater

Merops ornatus

210–240mm

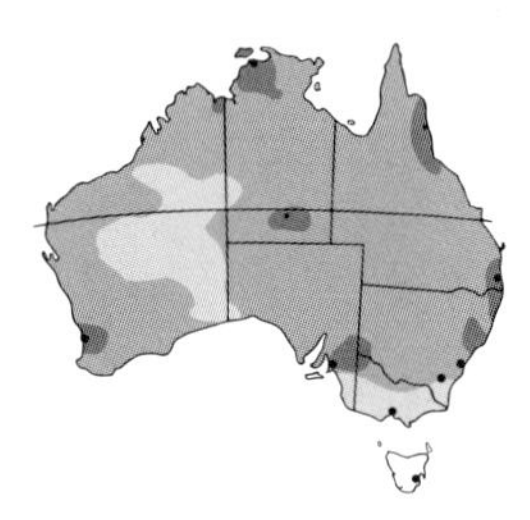

Sexes similar. Named for its plumage, containing the colours of a rainbow. Plumage is mostly bright green with a sky blue rump. Crown: Golden head with broad black eye stripe and bright blue stripe below. Throat: Orange-gold with broad, black band below. Breast: Green. Belly: Blue. Tail: Black with thin, streaming, central feathers. Bill: Black, long, curved. Underwings: Bright orange.
Female: Similar with shorter tail streamers. Immature: Indistinct throat band, duller, without tail streamers.

Habitat: Open country, woodlands, savannahs, shrubland, semi-cleared areas, usually near water. Avoids forests, rainforests, treeless areas.

Feeding: Mostly bees, wasps and other insects taken in flight. Sometimes from the ground or hovering over water. Plunges for aquatic life. Prey is taken back to the perch, battered then eaten.

Voice: Musical trilling 'ptrrrp-ptrrrp-ptrrrp'.

Status: Moderately common. Seasonal migration north to offshore islands, New Guinea.

Breeding: Nov.–Jan. in south. Sept.–Oct. and May in north. Tunnel drilled up to 1m long in a sandy embankment leading to nest chamber / 3–7 eggs.

Left: Male at nest chamber entrance. Right: Female (on left) and male.

Dollarbird

Eurystomus orientalis

260–300mm

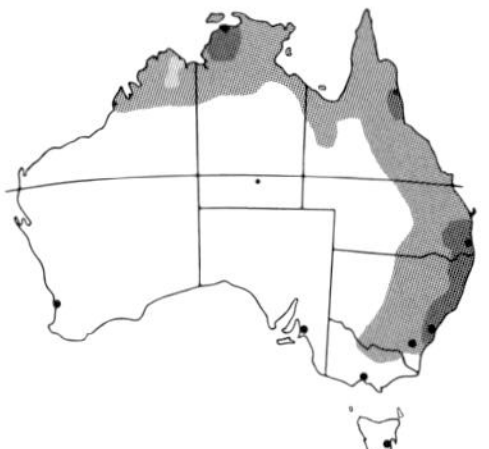

Sexes similar. Overall dull green appearance. Head, neck, breast: Dark brown. Bill: Bright red tipped black. Throat patch, undertail coverts: Bright blue. Wings: Deep blue-green with deep green coverts. Underwings: Silver-blue marking visible in flight.
Female duller. Immature: Duller, lacks blue throat patch. Bill, feet: Blackish.

Agile, tumbling, swooping, diving flight.

Habitat: Edges of tropical rainforests and open woodlands, nest hollows often along watercourses.

Feeding: Almost entirely large flying insects, pursued and snatched mid-air, returning to perch to batter and eat. Rarely ground feeding.

Voice: Harsh, nasal squawking.

Status: Moderately common. Breeding migrant. Arrive: Sept. Depart: Mar.–Apr.

Breeding: Oct.–Feb. In a shallow tree hollow 10m or higher / 3–5 eggs.

Left: Immature. Centre: Male. Right: In flight, showing 'dollar' markings.

16 Family Menuridae, Atrichornithidae, Climacteridae, Pittidae

A grouping of four families that commence the great songbird order, Passeriformes, which contains over half of Australia's bird species.

Family Menuridae contains Lyrebirds.
There are two species of lyrebird, both unique to Australia. The male Superb Lyrebird is noted for its vocal repertoire, mimicry and shimmering 'lyrate' plumes. During its display dance on its specially constructed raised display platform, the tail plumes are raised, fanned and held forward over the head. Their mimicry can include mechanical or urban sounds such as car engines, chainsaws or dogs barking, but more commonly the calls of other birds and mammals. The sequence of calls is learned from older birds and passed down, young birds taking 12 months to learn the sequence. They were introduced to Tasmania in the 19th century, where the calls of the birds today can still include those sequences retained from mainland ancestors, in addition to the sounds found in their adopted state. After the 2009 Black Saturday fires in Vic., researchers noted that areas inhabited by lyrebirds were less affected by fire. Their habit of raking through leaf litter reduces the litter fuel load by an average of 25% and suppresses the growth of vines that carry the fire into the canopy.
The smaller Albert's Lyrebird has a less spectacular display and less developed lyrates and plumage. It is restricted to a small area near the NSW–Qld border.

Family Atrichornithidae contains two species of scrub-bird.
Unique to Australia, scrub-birds live their lives scurrying mouse-like under the cover of the dense understorey vegetation of damp forest areas. The Noisy Scrub-bird was thought to be extinct until it was rediscovered in 1983. DNA studies have suggested that these birds are living fossils whose ancestors evolved 97–65 million years ago. They evolutionarily separated from their closest relative, the lyrebird, 30–35 million years ago.

Family Climacteridae contains the treecreepers.
This is a small family of seven species, with one species found in New Guinea, the remainder confined to Australia. Treecreepers are acrobatic birds that hunt for insects on the trunks of trees, climbing in a distinctive way by moving one foot forward and sliding the lower foot up to the higher foot and leading again with the same foot, appearing to shuffle up the tree. Using their large feet and long claws, they can feed on the underside of tree limbs or hang vertically as they probe the bark crevices for insects. Some species will also forage on the ground, often for ants. They nest in tree hollows or crevices, well lined with grass, feathers and animal fur. Most treecreepers are communal breeders, with group members helping with incubation and raising the young.

Family Pittidae contains the pittas.
There are three brilliantly coloured representatives in Australia. Pittas use selected rock 'anvils' to noisily break the shells of large rainforest land snails, eating the flesh and leaving tell-tale shell middens scattered throughout the forest. Although colourful, they are well camouflaged in their environment.

Opposite: The Noisy Scrub-bird, one of Australia's rarest birds.

Superb Lyrebird

Menura novaehollandiae

Endemic

800–1000mm

including 450mm tail

Noted for its elaborate tail and superb mimicry, the Superb Lyrebird resembles a large brown pheasant. Dark brown upperparts with reddish brown throat. Underparts: Grey-brown. Bill, legs, feet: Black. Tail: Long, lacy, filamentous feathers, with 2 outer 'lyre'-shaped feathers. Female: Smaller; tail lacks 'lyre' feathers. Immature: Similar to female.

Elusive and usually solitary, the male Superb Lyrebird has an elaborate courtship display and promiscuous mating system. He attracts potential mates by dancing on a display mound singing his song repertoire for up to 20 minutes, while throwing his ornate quivering tail forward over his body, creating a shimmering silvery white canopy of 'lyrate' plumes. Females secure their own territory, build their bulky dome-shaped stick nest on a rock ledge or in a hollow stump, and may visit several performing males before mating.

Display performances occur on prepared earth platforms about 1m in diameter and raised 150mm, the male building up to 15–20 in his territory.

Habitat: Dark, damp tree fern–lined gullies in wet eucalypt forests.

Feeding: Scratches in leaf litter for worms, spiders and insects.

Voice: Resounding loud, clear song, rattling, metallic whirring, thudding whistle and expert mimicry including farm and forest sounds. Their song can include the songs of up to 20 other bird species, camera shutters, chainsaws, etc. Unusual in most species, lyrebirds breed in the middle of winter, the male bird beginning the day singing from half an hour before sunrise from a high roosting perch. They sing less at other times of the year but can often be heard on a foggy cold morning. Females also sing but are less resounding.

Status: Sedentary, moderately common. Introduced Tas.

Breeding: Apr.–Sept. Bulky, dome-shaped nest well camouflaged with ferns, moss and rootlets, to 3m high / 1 egg.

Royal NP, NSW. Dandenong Ranges NP, Vic.

Top: Male in display. Centre: Female.

Albert's Lyrebird

Menura alberti

Endemic

850mm including 400mm tail

Sexes similar. Large ground-dwelling songbird. Rich, reddish brown throat and upperparts. Underparts: Grey-buff. Tail: Long, glossy black, lace-like tips. In display the male fans and quivers his silvery tail forward, enveloping the body, while calling, singing and prancing. Female: Shorter tail and lacks 'lacy' feathers. Immature: Similar to female.

Terrestrial, extremely wary and rarely seen.

Habitat: Rainforests and wet sclerophyll forests in a small area in the McPherson and Tweed Ranges area on the Qld–NSW border.

Feeding: Scratching for insects and insect larvae, snails and invertebrates in moist leaf litter.

Voice: Powerful voice with own metallic song and expert mimicry. During the breeding season, males sing for up to 4 hours a day.

Status: Vulnerable, through loss of rainforest habitat and predation by foxes and cats.

Breeding: June–July. Large dome-shaped nest usually placed on a rock ledge, stump or ground.

Lamington NP, NSW.

Opposite: Female.

Rufous Scrub-bird

Atrichornis rufescens

Endemic

165–180mm

Sexes similar. Stocky, ground-dwelling songbird. Dark reddish brown upperparts with fine black barring and long barred tail. Throat: Rufous-buff, mottled black with white streaks to sides. Underparts: Reddish brown streaked white on sides. Female: Paler breast markings and white throat. Immature: Duller, greyish throat. Chestnut underparts.

Shy. Rarely flies, runs mouse-like, actively foraging in leaf litter.

Habitat: Deep, tangled, leaf-littered understorey in moist highland forests.

Feeding: Small invertebrates, including snails, spiders, insects.

Voice: Loud, penetrating territorial song in the breeding season. Ringing, chattering and mimicry. 2 races with markedly different vocalisations.

Status: Rare, vulnerable, through habitat loss, logging and increased frequency of bushfires.

Breeding: Sept.–Dec. Domed nest with tunnel entrance placed close to ground / 2 eggs.

Lamington NP, Qld, and Barrington Tops NP, NSW.

Left and right: Female.

Noisy Scrub-bird

Atrichornis clamosus

Endemic Endangered

220–230mm

Sexes similar. One of Australia's rarest birds. Believed extinct, until a population was found in the 1960s. Distinctive white throat and chin with a blackish triangle-shaped patch. Breast: Blackish grading to off-white. Belly, flanks: Buff-rufous. Head, upperparts: Brown with fine black barring. Female: Smaller, lacks black markings. Immature: Similar to female, dark grey rump, tail and yellow gape.

Shy, difficult to see. Runs quickly through undergrowth. Flight is feeble.

Habitat: Moist areas of unburnt, densely vegetated mountain gullies, swampy areas, shrubby heathland, with deep leaf litter.

Feeding: Forages on the ground in leaf litter for insects and small invertebrates, occasionally seeds.

Voice: Male has piercing staccato notes that travel over 1km.

Status: Endangered, through habitat loss, logging and increased frequency of bushfires.

Breeding: Apr.–Oct. Well-hidden, rounded nest with side entrance, lined with decaying wood, usually near ground / 1 egg.

Two Peoples Bay NR and Waychinicup NP, WA.

Left and right: Female.

White-throated Treecreeper

Cormobates leucophaea

Endemic

140–170mm

Sexes similar. Mostly arboreal. Dark olive-brown upperparts and grey rump. Throat, breast: White. Belly, flanks: Striped. Undertail: White with black barring. Eyes: Brown. Race *minor*, far north Qld: Smaller, darker with greyish throat. Qld birds are generally darker. Female: All have a reddish orange spot below cheeks. Immature: Similar to female, but larger orange spot, plainer underparts. All immature females have bright chestnut rumps.

Mostly arboreal. Spirals up tree trunks gleaning insects. Solitary outside breeding season.

Habitat: Rainforests, eucalypt forests, woodlands, banksia scrub, mallee and brigalow.

Feeding: Prefers probing in rough bark for insects, invertebrates. Takes nectar and ants.

Voice: Quickly repeated piping whistles. Upward inflecting, loud musical ringing and trilling song.

Status: Sedentary, territorial.

Breeding: Aug.–Jan. Cup-shaped nest in a tree hollow / 2–3 eggs.

Eungella NP, Qld. Royal NP, NSW. Dandenong Ranges NP, Vic.

Left: Male. Right: Female.

White-browed Treecreeper

Climacteris affinis

Endemic

135–150mm

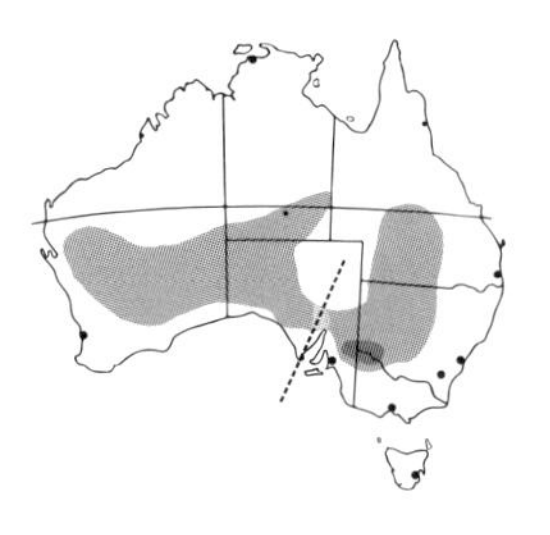

Sexes similar. Resembles the Red-browed Treecreeper but with a white eyebrow and heavier black and white streaked ear coverts and belly. Undertail coverts: Barred black and white. Nominate race: Grey-brown rump.
Race *superciliosa*, western SA, WA: Smaller. Reddish brown rump.
Female: All have a rufous eyebrow and fainter chest streaks, sometimes obscured. Immature: Similar, indistinct eyebrow.
Usually arboreal, sometimes searches fallen logs for prey.

Usually in pairs or family group.

Habitat: Semi-arid areas. White Cypress Pine, Mulga, brigalow and Belah woodlands.

Feeding: Forages spiralling up tree trunks and branches or among ground litter for insects, mostly for ants.

Voice: Sharp, metallic call and rapid trills.

Status: Uncommon.

Breeding: Aug.–Dec. Nest in tree hollow / 2–3 eggs.

Kunoth Well, Alice Springs, NT.

Left: Nominate male. Right: Nominate female.

Red-browed Treecreeper

Climacteris erythrops

Endemic

145–160mm

Rufous-brown face and broad red-rufous eyebrow. Upperparts: Dull brown. Crown, nape: Grey. Chin, throat: White. Breast, flanks: Olive-grey. Belly: Boldly streaked with white feathers edged with black. Undertail coverts: Barred black and white. Female: Similar with rufous-streaked chest. Immature: Grey face.

Prefers the upper canopy of smooth-barked trees. Communal, in family groups.

Habitat: Cool, dense, wet eucalypt forests with tall trees, along gullies and watercourses in mountainous areas.

Feeding: Almost exclusively on insects. Forages among peeling bark strips or sometimes on fallen timber.

Voice: High-pitched staccato chatter and squeaks.

Status: Uncommon, sedentary.

Breeding: Aug.–Jan. Bark- and fur-lined nest placed in tree hollow 6–30m high / 2–4 eggs.

Lamington NP, Qld.

Left: Male. Right: Female.

Black-tailed Treecreeper

Climacteris melanurus

Endemic

155–170mm

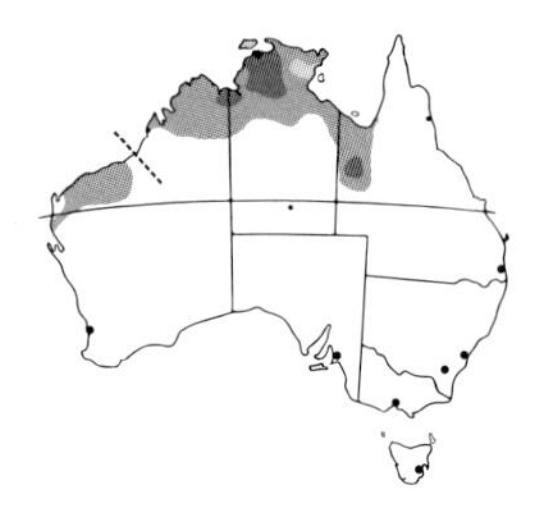

Sexes similar. Large, overall dusky brown treecreeper with faint buff streaks on ear covert and black and white streaked chin, throat, upper breast. Wings: Buff patch.
Isolated race *wellsi*, Pilbara region, WA: Smaller and paler. Female: Both races have white chin and throat. Upper breast: Rufous and white striped. Immature: Duller, throat white, edged black.

Forages on fallen timber or spirals up tree trunks probing for prey. Usually communal in groups of 2–6.

Habitat: Eucalypt woodlands and open forests.

Feeding: Ants and other insects.

Voice: Sharp, piping whistle call. Similar to White-throated Treecreeper.

Status: Common. Sedentary.

Breeding: Sept.–Jan. Grass nest in tree hollow 4–10m high / 2–3 eggs.

Carnarvon NP, Qld. Kakadu NP, NT.

Left: Male. Right: Female.

Rufous Treecreeper

Climacteris rufus

Endemic

150–170mm

Sexes similar. Most richly coloured treecreeper. Crown, neck: Grey. Upperparts: Ash brown. Wings: Greyish with pale rufous band. Brow, face, underparts: Bright rufous. Chest: Rufous washed buff-grey. Throat: Centre freckled black.
Female: Slightly smaller. Throat: Rufous freckled. Immature: Darker.

Appears to spend more time on ground probing for ants than probing tree bark for insects.

Habitat: Open eucalypt woodlands, open Jarrah forests and mallee where trees provide nesting hollows.

Feeding: Spirals up trees probing the bark for insects or forages on the forest floor.

Voice: Single, piping whistle call and rapid, piping, high-pitched whistled song.

Status: Moderately common.

Breeding: July–Nov. Grass and bark cup nest in tree hollow, up to 8m high. Sometimes in a hollow log / 1–4 eggs.

Porongurup NP and Dryandra SF, WA.

Left: Male. Right: Female.

Brown Treecreeper

Climacteris picumnus

Endemic

160–180mm

Sexes similar. Largest Australian treecreeper. Light to mid-grey-brown upperparts, pale grey face and whitish eyebrow. Upper breast: Centre patch finely streaked black. Breast, belly: Fine black lines. Lower flanks: Washed rufous. A distinctive, buff wing bar is visible in flight.
Race *melanotus*, far north Qld: Darker, more defined brow line.
Race *victoriae*, southeast coastal NSW, Vic.: More rufous. Most commonly seen on inland slopes and plains of Great Dividing Range.
Female: All have upper breast finely streaked rufous, often hidden. Immature: Darker, lower underparts washed rufous.

Unlike other treecreepers, the Brown Treecreeper spends most of its time foraging on the ground for insects and infrequently spirals up tree trunks.

Usually communal in small groups of up to 5.

Habitat: Open, dry eucalypt forests, woodlands with open ground and fallen timber. Mallee and stands of River Red Gum.

Feeding: Forages mostly in leaf litter and on fallen logs for ants and other insects.

Voice: Staccato 'whit-whit-whit'.

Status: Common. Sedentary.

Breeding: Co-operatively, with a breeding pair and sub-ordinate male 'helpers'. June–Jan. Nest in tree hollow 1–3m high / 2–3 eggs.

Glen Davis, NSW. Chiltern–Mt Pilot NP, Vic.

Centre row: Race *victoriae*.
Below left: Race *melanotus*. Below right: Nominate race.

Red-bellied Pitta

Pitta erythrogaster

160–180mm

Sexes similar. Upper breast and belly are richly coloured scarlet. The throat is black and the chest is bright iridescent blue, edged black. Upperparts, tail: Blue-green. Crown, frons: Brown. Nape: Reddish brown. Female: Duller. Immature: Olive-brown with white throat band, pale brown below.

Singularly or in pairs. Flight is close to ground.

Habitat: Rainforests, moist lowlands and scrub.

Feeding: Hops through leaf litter pecking for insects, invertebrates. Within its territory it has specific rocks, 'anvils' for shattering snail shells on.

Voice: Long, mournful whistles.

Status: Breeding summer migrant from New Guinea. Appears from Oct. to Mar. or Apr.

Breeding: Oct.–Dec. Twig and grass, dome-shaped nest, on ground or up to 3m high in dense vines / 2–4 eggs.

Wet season: Iron Range NP, Lockerbie scrub, tip of Cape York, Qld.

Left: Male. Right: Female.

Noisy Pitta

Pitta versicolor

170–200mm

Sexes similar. Short, stumpy green tail and green upperparts with a black head and throat. Crown: Reddish brown cap with a black centre. Rump: Iridescent turquoise. Underparts: Yellow-tan. Undertail coverts: Bright red. Shoulder: Yellowish stripe. Wings: Blue patch visible in flight. Leg, feet: Pinkish. Female: Undertail coverts: Pinkish red. Immature: Duller.

Shy, solitary, usually seen hopping along the forest floor.

Habitat: Rainforests, moist lowlands and scrub.

Feeding: Forages for large snails, smashing shells on a stone or timber 'anvil' at selected sites, which can be identified by the littered broken snail shells. Grubs, insects and berries are also eaten.

Voice: Loud, resonant whistle 'walk-to-walk' repeated twice.

Status: Uncommon.

Breeding: Oct.–Jan. Large, bulky, dome-shaped nest, well camouflaged on or near ground, often between the buttress roots of rainforest trees, with a 'doormat' of moist animal dung.

Iron Range NP, Broadwater near Ingham, and Lamington NP, Qld.

Left: Side view of adult. Right: Upperparts.

Rainbow Pitta

Pitta iris

Endemic

160–180mm

Sexes similar. Bright iridescent turquoise blue shoulder patch. Bright green, golden glossed upperparts are well defined against a black head and mostly black underparts. Crown: Cinnamon marked. Undertail coverts: Bright red. Flanks: Cinnamon. Tail: Short, green. Bill: Strong, black. Eyes: Dark brown. Female: Smaller, duller. Immature: Duller.

Mostly terrestrial. Territorial displays such as 'bowing' and wing flicking. Relatively tame.

Habitat: Coastal monsoon vine forests, bamboo thickets and mangroves.

Feeding: Foraging in leaf litter for earthworms, particularly in the breeding season. Also spiders, caterpillars, snails, small frogs, skinks, fruit. Uses selected 'anvils' for breaking snail shells.

Voice: 3 or 4 note territorial whistle.

Status: Moderately common. Sedentary.

Breeding: Oct.–Mar. Domed nest from ground to 3m high / 3–4 eggs.

Fogg Dam CR, Howard Springs NP, and East Arm, Darwin, NT.

Left: Side view of adult. Right: Back view of adult.

17 Family Maluridae

Fairy-wrens

Emu-wrens

Grasswrens

Family Maluridae is a small Australasian family containing the fairy-wrens, emu-wrens and grasswrens. Although called wrens they are not related to the 'true wrens' of the northern hemisphere, being incorrectly named by the early settlers.

There are nine endemic species of fairy-wren spread across Australia, with about five further species also found in New Guinea. Plus there are three endemic species of emu-wren and ten species of grasswrens.

Fairy-wrens live in small territorial communal groups made up of a permanently bonded breeding pair, their young and usually additional non-breeding males. The female builds a spherical nest with a side entrance, placed in a dense shrub or grass tussock usually about 1m above ground. The nest is built from grass stems woven with cobweb and lined with soft feathers and fine grass. The male protects the territory with 'song battles' with neighbouring fairy-wrens while the female incubates eggs and broods the young. All group members look after the young, allowing the breeding pair to raise two to four broods a season.

A 2017 genetic study found that the common ancestors of the superb and splendid fairy-wrens diverged approximately 4 million years ago. Thier common ancestor had diverged 7 million years ago from the genetic tree that produced the white-winged, red-backed and the white-shouldered (found in New Guinea) fairy-wrens.

Emu-wrens are among Australia's smallest birds. Fairy-wren-like, they are noted for their extremely long, thin, lacy tails of six filamentous feathers, thought to resemble emu feathers, and from which their name is derived. The birds are shy, and being small, brown and running mouse-like with their tails down, they are sometimes mistaken for bush mice as they quickly dart for dense cover. The three species of emu-wren live in distinctively different habitats.

Grasswrens are shy and elusive, with cryptic plumage, and mostly found in inland arid zones, where they are well camouflaged in areas containing mostly rocky spinifex or canegrass. Their range is restricted and inhospitable and may explain why they are not well known. For example, the Eyrean Grasswren inhabits clumps of Sandhill Canegrass in a small area of the Simpson Desert and had not been seen since 1874 when museum specimens were collected. Believed most likely extinct, it was not until the late 1970s that it was rediscovered.

Like all wren species, the female grasswren builds a bulky dome-shaped nest well hidden in a spinifex or canegrass tussock.

Opposite: Male Superb Fairy-wren in breeding plumage.

Superb Fairy-wren

Malurus cyaneus

Endemic

120–150mm

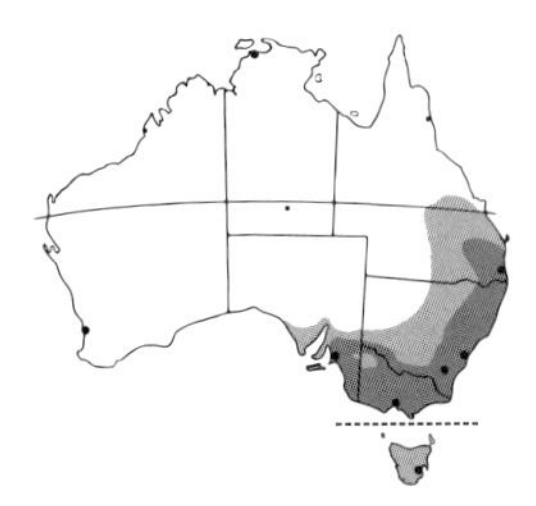

The familiar 'Blue-wren' of southeastern Australia. Easily recognised, with a blue-black head, nape, throat, chest, with iridescent bright blue crown, ear coverts and back. Underparts: Buff-white. Tail: Blue. Non-breeding: Olive-brown with dark blue tail. Dominant male retains its colour all year.
Nominate race: Endemic Tas. Largest and darkest. There are several similar mainland races.
Female: Upperparts: Light brown. Bill, lores, eye ring: Reddish. Throat, chest: Whitish. Belly: Fawn. Tail: Brown. Legs: Pale. Immature: Similar to female, but males develop a bluish tail and black bill.

Reside in family groups. Well adapted to urban areas.

Habitat: Varied. Shrubby, open forests and woodlands, coastal heaths, vegetation along watercourses, parks and gardens.

Feeding: Briskly hops through low thickets. Eats small insects, spiders, occasionally small fruits.

Voice: Vigorous, musical trilling warble, repeated.

Status: Common, sedentary.

Breeding: July–May. Often several broods. Dome-shaped nest to 6m high / 3–4 eggs.

Left: Adult male in breeding plumage.
Right: Non-breeding male.

Splendid Fairy-wren

Malurus splendens

Endemic

115–135mm

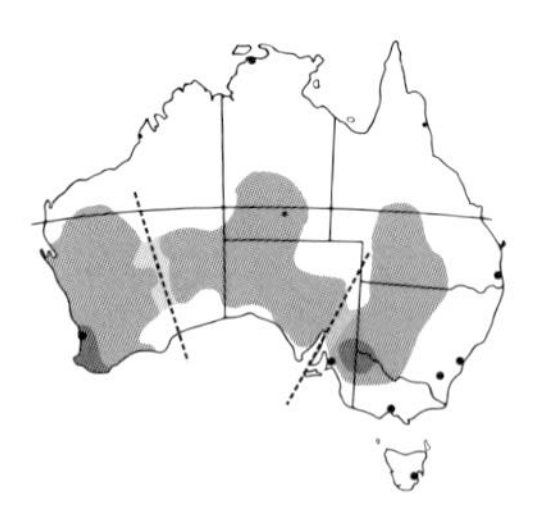

The familiar 'Blue-wren' of southwestern WA. Rich, incandescent violet-blue crown, upperparts and underparts. Back: No black on rump. Ear coverts: Sky blue. Lores, nape: Black. Tail: Long, slender, cobalt blue. Breast, throat: Cobalt blue washed with a violet sheen and wide black breast band. Bill: Black.

There are several similar races, including:
Nominate race, WA.
Race *musgravi*, Turquoise-wren central inland: Upperparts, ear coverts: Sky blue. Throat: Deep violet. Back: Wide black band across rump.
Race *melanotus*, Black-backed Fairy-wren, Qld, NSW, Vic. and eastern SA: Crown, mantle: Violet-blue. Ear coverts: Light blue. Throat: Blue, tinged violet. Narrow black breast band. Rump: Black.
Non-breeding males all races: Similar to female. Wings washed blue-grey and black bill.
Female all races: Upperparts: Brownish grey. Underparts: Whitish. Bill, lores, eye ring: Pale rufous. Tail: Light blue. Immature: Similar to females, but males develop bluish tail and black bill.

Weak flight. Resides in family groups, hopping through low shrubbery, generally feeding at higher levels than other fairy-wrens. Territorial.

Habitat: Southwestern forests with shrubby understorey, parkland, orchards. Avoids built-up areas. Inland habitat: Dense saltbush, mulga, mallee-spinifex areas and watercourse vegetation.

Feeding: Forages from the ground and higher for insects, larvae and spiders. Fruit, seeds also taken.

Voice: High-pitched, vigorous trilling, similar to Superb Fairy-wren.

Status: Locally common to uncommon.

Breeding: All year, mostly Sept.–Nov. Oval, domed nest with side entrance 2–3m high / 2–4 eggs.

Porongurup NP and Dryandra SF, WA. Kunoth Well, Alice Springs, NT. Hattah-Kulkyne NP, Vic.

Centre left: Nominate race, male in breeding plumage.Centre right: Female.
Below left: Race *musgravi*, Turquoise-wren, breeding male. Below right: Race *melanotus*, Black-backed Fairy-wren, breeding male.

Purple-crowned Fairy-wren

Malurus coronatus

Endemic

135–155cm

Deep purple crown with black centre and broad, black face mask extending around lores to nape.
Nominate race, Kimberley area WA, NT: Tail: Blue, finely tipped white. Upperparts: Brown. Underparts: Buff.
Non-breeding: Head: Grey, mottled black.
Race *macgillivrayi*, coastal Gulf of Carpentaria, NT, Qld: Tail: Greener. Upperparts: Grey-brown. Underparts: Whiter.
Female, both races: Deep reddish brown ear coverts, whitish eyebrow, eye ring and grey crown. Tail: Blue. Immature: Similar to female, no eye ring.

Rapidly hops in family groups of 5–9, usually in shady areas.

Habitat: Near fresh water. Pandanus, tall river grass, cane grass, dense grasslands with eucalypts, paperbark thickets and dense undercover.
Feeding: Hop-searches for insects, larvae, spiders.
Voice: Quiet 'chet' contact call. High-pitched reeling chirping. Harsh 'zit' alarm call.
Status: Sedentary. Moderately common.
Breeding: All year, mostly Sept.–Nov. Dome-shaped nest close to ground / 3 eggs.

Victoria River, NT. Lawn Hill NP, Qld.

Left: Breeding male. Right: Female with breeding male (below).

Variegated Fairy-wren

Malurus lamberti

Endemic

110–145mm

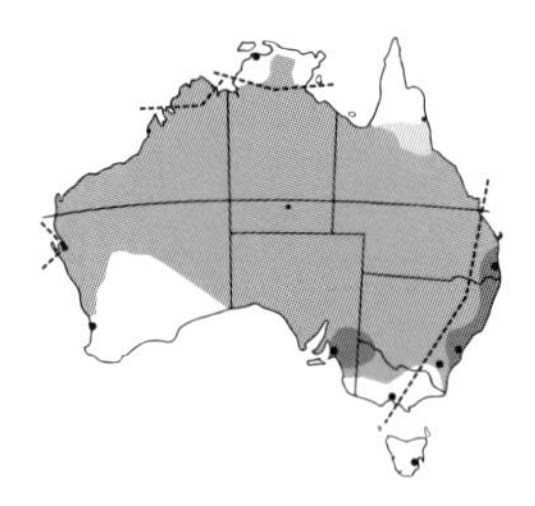

Best identified by geographic region. Distinctive, reddish tan shoulder patch. Crown, cheeks: Bright mid-blue. Back: Deeper blue extending to sides of breast. Bill, lores, nape, rump, upper breast: Black. Underparts: Buff-white. Tail: Long, dark blue, finely tipped white.
Non-breeding: Similar to female with black bill, lores. Blue tail.
Female: Bill, lores, eye ring: Tan. Upperparts: Light brown. Tail: Fawn, washed blue. Immature: Similar to female.
Nominate race found in eastern coastal south Qld, NSW, Vic.
Race *assimilis*, Purple-backed Fairy-wren, found across most of Australia, excluding eastern coastal area. Similar to nominate race, but has a darker blue cap and mantle.
Female: Like nominate race.
Race *rogersi*, Sandstone Fairy-wren, Kimberley area, WA: Rich chestnut shoulder patch. Crown deep purple-blue. Ear coverts: Bright blue. Breast: Black. Tail: broadly tipped white.
Female: Upperparts: Pale blue-grey. Underparts: White. Lores, eye ring, bill: Tan.
Both *rogersi* and *dulcis* interbreed with *assimilis* where their ranges overlap.
Race *dulcis*, Lavender-flanked Fairy-wren, Arnhem Land escarpment, NT, along watercourse vegetation. Similar to race *assimilis*. Flanks: Purplish.
Female: Similar to female race *rogersi*, but with white lores and eye ring.
Race *bernieri*, confined to Bernier and Dorre Is., Shark Bay World Heritage area, WA. Crown: Deep-violet.

Sprightly, bouncy hops using tail for balance.

Habitat: Nominate race, southeast Qld, NSW, Vic: Scrubby understorey of woodlands, forests, heaths, parks, gardens.
Inland and northern birds: rocky outcrops with dense cover; dense heaths, scrubby vegetation, protective spinifex, low shrubbery, sheltered gorges and watercourse vegetation.

Feeding: In family groups. Mostly insects, some fruit and seeds. Seldom out in the open.

Voice: High-pitched trilling, more metallic than Superb Fairy-wren.

Status: Common. Sedentary. Race *bernieri* vulnerable.

Breeding: Aug.–Jan. Oval, dome nest in low shrubs / 3–4 eggs.

Centre left: Nominate breeding male.

Centre right: Race *assimilis*, breeding male.

Below left: Nominate female. Below centre: Race *rogersi*, breeding male. Below right: Race *dulcis*, immature male.

Lovely Fairy-wren

Malurus amabilis

Endemic

120–130mm

Similar to the Variegated Fairy-Wren, but lighter blue with larger, rounded, bright blue cheek patches and more intense chestnut shoulder patch. Tail: Shorter, more rounded with a broad white tip. Underparts: White.
Non-breeding: Similar to female.
Female: Blue, washed grey back and head with brighter turquoise blue cheeks. Lores, eye ring, throat, underparts: White. Wings: Mid-grey-blue. Tail: Blue washed grey, tipped white. Bill: Black.
Immature: Similar to female, brown bill.

Reside in family groups of up to 7 birds.

Habitat: Edges of rainforests, mangroves, vine scrub and tropical coastal forests.

Feeding: Forages close to or on the ground in deep cover. Taking insects, some small fruits, seeds.

Voice: Musical trilling. Descending warble.

Status: Sedentary. Moderately common.

Breeding: Any time of year, mostly Sept.–Dec. Nest is low in dense cover / 2–3 eggs.

Iron Range NP, Qld.

Left: Breeding male. Right: Female.

Blue-breasted Fairy-wren

Malurus pulcherrimus

Endemic

125–150mm

Similar to Variegated Fairy-wren (race *assimilis*). Crown, mantle: Deep blue with a lilac sheen. Throat, upper breast: Black with a bluish sheen. Shoulder: Reddish tan patch. Tail: Long, smoky blue, faintly tipped white outer feathers.
Non-breeding: Similar to female with black bill, lores. Blue tail. White eye ring.
Female, immature: Bill, lores, eye ring: Tan. Upperparts: Light brown. Tail: Fawn, washed blue.

Shy and secretive. Difficult to observe.

Habitat: Open sand plains, heaths, scrubby mallee and swampy areas.

Feeding: Forages in small groups, usually in dense cover, for insects.

Voice: Soft, rippling warble.

Status: Sedentary. Removal of habitat in WA has made some populations locally extinct.

Breeding: Aug.–Sept. Untidy, dome-shaped nest, well hidden / 2–3 eggs.

Lincoln NP, SA. Porongurup NP. and Dryandra SF, WA.

Left: Breeding male. Right: Female (foreground) and breeding male.

Red-winged Fairy-wren

Malurus elegans

Endemic

135–155mm

Similar to the Blue-breasted Fairy-wren. Crown, cheeks: Mid-blue with silvery sheen. Mantle: Flecked blue and white. Throat, upper breast, rump: Deep blue-black. Shoulder: Broad, reddish tan patch. Tail: Long, blue-grey.
Non-breeding: Similar to female, black lores.
Female: Bill: Black. Lores, eye ring: Tan. Upperparts: Light grey with tan wash. Tail: Long, grey washed blue. Immature: Similar to female, duller.

Shy and secretive. Difficult to observe.

Habitat: Dense, wet undergrowth fringing freshwater creeks or swamps. Understorey in Jarrah–Karri forests, parks and gardens.

Feeding: Forages in small groups beneath undergrowth for insects, some fruit, seeds.

Voice: Rattling trill.

Status: Sedentary. Moderately common.

Breeding: Aug.–Jan. Large round nest, close to ground in swampy vegetation or dense cover near water / 2–3 eggs.

Two Peoples Bay NR, WA.

Left: Breeding male. Right: Non-breeding male.

Red-backed Fairy-wren

Malurus melanocephalus

Endemic

100–130mm

Overall glossy black with orange-red back and rump. Wings: Brownish black.
Non-breeding: Similar to female, retains blackish tail.
Race *cruentatus*, far north Qld, NT, WA: Similar, more intense crimson-orange back and rump.
Female, both races: Pale warm brown with buff-white underparts.
Immature: Similar to females, paler.
In display, male rump feathers are puffed out.

Habitat: Varied. Tropical, coastal to semi-arid inland with low, dense understorey. Thick grasslands with scattered trees and acacia thickets, rainforest fringe, swampy woodlands, open woodlands with tall grass, watercourse vegetation, spinifex, parks and gardens.

Feeding: Forages for insects early morning and evening in family groups, on or near ground.

Voice: Rippling warble.

Status: Sedentary. Common.

Breeding: Aug.–Jan in south. Nov.–Mar. in tropical north. Spherical bulky nest near water, often in pandanus / 3–4 eggs.

Cairns area, Qld. Kakadu NP, NT.

Left: Nominate race, females (on left and centre) and breeding male. Right: Nominate race, breeding male with feathers puffed out in display.

White-winged Fairy-wren

Malurus leucopterus

Endemic

110–135mm

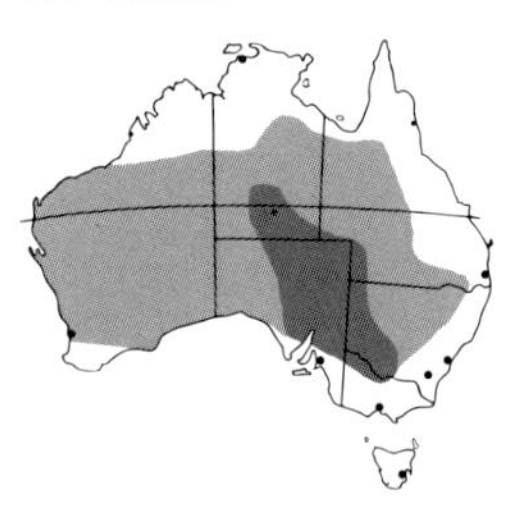

Intense cobalt blue with white wings. Tail: Deep blue.
Non-breeding: Similar to female with dark buff bill and bluer tail.
Female: Light brown-grey. Underparts: White. Tail: Blue-grey. Bill: Sandy brown.
Immature: Similar to female, browner.
2 races of black and white White-winged Fairy-wrens are found on Dirk Hartog Is. and Barrow Is., WA, respectively.

In small family groups. Flutters between shrubs with tail trailing.

Habitat: Open semi-arid to arid areas with tall grass; treeless areas; dense low shrubs; salt marshes; saltbush plains; coastal inlets. Offshore islands.

Feeding: Mostly in the outer foliage of shrubs or on the ground for insects.

Voice: Undulating, musical, metallic trill.

Status: Common to uncommon. Sedentary or locally nomadic.

Breeding: Aug.–Feb. Grass, spherical nest to 1m high / 2–4 eggs.

Left: Breeding male (foreground) and female. Right: Breeding male.

Southern Emu-wren

Stipiturus malachurus

Endemic

150–190mm

Distinctive tail comprising 6 long, brown, filamentous, trailing tail feathers, sometimes cocked.
Overall brown with heavily streaked black back, rufous crown and bright blue eyebrow, chin and throat. Underparts: Buff tan. Frons: Plain rufous-buff. Crown: Rufous-buff, streaked black. Belly: Buff-white.
Female: Similar, without blue markings.
Immature: Like female. Males have blue bib patch.

There are several races with subtle variations across range. Western birds more heavily streaked, more intense blue bib.

Timid. Runs mouse-like through undergrowth, bounces across clearings when disturbed. Feeble flight. In groups of 8–40 birds.

Habitat: Dense undergrowth, coastal dunes, heathlands, swampy areas.

Feeding: Forages under shelter for insects, spiders.

Voice: High-pitched trilling.

Status: Sedentary.

Breeding: Aug.–Jan. Oval nest, side entrance, in dense cover to 1m high / 2–4 eggs.

Barren Grounds NR, NSW. Croajingolong NP, Vic. Coorong NP, SA. Two Peoples Bay NR, WA.

Left: Male. Right: Female.

Mallee Emu-wren

Stipiturus mallee

Endemic Endangered

130–150mm

Similar to Southern Emu-wren, but bib is deeper blue and crown is plain rufous. Face, brow, throat, breast: Sky blue. Upperparts, wings: Grey-brown with black streaks. Tail: Dark, filamentous, shorter trailing feathers. Underparts: Buff. Female: Similar with streaked crown; lacks blue face, throat. Immature: Similar to female.

Shy, secretive. Usually in family groups or pairs.

Habitat: Old spinifex stands under low mallee trees and native cyprus, heathland with tea-tree and Broombush.

Feeding: Hop-forages on or near the ground for insects, some seeds and vegetable matter.

Voice: Thin, high-pitched trilling.

Status: Endangered.

Breeding: Sept.–Oct. Domed nest, close to ground, usually in spinifex / 2–3 eggs.

Hattah-Kulkyne NP, Vic.

Left: Female. Right: Male.

Rufous-crowned Emu-wren

Stipiturus ruficeps

Endemic

120–130mm

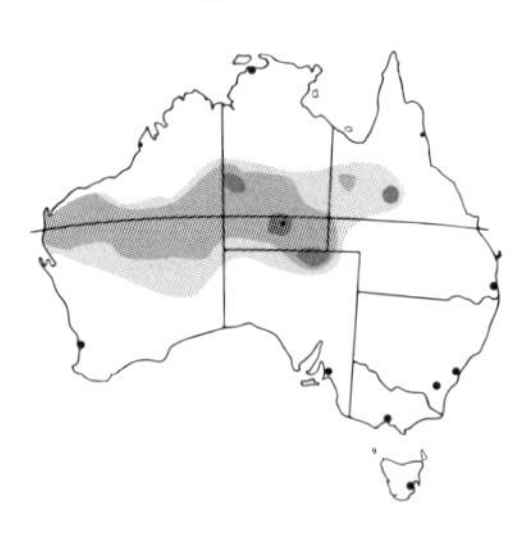

Similar to Mallee Emu-wren, but with plain, intense rufous crown and shorter, sturdier tail. Brow, face, throat: Deep, bright blue. Ear patch: Blue, fine black, white streaks.
Female: Lacks blue. Face, ear patches: Tan, finely streaked white. Lores, chin, throat: Whitish. Underparts: Buff-rufous. Immature: Similar to adults, paler.

Observes from spinifex stem then drops to move mouse-like through spinifex. Flies feebly.

Habitat: Spinifex-covered sand dunes and plains with scattered shrubs and low acacia.

Feeding: Inner branches of shrubs and spinifex for insects. In pairs or large feeding groups of 10–20.

Voice: High trilled chirp, repeated. Warbling trill.

Status: Moderately common. Sedentary.

Breeding: Aug.–Oct. Domed nest with side entrance in low shrubs or spinifex / 2–3 eggs.

Bladensburg NP, Qld. Ormiston Gorge, West MacDonnell NP, NT. Cape Range NP, WA.

Left: Male in spinifex. Right: Male.

Striated Grasswren

Amytornis striatus

Endemic

140–165mm

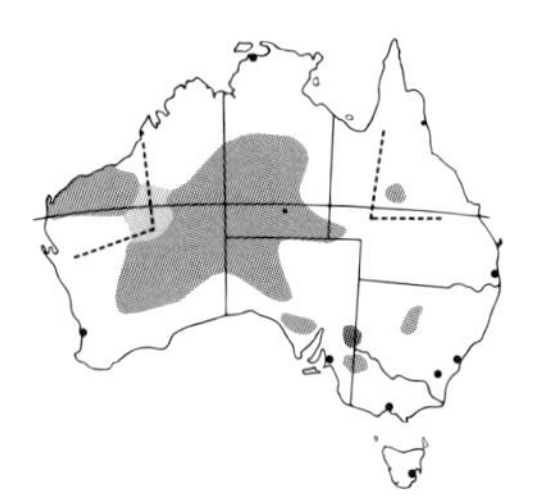

Sexes similar. Most widespread grasswren. Bold reddish brown and white streaked upperparts. Face: Bright orange-buff frons and long, black whisker stripe. Underparts: Buff-white. Tail: Long, dusky grey-brown. Nominate race most widespread and locally variable. Eastern birds are duller, greyer; western birds become brighter, more rufous.
Race *whitei*, Pilbara region, WA: Upperparts: Deep rusty rufous. Lower underparts: Buff-yellow-ochre.
Female: Rufous flanks.
Immature: Duller.

Stays close to shelter. Rapidly hops along ground when feeding.

Habitat: Low, stony hills, sandhills, sand plains with dense, mature spinifex, mallee, coastal scrub.

Feeding: Forages on the ground for insects, ants, seeds.

Voice: High squeaks and rippling, musical notes.

Status: Locally common. Sedentary. Usually in pairs or small groups.

Breeding: Aug.–Dec. Large domed nest in spinifex / 2–3 eggs.

Hattah-Kulkyne NP, Vic. Uluru–Kata Tjuta NP, NT.

Left: Female. Right: Male.

Short-tailed Grasswren

Amytornis merrotsyi

Endemic

150–160mm

Sexes similar. Similar to and once included with Striated Grasswren, it has a slightly heavier bill, shorter tail and longer legs. Facial markings are less prominent. Upperparts are brighter and richer colour. Female: Chestnut flanks.

Forges between clumps of spinifex running with a cocked tail. Often shelters in animal burrow.

Habitat: Sandstone ridges with tall spinifex and scattered trees, north Flinders and Gawler Ranges.

Feeding: Insects and seeds.

Voice: Similar to Striated Grasswren.

Status: Uncommon.

Breeding: Aug.–Dec. Large domed nest in spinifex / 2–3 eggs.

Stokes Hill Lookout, SA.

Opposite: Male.

White-throated Grasswren

Amytornis woodwardi

Endemic

200–220mm

Sexes similar. Similar to the Black Grasswren, but with a prominent white bib. Head, face, upper back: Black, streaked white. Lores, upper breast, chin, throat: White with black streaked breast band. Underparts, rump: Buff-tan. Tail: Grey-black. Female: Richer tan colouration. Immature: Similar to adults, darker.

Scurries through shrubby tussocks, with tail lowered. Calls from a vantage point.

Habitat: Arnhem Land escarpments, gullies, plateaus with spinifex, low shrubs and scattered trees.

Feeding: Forages in flocks for insects, some seeds and vegetable matter.

Voice: High-pitched trilling, rising and falling notes. Harsh alarm call.

Status: Vulnerable.

Breeding: Dec.–May. Bulky spherical nest usually in spinifex / 2 eggs.

Gunlom Falls, Kakadu NP, NT.

Opposite: Female.

Carpentarian Grasswren

Amytornis dorotheae

Endemic

160–180mm

Sexes similar. Similar to White-throated Grasswren, but smaller, more slender and paler. Crown, face: Black, finely streaked white, with black whisker stripe. Eyebrow: Rufous streak. Upperparts: Rufous, streaked with white feathers edged black. Tail: Rusty grey, tapered, usually upright. Chin, throat, breast: White. Underparts: Fawn. Female: Underparts: Rufous. Immature: Duller.

Usually seen in pairs or small family groups, running or hopping between clumps of spinifex. Flies poorly.

Habitat: Rugged sandstone outcrops with mature spinifex stands, boulders, sparse trees.

Feeding: Forages in small family groups of up to 5 for insects, spiders, seeds.

Voice: Cricket-like chirps; sweet trilling song.

Status: Vulnerable. Sedentary or locally nomadic.

Breeding: Aug.–Mar. or when conditions are favourable. Bulky nest, well hidden in clumps of spinifex / 2–3 eggs.

Borroloola area, NT. Mt Isa area, Qld.

Opposite: Male.

Western Grasswren

Amytornis textilis

Endemic

150–200mm

Sexes similar. Stout, finch-like bill. Upperparts: Drab grey-brown streaked with white. Underparts: Grey-buff, finely streaked chin and throat. Tail: Long, slender.
Female: Chestnut-washed flanks.
Race *myall*, northeast Eyre Peninsula, SA: Rarely flies.

Timid, elusive. Usually in pairs.

Habitat: In WA, coastal sand plains near Shark Bay. In SA, open saltbush, inland plains and dense acacia scrublands.

Feeding: Forages low or on the ground for insects, and seeds.

Voice: Soft cricket-like chirps. Loud alarm squeaks.

Status: Sparse to locally common. Vulnerable.

Breeding: July–Sept. Half-domed nest, often in saltbush. Incubated by male and female / 2–3 eggs.

Top left: Western Grasswren, female.

Thick-billed Grasswren

Amytornis modestus

Endemic

150–200mm

Far inland SA, NSW. Similar in appearance and behaviour to the Western Grasswren and best identified by range. (Map above.) Paler, with a longer tail and pale fawn, finely streaked white underparts.

Solitary or in pairs. Rarely flies. Runs and hops. Forages on ground and near cover.

Habitat: Open saltbush, bluebush, sandhill cane grass.

Feeding: Grass, seeds, insects.

Voice: Melodious, reedy, metallic trill. Loud alarm squeaks.

Status: Uncommon.

Breeding: Similar to Western Grasswren.

Top right: Thick-billed Grasswrens, male behind female.

Dusky Grasswren

Amytornis purnelli

Endemic

150–180mm

Sexes similar. Similar to the Thick-billed Grasswren. Thinner bill. Dusky brown head grading to dull rufous-brown upperparts with dark-edged white streaks. Breast: Russet-brown with white-streaks. Belly: Grey-brown. Tail: Dusky brown.
Female: Rufous flanks.
Immature: Duller.

Rarely flies. Scurrying, tail lowered, or hopping, tail cocked. Shy, but inquisitive.

Habitat: Tall spinifex clumps on rocky hills and gorges.

Feeding: Seeds, insects including grasshoppers, picked from crevices, low vegetation or ground.

Voice: High-pitched trill.

Status: Moderately common. Patchy, inland WA.

Breeding: Aug.–Oct. and Feb.–Apr. Full or semi-domed nest in spinifex / 2 eggs.

Mt Gillen and Simpsons Gap, Alice Springs area, NT.

Left: Female. Right: Male.

Kalkadoon Grasswren

Amytornis ballarae

Endemic

160–170mm

Sexes similar. Similar to the Dusky Grasswren but brighter-coloured plumage. Breast: Stiff, barb-tipped pale tan feathers. Belly, flanks: Grey.
Female: Rufous flanks and dark flecked belly.
Discovered in 1966. Rarely flies. Forages on ground close to spinifex cover.

Habitat: Rocky, spinifex-covered ridges around Mt Isa and Selwyn Ranges, Qld.

Feeding: Seeds and insects.

Voice: Weak, high-pitched peeps while feeding, high-pitched trill.

Status: Vulnerable.

Breeding: Feb.–Apr. Domed or semi-domed nest in spinifex clumps in sheltered gullies / 2–3 eggs.

Mica and Sybella Creeks, Mt Isa area, Qld.

Left: Female. Right: Male.

Black Grasswren

Amytornis housei

Endemic

180–210mm

Sexes similar. Stocky black and tan and white-streaked bird with a long, rounded black tail with very faint white barring. Head, mantle, underparts: Black with heavy white streaking. Back, wing coverts: Reddish brown. Female: Light reddish brown lower breast and underparts. Immature: Duller, fine white streaking to head, breast.

Runs with head and tail down, seeks shelter in rock crevices; calls from vantage point. In groups of 6–9 birds.

Habitat: Sandstone outcrops, ravines with tumbled rocks and spinifex.

Feeding: Forages in small parties for insects, seeds.

Voice: Wren-like. Metallic, slow, rattling trills.

Status: Near threatened. Sedentary.

Breeding: Apr.–Dec. Bulky spherical nest built into top of spinifex tussock / 2 eggs.

Mitchell Falls and Surveyors Pool, Mitchell River NP, WA.

Left: Female. Right: Male.

Grey Grasswren

Amytornis barbatus

Endemic

180–200mm

Sexes alike. Distinctive patterned black and white face, with strong black line through eye. Crown, upperparts: Cinnamon with black streaks. Underparts: White. Flanks: Buff. Tail: Long, brownish.
Nominate race, Qld, NSW.
Race *diamantina* is confined to Lake Eyre Basin. Larger and more rufous.
Immature: Plainer.

Low, swift flight. Hops with tail cocked in groups of up to 15.

Habitat: Dense clumps of lignum, spinifex and canegrass provide shelter, roosting, nest sites and food.

Feeding: Forages for insects and seeds.

Voice: Soft, 2–3-note constant contact call.

Status: Sedentary. Moderately common. Identified in 1967.

Breeding: July–Sept. Semi-domed nest in canegrass or lignum / 2–3 eggs.

Koonchera Dune, Birdsville Track, SA, in thickets of Sandhill Canegrass.

Opposite: Pair in Sandhill Canegrass.

Eyrean Grasswren

Amytornis goyderi

Endemic

140–165mm

Sexes similar. Distinctive, stout finch-like bill. Upperparts: Pale rufous streaked white. Underparts: White, buff flanks. Female: Buff-rufous flanks. Immature: Duller.

Hops with 1 foot in front of the other. In breeding season males sing loudly from a high perch.

Habitat: Simpson and Strzelecki Deserts. Sand ridges with hummocks of canegrass mixed with spinifex.

Feeding: Insects and seeds, especially canegrass.

Voice: High-pitched 2-note whistle.

Status: Uncommon. Nomadic, following availability of canegrass seed.

Breeding: Aug.–Sept. Semi-domed nest in centre of canegrass tussock / 2 eggs.

Opposite: Male singing.

Family Dasyoronithidae, Acanthizidae, Pardalotidae

A grouping of mostly small and often colourful insectivorous birds.

Family Dasyoronithidae contains the three species of Bristlebirds. Bristlebirds are named for the four or more stiff bristles at the base of the bill, visible at close range. The bristles fold back and are thought to protect their eyes as they run quickly through their dense heathland habitat. Bristlebird populations have been decimated by habitat destruction and fox and cat predation.

Family Acanthizidae contains the Pilotbird, Rockwarbler, Fernwren, Scrubwrens, Scrubtit, Heathwrens, Fieldwrens, Redthroat, Speckled Warbler, Weebill, Gerygones, Thornbills and Whitefaces.
Pilotbirds are close relatives of the Bristlebirds. They habitually follow in the wake of lyrebirds to prey on insects disturbed by the larger bird raking through leaf litter. Their loud call can be used to locate lyrebirds, hence their name Pilotbird.
The Rockwarbler, found only in NSW, is distinguished by its habit of building a dome-shaped, suspended nest inside a dark cave, often behind a waterfall.
The Fernwren, Scrubwrens and Scrubtit are small, dull and shy insectivores, found on the dark, gloomy floor of coastal rainforests and in dense shrubby undergrowth.
Heathwrens and Fieldwrens are shy and secretive birds of dense heathlands or mallee, usually quiet except during the breeding season when the male sings regularly from the top of a shrub.
The Redthroat and Speckled Warbler are both noted for their melodious song and mimicry of parts of other birds' songs.
The Weebill has the distinction of being Australia's smallest bird. It is sedentary in an established feeding territory in groups of up to ten. Gerygones were formerly known as warblers due to their loud, melodious song. However, they are not related to the 'true warblers' and were renamed 'gerygone', (pronounced 'jer-ig-on-nee'). Gerygones are noted for their beautifully built dome-shaped nests, which have a long 'tail'. The exception is the Large-billed Gerygone, which builds an untidy, well-camouflaged nest over water that resembles flood debris.
Thornbills are communal, acrobatic leaf gleaners that can be found in most habitats. Their nests are large dome-shaped grass structures with a side entrance and usually placed in a protective, prickly shrub.
The three species of Whiteface are gregarious birds with distinctive facial patterns and heavy finch-like bills for grinding grass and acacia seeds. The Banded Whiteface lives in the driest, most desolate inland regions and relies on insect prey for moisture.

Family Pardalotidae contains the four endemic pardalote species. Pardalotes are a brightly marked, endemic species of small, leaf gleaning birds with a distinctive 'bullet-like' flight. All species nest in a grass-lined nest chamber at the end of a 500–1500mm long tunnel, excavated in the side of an embankment. Some species will sometimes nest in tree hollows or nestboxes designed for pardalotes.

Opposite: Spotted Pardalote.

Rufous Bristlebird

Dasyornis broadbenti

Endemic

230–270mm

Sexes alike. Distinctive bright chestnut or russet-brown head with olive-brown upperparts, long tail and wings. Lores, malar line: White. Chin, throat: White, finely scalloped, flecked grey. Underparts: Grey-brown with scalloped pattern, darkest in eastern birds, whiter in west. Immature: Similar adult plainer. Weak flyer. Remains in dense scrub. Most active dawn and dusk.

Habitat: Coastal. Dense dune scrub, thickets of acacia, heaths. **Feeding:** Ground feeder, picking grubs, beetles, insects, larvae. **Voice:** Clear, ringing whistles. **Status:** Uncommon. Sedentary. **Breeding:** Sept.–Dec. Domed grass nest near ground / 2 eggs.

Port Campbell NP, Vic

Left and right: Adult.

Eastern Bristlebird

Dasyornis brachypterus

Endemic Endangered

200–220mm

Sexes alike. Rich olive-brown upperparts with chestnut-washed wings and rufous-brown rump. Tail: Long, faintly barred, brown. Underparts: Pale brown, light mottled chest. Belly, flanks: Grey. Eyes: Red-brown, pale eyebrow. Bill: Dusky.
Race *monoides*, eastern NSW–Qld border area: Darker back, plainer, olive-washed breast and belly.
Solitary, shy. Skulks in dense undergrowth. Runs rather than flies, tail partly cocked or fanned. Flies low, clumsy, short wings.

Habitat: Dense coastal heathlands bordering taller woodlands, tall swamps, stream thickets, woodlands with tussocky undergrowth. **Feeding:** Quietly, scratching ground debris for insects. **Voice:** Loud, melodious, penetrating 4-part call. **Status:** Sedentary. Uncommon. **Breeding:** Aug.–Dec. Domed nest close to ground / 2 eggs.

Booderee NP, Barren Grounds NR, NSW. Croajingolong NP, Vic.

Left and right: Nominate race.

Western Bristlebird

Dasyornis longirostris

Endemic Endangered

180–200mm

Sexes alike. Similar to the Eastern Bristlebird. Smaller. Breast more scalloped, dusky edged and greyer. Eyes: Red-brown. Eyebrow: Pale buff. Rump, tail coverts: Rufous-brown. Wings: Short, rounded, rufous-brown. Underparts: Grey-brown.
Immature: Pale olive-brown with grey underparts.
Shy, secretive. Weak, short, low flight into cover. Runs swiftly, wings drooped, tail sometimes raised or fanned.

Habitat: Confined to low wet coastal heathy undergrowth. **Feeding:** Ground foraging for insects and seeds. **Voice:** 5-part harsh call. High-pitched alarm call. **Status:** Endangered. **Breeding:** Aug.–Sept. Dome nest, side entrance, placed near ground in shrubbery / 2 eggs.

Two Peoples Bay NR, WA.

Left: Immature. Right: Adult singing.

Pilotbird

Pycnoptilus floccosus

Endemic

170mm

Sexes similar. Plump and large headed with cinnamon frons, amber eyes and slender, pointed, dusky bill. Upperparts: Deep rufous-brown. Tail: Long, broad, wedge tipped. Throat, breast, upperbelly: Cinnamon, scalloped brown. Lower underparts: Dull white, brown flanks, rufous tail coverts. Large, strong feet.
Race *sandlandi*, mid-NSW to Melbourne, Vic.: Plainer, paler. Female: Red eyes. Immature, grey-brown eyes.
Territorial. Feeble flight. Runs swiftly.

Habitat: Alpine areas. Race *sandlandi*: Wet eucalypt, temperate rainforests to subalpine areas. **Feeding:** Insects, invertebrates, some berries. Follows lyrebirds feeding in the disturbed litter. **Voice:** Loud, ringing, whistled, sweet song ending in a rising whipcrack. **Status:** Uncommon. Sedentary. **Breeding:** Aug.–Jan. Bulky domed nest hidden on ground / 1–2 eggs.

Barren Grounds NR, NSW.

Left: Race *sandlandi*. Right: Nominate.

Rockwarbler

Origma solitaria

Endemic

120–140mm

Sexes alike. Plain dark brown upperparts with rufous-washed rump. Frons: Flecked cinnamon brown. Eyes: Red. Bill: Dusky, slender. Throat: Pale grey, lightly speckled. Underparts: Rich rufous. Large, strong feet. Immature: Duller.
Short, swift dashing flight. Flicks tail sideways. Territorial.

Habitat: Endemic to Greater Sydney region in Hawkesbury sandsone formations. Usually near water.

Feeding: Forages over rocks, up vertical faces, hangs off rock overhang picking insects, seeds.

Voice: Melancholy shrill 'good-bye' 3–4 times. Staccato 'pink'.

Status: Moderately common. Sedentary.

Breeding: Aug.–Dec. Suspended grass dome nest, side entrance, often behind a waterfall / 2 eggs.

Morton, Dharug and Royal NPs, NSW.

Left: Adult upperparts. Right: Adult underparts.

Fernwren

Oreoscopus gutturalis

Endemic

120–140mm

Sexes similar. White throat, eyebrow and partial eye ring. Frons: Speckled white. Upperparts: Brown-green, appearing very dark in shadows. Breast: Large black crescent. Underparts: Lighter olive-brown. Eyes: Yellow. Bill: Long, black. Legs: Pinkish brown.
Female: Paler crown and brown eyes. Immature: Face, throat: Brown. Belly: Scalloped buff.

Constantly moving, tail flicking, bowing and bobbing head.

Habitat: Dense shady highlands and rainforests above 600m.

Feeding: Constantly flicking leaves and fossicking in the damp leaf litter for insects, snails, spiders. Burrows under thick mulch; follows larger birds picking in the disturbed mulch.

Voice: High-pitched whistles, chattering. Scolding calls.

Status: Moderately common. Sedentary.

Breeding: Aug.–Feb. Bulky dome nest near ground / 2 eggs.

Mt Hypipamee NP, Qld.

Left and right: Female calling.

White-browed Scrubwren

Sericornis frontalis

Endemic

110–140mm

Sexes similar. Small, constantly scolding, mostly olive-brown, ground-dwelling bird. Distinctive, long white eyebrow, silvery white whisker mark and blackish face mask with cream-yellow eyes. Rump, flanks: Dull rufous. Underparts: Soft buff-white with lightly streaked grey throat. Wings: Coverts tipped white. Several races with subtle plumage variations across range: Race *laevigaster*, Qld: Overall brightest. Ear coverts, eye stripe: Black. Flanks, belly: Buffy yellow. Tail: Black sub-terminal band. Race *maculatus*, Spotted Scrubwren from Adelaide, SA, to WA: Small white line below eye. Breast: Streaked. Tail: Pale tipped.
Female: Paler. Immature: Duller, darker.

Active, fearless and noisy. Pairs or small family groups.

Habitat: Dense, shrubby undergrowth, usually moist areas along creeks, dense sand plain heaths, mallee scrub, parks and gardens.

Feeding: Briskly hop-searches under cover over ground or among low branches. Insects, insect larvae, moths and the like.

Voice: Soft continuous chattering. Harsh metallic chirring when alarmed.

Status: Common. Sedentary.

Breeding: July–Jan. Communal. Domed grass nest, well hidden near ground / 2–3 eggs.

Below centre left and right: Nominate race.
Below left: Race *maculatus*, Spotted Scrubwren. Below right: Race *laevigaster*, Qld.

Tasmanian Scrubwren

Sericornis humilis

Endemic

120–150mm

Sexes alike. Similar to the White-browed Scrubwren but darker, plainer and larger. Face: Dull white eyebrow fades away at eye. Upperparts: Olive-brown. Tail: Plain. Immature: Similar, duller, indistinct facial pattern.

Rarely flies more than a few metres. Usually in pairs.

Habitat: Moist areas along creeks in dense, shrubby, gloomy undergrowth. Parks, gardens, Tas. offshore islands.

Feeding: Forages in the lower canopy and close to ground in dense cover for insects and some seeds.
Voice: Similar to the White-browed Scrubwren. Staccato alarm call.
Status: Common. Sedentary.
Breeding: Aug.–Dec. Domed nest, side entrance, well hidden near ground / 2–4 eggs.

Opposite: Adult.

Large-billed Scrubwren

Sericornis magnirostris

Endemic

110–125mm

Sexes similar. Brownish olive upperparts with rufous washed rump and tail. Face, lores: Bright, pale buff. Eyes: Dark red. Bill: Black, large. Underparts: Pale, buff-white.
Female: Duller, more olive. Frons, lores: Paler. Immature: Paler.

Agile, acrobatic, rapid, fluttering movements. Infrequently descends to ground.

Habitat: Damp rainforests, dense gullies, damp eucalypt forests.

Feeding: Loose parties of 5–10. Insects, small snails from foliage, bark, vines from ground to 17m.
Voice: Soft, tinkling twitter or harsh 'chew' contact call.
Status: Moderately common. Sedentary.
Breeding: July–Jan. Bulky domed nest in branch fork from near ground to 10m high. Sometimes in old nests of Yellow-throated Scrubwrens / 3-4 eggs.

Lamington NP, Qld. Dharug NP, NSW.

Left: Male. Right: Female.

Tropical Scrubwren

Sericornis beccarii

110–115mm

Sexes similar. Similar to the Large-billed Scrubwren but has black lores and frons patch. Eyes: Orange-red, short white eyebrow and thin, broken eye ring. Upperparts, face: Mid-russet-brown. Tail: Cinnamon. 2 double white wing bars. Underparts: Buff-white.
Race *dubius*, southern range: Indistinct facial and wing markings. Pale lores. Upperparts: Buff-cinnamon. Female: Paler with pale lores. Immature: Duller.

Pairs or small groups. Fossicks mostly on ground to mid-canopy.
Habitat: Tropical rainforests.
Feeding: Similar to the Large-billed Scrubwen, but more frequently comes to the ground.
Voice: Rapid, undulating musical warbling, 'wit-wit' contact call.
Status: Moderately common. Sedentary.
Breeding: Oct.–Dec. Similar to Large-billed Scrubwren / 2–3 eggs.

Iron Range NP, Qld.

Left: Male. Right: Female.

Atherton Scrubwren

Sericornis keri

Endemic

120–135mm

Sexes alike. Similar to the Large-billed Scrubwren, but overall darker. Upperparts: Brown to dark brown with rufous-brown rump. Eyes: Red with indistinct pale eyebrow. Underparts: Whitish with yellow tint. Bill: Slender, straight.
Immature: Similar, duller with brown eyes.

Mouse-like movements. Short flitting flight. Solitary or in pairs.

Habitat: Understorey mountain rainforests above 650m.

Feeding: Forages low in trees or shrubs or hop-searches over forest floor for insects and small snails.
Voice: Musical soft chattering.
Status: Moderately common. Sedentary.
Breeding: Aug.–Dec. Domed nest, well hidden in grass or ferns on ground / 2 eggs.

Mt Hypipamee and Mt Lewis NP, Qld.

Left: Adult upperparts. Right: Underparts.

Yellow-throated Scrubwren

Sericornis citreogularis

Endemic

120–140mm

Sexes similar. Ground dwelling. Distinctive black face mask and frons, dark eye and long, white grading to yellow eyebrow. Throat: Bright yellow. Crown, back: Olive-brown. Flanks: Light olive. Underparts: Grading to pale buff. Bill: Black. Legs: Long, pinkish. Races differ in voice, habits and nesting.
Race *cairsi*, far north Qld: Darker. Female: All have olive-brown face and frons. Immature: Similar to female, eyes brown, gape yellow.

Solitary or pairs.

Habitat: Gloomy forest floor. In north: Dense rainforest gullies usually above 600m. In south: Mountain and coastal rainforests.

Feeding: Exclusively on the ground. Hop-searches. Picks insects and occasionally seeds.

Voice: In south: Sharp 'tik' contact call, melodious whistling and mimicry. In north: Loud chattering, few whistling notes.

Status: Common. Sedentary.

Breeding: Aug.–Mar. Suspended, bulky domed nest with a hooded side entrance, hung from a small branch in south, from a vine tangle in north / 2–3 eggs.

Mt Hypipamee NP and Crater Lakes NP, Qld.

Opposite: Male.

Scrubtit

Acanthornis magna

Endemic

110–115mm

Sexes similar. Endemic to Tas. and King Is. Similar to the White-browed Scrubwren. Bill: Slightly down-curved, dusky. Crown, upperparts: Mid-grey washed russet. Face, ear coverts: Grey-brown. Eyes: Red-brown with white eye ring. Wings: Dark grey with white spot on coverts. Tail: Russet-grey, whitish tip. Throat, upper breast: Whitish. Underparts: Cream with russet washed flanks. Legs: Pinkish brown.
Immature: Similar with pale brown eyes, pale yellow underparts.

Active, small and secretive. Usually solitary.

Habitat: Forest undergrowth, ferny gullies and shady thickets.

Feeding: Foraging methodically, treecreeper-like on tree trunks and branches, poking and probing for small insects and spiders.

Voice: Churring, whistling and scolding chatter.

Status: Common. Sedentary. King Is. endangered.

Breeding: Sept.–Dec. Bulky dome nest near ground, often in dense ferns / 3–4 eggs.

Cape Bruny, Bruny Is., Tas.

Left: Adult. Right: Adult taking insect from bark.

Chestnut-rumped Heathwren

Calamanthus pyrrhopygia

Endemic

130–160mm

Sexes similar. Rufous rump, brown to brownish grey upperparts. Tail: Cocked, bright rufous coverts merging to dark brown with sub-terminal black band and grey tips; centre feathers rufous. Eyes: Yellowish brown with dull white eyebrow. Bill: Dark brown. Underparts: Dull white, heavily streaked dusky. Undertail: Mid-rufous.
Female: Buff eyebrow, lighter streaked, buff underparts.
Immature: Similar to female, plain buff underparts, washed russet breast.

Shy, secretive, skulking in dense cover. Swift, bouncy, fluttering flight into cover.

Habitat: Low, dense undergrowth of heathlands and forests.

Feeding: Loose groups forage low or on the ground for insects, some seeds taken.

Voice: Male begins to sing in winter. Melodic, canary-like song. Accomplished mimics with other birdcalls interspersed with their own song. Singing for up to 30 minutes at any time, sometimes in duets with female.

Status: Moderately common to vulnerable in SA. Sedentary.

Breeding: June–Nov. Domed nest, small spout entrance, well hidden close to ground / 2–3 eggs.

Royal NP, NSW. Deep Creek CP, SA.

Opposite: Male.

Shy Heathwren

Calamanthus cauta

Endemic

135–140mm

Sexes similar. Similar to the Chestnut-rumped Heathwren but more heavily streaked, richer colour and large white patch on wings. Eyes: Yellowish brown with long, deep, white eyebrow. Bill: Dark brown. Upperparts: Deep brown with bright chestnut rump and tail coverts. Tail: Bright chestnut with sub-terminal black band, white tips. Underparts: White, boldly streaked dusky. Female: Cream underparts, eyebrow. Immature: Duller.

Active, tail cocked. Ground dwelling.

Habitat: Mallee woodland with shrubby understorey.

Feeding: Hop-searches for insects, rarely seeds.

Voice: Sharp, musical sequence of trills and mimicry.

Status: Sedentary. Uncommon. Vulnerable in NSW.

Breeding: Aug.–Nov. Domed nest with small spout entrance, well hidden close to or on ground / 2–3 eggs.

Wyperfeld NP, Hattah-Kulkyne NP, Big Desert NR, Vic. Gluepot, SA.

Left and right: Male.

Striated Fieldwren

Calamanthus fuliginosus

Endemic

130–140mm

Sexes similar. 4 races with subtle variations. Overall body heavily streaked. Upperparts: Brown-buff with broad, heavy black streaks, rufous-washed frons and broad white eyebrow, lores. Underparts: Buff-white, heavily streaked with darker buff flanks and whiter throat. Tail: Cocked, dark sub-terminal band, white tips.
Female: Buff-yellow wash to throat, tawny eyebrows, lores. Immature: Similar to female.

Ground dwelling. Rarely flies more than a few metres. Usually in pairs.

Habitat: Damp coastal to alpine areas in moist heaths, wetland edges, Samphire flats and dunes.

Feeding: Mouse-like hop-searching for insects, seeds.

Voice: Male sings briefly from a perch before diving into cover. Succession of trills and chattering warbles.

Status: Sedentary, Uncommon.

Breeding: July–Dec. Domed nest, well hidden on or near ground / 3–4 eggs.

Green Cape Lighthouse, NSW. Werribee area, Vic. Strahan area, Mount Field NP, Tas.

Opposite: Male.

Rufous Fieldwren

Calamanthus campestris

Endemic

120–130mm

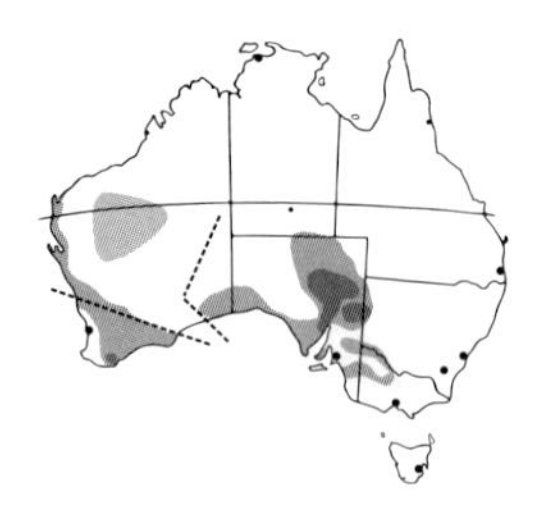

Sexes alike. Similar to the Striated Fieldwren but paler and more rufous. Upperparts: Overall grey-brown boldly streaked black. Lores, frons: White. Eyebrow: Pale. Underparts: Buffy white, streaked black. Tail: Erect, with dark band below white tip. Legs: Long, reddish brown. Several colour variations occur across its geographic range including:
Race *montanellus*, Western Fieldwren, southwest WA: Rufous-red upperparts and white, washed yellow underparts. Further north in WA, birds become more rufous-red. Immature: Duller.

Scurries quickly for cover when disturbed. Sedentary. Usually in pairs or small family groups.

Habitat: Arid and semi-arid regions; scrubby, sparse, low vegetation of saltbush, bluebush, spinifex, Samphire, salt marshes.

Feeding: Hop-searches for insects, some seeds.

Voice: Clear, high-pitched musical whistles, rattling trills.

Status: Uncommon.

Breeding: July–Nov. Domed nest hidden near ground / 2–3 eggs.

Arid Lands Botanic Garden, Port Augusta, SA. Stirling Ranges NP, Two Peoples NR and Cape Range NP, WA.

Left and right: Nominate race.

Redthroat

Pyrrholaemus brunneus

Endemic

115–120mm

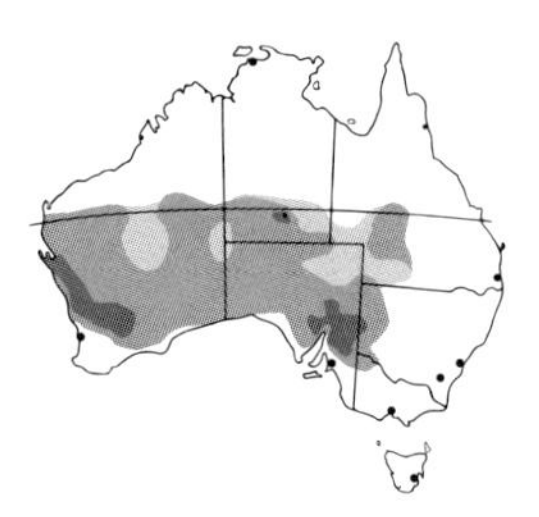

Sexes similar. Small, plain bird with a small patch of chestnut on chin and throat. Upperparts: Mid-grey-brown. Frons: White. Underparts: Dull white. Tail: Mid-brown with white outer tail feathers.
Female: Lacks throat patch. Immature: Similar to female but paler.

Shy. Usually in pairs or small family groups, sometimes in flocks of up to 30 birds.

Habitat: Arid and semi-arid areas in thick scrub, saltbush, mulga, Belah or mallee.

Feeding: Hop-searches with tail down, low in vegetation, taking insects and seeds.

Voice: Melodious, varied song incorporating mimicry of many other birds' song.

Status: Moderately common. Sedentary.

Breeding: Aug.–Dec. Domed nest to 1m above ground, sometimes in burrow, hollow fallen log or tree hollow / 3–4 eggs.

Wyperfeld NP, Vic. Arid Lands Botanic Garden, Port Augusta, SA. Carnarvon area, WA.

Opposite: Male.

Speckled Warbler

Pyrrholaemus sagittatus

Endemic

115–120mm

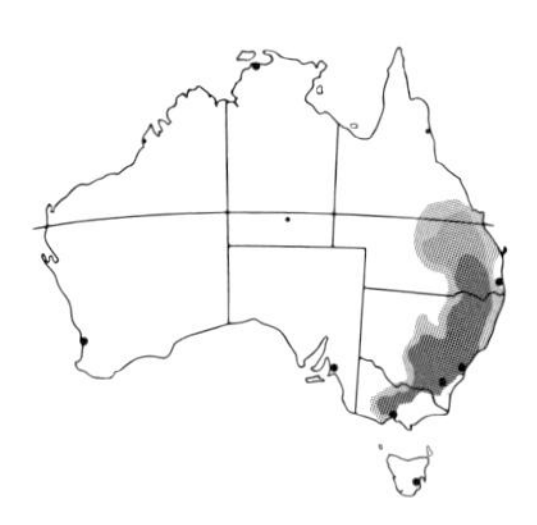

Sexes similar. Overall streaked, more heavily on the underparts. Black crown, buff streaked. Face: Grey-white streaked buff, white lores, long white eyebrow with broad, black upper-edge. Eyes: Dark red. Upperparts: Grey-brown with subdued streaking. Underparts: Cream with bold, heavy black streaks, yellow-washed flanks, vent. Tail: White tipped outer feathers, central feathers dark-tipped.
Female: Chestnut edge to eyebrow. Immature: Similar to female, less boldly streaked.

Mostly ground dwelling. Pairs or small family groups, often in mixed flocks with thornbills. Defends territory.

Habitat: Open, dry, eucalypt woodlands, northern vine scrub, mulga from western slopes of Great Dividing Range to inland, driest coastal areas.

Feeding: Briskly hops, probing leaf litter for insects and seeds.

Voice: Soft, rich trilling warble, mimicry of other birds' songs.

Status: Moderately common to declining. Sedentary.

Breeding: Sept.–Jan. Domed nest, near or on ground. Often parasitised by Black-eared cuckoos / 3–4 eggs.

Glen Davis and Mann River NR, NSW. Mulligans Flat NR, ACT. Chiltern–Mt Pilot NP, Vic.

Opposite: Male.

Weebill

Smicrornis brevirostris

Endemic

80–90mm

Sexes alike. Australia's smallest bird. Similar to the Yellow Thornbill but paler, plainer face. Distinctively short, stubby, pale brown bill. Head: Dull grey-brown with cream eyes and pale eyebrow. Upperparts: Yellowish olive. Underparts: Cream throat, breast grading to yellowish cream below. Tail: Outer feathers tipped white.
Nominate race, mid-Qld, NSW, Vic., eastern SA: Palest.
Race *occidentalis*, western SA to southern WA: Greyer.
Race *flavescens*, north Qld, NT, SA and north WA: Smaller and brighter.
Race *ochrogaster*, mid-WA, Pilbara area: Paler with little or absent streaking to throat.
Immature: Duller, eyes grey.

In pairs or small groups. Flits and hovers in the canopy, thornbill-like, while feeding.

Habitat: Dry eucalypt woodlands, open forests, mallee.

Feeding: Gleans scale, lerp and insects from outer foliage.

Voice: Short, clear, loud whistled 'willy-weet'.

Status: Common. Sedentary.

Breeding: Oct.–May in south. July–May in north. Domed nest with spout entrance, 1–10m high / 2 eggs.

Opposite: Nominate race.

Western Gerygone

Gerygone fusca

Endemic 100–110mm

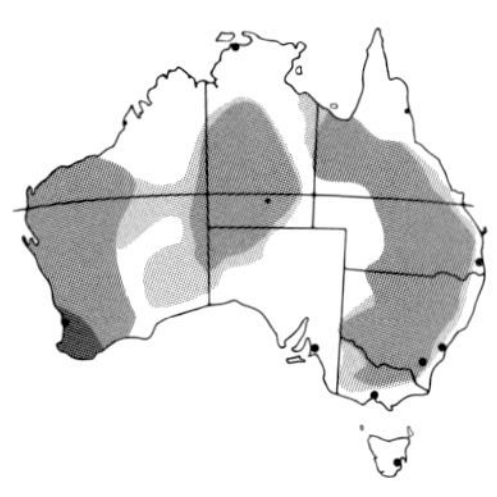

Sexes alike. Small arboreal bird with plain grey plumage washed olive-brown. Broad white band across base of tail distinguishes it from other gerygones. Eyes: Red. Eyebrow: Indistinct whitish extending into eye ring. Underparts: Whitish; throat, breast darker. Bill, legs: Blackish. Immature: Pale brown eyes. Underparts washed yellow. Solitary or pairs, rarely groups. **Habitat:** Variable. Drier inland open forests, woodland, mulga, mallee, Jarrah, wet Karri forests. **Feeding:** High in canopy, taking insects, aphids, or hovers above canopy for small flying insects. **Voice:** Clear, sweet, song whistled by male through day. **Status:** Common. **Breeding:** Sept.–Jan. Compact oval nest suspended from twigs, decorated with spider egg-sacs, 2–10m high / 2–3 eggs.

Left: Adult underparts. Right: Upperparts.

Brown Gerygone

Gerygone mouki

Endemic

90–110mm

Sexes similar. Plain, small bird. Pale-grey eyebrow, reddish eyes and brown upperparts. Underparts: Grey, buff tinted flanks. Tail: Centre feathers olive-brown, outer feathers dark sub-terminal band, pale tips. Nominate race, far north Qld. Race *amalia*, northern Qld. Race *richmondi*, southeast Qld, NSW, Vic.: Similar, pale grey face, throat olive-brown upperparts, black lores. Immature: Similar, brown eyes. Solitary, pairs or groups of 3–4 outside breeding times.

Habitat: Rainforests, mangroves and dense eucalypt forests. **Feeding:** Actively, methodically forages in foliage, along limbs, for insects, sometimes hawking. **Voice:** Incessant soft twittering, 'diddle-it-did-dit' rapidly repeated. **Status:** Moderately common. Sedentary. **Breeding:** Sept.–Feb. Suspended, slender dome nest with long tail, adorned with lichen, 1–15m high, often over water / 2–3 eggs.

Lamington NP, Qld.

Left, right: Race *richmondi*.

Large-billed Gerygone

Gerygone magnirostris

105–115mm

Sexes similar. Longer and thicker black bill than other gerygones. Face: Olive-brown. Eyes: Red-brown. Eye ring: White, indistinct brow. Upperparts: Olive-brown, browner wings. Underparts: White, buff sides to breast, flanks. Tail: Grey-brown, darker sub-terminal band and faint white spots on tips. Race *cairnsensis*, Qld: Smaller, warmer toned, rufous-edged wings. Tail: Large white spots on outer feathers. Immature. Brown bill, eyes. Flutters and hovers in outer foliage. Inconspicuous. **Habitat:** Riverine rainforests, paperbark forests, tall mangroves. **Feeding:** Gleans insects in outer foliage, usually upper canopy. **Voice:** Powerful sequence of rich notes, slowing and descending. **Status:** Common. Sedentary. **Breeding:** Sept.–Apr. Untidy nest with tail, over water, often near a wasp nest up 2–10m / 2–3 eggs.

Left: Race *cairnsensis*. Right: Nominate race.

Mangrove Gerygone

Gerygone levigaster

100–110mm

Sexes alike. Plain olive-brown upperparts, white eyebrows and red eyes. Underparts: Buff-white. Tail: Brown with large white spots on tips of outer feathers. Race *cantator*, Qld, NSW: Grey-white underparts, deeper brown upperparts. Immature: Similar, yellowish underparts. **Habitat:** Tidal mangroves, often in breeding season. Along waterways near scrubland, paperbark stands. Gardens.

Feeding: Extremely active foraging in outer branches, occasionally hovering for insects. **Voice:** Sustained, sweet, repeated, rising–falling whistled song, during most of the day. **Status:** Common. **Breeding:** Sept.–Apr. Compact, pear-shaped dome nest with side entrance, suspended from a branchlet 1–8m high / 2–3 eggs.

Left: Nominate race. Right: Race *cantator*.

Dusky Gerygone

Gerygone tenebrosa

Endemic
115mm

Sexes alike. Largest gerygone. Upperparts: Dusky grey washed pale olive-rufous. Eyes: Cream with long white eyebrow. Underparts: White with pale yellow-olive tint to throat, flanks. Southern birds, race *christophori*, larger, darker.
Immature: Similar. Underparts washed pale yellow and yellow-brown eyes.

Habitat: Coastal mangroves, swamplands and scrub along waterways.
Feeding: Gleans insects mostly from upper mangrove canopy.
Voice: Sweet, plaintive whistles, chattering contact calls.
Status: Common. Sedentary.
Breeding: Oct.–Mar. Domed, suspended nest in mangrove / 2 eggs.

Point Samson, WA.

Left and right: Adult foraging in mangroves.

Fairy Gerygone

Gerygone palpebrosa

100–110mm

2 races occur in Australia and interbreed where ranges overlap. Race *personata*, Black-throated Warbler, far north Qld: Head, throat, upper breast: Olive-black, broad white whisker line and 2 white spots on frons. Eyes: Red. Upperparts: Olive-grey. Underparts: Pale yellow. Wings: Grey-brown.
Race *flavida*, north Qld: Paler, lacks black chin. Tail: Black band and fine white tips.
Female: Both races paler, plainer, lack face marks.
Immature: Similar, buff throat.
Habitat: Vine thickets, eucalypt woodlands, scrubland, mangroves and rainforest fringes.
Feeding: Forages in the upper canopy for insects.
Voice: Scratchy chatter. Drawn-out ascending–descending cheerful song.
Status: Moderately common. Sedentary.
Breeding: Sept.–Mar. Domed nest, slender tail, suspended, near wasp nest 2–10m high / 2–3 eggs.

Iron Range NP and Centenary Lakes, Botanical Gardens Cairns, Qld.

Left: Nominate male. Right: Race *personata*, male.

White-throated Gerygone

Gerygone olivacea

100–110mm

Sexes alike. Plain, grey-brown washed olive upperparts and bright yellow underparts. Distinctive, bright white throat with 2 white frons spots.
Undertail coverts: White. Head, eyes: Bright red.
Race *rogersi*, NT, WA: Smaller, paler. Undertail, vent: Yellow.
Immature: Paler, throat yellow, eyes brown, indistinct frons.
Constantly flitting and hovering in outer floliage.
Habitat: Tall, open eucalypt woodlands, open forests, timbered waterways.
Feeding: Scale, lerp and insects.
Voice: Sweet, repeated trill of fast, whistled notes. Falling and rising tinkling.
Status: Locally common. Northern birds sedentary. Southern birds move north in winter.
Breeding: Aug.–Jan. Domed, hooded nest with slender tail, suspended from outer branches 2–15m high / 2–3 eggs.

Left: Nominate race. Right: At nest.

Green-backed Gerygone

Gerygone chloronota

90–100mm

Sexes alike. Green wash to back and wings. Head, face: Mid-grey. Eyes: Red. Bill: Black. Underparts: White, pale lemon flanks and undertail. Tail: Pale grey-brown.
Immature: Duller. Brown eyes, dark cream bill.
Hovers and flutters in the outer foliage in the mid-canopy zone. Solitary, pairs or small flocks.
Habitat: Eucalypt forests, rainforests, mangroves, paperbark-lined waterways, gardens.
Feeding: Actively gleans insects from outer foliage, rarely coming to ground.
Voice: Rapid, high-pitched warbling song at all times of day and throughout the year.
Status: Common. Sedentary.
Breeding: Apr.–Oct. Often near wasp nest, oval dome nest, side entrance, suspended in outer foliage 1–15m high / 2–3 eggs.

Buffalo Creek and East Point Reserve, Darwin, NT.

Left: Upperparts. Right: Underparts.

Slender-billed Thornbill
Acanthiza iredalei
Endemic
90–100mm

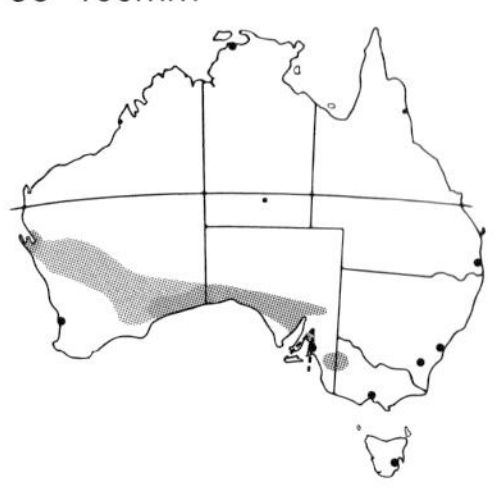

Sexes similar. Speckled-grey face and white eyes. Upperparts: Olive-grey. Wings: Grey-brown. Rump: Buff-yellow. Underparts: Creamy white, flecked olive-grey. Tail: Buff-yellow base, grey-brown with black band and grey tips to outer feathers.
Race *rosinae*, Gulf St Vincent, costal SA: Darker, more heavily streaked; vulnerable.
Race *hedleyi* inhabits dry heathland of SA, Vic border.
Immature: Similar.
Pairs or small, loose groups.
Low, bouncy, undulating flight.
Habitat: Samphire, saltbush, edge of salt lakes, mallee heath.
Feeding: Hop-gleaning, mostly in low shrubs for insects, spiders.
Voice: Constant chattering. High-pitched musical twitter in flight.
Status: Uncommon. Sedentary. Nomadic.
Breeding: July–Nov. Small domed nest, low in foliage / 3 eggs.

Little Desert NP, Vic. Chinaman's Creek, SA. New Beach, WA.

Left and right: Adult on Samphire.

Slaty-backed Thornbill
Acanthiza robustirostris
Endemic
100–110mm

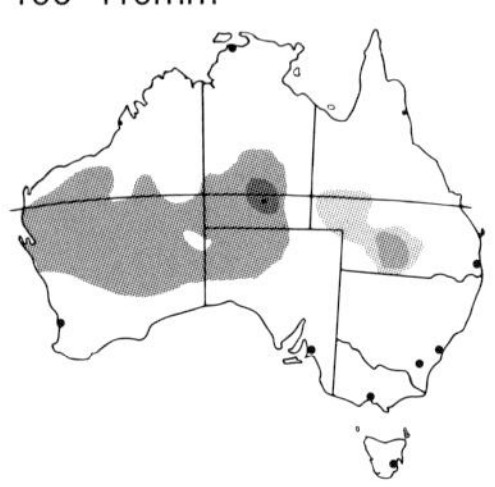

Sexes similar. Similar to the Inland Thornbill but distinguished by lack of striations to face and throat. Crown, frons: Grey with fine black streaks. Eyes: Red-brown. Bill: Black. Upperparts: Slate grey, rufous rump. Tail: Dull black with narrow white tip.
Immature: Similar, brown eyes.
In pairs or small groups with Inland Thornbills.
Low bouncing flight.
Habitat: Mulga, salt plains, salt lakes with low growth, woodlands.
Feeding: Forages for insects, spiders in the outer branches and foliage 1–5m above ground.
Voice: Loud, distinctive 'tiz-tiz'. Harsh 2-tone churr alarm call.
Status: Uncommon. Sedentary.
Breeding: July–Nov. subject to rainfall. Domed, roughish nest, placed low in foliage / 2–3 eggs.

Kunoth Well, Alice Springs, NT. Cooper Pedy area, SA.

Left: Adult. Right: Underparts.

Inland Thornbill
Acanthiza apicalis
Endemic
100–110mm

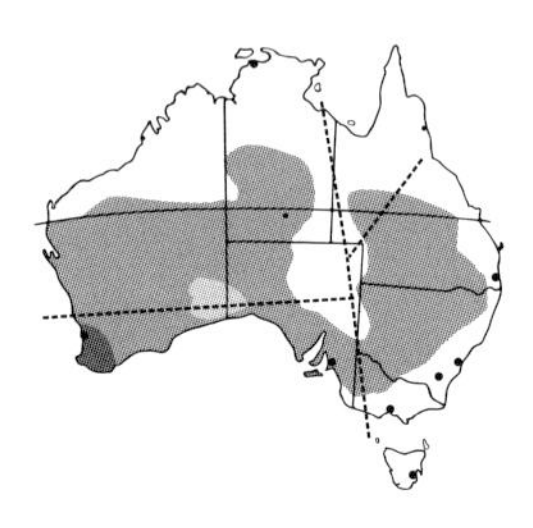

Sexes alike. Upperparts: Grey with rufous-brown rump and base of tail. Tail: Often cocked, outer feathers tipped white. Frons, breast: Heavily scalloped, speckled grey and white. Underparts: White, cinnamon-buff flanks. Eyes: Red. Bill: Black. Legs: Dusky brown.
Several similar races. Eastern birds have a more vibrant, rufous rump.
Race *albiventris*, eastern Qld, NSW, Vic.: Darkest, richest colour, heavier black scalloped frons and black-streaked breast.
Race *whitlocki*, NT, northern SA, WA: Paler.
Race *cinerascens*, far west Qld: Greyer.
Immature: Duller.
Usually solitary or in pairs.
Habitat: Inland dry scrub, woodlands.
Feeding: Low in trees or shrubbery for insects.
Voice: Strong song and mimicry.
Status: Common. Sedentary.
Breeding: July–Dec. Domed nest in low branches / 2–3 eggs.

Left and right: Nominate race.

Yellow Thornbill
Acanthiza nana
Endemic
90–100mm

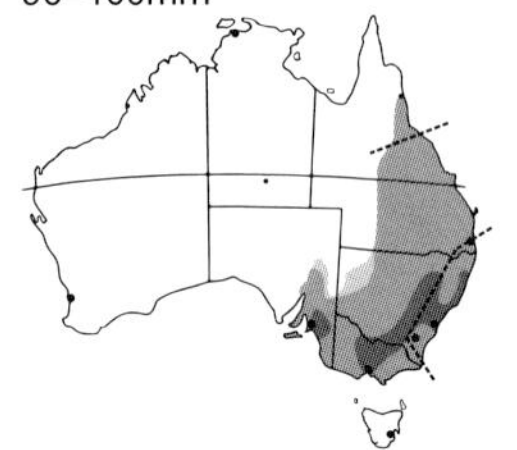

Sexes alike. Smallest and most yellow thornbill. Crown: Pale buff. Face: Grey, white streaked. Eyes: Brown. Lores: Indistinct white spots. Upperparts: Olive green. Underparts: Yellow, washed light buff on throat, flanks. Wings: Primaries edged olive-yellow. Tail: Olive-brown, dark tipped. Bill: Pinkish brown or black.
Race *flava*, Atherton area, Qld: Brightest.
Race *modesta*, Qld, inland NSW, Vic., SA: Paler.
Immature: Duller.
Usually in small family groups.
Habitat: Open forests, rainforests, mallee, mulga.
Feeding: Methodically hop-gleaning or hovering for insects.
Voice: Persistent, harsh, loud 'chips-chips' repeated. No song.
Status: Common. Sedentary.
Breeding: Aug.–Dec. Dome nest, entrance near top, suspended 3–12m high / 3 eggs.

Left and right: Nominate race.

Yellow-rumped Thornbill

Acanthiza chrysorrhoa

Endemic

110–115mm

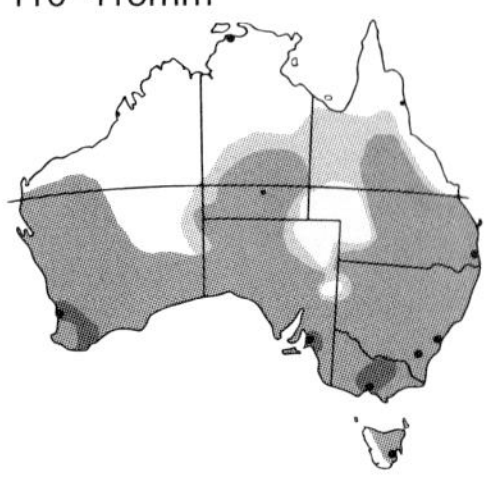

Sexes alike. Largest thornbill. Olive-brown back and bright yellow rump, cream underparts. Flanks washed olive-buff. Wings: Grey-brown. Crown: Speckled olive-brown. Face: Speckled, grey, white eyebrow and black frons. Eyes: Whitish. Bill: Black, slender. Tail: Short, black, outer feathers tipped white. Inland and northern birds slightly paler.
Bouncy flight. Communal.
Habitat: Clearings, edges of open woodlands, scrubland, farmlands, parks, gardens.
Feeding: Mostly on the ground, near cover, hop-searches for insects and occasional seeds.
Voice: Repeated, cheery undulating wh istles.
Status: Common. Sedentary.
Breeding: 4 breeding males and 1 breeding female. Up to 4 broods per season. Large, bulky domed nest with a false cup nest on top, 1–5m high / 3–4 eggs.

Left and right: Adult.

Striated Thornbill

Acanthiza lineata

Endemic

100mm

Sexes alike. Mostly olive-green upperparts with brownish wings and mid-brown rump. White and rufous-brown streaked crown. Cheeks, ear coverts: Olive-brown, heavily streaked. Bill: Dusky. Eyes: Grey-brown, white eyebrow. Tail: Olive-brown, dark sub-terminal band. Throat, upper breast: Whitish, dark streaked. Underparts: Yellow.
Race *alberti*, northern NSW, Qld: Greener.
Race *clelandi*, SA: More heavily streaked.
Immature: Duller.
Communal, flocks of 15–20 after breeding. Hovers, flits in upper canopy, often in mixed flocks.
Habitat: Open woodlands with understorey, dense eucalypt forests, mangroves, gardens.
Feeding: Insects, scale, lerp, psyllids, caterpillars, sometimes seeds and nectar.
Voice: Rapid, high-pitched staccato trill.
Status: Common, sedentary.
Breeding: July–Dec. Neat dome, suspended 5–10m high / 3 eggs.

Left: Race *clelandi*. Right: Nominate race, feeding in Sydney Golden Wattle.

Chestnut-rumped Thornbill

Acanthiza uropygialis

Endemic

100–110mm

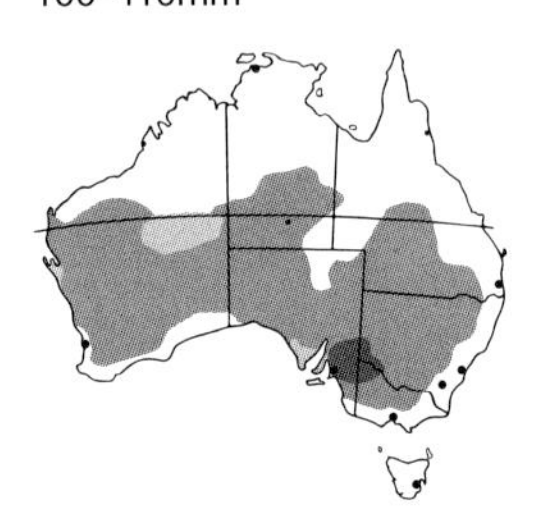

Sexes alike. Rich chestnut rump that is most obvious in flight. Upperparts: Brownish grey. Face: Speckled grey-brown, rufous-brown frons finely speckled pale grey. Eyes: White. Underparts: White, washed pale grey to sides of breast, flanks. Tail: Black with rufous base and white tips. Bill, legs: Dusky.
Immature: Brown eyes.
Usually in groups of 2–10. Often with Brown, Slaty-backed, Yellow and Yellow-rumped Thornbills.
Habitat: Arid mulga, mallee, Black Box and Belah woodlands, parks, gardens.
Feeding: Flits, hovers, hops, gleaning from all levels; poking bark or leaf litter.
Voice: Song, long chirping, twittering with mimicry mostly in breeding season. Quiet 'chips', repeated.
Status: Common. Sedentary.
Breeding: July–Dec. Dome nest in tree or post hollow or nest box 1–15m high / 2–4 eggs.

Left and right: Adult.

Western Thornbill

Acanthiza inornata

Endemic

90–100mm

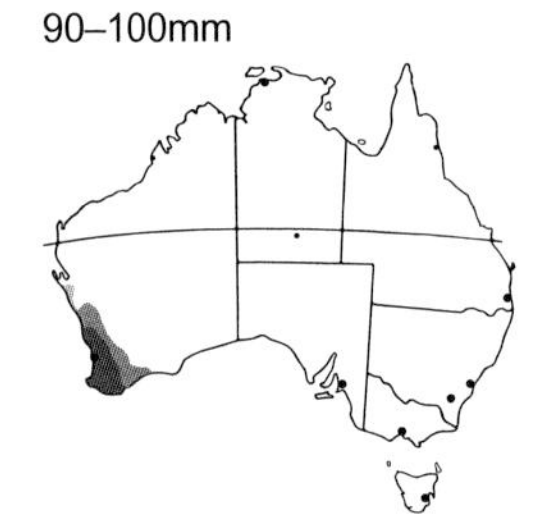

Sexes alike. Plain grey with a lightly speckled face. Eyes: White. Upperparts: Olive-grey with light brown rump. Underparts: Buffy white. Tail: Short, brown with broad, dusky, sub-terminal band. Bill, legs: Dusky.
Immature: Paler.
Communal, in groups of 5–15 birds, often in mixed flocks.
Habitat: Heathy woodlands, open shrubby forests, parks, gardens.
Feeding: Actively hop-searches the ground for beetles, bugs and in the lower foliage for insects and their larvae.
Voice: Rapid, soft, tinkling twitter and mimicry.
Status: Common. Sedentary or wanders locally.
Breeding: Aug.–Dec. Dome nest hidden close to ground, behind bark crevice, post or knot hole or under grass tussocks / 3–4 eggs.

Left and right: Adult.

Tasmanian Thornbill

Acanthiza ewingii

Endemic

100–110mm

Sexes alike. Similar to the Brown Thornbill, but distinguished by a brighter reddish rump, plainer, chestnut-tinted frons and chestnut-edged flight feathers. Breast: White, lightly mottled grey. Eyes: Red. Upperparts: Olive-brown. Underparts, undertail coverts: White. Bass Strait Island birds have deeper-toned upperparts. Immature: Paler. Brown eyes. Arboreal foraging at all levels.

Habitat: Cool temperate forests and wet sclerophyll shrubbery. **Feeding:** Insectivorous, occasionally seeds. **Voice:** Musical, whistled warbling and twittering. **Status:** Common. Sedentary. **Breeding:** Aug.–Dec. Compact, domed nest hung in outer foliage 2–4m high / 3–4 eggs.

Left: Adult feeding in Acacia. Right: Adult upperparts.

Buff-rumped Thornbill

Acanthiza reguloides

Endemic

100–110mm

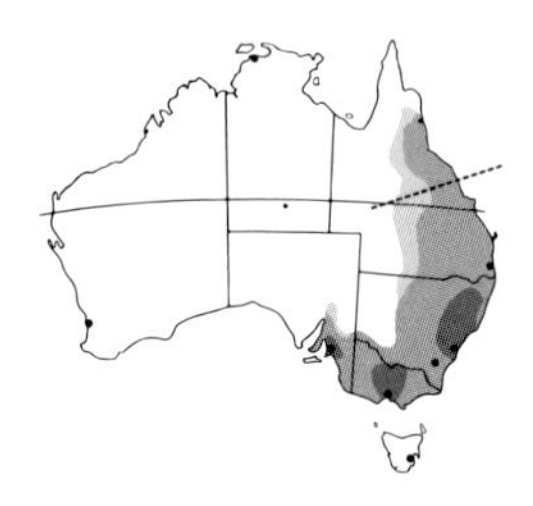

Sexes alike. Rufous-tinted, finely scalloped crown. Eyes: White. Upperparts: Olive-brown. Rump: Buff-cream. Underparts: Pale buff, flecked grey on throat and sides of breast. Tail: Dusky black, outer tips buff spots. Four similar races. Northern birds are brighter. Immature: Duller, plainer, brown eyes. After breeding, flocks of up to 20 birds form, usually with other small insectivores. **Habitat:** Open eucalypt forest.

Feeding: Hop-searching forest floor and lower canopy for insects, at times seeds. Far northern birds feed in the canopy where long grass inhibits ground foraging. **Voice:** Constant twittering, rapid, ringing staccato. **Status:** Moderately common. Sedentary. **Breeding:** Aug.–Jan. Dome nest hidden close to ground, behind bark crevice, post or knot hole or under grass tussocks / 3–5 eggs.

Left and right: Adult.

Brown Thornbill

Acanthiza pusilla

Endemic

95–105mm

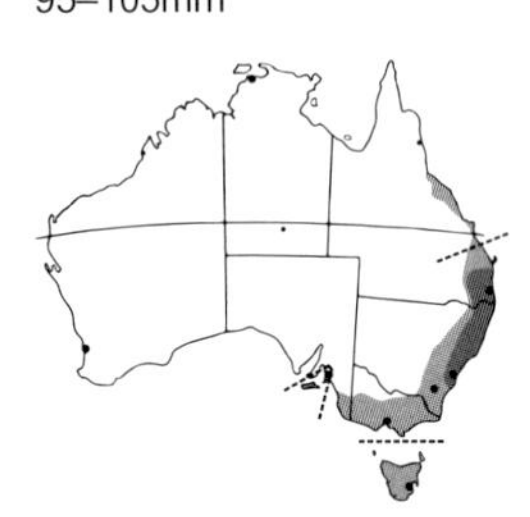

Sexes alike. Rufous and pale speckled crown. Face, throat, breast: Grey-brown with black streaking. Eyes: Dark red. Upperparts: Rich grey-brown with cinnamon rump. Underparts: Cream with pale yellow-buff flanks. Tail: Cinnamon base and blackish sub-terminal band tipped white. Subtle colour variation across range. Southern birds have white to buff flanks. Northern birds have deeper yellowish underparts. Immature: Less streaked, brown eyes.

Usually solitary or in pairs. **Habitat:** Forest and wooded areas with understorey. Gardens. **Feeding:** Forages in mid- to upper foliage, gleaning insects, their larvae and occasionally seeds and nectar. **Voice:** Musical notes, metallic trills and twitters, high-pitched 'seee'. **Status:** Sedentary. Common to locally uncommon. **Breeding:** June–Dec. Domed nest placed in low branches / 3 eggs.

Left: Adult at birdbath. Right: Adult.

Mountain Thornbill

Acanthiza katherina

Endemic

100mm

Sexes alike. Similar to Brown Thornbill. Eyes: Whitish cream. Upperparts: Olive-green to grey-green. Rump: Dull rufous. Frons: Buff scalloped. Chin, throat, upper breast: Cream. Underparts: Cream to lemon with buff flanks. Tail: Dark with white tip. Immature: Brown eyes. Flits between trees in short, undulating flight. Upper to middle canopy, constantly twittering.

Habitat: Mountain rainforests and adjacent stream vegetation. **Feeding:** Actively forages for insects in dense upper foliage. **Voice:** Loud, sweet warbling. Status: Common. Sedentary. **Breeding:** Sept.–Jan. Large domed nest, hooded side entrance, decorated with moss, suspended from a twig 5–15m high / 2 eggs.

Left and right: Adults showing distinctive almost white eyes.

Southern Whiteface

Aphelocephala leucopsis

Endemic

110–120mm

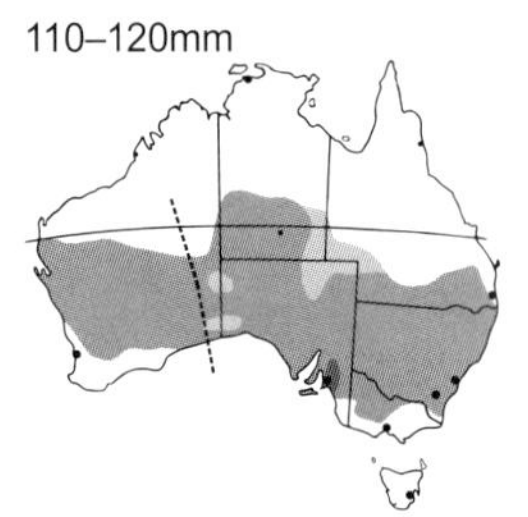

Sexes alike. Distinctive face with whitish, dark-edged frons and stubby black bill. Eyes: White. Crown, ear coverts, upperparts: Grey-brown, wings darker. Underparts: White with cream-grey wash to breast and brownish grey-flecked flanks. Tail: Dusky, outer feathers white tipped.
Race *castaneiventris*, WA: Chestnut washed flanks.
Immature: Paler.
Shy. Constant movement. Fast flight, dashes close to ground.
Habitat: Open woodlands, mallee, mulga with fallen timber.
Feeding: Uses stout bill to turn over litter, taking seeds, insects.
Voice: Melodic, bell-like whistled song. Twittering calls.
Status: Moderately common.
Breeding: July–Dec. Untidy domed nest in tree hollow, post hole, shrubbery, nest box / 2–5 eggs.

Left: Nominate race. Right: Race *castaneiventris*, WA.

Chestnut-breasted Whiteface

Aphelocephala pectoralis

Endemic

110–120mm

Sexes alike. Richly coloured. Frons: White with black band, merging into grey-brown crown. Eyes: White. Upperparts: Rufous-cinnamon. Wings: Grey-brown, flight feathers edged pale cinnamon. Underparts: Dull white with broad rufous-cinnamon band across breast and rufous-flecked flanks. Tail: Black, tipped white.
Immature: Similar, narrower, rufous, dusky-edged breast band.
Flies quickly to cover if disturbed. Low undulating flight.
Habitat: Semi-deserts, gibber plains.
Feeding: Hop-searches over ground picking up seeds, insects.
Voice: Loud, plaintive bell-like whistled song.
Status: Uncommon. Nomadic.
Breeding: Aug.–Sept., rain dependent. Rounded nest with side entrance placed low in a shrub / 3 eggs.

Left and right: Adults.

Banded Whiteface

Aphelocephala nigricincta

Endemic

110–120mm

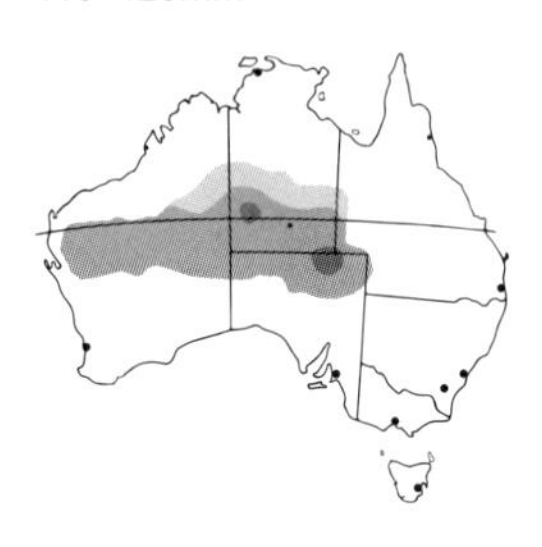

Sexes alike. Desert-adapted whiteface with a distinctive black band across its whitish underparts and white, edged black frons. Crown, face: Grey-brown streaked blackish. Upperparts: Cinnamon rufous. Wings: Brown-grey with pale rufous-edged flight feathers. Rump: Bright chestnut. Tail: Blackish, tipped white. Eyes: White.
Usually in pairs or small groups.
Habitat: Dry, almost treeless inland plains, saltbush plains, spinifex ridges and dunes. Relies on insect prey for moisture.
Feeding: Forages on open ground, hop-searches for seeds and insects.
Voice: Soft bell-like song. Staccato twittering.
Status: Moderately common. Wanders nomadically in pairs or small groups.
Breeding: Feb.–Aug. Bulky, domed nest, with long entrance spout lined with flowers, in prickly shrub 1–2m high / 2–4 eggs.

Uluru area, NT.

Left and right: Adults.

Forty-spotted Pardalote

Pardalotus quadragintus

Endemic Tas. Endangered

95–100mm

Sexes alike. Olive green with pale olive-yellow face and dark brown eyes. Underparts: Greyish white with dull yellow undertail. Wings: Black with prominent white spots (20 on each side). Rump: Dull olive. Bill: Short, notched, black. Legs: Fawn.
Immature: Paler.
Usually seen in small flocks.
Habitat: Mixed eucalypt forests where Manna Gums often dominate.
Feeding: Slow, methodical leaf-gleaning high in canopy, taking lerp, manna, insects, spiders.
Voice: Double-noted piping, second note lower pitched.
Status: Rare. Endangered. Sedentary in colonies.
Breeding: Aug.–Jan. Nests in tree hollow to 1–20m high or burrow nest / 3–5 eggs.

Bruny and Maria Is., Tas.

Opposite: Adult showing spots that give the bird its name.

Spotted Pardalote

Pardalotus punctatus

Endemic

80–100mm

Sexes similar. Small, brightly coloured birds. Nominate race occurs from southern Qld, NSW, partially Vic., SA to Tas. Jewel-like white spots on crown, wings and tail. Face: Flecked grey and white with grey-brown eyes and long, white eyebrows. Rump: Bright chestnut red. Tail: Black with golden yellow covert and sub-terminal glossy crimson and black band and white tips. Mid-back, scapulars, mantle: Brown, coarsely spotted buff. Wings, tail: Black tipped white. Throat: Dark yellow. Breast, belly: Creamy buff. Undertail: Bright yellow. Bill: Black.

Race *xanthopyge*, Yellow-rumped Pardalote: Far west NSW, Vic., SA and northern part of WA range. Back: Grey, finely spotted. Rump: Bright yellow. Interbreeds with nominate race and hybrids commonly occur in the overlapping range, most commonly in the eastern range.

Race *millitaris*, Queensland Spotted Pardalote: Isolated population in northeast Qld. Smaller. Russet rump and finely spotted back.

Female all races: Buff-yellow spots on crown. Back: Brown, spotted buff. Eyebrows: Buff. Throat: Cream-white. Undertail: Pale yellow. Immature: Similar female, paler, duller, grey crown.

Pairs or groups of 10–20 birds, usually in the higher canopy.

Habitat: Wet and dry sclerophyll forests, mallee, parks, gardens. Race *xanthopyge* almost exclusively inhabits open mallee woodlands.

Feeding: Leaf gleaning especially for psyllids and tiny sap-sucking lerp. Also manna and other insects.

Voice: Persistent soft calls, bell-like 'sleep-may-bee' call. Race *xanthopyge* higher pitched.

Status: Usually sedentary. Moderately common.

Breeding: June–Jan. Usually excavates tunnel in an earth embankment, 0.5–1.5m long, ending in domed nest in nesting chamber. Alternatively they sometimes use tree hollows or nest boxes, building a domed nest inside / 3–6 eggs.

Hattah-Kulkyne NP, Vic.

Top: Nominate male at entrance to nest tunnel.
Centre: Race *xanthopyge* Yellow-rumped Pardalote, male (left) and female (right).

Red-browed Pardalote

Pardalotus rubricatus

Endemic

100–110mm

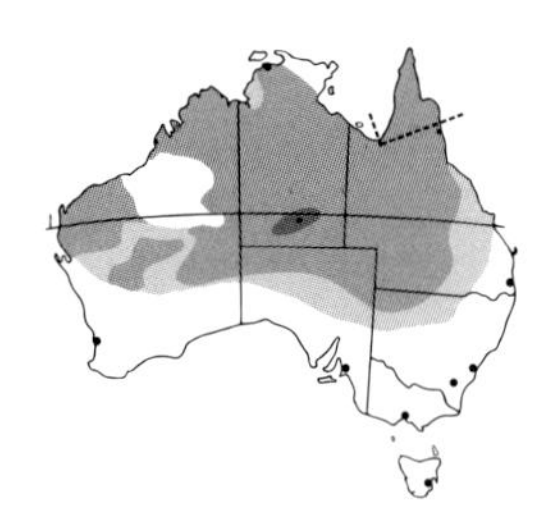

Sexes alike. Distinctive small orange-red eyebrow spot. Buff-yellow eyebrow over pale eyes. Crown: Black, spotted white. Upperparts: Grey-brown. Underparts: Pale fawn with pale yellow washed breast. Wings: Dusky with orange-yellow wash in a broad bar and flight feathers edged yellow and white. Rump: Pale yellow. Large bill.

Race *yorki*, Cape York, Qld: Smaller. Darker, more patterned upperparts and bright yellow undertail coverts.

Immature: Crown grey. Upperparts: Yellow-green. Underparts: Yellow.

Arboreal, preferring eucalypts.

Habitat: Drier eucalypt woodlands in the north. Eucalypt-lined waterways, mulga scrub, inland areas with desert bloodwood.

Feeding: Uses large 'scoop-shaped' bill for gleaning insects, especially lerp from the leaf surface. Also other insects, spiders. Often feeding in low shrubby eucalypts.

Voice: Mellow, high-pitched whistle. 5-note call, followed by higher-pitched more rapidly delivered notes, repeated, rising.

Status: Moderately common. Locally uncommon. Nomadic.

Breeding: July–Dec. or after rain in arid areas. Nest tunnel up to 80cm long, ending in a grass-lined chamber / 2–3 eggs.

Left and right: Adults. Nominate race.

Striated Pardalote

Pardalotus striatus

Endemic

100–115mm

Sexes alike. Found across most of Australia, absent only from the most arid regions. There are several geographic subspecies. Nominate race.Yellow-tipped Pardalote, Tas.: Prominent bright yellow throat and long, bright yellow to white eyebrow. Crown: Black, streaked white. Face: Grey, flecked black with a black line through eyes. Upperparts: Olive-grey with dull yellow rump. Underparts: White with yellow throat and buff yellow flanks. Wings: Black with bright yellow spot, thin white wing stripe and inner flight feathers edged rufous. Eyes: Light brown-grey. Bill: Stubby, black.
6 races, mostly differentiated by crown, colour of wing spot and width of white wing bars. Races hybridise and integrate where ranges overlap. The nominate race breeds in Tas., but makes regular trips to the mainland.
Race *substriatus*, Striated Pardalote, throughout the inland to central and southwest coastal WA. Crown, face: Black, streaked white. Wings: Broad white wing stripe, red or orange spot, secondary feathers buff edged. Rump: Dull ochre.
Race *ornatus*, Eastern Striated Pardalote, from mid-coastal eastern Australia south to SA. Crown: Black, streaked white. Wing spot: Red or orange with narrow, white wing stripe, secondary feathers buff edged. Rump: Dull ochre.
Race *melanocephalus*, Black-headed Pardalote, northeast NSW to northeast Qld. Crown, face: Black without streaks. Red wing spot, bright tan rump.
Race *uropygialis*, Kimberley, Top End, NT, and northwest Qld. Yellow-rumped form of the Black-headed Pardalote.
Race *melvillensis*, Melville Is., NT, Tan-rumped Pardalote.
Sexes alike, except for race *uropygialis*, where the female crown is fawn. Immature: Similar, paler with an olive-grey crown and cream eyebrows.

Flight is high, fast and direct from tree to tree.

Habitat: Widespread, occupying diverse habitats. Open eucalypt forests, woodlands, mallee, mulga, rainforests, mangroves, shrubby areas, River Red Gum–lined watercourses, parks and gardens.

Feeding: Forages quickly in upper canopy and outer foliage of mostly eucalypts. Leaf-gleaning, lerp, thrips, insects, bugs, flies.

Voice: Variable, between races and locally. Constant soft trill contact call. Clear, musical, sharp, loud 'chip-chip' repeated frequently.

Status: Sedentary in north. Southern birds move northward, inland or to lowlands in winter. Most Tas. birds migrate Mar.–Oct. to the mainland as far north as Qld.

Breeding: June–Jan. Northern birds generally build nest burrows. Other races will either make a nest burrow or select a tree hollow 10m or higher. Human-made structures with suitable hollows are also used / 3–5 eggs.

Top left: Nominate race, Yellow-tipped Pardalote. Top right: Immature, note olive-grey crown.

Top centre left and right: Race *substriatus*, Striated Pardalote.

Below centre: Race *ornatus*, Eastern Striated Pardalote, adult (left) and at entrance to nest (right).

Below left: Race *melanocephalus*, Black-headed Pardalote. Below right: Race *uropygialis*.

Family Meliphagidae

Honeyeaters

Chats

Gibberbird

A grouping of the diverse members of the large family containing honeyeaters, abberant honeyeaters and the unique Gibberbird.

Family Meliphagidae is one of Australia's largest and most diverse family groups. It contains 69 species of honeyeaters, 4 species of chats and the closely related Gibberbird.

Honeyeaters are a group of iconic birds that are major pollinators of many of Australia's unique native plants. Many of their adaptions reflect their co-evolution with Australia's endemic honey-flora. These adaptations include their generally down-curved bill to access the flower nectary, the bill shape reflecting the shape of their favoured flower types. The honeyeater's long tongue is split with four fine hair-like extensions at the tip, resembling a brush tip, allowing the honeyeater to 'lick up' the liquid nectar.

Feeding largely on nectar, they probe the mostly tubular flowers of native honey-flora plants including banksias, grevilleas, hakeas and waratahs. While honeyeaters are feeding on the honey-flora flowers, pollen is deposited on their head and transferred to the next plant they visit, thereby assisting cross-pollination.

As well as nectar, honeyeaters supplement their diet with insects, especially in the nesting season, with some species eating more insects than nectar. Some species supplement their nectar diet with small native fruits, especially from tropical and rainforest species. Generally species with long fine bills, such as spinebills, are mainly nectarivorous.The Painted Honeyeater is a mistletoe specialist, only foraging in the flowers of the parasitic mistletoe growing in trees throughout Australia.

Most honeyeaters are blossom nomads, following the cyclical flush of honey-flora, and time their nesting period to coincide with the peak flowering of their favoured plants. This saves time feeding themselves while hunting for insects to feed their young. Nests are typically neat, cup-shaped and constructed from thin bark strips and grass, closely bound with spider web, lined with soft plant matter and suspended by the rim in a leafy branchlet or placed in a horizontal fork. Honeyeaters will readily visit parks and gardens containing Australian honey flora, often taking up residence.

Australian chats are small, aberrant honeyeaters that have adapted to a terrestrial life in a wide range of habitats in the more arid regions of inland Australia. They have short honeyeater brush-tipped tongues; while being almost exclusively ground-feeding insectivores, they occasionally rifle nectar and pollen from flowering groundcover plants such as Sturt's Desert Pea or Emu-bush. Unusually, they spend most of their time on the ground walking around, rather than hopping. They rarely fly.

The Crimson and Yellow Chats are unusual among honeyeaters in that their seasonal colour changes, the males being much duller outside the breeding season. Typical chat nests are cup-shaped, built low in a small dense shrub, constructed from twigs and fine grass and lined with fine rootlets.

The Gibberbird, also known as the Desert Chat, is a distinctive species of chat and the only member of the genus *Ashbyia*. The bird's common name is a reflection of the gibber plains that it mainly inhabits, thinly vegetated stony desert areas of the inland.

Opposite: Eastern Spinebill in the nectar-rich flowers of 'Superb' Grevillea.

Yellow Wattlebird

Anthochaera paradoxa

Endemic

440–450mm

Sexes alike. Australia's largest honeyeater. Prominent pendulous, orange-yellow cheek wattles, brighter when breeding and lengthening with age. Crown and nape: White, streaked black. Face: Silvery white with dusky malar line, dark brown eyes and black bill. Throat: Grey-white bearded. Upperparts: Grey-brown, streaked white. Wings: Dark grey-brown, edged white. Tail: Long, grey-brown tipped white. Underparts: Grey-white with dusky streaks and yellow lower breast and belly. Legs: Pinkish brown.
Immature: Similar, duller, brownish belly, small wattles.

Strong, undulating flight. Gregarious.

Habitat: Prefers tall, mature eucalypt woodlands, parks, gardens and orchards.

Feeding: Forages from near ground to upper canopy, taking insects, nectar, manna, sometimes fruit.

Voice: Loud 'kuk-kukuk' accompanied by jerky head movements. Discordant gurgling.

Status: Common.

Breeding: July–Jan. Large, open twig nest / 2–3 eggs.

Left: Adult. Right: Adult showing wattles.

Red Wattlebird

Anthochaera carunculata

Endemic

330–370mm

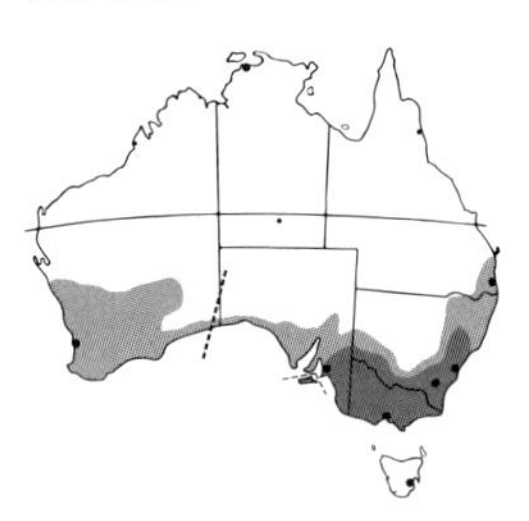

Sexes similar. Largest mainland honeyeater. Overall grey-brown and white streaked with distinctive red wattles that lengthen and darken with age. Crown: Black. Face: Silvery white. Eyes: Dark red. Underparts: Pale grey-brown heavily streaked white with yellow centre belly.
Race *woodwardi*, southwestern WA. Smaller with heavier streaking.
Race *clelandi*, restricted to Kangaroo Is. Deeper yellow belly.
Female: All are smaller.
Immature: Reddish eyes, small wattles, overall browner with bright white wing edges.

Habitat: Eucalypt woodlands, heathlands, mallee, parks and gardens.

Feeding: Mostly nectar. Large quantities of insects, manna and honeydew. Aggressive when feeding. Hops along the ground taking insects or hawks for flying insects.

Voice: Noisy. Harsh 'tchock'.

Status: Common. Nomadic, with groups of 5–100 birds following flowering events.

Breeding: July–Dec. Nest is a cup-shaped structure made from sticks and twigs lined with bark and grass / 2–3 eggs.

Left: Red Wattlebird race *woodwardi* on Firewood Banksia. Right: Nominate race feeding on Monga Waratah.

Little Wattlebird

Anthochaera chrysoptera

Endemic

270–350mm

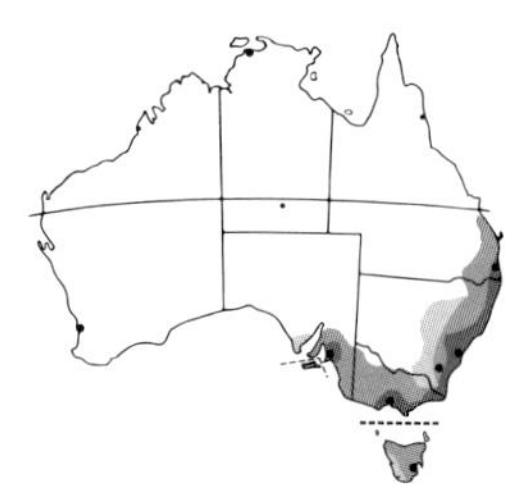

Sexes similar. Smallest wattlebird. Dark, brownish grey body streaked white and without 'wattles'. Faint silvery cheek patch. Eyes: Pale blue-grey. Tail, wings: Tipped white. Underparts: Grey, heavily streaked white. Large rufous wing patch visible in flight. Race *tasmanica,* Tas., and race *halmaturina*, Kangaroo Is., are darker.
Female: All are smaller.
Immature: Similar, less prominent streaks and dark brown eye.

Often darting almost vertically into the air in pursuit of flying insects.

Habitat: Coastal heathland, shrubby forests, parks and gardens.

Feeding: Nectar, manna and insects. Aggressive towards other birds around food source.

Voice: Constant, rasping squawk by male. High-pitched call by female. Soft, metallic duet sung by pairs.

Status: Common.

Breeding: July–Dec. Nest is an untidy cup-shaped structure of twigs, lined with shredded bark, placed in the fork of a banksia or eucalypt tree / 2 eggs.

Left: Nominate race. Note rufous wing patch. Right: Nominate race on Bull Banksia.

Western Wattlebird

Anthochaera lunulata

Endemic

290–330mm

Sexes similar. Similar to the Little Wattlebird but crown and nape are not streaked. Face, cheeks: No wattles, conspicuously silvery white streaks from gape fanning out down sides of neck. Belly: Grey and white streaked. Underwing: Reddish brown patch. Upperparts: Grey-brown with broad white bars on wings. Tail: Broadly tipped white. Bill: Slender, black. Eyes: Red. Female: Duller, smaller. Immature: Less streaked.

Birds usually feed alone or in small groups.

Habitat: Coastal heathlands, woodlands, parks and gardens with nectar-producing trees and shrubs.

Feeding: Wanders in small groups for nectar. Insects are also taken.

Voice: Bubbling, complex calls, cackles and twittering, often in pairs.

Status: Common.

Breeding: July–Oct. Nest is an untidy cup shape of twigs lined with shredded bark / 1–3 eggs.

Left: Feeding on Acorn Banskia. Right: Firewood Banksia. Note red eye.

Spiny-cheeked Honeyeater

Acanthagenys rufogularis

Endemic

220–270mm

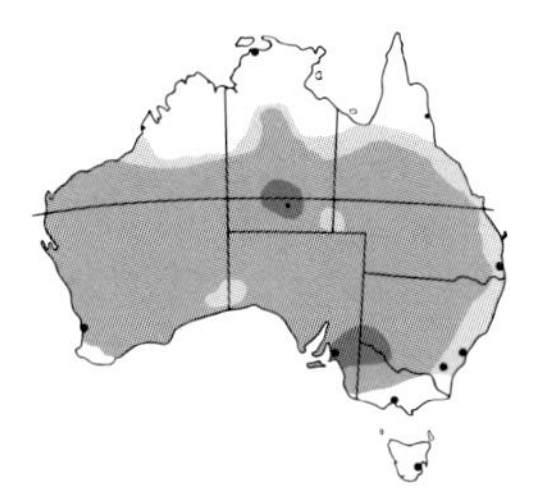

Sexes similar. Mid-sized honeyeater with a distinctive pink, black-tipped bill, pink gape and pink skin extending under eyes. Eyes are bright blue with a dusky broad line from lores to behind the eyes. Cheeks: Broad 'spiny' white and yellow streak to below the ears. Throat, breast: Cinnamon. Crown: Grey-brown, mottled with black. Upperparts: Brownish grey with darker wings edged cream-white and light grey rump. Tail: Dusky grey, tipped white outer feathers. Female: Smaller. Immature: Similar, brown eyes with pale pink skin and yellow cheek 'spines'.

Sociable, aggressive, often feeding in large flocks. Quick, undulating flight between trees or singing from a high perch and quickly gliding back into cover.

Habitat: Coastal woodlands, open arid and semi-arid scrubland, mallee and desert areas.

Feeding: Nectar, manna and small fruits. Insects gleaned or snapped mid-air.

Voice: Liquid trills and warbles. Contact, alarm call, sharp 'tok'.

Status: Common. Locally nomadic.

Breeding: July–Jan. in east and southeast. Irregularly elsewhere. Nest is a suspended deep cup shape of plant fibres and grass bound with spider web / 2 eggs.

Left: Adult at birdbath. Right: Adult upperparts.

Striped Honeyeater

Plectorhyncha lanceolata

Endemic

215–235mm

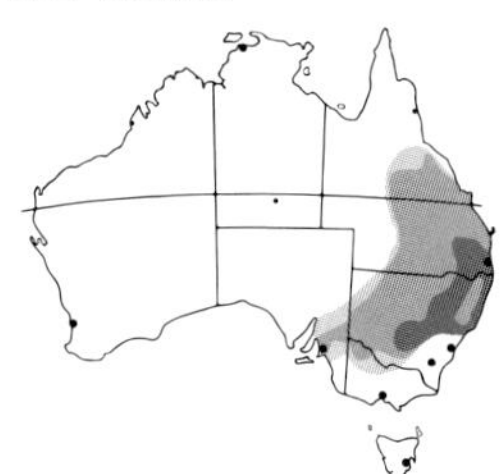

Sexes similar. Overall streaked appearance with long spiky throat and upper breast feathers. Head, nape: White, streaked black. Mantle: Grey-brown, streaked brown. Underparts: Whitish with less streaking. Tail: Black, brown, white tipped. Bill: Blue-grey dark tipped. Eyes: Brown. Female: Browner upperparts. Underparts: Greyer. Immature: Duller, less streaking.

Active, fearless. Dashing flight between trees.

Habitat: Woodland and inland scrubby mallee.

Feeding: In the mid-canopy or foraging on the ground. Mostly insects and also fruit, nectar.

Voice: Loud, bubbling, melodious song. Often in duets.

Status: Reasonably common. Usually sedentary. Inland birds locally nomadic in autumn.

Breeding: Aug.–Jan. Nest is a suspended cup shape structure of plant fibres and grass bound with spider web / 4–5 eggs.

Oxley Creek Common, Qld. Round Hill NR, NSW.

Left: Female nest building. Right: Male upperparts.

Silver-crowned Friarbird

Philemon argenticeps

Endemic

250–320mm

Sexes similar. Large honeyeater with prominate knob on bill, a silvery white crown and neck, and black patch of bare facial skin. Dusky, furry ear patch and red eyes. Chin, throat, upper breast: Silver grey gorget feathers. Upperparts: Olive-brown, darker on wings and tail. Underparts: Fawn.
Female: Smaller. Immature: As adults but washed yellow breast. White scalloping to upperparts. Small knob on forehead and brown eyes.

Aggressive to other birds.

Habitat: Tropical eucalypt forests, melaleuca woodlands, gardens near water.

Feeding: Feeds on nectar and insects, often with other honeyeaters and lorikeets.

Voice: Harsh raucous calls.

Status: Locally common. Nomadic following flowering honey flora.

Breeding: Sept.–Dec. Cup-shaped nest of plant fibres and grass suspended by rim, 2–11m high / 1–2 eggs.

Judbarra NP and Escarpment Walk, Cape Crawford, NT. Mitchell Falls, WA.

Left and right: Adult, note prominent bill knob.

Helmeted Friarbird

Philemon buceroides

320–360mm

Sexes similar. Mid-sized with a silver crown and a sloping knob on its bill. 3 similar races in Australia.
Race *yorki*, northeastern Qld. Short, tufty grey feathers, crown and nape. Face: Slate grey facial skin with dark furry ear coverts, dark red eye. Bill: Black, sloping knob. Upperparts: Mid-brown. Underparts: Light grey-brown with white-grey gorget feathers to throat and breast.
Race *ammitophila*, Sandstone Friarbird. Top End, NT. Smallest knob. Prefers sandstone country.
Race *gordoni*, Melville Island Friarbird. Smaller with a rounded knob. Prefers mangrove areas.
Female: All smaller. Immature: Brown eyes, smaller casque.
Aggressive to other birds.

Habitat: Edge of rainforests, mangroves, paperbark swamps, always close to rainforest and monsoon vine thickets for roosting. Parks and gardens.

Feeding: Usually upper canopy, individually or groups of 10–30, noisy, taking nectar, fruit, insects.

Voice: Extremely noisy, harsh and repetitive calls.

Status: Locally common.

Breeding: Sept.–Apr. Nest is a deep cup shape from plant fibres and grass bound with spider web, slung from a forked branch to 14m high / 1–5 eggs.

Cairns area, Qld and Gunlom Falls, Kakadu NP, NT.

Left and right: Adult feeding on Golden Tree.

Noisy Friarbird

Philemon corniculatus

Endemic

300–380mm

Sexes alike. Distinctive bare black head and upper neck with a prominent knob on a strong bill. Neck collar, centre breast: Long silvery feathers. Back, rump: Brown. Underparts: Off-white. Wings: Mid-grey. Tail: Brown, white tips. Eyes: Red.
Race *monachus*, north Qld. Larger.
Immature: Small knob, lacks gorget.

Gregarious, noisy. Aggressively defends feeding area.

Habitat: Eucalypt forests, open woodlands, parks and gardens.

Feeding: Nectar, insects, native fruits.

Voice: 'Four-o-clock' call with a cacophony of noisy cackles and chuckles.

Status: Common. In winter, southern birds migrate northward and highlands birds move to lower altitudes.

Breeding: Aug.–Jan. Nest is large and untidy deep cup shape made from plant fibres and grass bound with spider web, slung from a forked branch / 2–3 eggs.

Left: Adult nominate race. Right: Feeding in Red-flowering Gum.

Little Friarbird

Philemon citreogularis

250–290mm

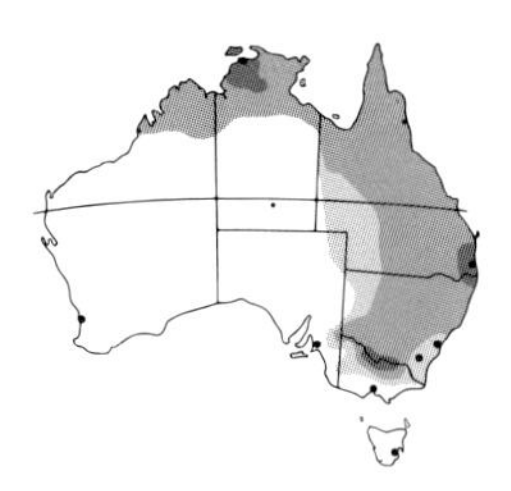

Sexes alike. Smallest friarbird, lacks 'knob' on bill. Bluish black facial skin. Eyes: Dark. Shoulders, back: Grey-brown with bluish tinge. Nape: Whitish. Chin: Fine, silky white feathers. Underparts: Pale grey. Tail: White tipped.
Immature: Paler facial skin. White scalloping to back, rump. Breast, throat: Washed yellow. Constantly chases other honeyeaters from food source.

Habitat: Open eucalypt woodlands, forests, orchards. In the north, melaleuca forests, mangroves. Wetlands. Arid and semi-arid areas along watercourses. Parks and gardens, particularly with water.
Feeding: Prefers the outer foliage, foraging for nectar. Gleans and hawks for insects.
Voice: Raucous 'ar-coo' repeated monotonously.
Status: Locally common. Migrates locally following food. Many southern birds move northward to Qld for winter, returning in spring to nest.
Breeding: Aug.–Apr. Flimsy deep cup, suspended by rim. Always above or near water / 2–3 eggs.

Cairns area, Qld. Kakadu NP, NT.

Left: Gleaning insects. Right: Taking nectar from grevillea flowers.

Blue-faced Honeyeater

Entomyzon cyanotis

260–320mm

Sexes alike. Golden olive green upperparts and striking 2-tone turquoise and cobalt skin patch around its yellow eye. Crown, face, hind neck: Black with broken white nape crescent. Tail: Tipped white. Underparts: White, with black throat and long black breast bib and white malar line joining white breast. Bill, legs: Blue-grey.
Race *griseigularis*, Cape York, Qld. Smaller.
Race *albipennis*, Top End, NT, WA. Deeper blue facial skin, large white wing windows.
Immature: Brownish with grey chin and bib and yellow-green facial skin.
In pairs or small flocks. Aggressive towards other birds.

Habitat: Rainforest fringes, open woodlands, mangroves, eucalypt, paperbark, pandanus.
Feeding: Mostly insects, also nectar and fruit.
Voice: Loud monotonous 'ki-owt' at daybreak and piping calls. Soft contact calls in flight.
Status: Locally common.
Breeding: June–Jan. Builds nest mostly on top of abandoned nest, commonly that of the Grey-crowned Babbler / 2–3 eggs.

Chiltern–Mt Pilot NP, Vic. Burrendong Botanic Garden, NSW. Cairns area, Qld. Nitmiluk NP, NT.

Left: Adult on Robyn Gordon Grevillea. Right: Immature in Gorge Hakea.

Regent Honeyeater

Anthochaera phrygia

Endemic Endangered

200–240mm

Sexes similar. Striking, black and yellow plumage. Black head with yellowish or pink warty skin around eyes and black upper breast and neck. Upperparts: Lemon yellow scalloped black, grading to whitish rump. Lower breast, belly: Lemon yellow scalloped black grading to pale yellow undertail coverts. Wings: Mostly black with prominent bright yellow. Tail: Black with outer feathers edged bright yellow and white below.
Female: Smaller. Bare yellow skin only under eyes and smaller black breast area. Immature: Brown. Flight and tail feathers edged yellow.
Swift flight. Follows flowering of eucalypts and other trees.

Habitat: Box–ironbark forests and woodlands.
Feeding: Prefers outer foliage, taking nectar, manna, lerp and fruit. Insects are prised from bark or hawked mid-air.
Voice: Melodic bell-like call. Clinking metallic call.
Status: Endangered. Rare. Nomadic, moving north in autumn, returning south to breed.
Breeding: Aug.–Jan. Thick cup nest placed in tree fork 1–9m high / 2–3 eggs.

Chiltern–Mt Pilot NP, Vic. Glen Davis, NSW.

Opposite: Adult feeding in Yellow Box. Note warty face skin around eye.

Noisy Miner

Manorina melanocephala

Endemic

240–280mm

Sexes alike. Mostly grey with white-tipped tail. Crown, ear coverts, chin: Black. Frons, lores: White with bare bright yellow skin patch behind the eyes. Bill, legs: Yellow. Underparts: Pale grey with scalloped grey throat and breast.
Race *crassirostris*, northern Qld. Paler.
Immature similar.

Vigorously defends feeding area. Living in loose colonies, they unite to drive away intruders if danger threatens.

Habitat: Open forests, woodlands, remnant bushland, parks, gardens.

Feeding: In large groups foraging in trees or on the ground, taking nectar, fruits, insects, lerp, honeydew.

Voice: Noisy, raucous 'pee-pee-pee'.

Status: Locally common. Sedentary.

Breeding: June–Dec. Suspended, deep cup-shaped nest from twigs and shredded bark, to 20m high / 2–4 eggs.

Left: Nominate adult. Right: Adult feeding in 'Peaches and Cream' Grevillea.

Yellow-throated Miner

Manorina flavigula

Endemic

220–280mm

Sexes alike. Similar to the Noisy Miner but has pale yellow frons, yellow wash to sides of throat and a distinctive white rump. Eyes: Brown with bare yellow skin spot. Upperparts: Grey with yellow-green edged wings. Tail: Tipped white. Underparts: White with grey mottled throat and breast. Geographic changes in plumage across several subspecies.
Nominate race, eastern Australia.
Race *wayensis*, WA, southern NT, SA, far inland NSW, Qld. Paler.
Race *obscura*, Dusky Miner, southwest WA. Overall darker. Narrower face mask.
Race *lutea*, Top End, NT, WA. Palest, smallest.

Noisy and sociable. Aggressive to other birds.

Habitat: Dry open woodland, mallee, arid scrubland west of the Great Dividing Range.

Feeding: Mostly insects, also nectar, fruit and seeds.

Voice: Harsh, similar to Noisy Miner.

Status: Sedentary. Locally nomadic.

Breeding: July–Dec. Communally.Typical nest, bulky, similar to the Noisy Miner, 3–6m high / 3–4 eggs.

Left: Race *wayensis,* pair at Alice Springs, NT. Right: Nominate race.

Black-eared Miner

Manorina melanotis

Endemic Endangered

230–260mm

Sexes alike. Similar to the Yellow-throated Miner, but overall darker, with a larger black face mask from bill to ear coverts and a small yellow patch behind eyes. Upperparts: Dark grey, darker on crown and rump. Underparts: Mid-grey, darker mottled throat, breast. Underparts: Plain and paler.

Couples pair for life. Co-operative breeders in groups of 8 to 40 birds with up to 12 helpers at nest. Shyest miner, unsually quiet outside breeding.

Habitat: Restricted to undisturbed mature mallee, eucalypt forests and woodlands.

Feeding: Mostly insects and lerp gleaned from leaves, bark or taken in the air or from the ground. Also nectar.

Voice: Similar to Noisy Miner.

Status: Endangered. Sedentary. Threatened by habitat loss and interbreeding with the Yellow-throated Miner.

Breeding: Aug.–Jan. Communal, typical nest similar to Noisy Miner / 2–3 eggs.

Gluepot NR, SA.

Left: Adult with insect prey. Right: Adult Upperparts.

Bell Miner

Manorina melanophrys

Endemic

175–200mm

Sexes alike. Commonly called 'Bellbird' in reference to their distinctive continuous, bell-like call. The smallest of the miner birds, they are mostly dark olive green with a yellowish belly. Eyes: Brown, with orange-red eye patch skin. Lores, malar line: Blackish. Bill: Yellow. Underparts: Yellow-green. Legs: Orange-yellow.
Immature: Brownish. Eye skin: Pale yellow.

Hopping and fluttering through the mid-canopy. Pugnacious and aggressive, driving other birds from feeding territory. Reside in small, complex, social groups within a larger colony.

Habitat: Monopolises pockets of wet eucalypt forests with dense understorey, usually near water.

Feeding: In colonies in the upper canopy of eucalypts, primarily on psyllids and lerp, also insects, manna and nectar.

Voice: Sharp, high-pitched far-carrying contact 'ping' calls of colony ring out through the day.

Status: Common. Sedentary.

Breeding: July–Feb. Well-hidden, typical nest similar to Noisy Miner 4–5m high / 1–3 eggs.

Long Island in Melbourne Botanic Gardens, Vic.

Left: Adult. Right: Pair feeding.

Lewin's Honeyeater

Meliphaga lewinii

Endemic

190–215mm

Sexes alike. Overall dark olive green with a conspicuous, pale yellow, crescent-shaped ear patch. Bill: Black with thin yellowish gape extending under eye. Eyes: Blue-grey. Frons, crown: Dark grey. Wings: Dark olive green. Underparts: Pale olive, faintly streaked grey.
Race *amphochlora*, far north Qld. Pale olive-yellow and longer bill.
Race *mab*, central north Qld. Smaller.
Immature: Similar, brown eyes and paler crown.

Fearless and inquisitive. Aggressively chases other birds away from feeding area. Strong, undulating flight, audible wingbeats. Largest and darkest of the 3 look-alike honeyeaters.

Habitat: Wet eucalypt forests, rainforests, heathlands and gardens.

Feeding: In mid- to upper canopy for nectar, fruit. Spirals up trees, picking insects from bark crevices.

Voice: High-pitched staccato. Single harsh call.

Status: Locally common.

Breeding: Oct.–Jan. Large, untidy cup nest of bark, grass and moss 2–6m high / 2–3 eggs.

O'Reilly's, Lamington NP, Qld.

Left: Nominate race feeding on Grevillea flowers. Right: Adult upperparts.

Graceful Honeyeater

Meliphaga gracilis

150–170mm

Sexes alike. Similar to the Yellow-spotted Honeyeater but smaller with a more graceful appearance. Smaller, crescent-shaped, yellow ear patch with dusky green lores and frons. Bright yellow gape and line extending under brown eye. Bill: Fine, more strongly down-curved. Upperparts, head: Olive-brown. Underparts: Pale grey-green.
Immature similar.

Acrobatic when feeding, fluttering in the outer foliage. Upright stance when perched.

Habitat: Coastal rainforests, mangroves, lowland wet forests and tropical gardens.

Feeding: Forages high in the canopy for nectar, insects and fruit. Visits gardens with nectar-producing plants.

Voice: Sharp 'tick' or 'tuck' contact call, repeated constantly.

Status: Common. Sedentary.

Breeding: Oct.–Feb. Typical, attractive, suspended nest, 2–6m high / 2 eggs.

Cairns area including Centenary Lakes Botanic Gardens, Qld.

Opposite: Adult feeding in Cliff Bottlebrush.

Yellow-spotted Honeyeater

Meliphaga notata

Endemic

170–200mm

Sexes alike. Similar to Lewin's Honeyeater but has smaller, more rounded yellow ear patch. Bill: Black with thin bright yellow gape and line extending under brown eye.
Race *mixta*, southerly range, are slightly darker.
Immature: Paler with underparts washed ochre.

Direct, swift and undulating flight.

Habitat: Coastal rainforests, mangroves, lowland wet forests and tropical gardens. Resident on Barrier Reef islands.

Feeding: Noisy and aggressive when feeding, taking nectar, insects and fruit, favouring native raspberry.

Voice: Harsh 'tchu-tchu-chua' descending in pitch.

Status: Common.

Breeding: Sept.–Feb. Cup nest, from palm fibre and bark, bound with spider web, 1–3m high / 2 eggs.

Left and right: Adult nominate race.

White-gaped Honeyeater

Stomiopera unicolor

Endemic

170–210mm

Sexes alike. Plain grey bird with a conspicuous creamy white half-moon on bare gape. Upperparts: Dull olive-grey with darker crown and yellow-edged tail and wings. Underparts: Mid-grey. Eyes: Brownish green.
Immature: Yellow gape.

Usually in pairs or small groups. Defensive, aggressively chases other birds away from feeding area. Noisy group display with errect tail and wings flicking.

Habitat: Mangroves, riverine thickets, moist lowlands, commonly along streams, also gardens.

Feeding: Hopping through foliage, often with partial cocked tail, taking nectar, insects, fruit and seeds.

Voice: Rollicking, trilled whistles, often in duets advertising territory or feeding area.

Status: Common. Sedentary.

Breeding: Sept.–May. Cup-shaped nest of palm fibre and bark, bound with spider web, often over water / 2 eggs.

George Brown Botanic Gardens and Middle Arm, Darwin, NT.

Left: Adult. Right: Adult feeding on Grass Tree.

White-lined Honeyeater

Meliphaga albilineata

Endemic

170–205mm

Sexes similar. Mid-sized, slim honeyeater. Deep brown upperparts and a dark grey face with a yellow gape and a white gape line ending in a small ear tuft. Tail, wings: Deep brown edged pale lemon. Eyes: Bluish. Underparts: Light grey with throat and breast mottled grey-brown. Female: Smaller. Immature: Greyer, ear patch yellow, wings washed rufous.

Forages high in the canopy. Rarely seen.

Habitat: Monsoon forests in rocky gorges and sandstone escarpments.

Feeding: Aggressive and active, foraging for insects, fruit, nectar.

Voice: Calls throughout the day from a shady perch. Loud, echoing rising–falling whistles.

Status: Uncommon to locally common. Sedentary.

Breeding: Aug.–Feb. Typical cup nest slung in the outer foliage, 1–5m high / 2 eggs.

Gunlom Falls and Nourlangie Rock, Kakadu NP, NT.

Left: Adult feeding in White Bottlebrush. Right: Adult underparts.

Kimberley Honeyeater

Meliphaga fordiana

Endemic

170–205mm

Sexes alike. Similar to White-lined Honeyeater, but has plain, paler belly and lacks lemon edge to flight and tail feathers. Eyes: Grey. Generally a little-known species, best identified by location and song.

Habitat: Endemic to the Kimberley region, WA. Frequents sandstone gorges, monsoon rainforests, eucalypt woodlands and paperbark forests.

Not much is known about this species. The Kimberley Honeyeater is morphologically similar to White-lined Honeyeater, however, research by E.T. Miller and S.R. Wagney in 2013 shows large difference in songs of the 2 species, justifying separation of the 2 species.

Feeding: Forages mostly for insects but also for nectar and fruits.

Voice: 10–12 rapid up and down chirping whistles.

Status: Uncommon. Isolated.

Breeding: Aug.–Jan. Deep cup nest slung in the outer foliage / 2 eggs.

Mitchell Falls, Mitchell River NP, WA.

Left: Adult. Right: Adult upperparts.

Bridled Honeyeater

Bolemoreus frenatus

Endemic

200–210mm

Sexes similar. Dark brown head. Distinctive black bill with bright yellow base and gape and a bright yellow malar line extending from bill and curving up to behind eye. Yellow line extending from gape to pink skin spot below eye, finishing with a white upward-fanned 'bridle' behind eye. Eyes: Blue-grey. Wings, tail: Dusky brown edged lemon. Underparts: Grey-brown, mottled grey, grading to pale brown undertail coverts.
Female: Smaller. Immature: Similar, greyer.

Usually solitary, occasionally pairs or small groups. May gather in noisy flocks at flowering or fruiting trees. Aggressive, particularly in breeding season.

Habitat: Wet tropical rainforests above 300m, open lowlands in winter.

Feeding: Usually in the mid-canopy. Mostly nectar feeding, also taking insects, native fruits.

Voice: Melodious, descending 5-note whistled song and harsh feeding call.

Status: Moderately common. Usually solitary, sometimes pairs or small groups. Wanders locally.

Breeding: Sept.–Jan. Aggressive during breeding season. Typical nest of fine twigs 1–3m high / 2 eggs.

Lake Eacham picnic ground, Crater Lakes NP, Julatten area, and Mt Hypipamee NP, Qld.

Left: Adult. Right: Adult.

Eungella Honeyeater

Bolemoreus hindwoodi

Endemic

180–200mm

Sexes similar. Similar to the Bridled Honeyeater but smaller, greyer. Face: Dusky olive ear coverts, white stripe below eyes, white spot above eyes and white plume patch to sides of neck. Eyes: Blue-grey. Bill: Black, brownish pink gape. Upperparts: Dark olive-grey, lightly mottled. Underparts: Light, streaky grey grading to grey-brown undertail, edged whitish.
Female: Smaller. Immature: Duller.

May gather in noisy flocks around flowering or fruiting trees. Usually inconspicuous.

Habitat: High rainforest ridges, open lowlands in winter.

Feeding: Similar to Bridled Honeyeater. Nectar, insects and native fruits.

Voice: Distinctive, high-pitched, loud, joyous up and down whistle.

Status: Uncommon.

Breeding: Sept.–Jan. Typical nest deep, concealed with moss in outer foliage / 2 eggs.

Iluka Park area, East Mackay, and Eungella NP, Qld.

Left: Adult. Right: Adult underparts.

Yellow-faced Honeyeater

Caligavis chrysops

Endemic

160–170mm

Sexes alike. Mostly olive-grey honeyeater with a distinctive, curving yellow stripe under eye, bordered black above and below and a small white ear tuft and white spot above the eye. Upperparts, wings, tail: Grey-brown. Underparts: Whitish, faintly streaked grey-brown. Eyes: Blue-grey. Bill: Black. Race *barroni*, far north Qld. Confined to highland areas. Immature: Duller.

Conspicuous in autumn, in large migrating flocks, moving north for winter along eastern ranges, returning south in spring in small inconspicuous groups.

Habitat: Coastal open woodlands, mangroves, parks and gardens.

Feeding: Nectar. Also takes manna, honeydew and insects, particularly in summer.

Voice: Cheerful, sharp, musical descending 'calip-calip-calip' and 'chick-up' call repeated.

Status: Common. Vagrant Tas.

Breeding: July–Mar. Small cup, neatly woven nest slung in foliage to 7m high / 2–3 eggs.

Opposite: Adult feeding in Red Spider Flower.

Singing Honeyeater

Gavicalis virescens

Endemic

170–210mm

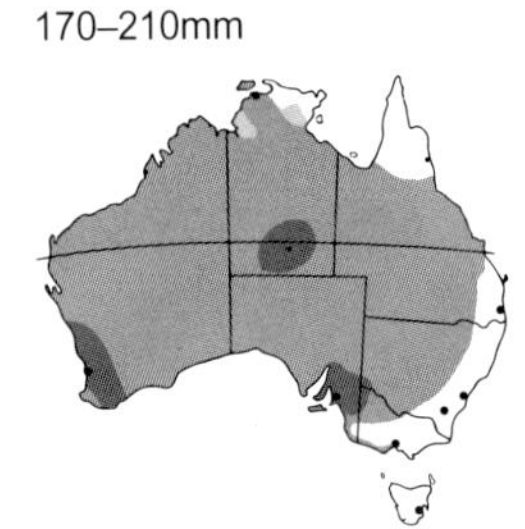

Sexes alike. Most widespread honeyeater, with geographical colour variations across range. Distinctive face with black streak from bill through eyes and down sides of neck, edged below with a bright yellow and white streak. Crown, upperparts: Grey-brown. Wings, tail: Brown, edged yellow. Throat, underparts: Streaked white, yellow, grey. Bill: Black. Eyes: Reddish brown. Immature: Similar, paler face.

Usually solitary, sometimes in loose flocks following flowering events.

Habitat: Well adapted to dry areas. Scrubby heathlands, arid scrub, open woodlands with acacia, sand plains, dune thickets, parks and gardens.

Feeding: Foraging low in the canopy, in shrubbery or on the ground taking nectar, fruit and insects.

Voice: Lively, musical, resonant 'prrrip-prrrip', often in duets.

Status: Common. Localised movement in response to conditions.

Breeding: July–Feb. Typical cup nest, untidy and flimsy / 2–3 eggs.

Cairns Esplanade and Daintree River, Qld.

Left: Adult. Right: Immature.

Varied Honeyeater

Gavicalis versicolor

180–210mm

Sexes alike. Similar to Singing Honeyeater, but larger and occupying different habitat. Face: Broad, black streak from bill through eye, edged below with a bright yellow and white stripe. Black line continues, edged broadly white, down sides of neck and shoulder. Crown, upperparts: Mottled brown. Wings, tail: Brown, mottled olive-brown edged olive-yellow. Throat, underparts: Yellow, streaked brown. Bill: Black. Eyes: Blue-grey.
Immature: Similar, duller, lighter face and crown.

Dashing, undulating flight. Territorial, communal groups of 7–15 birds. Conspicuous, loud. Aggressively chases other birds from feeding area.

Habitat: Coastal forests, mangroves, open woodlands, dunes, thickets, parks, gardens.

Feeding: Nectar and insects.

Voice: Loud, rollicking trills. Similar to Singing Honeyeater, but louder with more variation.

Status: Moderately common.

Breeding: Apr.–Dec. Typical cup nest, usually in mangroves 1–5m high / 2–3 eggs.

Cairns Esplanade and Daintree River, Qld.

Left and right: Adult foraging in mangroves.

Mangrove Honeyeater

Gavicalis fasciogularis

Endemic

195mm

Sexes alike. Similar to the Varied Honeyeater. Interbreeds where ranges overlap. Upperparts, tail: Brownish olive. Throat: Scaly brown and yellow. Underparts: Whitish with olive-brown streaks, darker grey-brown over breast. Eyes: Blue, with black band through eye, bordered by a yellow streak below.
Immature: Browner upperparts, paler underparts.

Gregarious, gathering in small, noisy groups.

Habitat: Mangroves and adjacent coastal areas, paperbark swamps, gardens and nearby islands.

Feeding: Mostly forages in and around mangroves. Nectar from mangrove blossoms, insects, marine snails and small crabs.

Voice: Varied, resonant, musical whistling calls. Scolding chatter.

Status: Common in Qld. Vulnerable in NSW.

Breeding: Apr.–May in north. Aug.–Dec. in south. Typical nest, usually in a fork in dense mangroves over water / 2 eggs.

Opposite: Adult feeding on Grevillea.

Yellow Honeyeater

Stomiopera flava

Endemic

170–190mm

Sexes alike. Uniformly yellow-olive upperparts. Face, lores, ear coverts: Olive-grey with yellow strip above and below grey eyes. Underparts: Lemon yellow. Bill: Brownish black.
Race *addendus*, southern range, Larger and greenish tinged.
Immature: Duller, pale bill.

Usually in pairs. Constant jerky movement, hopping and quick flight, chasing away competitors.

Habitat: Open forests, mangroves, riverine vegetation, parks and gardens with shrubby understorey.

Feeding: Nectar, insects and fruit taken from understorey to upper canopy vegetation.

Voice: Loud, varied, metallic whistles. Scolding chatter.

Status: Moderately common. Sedentary or locally nomadic.

Breeding: Oct.–Mar. Typical nest, shallow, 1–10m high / 2 eggs.

Townsville and Cairns area, Qld.

Left: Adult nominate race. Right: Adult feeding on Lotus Lily.

White-eared Honeyeater

Nesoptilotis leucotis

Endemic

165–210mm

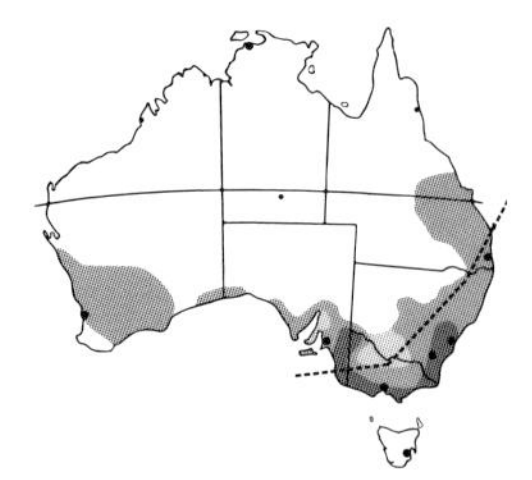

Sexes similar. Conspicuous white ear patch and relatively short bill. Face, neck, upper breast: Black. Crown, hind neck: Dark grey, streaked black. Upperparts: Bright olive green. Belly: Yellow-green. Eyes: Blue-grey.
Nominate race, coastal southern Qld, NSW, Vic. to southeast SA. Race *novaenorciae*, Qld, inland NSW to SA, WA. Smaller, duller.
Female: Smaller. Immature: Duller, browner with cream ear patch.

Usually solitary, often hopping sideways along a branch, probing crevices for insects.

Habitat: Dry eucalypt forests and woodlands with complex understorey. Heathlands, mangroves, mallee, inland scrub. Alpine to lowland areas. Parks, gardens. Avoids open spaces.

Feeding: Usually in the lower canopy. Manna, lerp, honeydew and insects. Some nectar taken by puncturing the flower base.

Voice: Loud, ringing 'chock-up, chock-up'.

Status: Common to locally common.

Breeding: July–Apr. Open, thick-walled cup nest lined with hair, low to 3m high / 2 eggs.

Barren Grounds, NSW. Girraween NP, Qld. Wyperfeld, Hattah–Kulkyne and Murray–Sunset NP, Vic.

Left: Adult nominate race. Feeding in Grass-leaf Hakea. Right: Adult upperparts.

Yellow-throated Honeyeater

Nesoptilotis flavicollis

Endemic

200–230mm

Sexes similar. Similar to the White-eared Honeyeater, but with a conspicuous yellow chin and throat above dusk grey underparts. Face: Lores, cheeks are dark grey fading to silvery with a small yellow tuft. Crown, hind neck: Grey flecked black. Upperparts: Bright olive-green. Eyes: Deep red-brown. Bill: Black.

Female: Smaller. Immature: Duller, paler throat and green-grey eyes.

Habitat: Endemic Tas. Wide ranging, including wet or dry forests, heathlands, rainforests, parks and gardens.

Feeding: In foliage and bark crevices in all levels of the canopy. Mostly insects, some nectar, manna and honeydew. Occasionally fruit and seeds.

Voice: Varied warbling. Rapid loud 'tonk, tonk', repeated frequently.

Status: Common. Sedentary. Males hold territory, chasing female and young away after breeding.

Breeding: July–Jan. Open cup-shaped, thick-walled nest, low, often in tussocks or up to 10m / 2–3 eggs.

Left: Adult. Right: Adult underparts.

Yellow-tufted Honeyeater

Lichenostomus melanops

Endemic

170–230mm

Sexes alike. Black face mask from bill to ear coverts with yellow crown, yellow sides of throat and bright yellow tufts that are raised when alarmed. Upperparts: Olive-brown. Underparts: Olive-yellow, lightly streaked.

Race *meltoni*, inland slopes of Great Dividing Range from mid-northern Qld, just reaching SA. Smaller and greyer.

Race *cassidix*, 'Helmeted Honeyeater'. Larger and brighter with longer tufts on frons. Restricted to Yellingbo area, Vic. Strongly associated with Mountain Swamp Gum.

Noisy, active and aggressive towards food competitors. Colonies of a few to 100 birds.

Habitat: Shrubby coastal eucalypt forests, woodlands. Race *meltoni*: Drier woodlands with open understorey, often along watercourses.

Feeding: Mostly nectar from eucalypts, also insects, lerp. Aerially pursues insects.

Voice: Loud, nasal 'tchurr' contact call. Occasional sweet whistling.

Status: Moderately common. Race *cassidix*, endangered.

Breeding: July–Jan. Typical nest, large, tightly woven, 0.3–4m above ground / 2–3 eggs.

Chiltern–Mt Pilot NP, Vic.

Left: Adult nominate. Yellow-tufted Honeyeater. Right: Race *cassidix*, 'Helmeted Honeyeater'.

Purple-gaped Honeyeater

Lichenostomus cratitius

Endemic

160–170mm

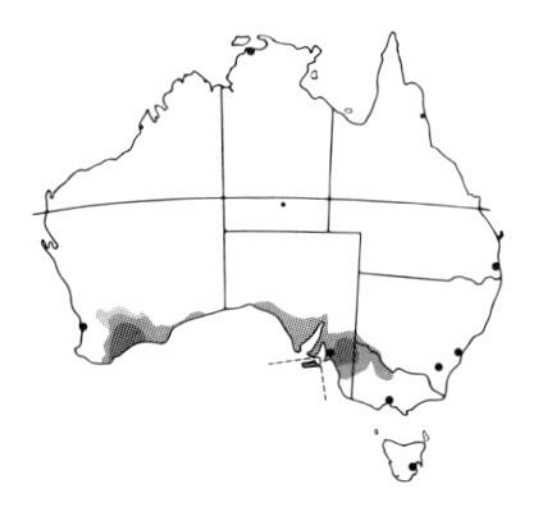

Sexes similar. Mostly grey-olive upperparts. Prominent black eye mask and distinctive purple gape above a bright yellow malar line. Small, bright yellow ear tufts. Underparts: Yellow-grey. Eyes: Brown. Bill: Black.

Nominate race confined to Kangaroo Is.

Race *occidentalis*, mainland, is brighter.

Female: Smaller. Immature: Duller, yellow gape.

Shy. Usually solitary or loose groups of 5–6 birds.

Habitat: Mallee, heathlands, open woodlands, gardens. Strongly associated with Whipstick Mallee.

Feeding: Nectar, some honeydew and pollen. Hawks for insects or gleans from bark and foliage.

Voice: Loud, harsh chattering song. Sharp chirps and whistles.

Status: Sedentary, moderately common to locally uncommon. Vulnerable NSW. Wanders locally following flowering events.

Breeding: Aug.–Dec. Tightly woven cup nest hung in a shrub, often Broombush or eucalypt, low to 2m high / 2 eggs.

Cape Gantheaume, Kangaroo Is., Big Desert WP and Little Desert NP, Vic. Innes NP, SA. Stirling Ranges NP, WA.

Left: Adult race *occidentalis* with pollen-dusted frons. Right: Adult race *occidentalis*.

Grey-headed Honeyeater

Ptilotula keartlandi

Endemic

130–160mm

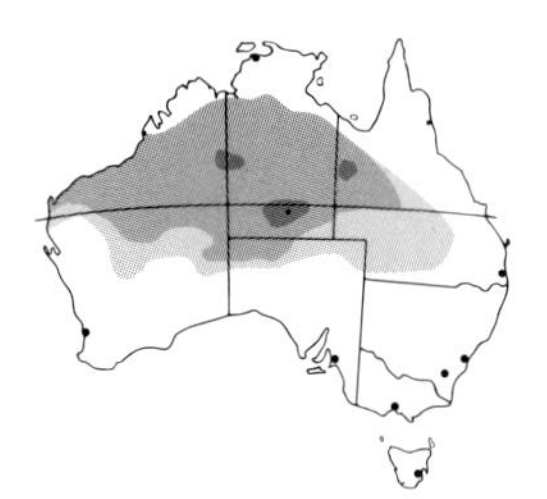

Sexes similar. Similar to the Purple-gaped Honeyeater but located further north. Grey crown and grading fawn-grey upperparts. Face: Black ear covert underlined yellow with up-curved yellow ear tuft. Tail, wings: Grey-brown edged yellow. Underparts: Greyish white with dark streaks.
Female: Smaller. Immature: Duller.

Usually seen in small groups of 5–6 birds flitting in the outer foliage of small eucalypts.

Habitat: Rocky gorges, eucalypt woodlands, mulga scrub and shrubby desert dunes.

Feeding: In the canopy, mostly taking nectar. Also insects, manna and honeydew.

Voice: Cheerful, chattering. Loud 'chek-chek' repeated.

Status: Moderately common. Mostly sedentary, locally nomadic.

Breeding: July–Nov. Typical nest usually in low scrub, 1–5m high / 2 eggs.

Simpsons Gap and Ormiston Gorge, NT.

Opposite: Adult.

Yellow-plumed Honeyeater

Ptilotula ornata

Endemic

140–165mm

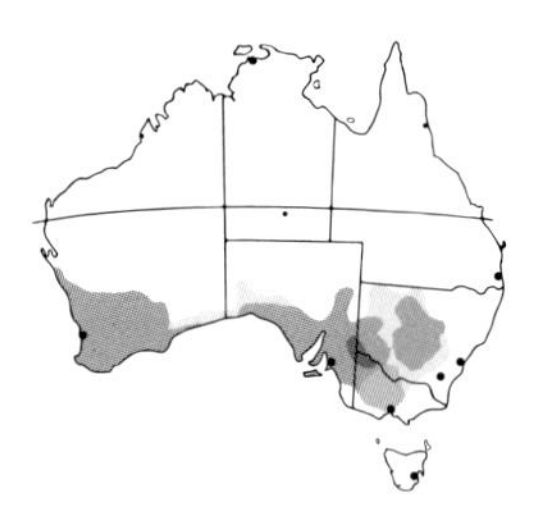

Sexes alike. Similar to the Grey-fronted Honeyeater. Upperparts: Olive brown. Neck: Narrow yellow upswept plume finely edged black along top. Underparts: Pale fawn, heavy brown streaking. Ear coverts: Dark brown. Bill: Black, yellowish base. When breeding base and gape are black. Eyes: Dark brown.
Immature: Duller. Eye ring, gape, base of bill: Yellow.

Aggressive, quarrelsome, gathering in small flocks to defend territory. Communal wing-quivering display.

Habitat: Arid and semi-arid mallee or coastal wet mallee. Tall, temperate eucalypt woodlands.

Feeding: Forages in pairs or small flocks, mostly in low outer foliage, sometimes coming to ground. Forages for insects, lerp, honeydew, manna and nectar.

Voice: Loud 'chick-oweee'. Sharp, alarm and flight call.

Status: Common to moderately common. Usually sedentary.

Breeding: Aug.–Dec. Tightly woven cup nest hung in small shrub, low / 1–3 eggs.

Wyperfeld, Hattah–Kulkyne and Murray–Sunset NP, Vic. Gluepot SA.

Left and right: Adult.

Grey-fronted Honeyeater

Ptilotula plumula

Endemic

150–160mm

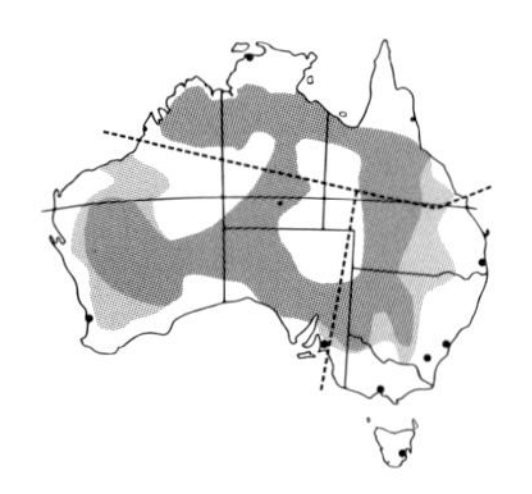

Sexes similar. Similar to the Yellow-Plumed honeyeater. Upperparts, wings, tail: Olive-grey. Face, crown: Olive-yellow with dark grey frons. Neck: Yellow, upswept plume finely edged black stripe below. Underparts: Buff grey with faint streaking. Bill: Black with yellow base and gape.
Breeding: Bill, gape is black.
Race *planasi*, tropical north. Pale grey back and frons patch.
Race *graingeri,* southern Qld, NSW, Vic. to SA. Larger, darker.
Female: Smaller. Immature: Duller, plainer. Gape and base of bill yellowish.

Noisy, quarrelsome groups gather to chase off intruders.

Habitat: Arid spinifex deserts, dunes, mulga, mallee scrub and eucalypt woodlands.

Feeding: Nectar from eucalypt flowers, also insects, lerp, manna and honeydew.

Voice: Penetrating, sharp 'clit'. Occasionally clear penetrating melodic song, a rattling 'wirt wirt wirt'.

Status: Moderately common. Locally nomadic.

Breeding: Aug.–Jan. or after good rain. Cup nest of bark fibre, spider web and egg sacs, hung in small shrub 1–3m high / 2–3 eggs.

Left: Non-breeding adult in Desert Grevillea. Right: Breeding. Note black bill.

Fuscous Honeyeater

Ptilotula fusca

Endemic

130–165mm

Sexes similar. Mid-sized, mostly olive-brown, paler below with wings and tail edged olive-yellow. Plain face with a small ear tuft edged black. Eyes: Lightly smudged around dark brown eye with yellow eye ring. Bill: Black with yellow base and gape. Breeding plumage: Eye ring and base of bill are dark.
Female: Smaller. Immature: Browner, brown bill.

Active, agitated. Foraging in pairs or flocks of 6–12.

Habitat: Open forests and grassy woodlands, mostly on western slopes of the Great Dividing Range.

Feeding: Mostly insects, lerp, honeydew, manna, also nectar.

Voice: Rolling chitter. Flight song 'chichowee, chickawee' repeated.

Status: Common. Colonial. Sedentary, wandering locally for flowering events.

Breeding: July–Mar. Dainty cup nest of bark fibre, grass and spider web hung to 20m high / 1–3 eggs.

Glen Davis, NSW. Chiltern–Mt Pilot NP, Vic.

Opposite: Breeding birds. Male (left) and female.

Yellow-tinted Honeyeater

Ptilotula flavescens

130–165mm

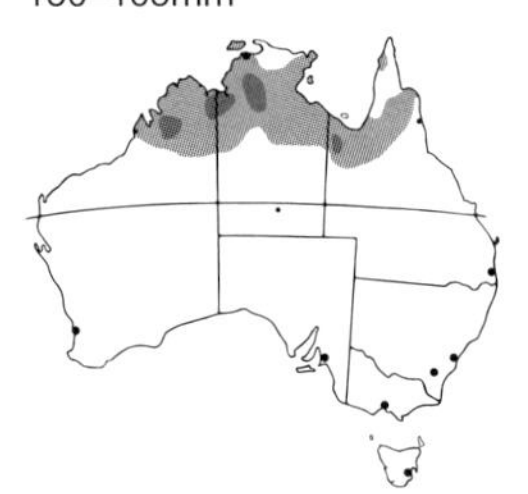

Sexes alike. Similar to Fuscous Honeyeater but overall brighter and more yellowish. Black crescent-shaped stripe above a bright yellow plume behind ear coverts. Face, frons: Yellow. Crown, nape: Fawnish yellow. Upperparts, tail: Olive-fawn edged yellow. Eyes: Black. Underparts: Yellow to cream with darker streaks.
Immature: Duller. Bill: Yellow base.

Pugnacious. Residing in small colonies, constantly chasing one another with fast, aerial acrobatics or dashing from tree to tree.

Habitat: Open eucalypt woodlands, scrublands, often near water. Occasionally mangroves.

Feeding: Foraging in the mid- to upper canopy, nectar, insects, lerp, honeydew and manna.

Voice: High-pitched whistled 'chip'. Harshly repeated when alarmed. Rattling, repeated song similar to White-plumed Honeyeater.

Status: Common. Sedentary, wanders locally for blossoms.

Breeding: July–Nov. Cup nest of bark fibre, grass, spider web and egg sacs, hung in shrub or tree 2–12m high / 2–3 eggs.

Left: Adult pair at waterhole. Right: Adult underparts.

White-plumed Honeyeater

Ptilotula penicillata

Endemic

135–175mm

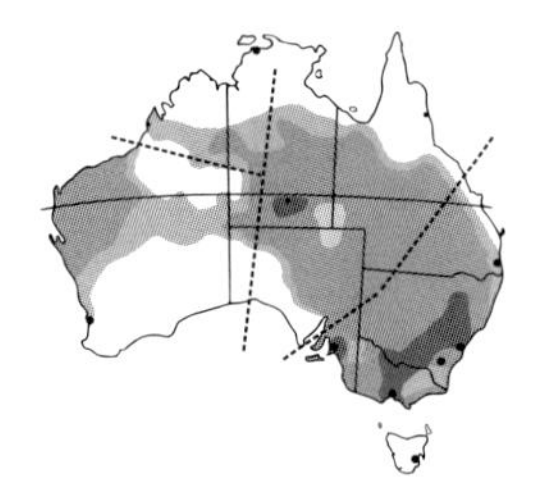

Sexes alike. The only honeyeater with a white neck plume. Back and wings are dark olive-brown. Underparts: Pale greyish brown. Face: Yellowish. Bill: Black when breeding, nonbreeding the base is yellowish.
Nominate race, southern Qld, NSW, Vic. to eastern SA.
Race *leilavalensis*, central inland. Smaller, brighter yellow.
Race *calconi*, northern WA. Smallest, palest.
Race *carteri*, south WA. Bright yellow face and crown. Underparts: Brighter, paler.
Immature: Duller.

Well adapted to living in an urban environment. In colonies. Fearless and aggressive, mobbing much larger birds.

Habitat: Tall eucalypt woodlands, parks and gardens. Inland areas near water and River Red Gums.

Feeding: Energetically forages at all levels in the outer foliage, mostly in eucalypts for nectar, insects, lerp, honeydew, manna.

Voice: Musical 'chick-o-wee' and repeated 'chip'.

Status: Common to moderately common.

Breeding: July–Jan. or after rain in inland areas. Deep cup nest of bark fibre, grass, spider web, egg sacs, slung in outer branches 1–20 m high / 2–3 eggs.

Left: Nominate adult. Feeding in Nepean Spider Flower. Right: Adult race *calconi*.

Strong-billed Honeyeater

Melithreptus validirostris

Endemic

150–170mm

Sexes alike. Black head with bright white crescent around nape. Dark brown eyes with greenish white skin crescent over eye. Bill: Large, almost straight, sharp pointed. Upperparts: Grey-brown with light yellow wash to centre. Underparts: Pale grey-brown with a white throat. Immature: Head: brownish. Eye skin, bill: Yellow-orange.

Falling debris from feeding birds usually indicates their presence.

Habitat: Endemic Tas. Mature wet forests favouring gullies; dryer forests and slopes in winter. Also wet coastal scrub and occasionally gardens.

Feeding: Noisy. Forages from ground level to upper canopy, tearing away or digging in bark for insects, larvae and spiders. Occasionally nectar, fruit taken.

Voice: Loud 'cheep' repeated. Harsh 'churring' alarm call.

Status: Common. Wander in nomadic groups after breeding.

Breeding: July–Dec. Deep cup nest of bark fibre, grass, spider web, egg sacs, slung in outer foliage 2–25m high / 2–3 eggs.

Left: Immature birds along log with adult sitting at far end. Right: Adult.

Black-chinned Honeyeater

Melithreptus gularis

Endemic

140–160mm

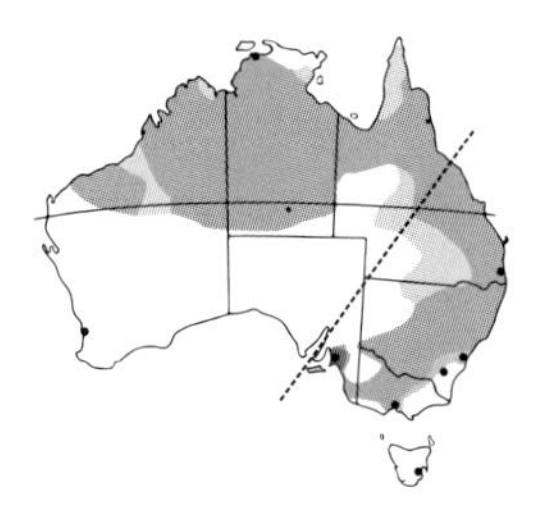

Sexes alike. Prominent crescent-shaped blue eye skin over deep brown eyes. Black head with a bright white crescent around nape. Upperparts: Olive green. Chin: Black. Throat: White. Underparts: Light grey. Bill: Black. Legs: Orange-red.
Race *laetior*, Golden-backed Honeyeater, far west Qld to WA. Back: Bright olive-gold. Eye skin: Green-yellow.
Immature: Duller, browner. Yellow-orange bill, dull blue eye skin.

Usually in small flocks.

Habitat: Savannah woodlands, dry eucalypt forests, spinifex scrub, along watercourses. Occasionally gardens.

Feeding: Usually in the outer foliage of upper canopy or probing limbs or bark. Nectar, insects, honeydew in summer. Often drinks at midday.

Voice: High-pitched 'chirrup' and 'churring' calls.

Status: Moderately common. Sedentary. Locally nomadic. Vulnerable NSW, SA.

Breeding: July–Dec. Fragile cup nest of bark fibre, grass and spider web, slung in outer foliage 1–15m high / 1–2 eggs.

Chiltern–Mt Pilot NP, Vic. Glen Davis, NSW. Mt Isa area, Qld. Cape Crawford, NT.

Left: Race *laetior*. Golden-backed Honeyeater. Right: Nominate race.

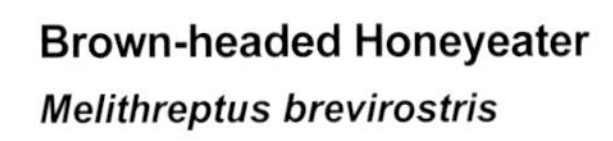

Brown-headed Honeyeater

Melithreptus brevirostris

Endemic

115–140mm

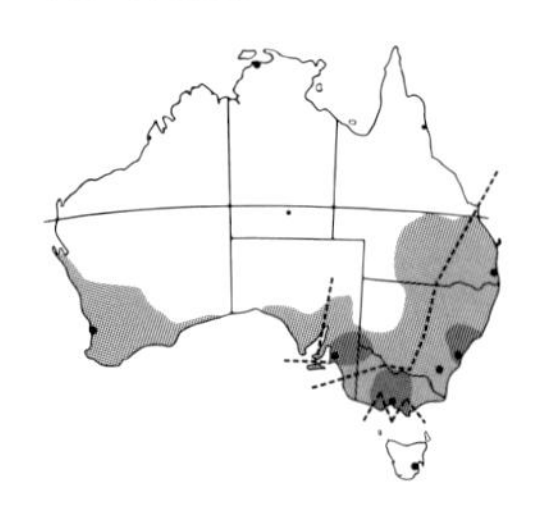

Sexes similar. Dark brown head with cream tinged green, crescent skin over deep brown eyes and a fawn crescent around nape. Upperparts: Olive-brown. Underparts: Light grey. 5 races with increasingly darker heads graduating westward.
Race *pallidiceps*, inland Qld, NSW, SA. Smaller, paler.
Race *leucogenys*, SA, WA. Paler with orange-yellow eye skin.
Race *magnirostris*, Kangaroo Is. Darker.
Race *wombeyi* confined to Otway, Strzelecki and Hoddle Ranges, Vic. Largest. Upperparts: Dark olive.
Immature: Paler. Bluish eye skin.

Communal, sociable groups of 10–20 birds, feeding, roosting, preening and caring for young.

Habitat: Open eucalypt forests, mallee scrub, sub-alpine woodlands, parks and gardens.

Feeding: Probes bark for insects. Lerp, manna and honeydew particularly in winter.

Voice: High-pitched call 'chirrip' and sharp chattering.

Status: Common. Sedentary or locally nomadic.

Breeding: Aug.–Dec. Pairs or communal groups. Deep cup nest slung in the outer foliage / 2–3 eggs.

Left: Nominate race. Right: Nominate race in Red Ironbark.

White-throated Honeyeater

Melithreptus albogularis

Endemic

130–140mm

Sexes alike. Similar to White-naped Honeyeater, with black head and distinctive bright white crescent around the nape. Eyes: Red-brown with crescent-shaped, bluish white skin above eyes. Upperparts: Olive-yellow. Underparts: White. Race *inopinatus*, southern Qld. NSW. Larger and duller. Immature: Brownish, incomplete nape band.

Usually in pairs or small groups.

Habitat: Eucalypt woodlands and forests, paperbark swamps, occasionally mangroves, parks and gardens.

Feeding: Forages in the outer foliage of the upper canopy for insects, manna, honeydew and nectar. Often in mixed flocks.

Voice: Sharp 'tsip', harsh hissing, rapid piping song.

Status: Common. Sedentary.

Breeding: Jan.–Oct. Delicate cup nest of plant fibres and grass bound with spider web and slung in outer foliage 5–9m high / 2 eggs.

Left: Nominate adult. Feeding in Scarlet Honey Myrtle. Right: On Grass Tree flower spike.

White-naped Honeyeater

Melithreptus lunatus

Endemic

140–150mm

Sexes alike. Black head with a short white nape band. Orange-red crescent-shaped skin patch over eye. Bill: Black with orange gape. Upperparts: Olive green. Underparts: White. Immature: Crown, upperparts: Brownish with incomplete nape band, orange eye crescent.

Undulating flight. Noisy flocks of 30–40. Southeast highland birds partially migrate northward in autumn, returning in spring.

Habitat: Open forests, tall eucalypt woodlands, gardens.

Feeding: Upper canopy for manna and honeydew, also nectar, insects, lerp.

Voice: Soft, trilled piping. Harsh hissing. No song.

Status: Common.

Breeding: July–Jan. Deep, cup-shaped nest in outer foliage 5–20m high / 2–3 eggs.

Left: Adult at birdbath. Right: Adult foraging.

Gilbert's Honeyeater

Melithreptus chloropsis

Endemic

140–150mm

Sexes alike. Also known as Western White-naped Honeyeater or Swan River Honeyeater. Similar to White-naped Honeyeater but with a distinctive white or pale green eye skin patch, bolder white crescent-shaped patch on nape, with a longer bill and darker chin. Upperparts: Olive green. Head, nape, throat: Black. Underparts: White. Eyes: Dull red. Bill: Brownish black.

Habitat: Dry sclerophyll forests, Jarrah forest belt, coastal heaths, Dryandra Woodland southeast of Perth.

Feeding: Upper canopy for manna and honeydew, also nectar, insects and lerp.

Voice: Soft, trilled piping. Harsh grating hissing.

Status: Common.

Breeding: Nov.–Jan. Deep cup-shaped nest of bark, grass and rootlets in outer foliage 5–20m / 2–3 eggs.

Left: Adult at birdbath. Right: Adult upperparts.

Black-headed Honeyeater

Melithreptus affinis

Endemic

145mm

Sexes alike. Tas. endemic, similar to the mainland White-naped Honeyeater. It has a black head and throat with a white-tinted turquoise skin crescent over brown eyes. Upperparts: Dull olive green. Underparts: White, grading to grey with black mark on sides of breast. Immature: Head: Completely brownish. Underparts: Tinged yellow. Bill: Yellow gape.

Gregarious, often in small flocks. Aggressive towards other birds. Undulating, rapid flight.

Habitat: Mature forests and dense vegetation, coastal heaths, low shrubland. Occasionally parks and gardens.

Feeding: Foraging in pairs or small flocks, mostly in the upper canopy, for insects, honeydew, nectar, fruit.

Voice: Soft 'chip' contact call. High-pitched 2-note whistle.

Status: Common. Nomadic in winter.

Breeding: Oct.–Jan. Nest is similar to White-naped Honeyeater, 5–20m high / 2–3 eggs.

Left: Adult at waterhole. Right: Adult foraging in Manuka.

New Holland Honeyeater

Phylidonyris novaehollandiae

Endemic

160–180mm

Sexes alike. Mostly black and white streaked with a large bold yellow wing patch and white tufts above and below stark white eyes. Underparts: White streaked black, more finely streaked on the throat. Wings: Grey-black with wide yellow patch. Tail: Black with yellow edge, white tip. Several similar races.
Race *longirostris*, southwest WA. Longer bill and smaller malar line.
Immature: Brownish, eyes grey.

Usually in large groups. Low, dashing, undulating and erratic flight. Rises vertically from shrubs and sings briefly before diving back to cover. Chases other birds from its nectar source.

Habitat: Coastal heaths, forests, woodlands, parks and gardens particularly with banksias.

Feeding: Actively forages in the lower strata, mostly reliant on nectar. Also manna, honeydew, fruit and insects gleaned or caught in mid-air.

Voice: Loud, high-pitched, metallic call. Harsh contact call. High-pitched single alarm whistle.

Status: Common.

Breeding: Mar.–May and July–Dec. Dependent on food supply. Coarsely woven cup nest, low to 5m high / 1–3 eggs.

Left: Nominate adult. Feeding in Coast Banksia. Right: In Heath-leafed Banksia.

White-cheeked Honeyeater

Phylidonyris niger

Endemic

160–180mm

Sexes alike. Similar to New Holland Honeyeater but with a larger white cheek patch and broad white eyebrow above a brown eye. Head, upperparts: Black with light white streaks to back. Wings: Grey-black with wide yellow patch. Tail: Black, yellow edge, white tip. Underparts: White, streaked black. Bill: Black.
Race *gouldii*, southwestern WA. Similar but longer bill and smaller cheek patch.
Immature: Brownish.

Agile, active, squabbling groups, often chasing one another. Often with New Holland Honeyeaters.

Habitat: Coastal heathlands, forests, woodlands with a dense understorey of nectar-rich plants. Parks and gardens.

Feeding: Forages mostly for nectar, also insects gleaned from foliage, bark or taken mid-air.

Voice: Protracted musical warbling in flight display. Rapid, loud, prolonged 'chippity-chip' whistle.

Status: Common. Sedentary or locally nomadic, following flowering events.

Breeding: Aug.–Nov. Cup nest of plant fibres, grass and spider web, slung in low, dense vegetation / 1–3 eggs.

Left: Nominate race. Right: Race *gouldii* feeding in Parrot Bush.

White-fronted Honeyeater

Purnella albifrons

Endemic

150–160mm

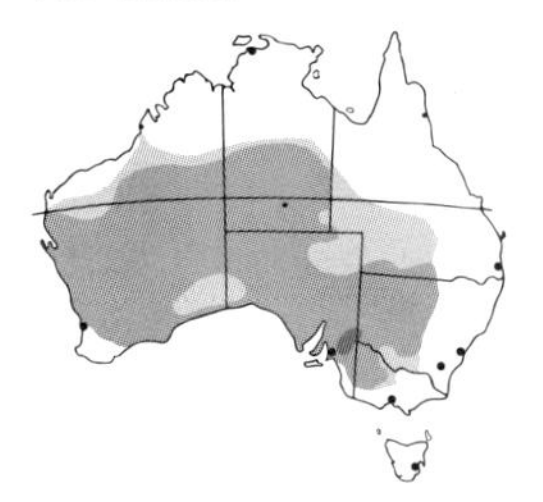

Sexes similar. Distinctive face with white lores, frons, malar line, eye ring and a small red wattle behind the eyes. Eyes: Red-brown. Ear coverts: Grey with small white plume. Bill: Black, strong. Crown: Black speckled white. Upperparts: Brown-black with olive-yellow wing patch and rufous brown rump. Throat, breast: Black-brown. Underparts: White with brown-black streaked breast and flanks.
Female: Browner. Immature: Plainer, browner, mottled breast. Lacks red wattle spot.

Swift, undulating flight, short sallies for flying insects.

Habitat: Arid and semi-arid mallee, scrublands and coastal heathlands. Occasionally dry open woodlands, parks, gardens.

Feeding: Arboreal, foraging mostly for nectar, also insects and honeydew.

Voice: Metallic, loud, canary-like melodious song.

Status: Moderately common. Highly nomadic, following flowering mallee eucalypts.

Breeding: Aug.–Nov. Communal. Cup nest, low in shrubbery or spinifex clumps / 1–3 eggs.

Hattah–Kulkyne NP, Vic. Arid Lands Botanic Garden, SA.

Left: Adult feeding in Pear-fruited Mallee. Right: Female.

Painted Honeyeater

Grantiella picta

Endemic

140–150mm

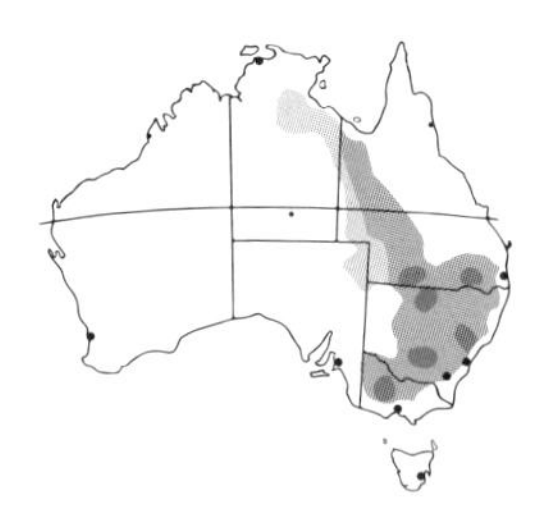

Black head and upperparts, bright pink bill and a small white tuft on the neck. Wings: Black, broadly edged yellow. Tail: Black, edged yellow and tipped white. Underparts: White, with dark flecks on flanks. Eyes: Reddish brown.
Female: Smaller, browner upperparts and head. Immature: Similar to female. No flecks on flanks.

Usually solitary, sometimes pairs or small groups. Males establish breeding territory with vertical song flight and singing from exposed perch. Defends territory.

Habitat: Open forests of ironbark, box, woodlands with mistletoe, gardens with large eucalypts.

Feeding: Specialist feeder on mistletoe fruits, particularly from *Amyema* spp. Also takes nectar, insects.

Voice: Distinctive, undulating, double whistle, 'georgie-georgie'.

Status: Uncommon. Vulnerable NSW, Vic. Migratory. Overwinters in the northern part of its range. Spring, summer breeding visitor in its southern range.

Breeding: Oct.–Mar. Frail cup nest of plant fibres and grass bound with web 3–20m high / 2 eggs.

Dunach CR, Clunes, Vic.

Opposite: Male.

White-streaked Honeyeater

Trichodere cockerelli

Endemic

160–180mm

Sexes alike. Unique honeyeater with no close relatives. The throat and breast have grey ruffled 'beard' feathers, edged white. Crown, face, lores: Grey with short yellow malar lines and yellow ear tufts. Eyes: Red. Upperparts: Dark brown with dusky brown wings broadly edged yellow. Tail: Dusky grey edged yellow. Underparts: Whitish. Bill: Black with blue-grey gape.
Immature: Browner, plainer, brown eyes and yellow gape. Lacks throat plumes.

Active, moving rapidly through the upper canopy.

Habitat: Paperbark, eucalypt woodlands, tropical heathlands, riverine forests, mangroves, edges of rainforests.

Feeding: Often in flocks, in the upper canopy, taking insects and nectar. Follows the blossoming sequence of paperbarks.

Voice: Melodic 4-note whistling. Metallic, scolding 'churr'.

Status: Moderately common. Locally nomadic.

Breeding: Jan.–May. Cup nest of plant fibres, grass and spider web in low fork, often over water in a paperbark / 2 eggs.

Iron Range NP, Qld.

Opposite: Adult.

Crescent Honeyeater

Phylidonyris pyrrhopterus

Endemic

150–160mm

Small to medium-sized honeyeater with a distinctive dark crescent from shoulder to sides of breast. Upperparts: Dark grey with bright yellow wing patch. Eyes: Deep red, white stripe behind. Throat: White, lightly streaked. Breast: White, flanked by black crescents bordered white.
Female: Olive-brown, faded yellow wing patch. Immature: Duller, indistinct, faint crescents.

Breeding pairs often remain in the occupied territory for several years.

Habitat: Moist forests with dense understorey, heathlands, parks and gardens.

Feeding: Arboreal. Mostly nectar, also honeydew, insects and fruit.

Voice: Sharp 'egypt-egypt' call by male in defence of territory, often repeated. Twittering.

Status: Common in Tas. Moderately common elsewhere. Some autumn migration from high altitude to lower altitude.

Breeding: July–Jan. Typical nest, low to 2m, often near water / 2–4 eggs.

Left: Male in Silver Banksia. Right: Female.

Tawny-crowned Honeyeater

Gliciphila melanops

Endemic

150–170mm

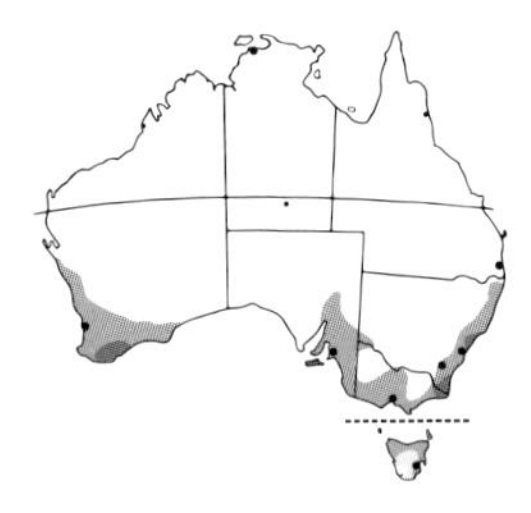

Sexes similar. Distinctive tawny crown. Face: Dark crescent from gape to eyes and down sides of breast, with a small narrow white crescent-shaped down-swept line. Upperparts: Grey-brown. Eyebrow, narrow ear crescent, chin, throat: White. Breast: White grading to lightly mottled.
Race *chelidonia*, Tas. More rufous brown.
Female: Smaller. Immature: Brownish, paler, yellow throat patch.

Dramatic flight display, rising vertically, spiralling downwards with outstretched wings, singing. Often seen hopping on the ground feeding.

Habitat: Coastal heaths, mallee scrub, occasionally open forests.

Feeding: Prefers low, shrubby nectar-rich plants. Insects are hawked mid-air.

Voice: Often sings from a high perch. Flute-like, high-pitched, ascending call. Metallic, ascending whistles in flight.

Status: Moderately common.

Breeding: July–Mar. Bulky cup nest of stringy bark, grass and spider web, in a low dense shrub 2–3m high / 2 eggs.

Royal NP, NSW. Lincoln NP, SA. Stirling Ranges NP, WA.

Left: Adult in Netbush. Right: Immature.

Banded Honeyeater

Cissomela pectoralis

Endemic

110–130

Sexes alike. Distinctive black and white honeyeater. Crown, back, wings, tail: Black. Underparts: White with black breast band. Bill: Black. Eyes: Brown. Non-breeding adults have a brown and black streaked mantle and yellow gape and base of bill and white 'armpits' seen in flight as white flashes.
Immature: Crown: Brown. Back: Reddish brown. Ear covert: Yellow. Gape: Yellow. Underparts: Buff with prominent brown breast band.

Often seen flying high in flocks of up to 50 birds in search of flowering trees.

Habitat: Tropical eucalypt woodlands, forests, mangroves, paperbark swamps and riverside vegetation.

Feeding: Feeding in large groups and mixed flocks, relying mostly on nectar. Insects are also taken.

Voice: Cheerful chattering. Descending song. Constant 'chip' flight contact call.

Status: Moderately common. Highly nomadic, following flowering events.

Breeding: Oct.–Apr. Flimsy cup nest typical of honeyeaters to 3m high / 2 eggs.

Cape Crawford and Gunlom Falls, Kakadu NP, NT.

Left: Breeding adult. Right: Immature.

Eastern Spinebill

Acanthorhynchus tenuirostris

Endemic

150–160mm

Identified by its very long, fine down-curved bill and rapid, erratic flight with whirring wings and flashing white outer tail feathers. Distinctive white throat with centre tan patch and white breast flanked by wide, black crescents. Black head and white throat are sharply defined. Upperparts: Grey-brown with rich tan nape and grey rump. Wings: Black. Tail: Black, outer tail feathers broad-tipped white. Underparts: Tan. Eyes: Red. Race *dubius*, Tas. Duller, browner.
Female: Dark grey crown.
Immature: Upperparts: Grey-olive. Underparts: Fawn.

Often hovers while feeding on nectar from flowers.

Habitat: Generally east of the Great Dividing Range in heathlands, forests with heath understorey and well-vegetated parks and gardens.

Feeding: Nectar. Specialised bill to probe tubular flowers. Occasionally takes insects from foliage or by sallying mid-air.

Voice: High-pitched, rapid piping.

Status: Common.

Breeding: Aug.–Mar. Nest is a small cup of twigs, grass and spider web placed in a bushy shrub 1–5m high / 2–3 eggs.

Left: Male. Right: Immature.

Western Spinebill

Acanthorhynchus superciliosus

Endemic

150–160mm

Similar to Eastern Spinebill in appearance and habits, but with a broad chestnut collar over nape and throat and a white breast band with black breast band below. Black face mask, red eyes, white eyebrows, malar lines. Back: Olive-grey. Head: Black. Bill: Black, long, thin, down curved. Underparts: Buff-grey.
Female: Duller. Olive-grey upperparts, buff brown throat and pale tan nape. Immature: Like females, duller.

Flight is rapid, erratic, with distinctive whirring from wings and tail flashing white.

Habitat: Heathlands, banksia woodlands, parks and gardens.

Feeding: Usually feeds in low shrubs, bill specialised for taking nectar from tubular flowers. Occasionally takes insects from foliage or in mid-air.

Voice: Sharp, staccato piping 'kleet-kleet'. Fluted, metallic whistle.

Status: Common.

Breeding: Sept.–Jan. Nest is a small cup of twigs, grass and spider web, low, usually in a bushy shrub / 1–2 eggs.

Left: Male feeding in *Grevillea pinaster*. Right: Immature feeding on Scarlet Banksia.

Rufous-banded Honeyeater

Conopophila albogularis

120–140mm

Sexes alike. Distinctive white chin and throat and broad reddish brown breast band. Head: Grey. Upperparts: Buffy brown. Wings, tail: Grey-brown edged deep yellow, pronounced on flight feathers. Underparts: White with brownish flanks. 2 isolated populations.
Immature: Browner head and indistinct or absent breast band.

Active, lively, hopping and fluttering in foliage; dashing or sallying after insects.

Habitat: Coastal monsoon vine scrub, paperbark swamps, mangroves.

Feeding: Noisily forages in small groups for nectar and insects in low- to mid-strata scrub. Often along watercourses.

Voice: Sharp, melodious 'zit, zit'.

Status: Locally common. Semi-nomadic, following flowering events.

Breeding: Sept.–Apr. Deep 'hammock' nest of paperbark strips, grass and spider web, often over water to 7m high / 2–3 eggs.

Holmes Jungle NP, Darwin, NT.

Left and right: Adult.

Pied Honeyeater

Certhionyx variegatus

Endemic

150–180mm

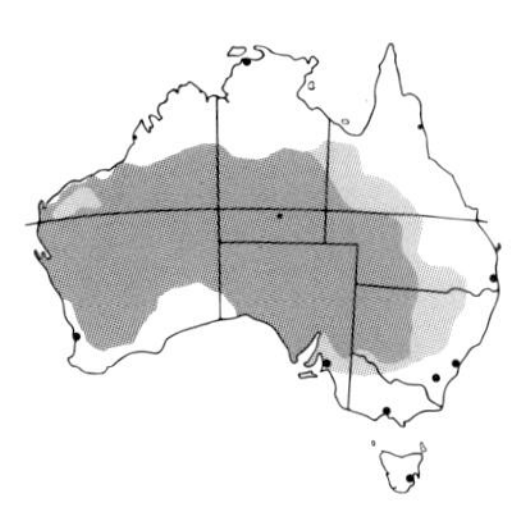

Distinctive black and white bird with a small bright blue skin patch below eyes. Head, throat: Black. Upperparts: Black with white rump. Wings: Black with broad white stripe. Underparts: White, undertail black tipped. Bill: Black, down-curved.
Female: Upperparts: Mottled, mid-brown. Underparts: Light brown with streaked throat and breast. Eye patch: Less conspicuous blue.
Immature: Similar to female, yellow gape.

Male territorial display flight includes vertical climb and gliding descent while singing.

Habitat: Varied. Arid and semi-arid areas with acacia, spinifex, mallee scrub. Coastal eucalypt woodlands.

Feeding: Mostly nectar. Probes tubular flowers, especially Emu-bush, grevilleas and eucalypts. Also takes insects and occasionally fruit and seeds.

Voice: Usually quiet, except when breeding. Piercing 'tr-tee-tee-tee' whistle.

Status: Locally common to rare. Nomadic following flowering events.

Breeding: Sept.–Feb., or following rain. Untidy cup of twigs, grass and spider web low in a bushy shrub / 2–4 eggs.

Francois Peron NP, Cape Range NP. and Canning Stock Route, WA.

Left: Male feeding in Berrigan. Right: Immature.

Black Honeyeater

Sugomel nigrum

Endemic

100–120mm

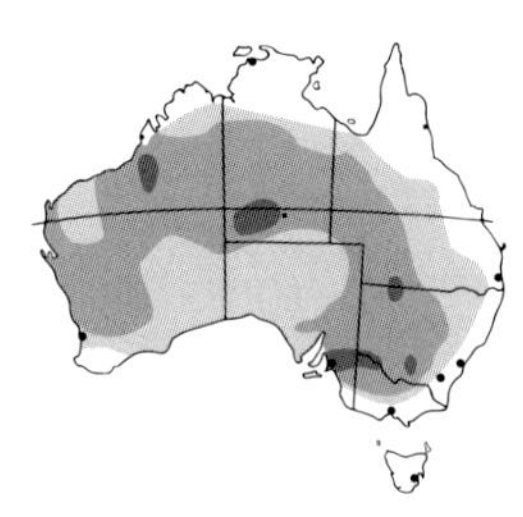

Black head and upperparts. Throat: Blackish extending in a tapering line down over white breast and belly. Tail, legs: Black. Underparts: White.
Female: Upperparts: Light brown with buff-white stripe behind eyes. Throat, breast: White, washed brown. Underparts: White. Immature: Similar to female, plainer.

Male territorial display flight includes vertical climb and gliding descent while singing.

Habitat: Mulga, mallee and spinifex scrubs.

Feeding: Mostly nectar taken almost exclusively from Emu-bush. Hovers over flowers and also hawks for insects.

Voice: Sparrow-like chirping. Thin whistled 'preee'.

Status: Highly nomadic, following flowering events. Southern birds move north to overwinter.

Breeding: Oct.–Nov., or following rain. Shallow cup nest of twigs, grass and spider web, in a low shrub, in loose colonies / 2–3 eggs.

Left: Male. Right: Female taking nectar from Emu-bush.

Rufous-throated Honeyeater

Conopophila rufogularis

Endemic

120–140mm

Sexes alike. Deep, reddish brown throat feathers that are puffed out in an aggressive display together with an upright pose. Golden yellow bands under wings are also used in aggression display. Breast: Buff wash. Underparts: Light grey. Upperparts: Mid-grey brown with bright yellow wing patch and bright yellow edged tail. Eyes, bill: Grey-brown.
Immature: Similar, throat white.

Active, relatively tame.

Habitat: Tropical eucalypt forests and woodlands.

Feeding: Noisily forages in upper branches for nectar. Sallies and hovers for insects. During the day drinks and bathes frequently. Aggressive when feeding.

Voice: Rasping chattering notes, often in duets. Loud 'zit-zit' alarm call.

Status: Common. Partially nomadic in dry season.

Breeding: Sept.–Apr. Deep cup nest of paperbark, grass and spider web placed 1.5–7m high / 2–3 eggs.

Left: Adult. Right: Female constructing nest.

Green-backed Honeyeater

Glycichaera fallax

120mm

Sexes alike. Small, nondescript, tropical honeyeater with dull olive upperparts and grey-green frons, crown and nape. Eyes: Light blue-grey with faint chalky white eye ring. Underparts: Dull lemon. Legs: Bluish-grey.
Immature: Duller. Gape: Pale yellow. Eyes: Dark, indistinct eye ring.

Constantly active in the upper rainforest canopy. Fast, direct flight and fluttering, hovering, acrobatic or hopping.

Habitat: Tropical rainforests, vine scrub and adjacent eucalypt woodlands when flowering.

Feeding: Forages from near ground level to the upper canopy on edge of rainforest, hovers and flutters to pick insects from the outer foliage and occasionally chases insects in aerial pursuit. Occasionally takes nectar from palm and eucalypt flowers.

Voice: Pleasant, soft twitters. Harsh scolding notes.

Status: Moderately common.

Breeding: Time of year unknown. Cup nest / 2 eggs.

Iron Range NP, Qld.

Opposite: Adult.

Grey Honeyeater

Conopophila whitei

Endemic

100–130mm

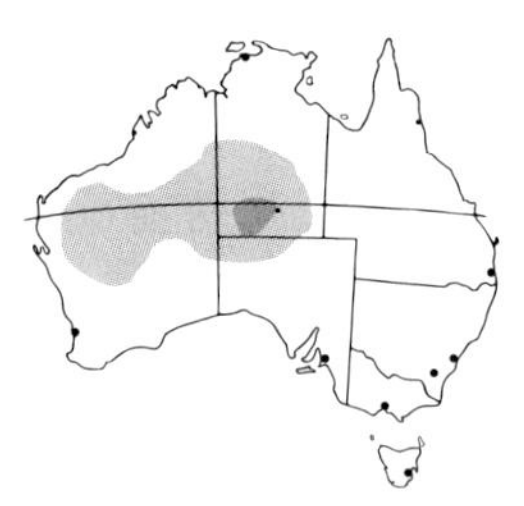

Sexes alike. Tiny, very plain honeyeater with grey-brown upperparts, olive washed wings and a short, grey, almost straight bill. Eyes: Brown with ash grey eye ring. Tail: Dark brown, tipped white. Underparts: Off-white with greyish throat, breast. Immature: Cheek, throat: Pale yellow.

Inhabits remote inland areas and is one of Australia's least known birds. Often seen with thornbills and gerygones.

Habitat: Remote mulga and spinifex shrubs. Native gardens in the Alice Springs area.

Feeding: Mistletoe fruits, insects. Pierces tubular flowers for nectar, particularly Emu-bush.

Voice: High-pitched harsh call. Twittering trills, repeated.

Status: Uncommon to rare. Nomadic.

Breeding: Aug.–Nov. or in response to rain. Fragile cup nest of grass bound with spider web about 3m high / 2 eggs.

Olive Pink Botanic Garden and Kunoth Bore, Alice Springs, NT. Yalgoo area, WA.

Left: Adult in Hakea. Right: Adult.

Brown-backed Honeyeater

Ramsayornis modestus

120–130mm

Sexes alike. Small, drab honeyeater with plain brown upperparts and faint, dark flecks to crown and nape. Wings, tail: Grey-brown. Underparts: Chalky white with faint, pale brown barring on breast. Face: Brown, darker from lores to ear coverts with white line from gape to under eyes. Eyes: Red-brown. Bill, legs: Pinkish.
Immature: Breast heavily streaked dark brown, yellow face line.

Lively, active, often short sallies for insects.

Habitat: Paperbark swamps, wet riverside vegetation, mangroves and nearby eucalypt woodlands.

Feeding: In low shrubbery to the high canopy. Often feeds in mixed flocks. Insects taken from foliage, under bark or mid-air. Sometimes nectar.

Voice: Sharp, repeated 'chit, chit' contact call. Constant soft 'mik-mik-mik' in flight. Chattering song.

Status: Locally common to uncommon. Nomadic, migratory. Overwintering in Torres Strait islands and PNG.

Breeding: Aug.–Mar. Deep, cup-shaped, domed nest, often suspended over water, 2–4m high / 2–3 eggs.

Centenary Lakes Botanic Gardens, Cairns, Qld.

Left: Adult with prey. Right: Adult collecting nesting material.

Macleay's Honeyeater

Xanthotis macleayanus

Endemic

200mm

Sexes alike. Inconspicuous honeyeater in the upper foliage with brown-black frons and crown, a brown-black and white speckled nape. Distinctive yellow skin patch below the eyes and golden ear tufts. Back: Brown, heavily streaked with yellow. Breast: Olive and white streaked. Underwing coverts: Buff. Legs: Grey, bluish tinged.
Immature: Duller, greyer.

Acrobatic amongst blossoms. Climbs trunks and along limbs in search of insects. Usually inconspicuous, but aggressive towards other honeyeaters.

Habitat: Upper layer of lowland rainforests and mangroves. Parks and gardens.

Feeding: Insects make up the bulk of the diet, followed by fruit and nectar.

Voice: Piercing, musical 'to-wit-too-wee-twit'.

Status: Moderately common. Sedentary.

Breeding: Oct.–Dec. Cup nest of palm fibre, bark strips and spider web, from 2m to high in the canopy / 2 eggs.

Opposite: Adult feeding in 'Honey Gem' Grevillea.

Bar-breasted Honeyeater

Ramsayornis fasciatus

Endemic

130–140mm

Sexes alike. Distinctive white breast with pronounced black-grey barring. Crown: Black with fine white scalloping. Face: White with black malar line and red-brown eyes. Upperparts: Brown with black streaking. Belly, undertail: White with sparse fading streaks and black streaked flanks. Bill, legs: Pinkish brown.
Immature: Plain brown head. Breast streaked, not barred.

Relatively tame. Inconspicuous. Solitary or in pairs. Often feeds with other honeyeaters.

Habitat: Tropical rainforests, paperbark and mangrove swamps, forests along streams.

Feeding: In all levels of vegetation. Primarily nectar. Insects also taken.

Voice: Repeated shrill, piping alarm call. Piercing metallic chattering song.

Status: Moderately common. Nomadic, following flowering trees.

Breeding: Aug.–Jan. Deep cup-shaped nest with roof, side entrance, usually over water, 1–5m high / 2–3 eggs.

Left: Adult feeding on the flowers of White-flowered Black Mangrove. Right: Immature.

Brown Honeyeater

Lichmera indistincta

110–150mm

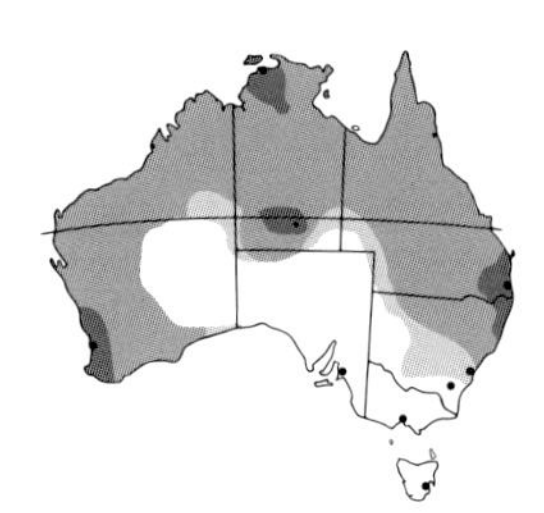

Sexes similar. Mid-sized, plain olive-brown honeyeater with a longish curved bill and a small light yellow tuft behind the eyes. Wings: Dull yellow patch. Throat, upper breast: Pale olive-brown grading to off-white underparts. Breeding male: Black gape. Non-breeding: Yellow gape.
Female: Olive-tinged plumage, pale yellow gape. Immature: Yellow gape, indistinct tuft behind eye.

Actively defends feeding area.

Habitat: Varied. Understorey along watercourses, woodlands, edge of mangroves, heathlands, mulga and spinifex scrub, edges of tropical rainforests.

Feeding: Forages at all strata levels in trees and shrubs, usually in mixed flocks. Mostly nectar, also insects.

Voice: Noted for its pleasant, melodious, warbled, metallic notes.

Status: Common. Locally nomadic, following flowering events. Southern birds migrate northward to overwinter.

Breeding: June–Jan. Cup nest of rootlets, grass and spider web, low to 2m high, often over water / 2 eggs.

Left: Adult feeding on Firewood Banksia. Right: Adult feeding on Gold-tipped Bottlebrush.

Dusky Honeyeater

Myzomela obscura

130–150mm

Sexes alike. Plain, dusky brown upperparts with slightly darker head and breast. Underparts: Paler dusky brown. Wings, tail: Grey-brown. Eyes: Dark brown. Bill: Black, long, thin down-curved.
Race *harterti*, Qld. Darker, warmer brown.
Immature: Warmer brown, reddish wash to face and yellow gape.

Singular or pairs. Occasionally small groups. Inquisitive. Active, acrobatic feeders, flitting rapidly.

Habitat: Dense wet tropical rainforests, dense coastal woodlands and mangroves, parks and gardens.

Feeding: Often in mixed flocks, hovering or hanging from flowers foraging for nectar. Also insects.

Voice: Squeaky chattering. Trill calls.

Status: Common. Locally nomadic.

Breeding: Aug.–Jan. Cup nest of rootlets, grass and spider web from 5m to upper canopy / 2 eggs.

Centenary Lakes Botanic Gardens Cairns, Qld.

Left: Nominate race adult. Feeding in *Grevillea longistyla*. Right: In 'Misty Pink' Grevillea.

Tawny-breasted Honeyeater

Xanthotis flaviventer

170–210mm

Sexes alike. Similar to Macleay's Honeyeater but with distinctive bare skin around the eyes, a thin white streak under and behind the eyes and a small yellow ear tuft. Upperparts: Dark olive-brown with light speckles on the nape. Bill: Black. Throat, side of neck: Light grey. Breast: Tawny, faintly streaked. Legs: Blue-grey.
Immature: Plainer.

Noisy, aggressive, arboreal feeders, often in small groups.

Habitat: Tropical rainforests, mangrove and vine scrub along waterways, and adjacent areas when eucalypts are flowering.

Feeding: Methodically searches in the middle to upper canopy, primarily taking insects from foliage or bark crevices. Some nectar taken.

Voice: Repeated whistles, up and down in pitch. Usually quiet except when feeding.

Status: Moderately common.

Breeding: Nov.–Feb. Cup-shaped nest of bark shreds, paperbark and spider web, 2–20m high / 2 eggs.

Left and right: Adults.

Red-headed Honeyeater

Myzomela erythrocephala

100–130mm

Distinctive, small and agile, with a striking glossy scarlet head with black lores, eye ring and long black down-curved bill. Sides of neck, throat, rump: Glossy scarlet. Back, wings, upper breast: Brownish grey. Underparts: Light brownish grey. Eyes: Reddish brown.
Female: Smaller. Upperparts: Brown-grey with crimson wash to face, with no red behind the eyes. Underparts: Whitish. Flight feathers edged whitish.
Immature: Similar to female, yellow lower bill.

Constantly moving, flitting. Undulating, fast, buoyant flight.

Habitat: Tropical mangroves and often paperbark forests. Occasionally in urban gardens.

Feeding: Arboreal, active, calling when feeding, preferring the outer foliage, probing flowers for nectar and gleaning or taking insects mid-air.

Voice: Metallic twittering. 'Chirrup' calls.

Status: Locally common. Also in New Guinea.

Breeding: Mar.–Sept. Cup nest of bark shreds and spider web, usually low in a mangrove over water / 2 eggs.

Middle Arm and Buffalo Creek boat ramp area, Darwin, NT. Broome, WA.

Left: Adult male. Right: Adult female.

Scarlet Honeyeater

Myzomela sanguinolenta

100–110mm

Jewel-like honeyeater with a glossy bright scarlet head, upper body and breast, and contrasting black lores, bill, tail. Black wings are edged white. Underparts: Whitish grey. Eyes: Brownish black.
Female: Upperparts: Olive-brown. Underparts: Whitish with variable red wash on chin. Immature: Similar to female.

Constantly active, acrobatic, hovering while feeding. Fast bouncy flight and sallies for flying insects. Usually in pairs or small flocks. Sings while feeding.

Habitat: Open coastal forests, woodlands, mangrove thickets, parks and gardens.

Feeding: Prefers the upper canopy, primarily taking nectar. Occasionally fruit and insects.

Voice: Sweet, clear, melodic, warbling by male. Squeaks by female.

Status: Moderately common. Locally nomadic, following flowering events. Also in New Guinea.

Breeding: July–Jan. Delicate cup of bark shreds and spider web, 1–15m above ground, up to 3 broods per season / 2 eggs.

Royal NP, NSW. Lamington NP, Qld.

Left: Adult male. Right: Adult female feeding in broad-leaved paperbark.

Crimson Chat

Epthianura tricolor

Endemic

110–120mm

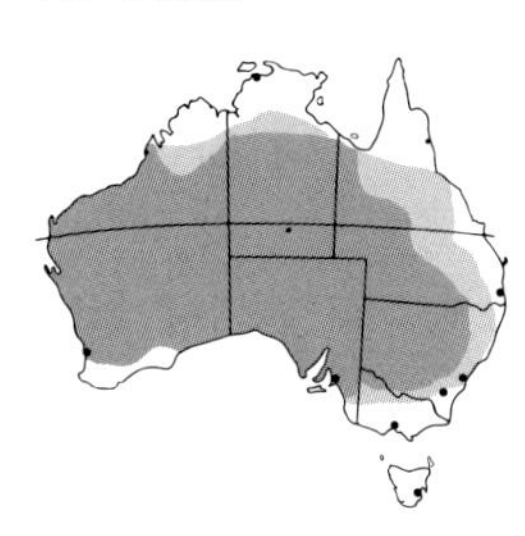

Brilliant crimson crown, rump, breast and underparts. Eyes: Cream-yellow. Chin, throat: Bright white. Upperparts: Deep brown. Undertail coverts: White. Tail: Blackish-brown, tipped white.
Non-breeding: Duller, greyer face mask with patchy crimson on crown and underparts.
Female: Upperparts, head: Greyish brown. Breast, rump: White, patchy red. Throat, belly: White. Non-breeding: No red in underparts or crown. Immature: Similar to female, without any red plumage.

Unlike most other birds it walks rather than hops. Bouncing flight.

Habitat: Semi-arid mulga, mallee, grasslands and salt plains.
Feeding: Usually ground feeding, quickly running, taking insects. Sometimes nectar from low flowers.
Voice: High-pitched tinkling notes, repeated.
Status: Generally common, but nomadic following erratic rains. Migrate seasonally within their range – southward in summer and northward in winter.
Breeding: Aug.–Oct. in south, elsewhere in good seasons. Often in colonies. Deep cup nest of twigs and grass to 1m high / 3–4 eggs.

Kunoth Well, Alice Springs, NT.

Left: Breeding male. Right: Breeding female.

Orange Chat

Epthianura aurifrons

Endemic

105–120mm

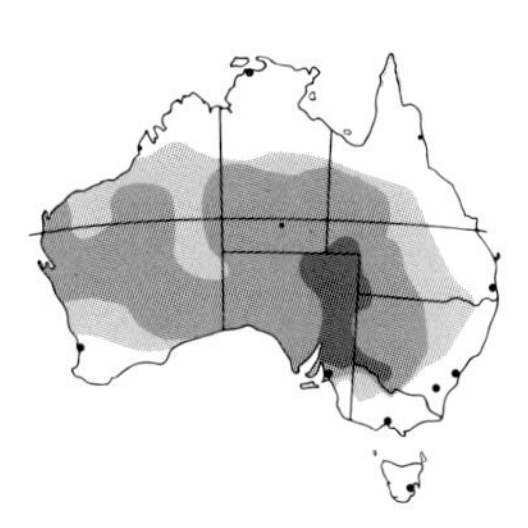

Orange-yellow head, rump and underparts. Frons and breast more luminous orange-yellow. Lores, throat: Black. Eyes: Deep red to orange-brown. Back: Olive-yellow with black streaks. Tail: Blackish with white tips and yellow uppertail coverts.
Non-breeding: Duller, paler.
Female: Upperparts: Mottled grey-brown. Underparts: Pale-yellow. Breast: Grey-brown wash. Immature: Similar to female.

Flight is low, fluttering to low vantage perch.

Habitat: Low saltbush plains and Samphire around salt lakes and shrubby gibber plains.

Feeding: Forages exclusively for insects on or close to ground in low foliage.

Voice: Metallic, musical notes. 'Cheep-cheep' flight call.

Status: Moderately common. Nomadic.

Breeding: Aug.–Sept. In colonies, cup nest of grass in saltbush / 3–4 eggs.

Deniliquin area, NSW. New Beach, WA.

Left: Breeding male. Right: Female.

Yellow Chat

Epthianura crocea

Endemic

110–120mm

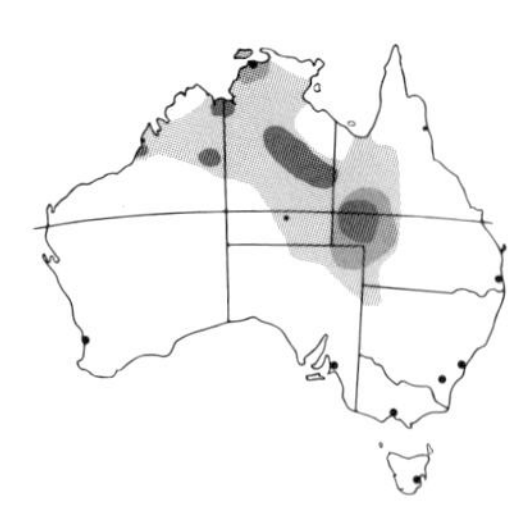

Least known chat. Frons, face, underparts, rump: Bright yellow. Breast: Black crescent band. Back, crown: Grey-brown, tinged yellow with darker mottling. Eyes: Cream with thin dark eyeline. Tail: Grey-brown tipped white. Non-breeding: Paler, greyer, crown streaked olive, breast band thinner. Female: Similar, paler, no breast band, tan tinted eyeline. Immature: Similar to female, paler, pale bill, tan eye.

Usually in groups of 2–10.

Habitat: Open, well-grassed coastal or inland seasonal swamps, bore-drains, fringing beds of reeds or low vegetation.

Feeding: Runs between vegetation clumps, feeding on the ground for insects.
Voice: High, piping 3-note musical call.
Status: Rare to uncommon. Nomadic.
Race *macgregori*, Qld, and race *tunneyi*, far north NT, endangered.
Breeding: Sept.–Nov., Mar.–Jun. after rain. Cup nest of grass in low reeds, often over water / 3 eggs.

Broome Bird Observatory, WA. Barkly Tableland, NT.

Left: Female. Centre and right: Breeding male.

White-fronted Chat

Epthianura albifrons

Endemic

120–130mm

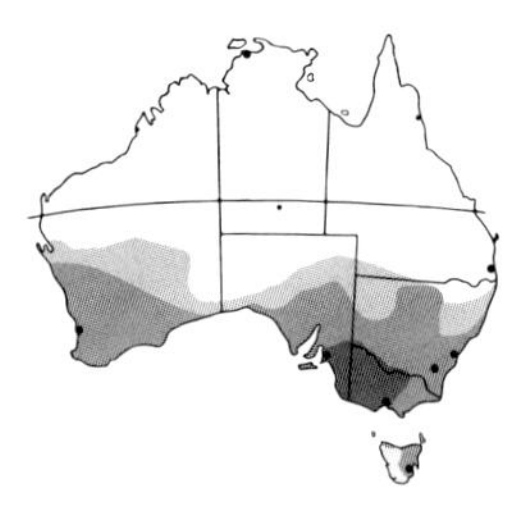

Distinctive black breastband extending around the nape and a black crown. Face, throat, underparts: White. Eyes: Orange or buff-white. Back: Mid-grey. Wings: Brownish grey. Tail: Dark brown with white tips prominent when tail is spread. Female: Head, upperparts: Grey-brown with a thinner browner breast band. Immature: Similar to female, browner, indistinct or absent breast band.

Wary. Gregarious. Perches prominently. High, bouncy, undulating flight or runs along the ground.

Habitat: Low, shrubby damp areas around inland salt lakes, swamps, open fields and heaths.

Feeding: Insects taken from the ground and low bushes. Nectar possibly taken.

Voice: Finch-like metallic 'tang' call.

Status: Common. Sedentary in southern wetter areas. Partially nomadic inland, in flocks of 5–50 birds.

Breeding: July–Jan. Nests near ground in low shrub / 2–4 eggs.

Left: Male. Right: Female in Black-seeded Samphire.

Gibberbird

Ashbyia lovensis

Endemic

125–130mm

Sexes similar. Crown, nape, back, wings: Mottled, buff-grey-brown. Face, eyebrow, lores: Yellow. Eyes: Yellowish cream. Breast, flanks: Yellowish grey-brown. Tail: Black. Female: Browner upperparts, underparts paler with brown washed breast. Immature: Similar to female. Bill: Cream.

Similar to chats. Relatively tame. Pairs or small groups. Perches with an upright posture on elevated stones. Runs with a swagger and wags tail. Short, bouncy, undulating flight, but mostly terrestrial.

Habitat: Sparsely vegetated stony gibber plains of Lake Eyre Basin.

Feeding: Searches the ground for insects.

Voice: Musical chattering. Male has piercing 'wheet, wheet, wheet' breeding display call, as it rises to about 30m then drops vertically.

Status: Mostly sedentary. Locally common.

Breeding: June–Dec. Well constructed deep nest, formed in a deep depression in the ground, lined with twigs, fine grass and rootlets, often under a shrub / 2–4 eggs.

Sturt NP, NSW. Strzelecki Track, SA.

Left: Female. Centre and right: Male.

Babblers

Flyrobins

Robins

Scrub-robins

Chowchilla

Logrunner

A group of three families, with similarities in appearance and behaviour.

Family Pomatostomidae contains the babblers, mostly endemic to Australia, with one, the Grey-crowned Babbler, also found in New Guinea.

Babblers are typically found in arid and semi-arid areas. They are gregarious, medium-sized birds with strongly down-curved bills. Their tails are long, often held fanned out, and they have long legs and strong feet for running and scratching while foraging on the ground. Communal birds in groups of 12 to 20, they are territorial and build several large communal roosting nests from strong twigs and grass and one breeding nest within their territory. Group members attend to the breeding pair.

Family Orthonychidae contains the Australian Logrunner and the Chowchilla.

The Logrunner and Chowchilla are small, robust, ground-dwelling birds of the rainforest. They have strong spiny tail feathers that they use as a prop while they forage, scratching while spinning with one foot in the leaf litter, leaving a trail of distinctive small, cleared circles behind on the forest floor. The female builds a dome-shaped nest with a side entrance from sticks covered with moss and lichen, either on or near the ground. The female incubates the eggs and feeds the young insects caught by the male.

Family Petroicidae, contains the scrub-robins, robins and flycatchers.

Australian robins were named by the early settlers after the robins of the northern hemisphere but are unrelated to 'true robins'. Mostly endemic to Australia, some species are also found on nearby Pacific islands.

These small, graceful and unobtrusive flycatcher-like birds are mostly insectivorous. Perched motionless while scanning the ground, they pounce on prey or sally into the air to catch airborne insects. A beautiful cup-shaped nest is built by the female, from fine bark strips and plant fibre bound with spider web and decorated with moss and lichen or bark for camouflage, and is usually placed in the fork of a tree. The female incubates the eggs and is fed by the male. Both parents feed the young.

Opposite: Male Red-capped Robin perched below an immature offspring.

Grey-crowned Babbler

Pomatostomus temporalis

290mm

Sexes alike. White crown with narrow central grey stripe, broad white eyebrow and dark grey eye stripe. Upperparts: Greyish brown. Wings: Brown, chestnut wing patch seen in fight. Eyes: Pale yellow. Bill: Down-curved. Chin, throat, upper breast: White: Underparts: Buff-grey with variable rufous-brown belly. Tail: Long, black-brown, tipped white. Race *rubeculus*, Red-breasted Babbler, far west Qld, NT, WA. Smaller, similar. White frons, rufous breast. Underparts: Dark brown. Tail: Narrow white tip. Immature: Paler, brown eyes.

Highly gregarious. In noisy family groups of about 15 birds.

Habitat: Inland plains, scrubby dry woodlands, roadsides, farms.

Feeding: Rummages through leaf litter, turning over debris, picking up insects and spiders.

Voice: Female 'ya'. Male answers 'ahoo'. Flock join in to make a melodious cacophony.

Status: Sedentary. Moderately common. Rare in Vic., SA. Race *rubeculus* common in northwest.

Breeding: July–Feb. Roosts and nests communally, domed nest 4m high / 2–3 eggs.

Capertee Valley, NSW. Kakadu NP, NT.

Left: Nominate race. Right: Race *rubeculus*.

White-browed Babbler

Pomatostomus superciliosus

Endemic

180–210mm

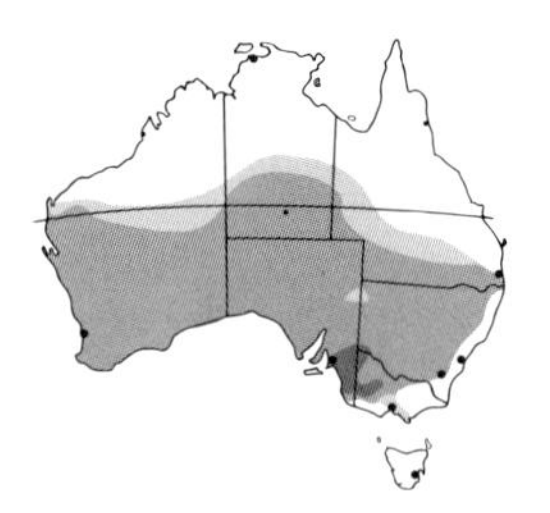

Sexes alike. Dusky brown head with a prominent, white, dusky edged eyebrow. Underparts: White grading to grey-brown belly. Bill: Long, narrow down-curved. Upperparts: Buff-brown, darker rump. Tail: White tipped. Immature: Duller. Throat, breast: Buff-white.

Noisy, social, groups of 3–10 birds, roosting together at night.

Habitat: Open forests with understorey, woodlands, mallee and mulga scrub.

Feeding: Forages mostly on the ground among leaf litter and fallen timber for insects, invertebrates, small frogs and lizards. Fruits and seeds are also eaten.

Voice: High-pitched, chattering notes and whistles.

Status: Moderately common. Sedentary.

Breeding: July–Dec. Nests communally. Domed nest 1–6m high in small tree / 2–3 eggs.

Hattah–Kulkyne NP. Vic.

Opposite: Adult.

Hall's Babbler

Pomatostomus halli

Endemic

200–220mm

Sexes alike. Similar to the White-browed Babbler but mostly dark brown. Distinctive white bib, contrasting with dark brown breast and belly. Eyebrow: Long, white, widest at nape. Eyes: Brown to red-brown. Tail: Broad white tip. Bill: Grey-black, down-curved.
Immature: Duller.

Gregarious, noisy, in small groups. Roosts communally.

Habitat: Lightly timbered mulga and dry, rocky ridges with acacia.

Feeding: Mostly ground feeding in family groups, turning over stones and twigs, picking up insects, spiders.

Voice: Cacophony of chuckles, growls and musical notes. Similar to White-browed Babbler.

Status: Moderately common. Locally nomadic.

Breeding: Sept.–June. Domed nest, side entrance, about 5m high. Communal roosting and nest building / 1–2 eggs.

Left: Adult. Right: Small group.

Chestnut-crowned Babbler

Pomatostomus ruficeps

Endemic

210–230mm

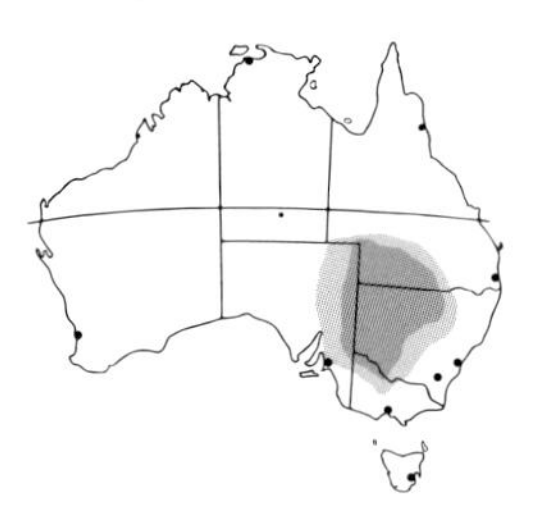

Sexes alike. Slim bird with rich chestnut crown and nape bordered by long white eyebrow. Brown eyes and face mask above a white throat and breast merging to a grey belly. Upperparts: Mottled grey-brown. Flanks: Dark brown. Tail: Black-brown, white tipped. Immature: Duller. Crown: Brown. Eye stripe, wing bars: Tannish.

Noisy, social groups of about 6–15 birds. Shy and elusive to approach. Roost in communal 'dormitory' nest.

Habitat: Dry inland, low open scrub, mallee, mulga, casuarina woodlands and edges of salt lakes.

Feeding: In groups on the ground, probing for insects, spiders, small amphibians and reptiles. Sometimes fruit, seeds.

Voice: Penetrating whistle, repeated chattering.

Status: Locally common to scarce. Sedentary.

Breeding: June–Oct. Large domed nest, 5–10m high / 3–5 eggs.

Hattah–Kulkyne NP, Vic.

Opposite: Adult.

Lemon-bellied Flyrobin

Microeca flavigaster

120–140mm

Sexes alike. Also called Lemon-bellied Flycatcher. Upperparts: Olive-brown, rump olive-yellow. Underparts: Washed lemon with white throat. Bill: Dark above, pale below. Lores: Grey streak. Eye ring: Pale. Tail: Shallow fork. Legs: Black. 4 races.
Nominate race, Top End, NT. Moonsoon forests.
Race *laetissima*, central coastal wet tropical Qld. Larger. Pale yellow breast.
Race *flavissima*, far northern Qld. In open forest or woodland. Breast, belly: Bright yellow. Upperparts: Yellow-olive.
Race *tormenti*, Kimberley Flycatcher, Northern WA. Usually in mangrove thickets. Breast, belly: White. Upperparts: Grey-brown.
Immature all races: Brown streaked.

Usually solitary.

Habitat: Tropical rainforest margins, paperbark swamps, monsoon and vine scrub, mangroves, savannah woodlands.

Feeding: Hawks insects from above the canopy or from foliage, returning to the same perch. Race *tormenti* takes small crustaceans.

Voice: Clear, sweet, repeated, varied, rising–falling musical notes.

Status: Common to locally uncommon. Sedentary.

Breeding: Aug.–Mar. Tiny cup nest 1.5–10m high / 1 egg.

Iron Range NP and Julatten area, Qld. Fogg Dam, NT. Derby boat ramp mangroves, WA.

Left: Race *tormenti*. Centre: Race *laetissima*. Right: Nominate race.

Yellow-legged Flyrobin

Microeca griseoceps

120mm

Sexes alike. Also called Yellow-legged Flycatcher. Similar to the Lemon-bellied Flyrobin with distinctive orange-yellow legs. Upperparts: Dull green. Head: Mid-grey. Throat: White. Breast: Light grey. Belly, tail coverts: Pale lemon. Wings, tail: Grey-brown, edged olive. Bill: Small, dark above, pale yellowish below. Immature: Duller.

Often raises tail when perched. Short fluttering flight.

Habitat: Outer canopy of tropical rainforests and adjacent edges of eucalypt woodlands, paperbark swamps.

Feeding: Gleans insects from foliage or hawks for them mid-air.

Voice: Repeated 'zzt-zzt' calls. Strong, piped descending call.

Status: Uncommon. Solitary. Wanders locally.

Breeding: Nov.–Jan. Small, cup-shaped nest, 10–15m high / 2 eggs.

Iron Range NP, Qld.

Left and right: Adults.

Jacky Winter

Microeca fascinans

120–140mm

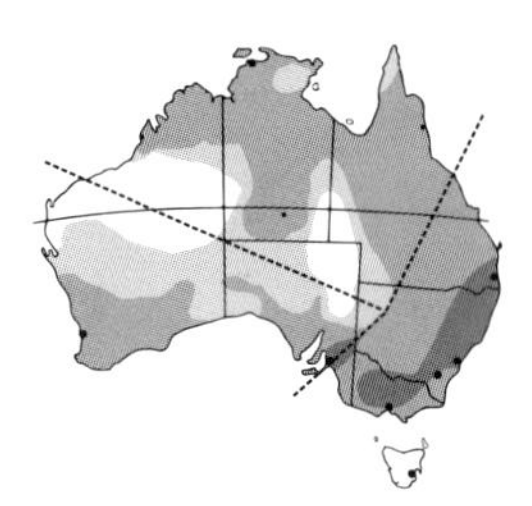

Sexes alike. Grey-brown upperparts with a faint white eyebrow. Underparts: White with faint brown-washed breast. Wings: Darker, edged white. Tail: Blackish brown with white outer feathers.
Nominate race, eastern Qld to SA.
Race *assimilis*, southwest SA, WA. Darker. Outer tail feathers darker towards base. Breast, flanks: Greyish.
Race *pallida*, Cape York, Qld, Top End, NT. Smaller, sandy brown, becoming paler further inland.
Immature: Upperparts: Pale streaked. Underparts: Mottled.

Relatively tame. Tail often fanned while hovering.

Habitat: Varied. Open woodlands, forest edges, mallee, scrublands, paddocks, parks and gardens.

Feeding: Insects taken from foliage or sallies from perch or hawks insects mid-air. Hovers low over grassy areas.

Voice: Clear, sweet, ringing, whistling song.

Status: Moderately common. Sedentary or nomadic. Disperses in winter.

Breeding: July–Aug. or after rain in arid areas. Small, shallow cup nest, on horizontal branch or fork 0.5–20m high / 2 eggs.

Left: Nominate race displaying white outer tail feathers. Right: Female on nest.

Scarlet Robin

Petroica boodang

Endemic

120–130mm

Bright scarlet breast with black upperparts, head and throat and a large white spot on frons. Wings: Boldly marked white. Tail: Outer feathers edged white. Lower belly, undertail coverts: White.
Female: Upperparts, head: Grey-brown, throat paler. Breast: Pale scarlet wash. Frons, wing marks: Smaller, buff-white. Immature: Similar to female, lacks any red.

Solitary or pairs. Relatively tame. Does not form winter flocks.

Habitat: Dry eucalypt forests, woodlands with light understorey and fallen timbers. Winter habitat: Open grassy areas, River Red Gum forests, gardens.

Feeding: Perch and pounce taking prey from the ground. Insects and larvae. Sometimes sallying.

Voice: Sweet, high trilling notes, repeated.

Status: Moderately common. Disperses in winter.

Breeding: July–Aug. Dec.–Jan. Open-cup nest, often on a dead branch, 2–20m high / 3 eggs.

Left: Female. Right: Male.

Red-capped Robin

Petroica goodenovii

Endemic

110–115mm

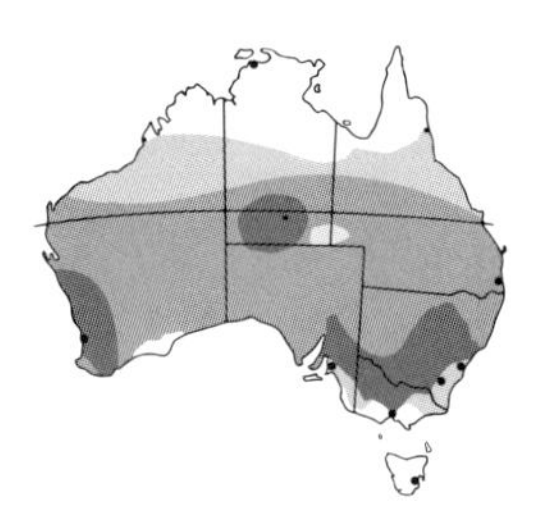

Vibrant red cap and breast contrasting with a brownish black head, throat and upperparts. Underparts, flanks: White. Wings: Black, boldly marked white. Tail: White outer feathers. Female: Dull red cap. Upperparts: Pale grey-brown. Underparts: Buff-grey. Wings, tail: Buff-white marks. Immature: Similar to female, males have faint red wash to breast.

Male defends breeding territory by singing from perimeter perches. Often tame, confiding.

Habitat: Open drier woodlands, semi-arid mulga, inland scrub, mallee and farms.

Feeding: Perches near ground and pounces on insects, grasshoppers, ants, moths, caterpillars, other invertebrates.

Voice: Soft, abrupt, metallic 'tick, tick, tick' call. Rattling trill. Scolding defence 'kek-kek-kek' repeated.

Status: Moderately common. Sedentary. Southern birds disperse in winter, moving to more open habitat or towards the coast in WA.

Breeding: July–Jan. Tiny cup nest, camouflaged with lichen or bark, 5–10m high / 3 eggs.

Left: Female. Right: Male.

Flame Robin

Petroica phoenicea

Endemic

125–135mm

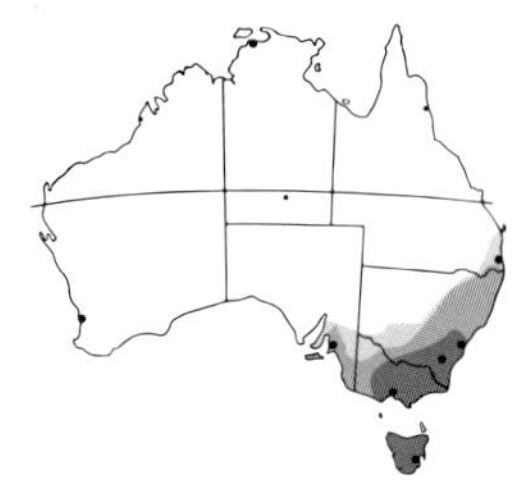

Bright orange-red throat, breast and lower belly with slate grey upperparts and large white patch on frons. Vent, undertail: White. Wings: Conspicuous white bar. Tail: White outer feathers. Female: Pale brown upperparts; small white, indistinct frons spot. Breast: Buff-brown. Wings: Buff patches. Tail: Brown, white outer shafts. Immature: Similar, often orange-washed breast.

Solitary, but after breeding often gather in flocks of 5–30, usually with mixed feeding flocks.

Habitat: Tall, moist eucalypt woodlands and forests with open understorey and clearings. Winter habitat: Lowland, drier open woodlands and forests, native grasslands, gardens.

Feeding: Perches, pounces from a low branch. Insects, larvae, spiders. Sallies for flying insects or probes bark or foliage. In winter, often scans from a prominent perch.

Voice: Soft, descending, piping trill.

Status: Moderately common. Some Tas. birds overwinter on the mainland.

Breeding: Aug.–Sept. Cup nest in a shallow tree hollow, stump, rock crevice or eucalypt fork / 3–4 eggs.

Left: Female. Right: Male.

Rose Robin

Petroica rosea

Endemic

115–125mm

Small, long-tailed robin with a deep rose pink breast, white belly and small white frons spot. Head, throat, upperparts: Slate grey. Wings: Edged grey-brown. Female: Upperparts: Brownish grey. Underparts: Pale grey with pink flush to breast. Wings: 2 buff bars. Tail: Brown with white outer feathers. Immature: Similar to female without pink flush.

Acrobatic, flits flycatcher-like with tail fanned. Flicks tail when perched.

Habitat: Wet eucalypt forests or rainforests with dense gullies and creeks with tree fern and acacia understorey.

Feeding: Takes insects from foliage or ground. Pursues flying insects.

Voice: Musical, cheery, trilling song.

Status: Uncommon. Solitary or in pairs. Disperses to lower altitudes and northward for winter.

Breeding: Sept.–Jan. Dainty cup nest, on outer lichen-covered branch or fork / 2–3 eggs.

Lamington NP, Qld. Barrington Tops NP, NSW.

Left: Female. Right: Male.

Pink Robin

Petroica rodinogaster

Endemic

120–130mm

Small robin with sooty black upperparts, head and throat and contrasting lilac-pink breast and belly. Small, white frons spot. Wings: Faint buff bars. Tail: Plain sooty black.
Female, immature: Upperparts: Warm brown. Underparts: Buff-cinnamon brown. Breast: May have a pink wash. Wings: Buff wing patches.

Pairs or family groups.

Habitat: Fern gullies and shady understorey of tall eucalypt forests and temperate rainforests.

Feeding: Perches and pounces mostly for insects and spiders. Often forages on the ground.

Voice: Soft 'tick' call and warbling song.

Status: Moderately common. In winter, mainland birds disperse north and west into drier and more open regions.

Breeding: Sept.–Dec. Deep, cup-shaped nest, camouflaged with lichen in forked branch in dense cover, low to 6m high / 3–4 eggs.

Mt Wellington, Tas.

Left: Female. Right: Male.

Hooded Robin

Melanodryas cucullata

Endemic

140–170mm

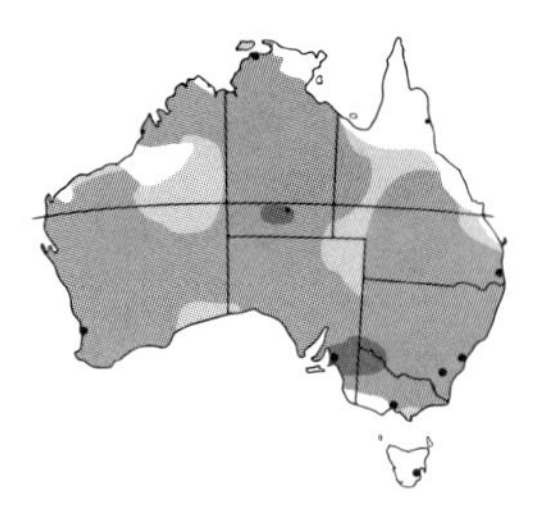

Named for the black 'hood' that covers its head, extending to the upper breast and back. Underparts: White, washed grey. Shoulder: Bold white bar. Wings: Black with white stripe. Female: Grey-brown. Wings: Dark brown, white bar and margins. Breast and underparts: Light brownish-grey. Immature: Similar to female.

Wary and unobtrusive. Usually in pairs or small groups.Swift, undulating flight in short bursts.

Habitat: Open dry forests with scattered trees and understorey with open grassy patches and woody debris. Eucalypt forests, mallee, mulga, native pine, casuarina, and drier coastal areas of banksia heathlands. Farmland with scattered trees and suitable understorey.

Feeding: Perch and pounce from a low vantage point or on the ground among woody debris. Mostly insects, spiders.

Voice: Penetrating, descending and fading whistle. Often quiet.

Status: Uncommon to moderately common in the southeast of their range. Vunerable in NSW. Adults sedentary. Immature disperse.

Breeding: July–Dec. in south. Mar.–Apr. in north. Communal. Cup nest placed in a crevice or hole 0.5–10m high / 2–3 eggs.

Left: Female. Right: Male.

Dusky Robin

Melanodryas vittata

Endemic

160–170mm

Sexes alike. Found exclusively in Tas. Sombre plumage with plain olive-brown upperparts with a fine, brown eyeline. Wings: Indistinct pale buff patch and stripe. Tail: Outer feathers edged buff. Underparts: Pale brown. Immature: Similar to adult, indistinct markings.

Solitary or in pairs. Small flocks form in winter.

Habitat: Edges of temperate rainforests, open woodlands with undergrowth, newly cleared areas, sometimes gardens and parks.

Feeding: Sits motionless surveying the ground, pouncing on prey, usually insects and invertebrates. Sometimes forages in low shrubbery or catches insects mid-air.

Voice: Low, monotonous, plaintive double whistle, second note lower.

Status: Moderately common. Sedentary.

Breeding: July–Dec. Untidy, grass cup-shaped nest placed in a cavity, crevice or tree fork, 1–6m high / 3 eggs.

Opposite: Adult.

Pale-yellow Robin

Tregellasia capito

Endemic

120–135mm

Sexes alike. Inconspicuous small robin with olive-grey upperparts. Chin, throat, lores: White. Underparts: Pale yellow, darker on sides of breast. Tail: Grey-brown. Legs: Pale orange-yellow. Race *nana*, northeastern Qld. Smaller. Lores, eye ring: Rufous-buff. Immature: Similar, browner and whitish streaked.

Quiet, unobtrusive, preferring gloomy dense vegetation. Relatively tame.

Habitat: Coastal and upland rainforests and moist eucalypt forests with dense understorey, vine thickets.

Feeding: Hangs sideways, perches, pounces mostly on insects from the ground to the mid-canopy. Sometimes seeds.

Voice: Penetrating whistle, repeated squeaks. One of the first calls to be heard at dawn.

Status: Moderately common, sedentary. Usually seen in pairs.

Breeding: July–Jan. Cup-shaped nest in a fork of a tree or often in Lawyer Vine / 2 eggs.

Left: Race *nana*. Right: Nominate race at nest in Lawyer Vine.

White-faced Robin

Tregellasia leucops

120–130mm

Sexes alike. Distinctive, 'owl-like' white face. Head: Olive-black. Eyes: Black. Chin: White. Upperparts: Deep olive. Underparts: Yellow with olive washed flanks. Bill: Black with white base.
Immature: Similar, duller.

Relatively tame. Usually in pairs or loose feeding flocks. Quiet and unobtrusive.

Habitat: Tropical rainforests or moist lowland forests.

Feeding: Perches and pounces. Sits motionless, then darts to ground for insects. Probes bark for insects.

Voice: Rising, whistled 5-note trill. Harsh 'chee-chee' call.

Status: Moderately common.

Breeding: July–Jan. Cup-shaped nest, usually in Lawyer Vine / 2 eggs.

Iron Range NP, Qld.

Left and right: Adult.

Eastern Yellow Robin

Eopsaltria australis

Endemic

145–155mm

Sexes similar. Bright citrus yellow underparts. Mid-grey head and upperparts with an olive-yellow rump conspicuous in flight. Chin: White. Eyebrow: Faint, pale. Bill: Blackish. Wings, tail: Dark brown-grey. Legs: Brownish. Race *chrysorrhoa*, northern NSW, Qld. Bright yellow rump. Female: Smaller. Immature: Rufous-brown and white streaked.

Inquisitive and relatively tame. Usually in pairs or small family groups. Flits from perch to perch. Often clings to the side of a tree trunk. Flicks tail and wings.

Habitat: Coastal mountain forests, woodlands with undergrowth, banksia heaths, rainforests, mallee, mulga, parks and gardens.

Feeding: Perches and pounces. Insects, spiders, invertebrates, usually taken from the ground.

Voice: Clear piping whistles. One of the first birds heard in the morning and last in the evening.

Status: Common, sedentary. Winter migration from high altitude to lower altitudes.

Breeding: July–Dec. Camouflaged, cup nest in tree fork 1–7m high / 2 eggs.

Opposite: Nominate race.

Western Yellow Robin

Eopsaltria griseogularis

Endemic

145–155mm

Sexes alike. Behaviour and appearance similar to the Eastern Yellow Robin. Grey head with white throat, grey breast and bright yellow underparts. Upperparts: Grey with darker grey wings with concealed white wing bars and white-tipped tail. Eyes, lores: Dark brown-black. Nominate race found southwestern corner WA.
Race *rosinae*, elsewhere in WA and Eyre Peninsula, SA. Darker with greyer breast and olive-yellow rump.
Immature: Dark brown with pale streaks.

Habitat: Diverse habitat. Dense forest canopies, open woodlands and mallee with shrubby undergrowth; areas with leaf litter and fallen timber.

Feeding: Perches and pounces. Insects, spiders, invertebrates, usually taken from the ground.

Voice: Repeated, slow piping whistle, trilled whistles at dawn.

Status: Common to uncommon.

Breeding: July–Dec. Camouflaged cup-shaped nest 1–8m high / 2 eggs.

Left and right: Nominate race.

White-breasted Robin

Eopsaltria georgiana

Endemic

145–155mm

Sexes alike. Blue-grey head with black lores and pale eyebrow. Underparts: White with grey washed breast. Upperparts: Blue-grey with dark grey wings with concealed pale bar seen in flight. Tail: Dark grey, outer feathers tipped white. Eyes: Dark brown. Bill: Black.
Immature: Browner, mottled.

Pairs or small groups occupy a territory.

Habitat: Shady areas. Moist vegetation, along watercourses, gullies, tall coastal heaths and wet woodlands with substantial leaf litter.

Feeding: Pounces and perches, taking insects, ants and invertebrates from the ground.

Voice: Harsh 'chit-chit' alarm. Liquid chirruping song.

Status: Uncommon. Sedentary, some seasonal movement.

Breeding: July–Dec. Co-operative. 'Helper' birds assist in rearing the young. Loose, cup nest in cover, 1–6m high / 2 eggs.

Two Peoples Bay NR, WA.

Left and right: Adult.

Mangrove Robin

Peneonanthe pulverulenta

124–160mm

Sexes alike. 3 similar races occur in Australia. Slate-backed robin with an unusually long bill. Race *leucura*, northern Qld. Largest. Upperparts: Slate grey with dusky crown. Lores, eye stripe: Black. Bill: Long, straight, black. Rump: Black. Tail: Black with outer feathers partly white. Chin, throat, underparts: White with blue-grey washed breast. Wings: Pale bar in flight.
Race *alligator*, NT. Darker. Breast: Greyer.
Race *cinereiceps*, northwest WA. Upperparts: Ash grey. Underparts: Whitish.
Immature: Streaked brown above, mottled grey below.

Relatively tame. Pairs or small groups in a small territory. Usually in cover.

Habitat: Low coastal mangrove thickets along tidal creeks, estuaries.

Feeding: Sits motionless, picks insects, crustaceans from mud, among mangrove roots or foliage.

Voice: Soft, plaintive 2-note whistle. Lively chattering.

Status: Moderately common. Sedentary.

Breeding: Apr.–Sept. Shallow, camouflaged nest, often in mangrove fork or Lawyer Vine / 2–3 eggs.

Daintree River, Qld. Charles Darwin NP, NT.

Left and right: Race *leucura*. Adult.

Grey-headed Robin

Heteromyias cinereifrons

Endemic

160–180mm

Sexes alike. Dark grey crown, nape and lores with eye streak and white eyebrow, white under-eye line joining a white throat and chin. Bill: Black with pale tip. Eyes: Brown-black. Upperparts: Tortoiseshell colouring with rufous rump and blackish wings with uneven white bars. Underparts: Pale, buff-rufous with pale grey wash to breast. Legs: Long, pinkish.
Immature: Upperparts: Patchy brown. Underparts, head: Mottled orange-brown.

Solitary, pairs or family groups. Hops briskly with an upright posture. Sits motionless or clings sideways on tree trunk.

Habitat: Lower elevation of mountain rainforests and coastal highlands.

Feeding: Hops through low cover or perches and pounces from low branch. Takes insects, spiders, snails from the ground.

Voice: Soft, short piping whistle. Loud, high piping followed by 2–3 lower notes, repeated.

Status: Moderately common. Sedentary.

Breeding: Aug.–Jan. Pairs build a shallow cup nest of moss and tendrils in a vine tangle or tree, 2–3m high / 1–2 eggs.

Mt Hypipamee NP, Qld.

Left and right: Adult.

White-browed Robin

Poecilodryas superciliosa

Endemic

140–180mm

Sexes alike. Distinctive face pattern with brown eye stripe between long, white eyebrow and malar line. Eyes: Dark, underlined with thin white crescent. Crown, upperparts: Olive-brown. Wings: Dark, with 2 white bars and white tips. Underparts: Pale grey with white belly. Bill: Long, heavy. Immature: Similar, duller.

Tail is often cocked, wren-like. Usually in pairs.

Habitat: Rainforests, vine scrub, coastal woodlands, mangroves and watercourse vegetation.

Feeding: Sallies from a low perch or hops on the ground for insects, grasshoppers, caterpillars, crustaceans.

Voice: Clear, piping whistle, repeated 3–4 times.

Status: Moderately common. Sedentary.

Breeding: July–Mar. Fragile camouflaged nest 1–3m high / 2 eggs.

Big Mitchell Creek, north of Mareeba, Qld.

Left and right: Adult.

Buff-sided Robin

Peocilodryas cerviniventris

150–170mm

Sexes alike. Similar to White-browed Robin but larger. Head, upperparts: Rich dark brown. Lores: Black. Eyebrow: Wide, long, white. Malar line: White, short. Wings: 2 broad white bars and white tips. Underparts: Pale grey. Flanks, undertail coverts: Tawny rufous. Tail: White tipped. Immature: Similar, darker.

Tail is often cocked. Flicks wings when perched. Inquisitive, unobtrusive.

Habitat: Damp forests along watercourses. Monsoon rainforests, vine scrub and pandanus.

Feeding: From a low perch, scans the ground for prey or sallies for insects mid-air.

Voice: 3 clear sweet notes, repeated. Chattering.

Status: Locally common to uncommon. Sedentary.

Breeding: Aug.–Oct. Cup-shaped, loose, camouflaged nest 1–3m high / 1–3 eggs.

Gunlom Falls, Kakadu NP, NT.

Opposite: Adult.

Northern Scrub-robin

Drymodes superciliaris

210–220mm

Sexes alike. Thrush-like bird with a distinctive, long, black vertical eye mark, long legs and long tail tipped white. Face, chin, throat: Whitish. Crown, upperparts: Rich cinnamon. Wings: Black with 2 white bars. Rump: Bright cinnamon. Underparts: Cream-buff. Flanks: Tawny buff. Immature: Similar, duller, dark flecks to throat and breast.

Solitary or in pairs. Curious, relatively tame and well camouflaged on the forest floor. Runs briskly, hops, tail bobbing and fanning.

Habitat: Vine scrubs, rainforests with deep leaf litter.

Feeding: Forages on the ground, probing and picking through leaf litter for insects, invertebrates.

Voice: Lisping, drawn-out rising whistle.

Status: Uncommon, sedentary.

Breeding: Oct.–Jan. Depression in ground, leaf-lined with a perimeter of twigs / 2 eggs.

Iron Range NP, Qld.

Left and right: Adults.

Southern Scrub-robin

Drymodes brunneopygia

Endemic

210–230mm

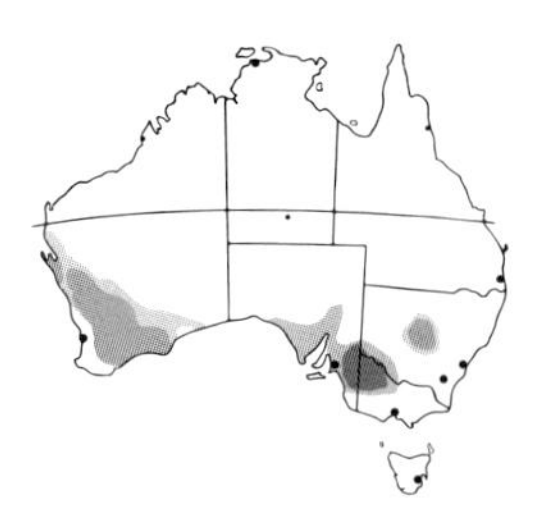

Sexes similar. Dark grey-brown upperparts and pale grey face with short, vertical, dusky mark through eye. Throat, breast, underparts: Grey merging to white belly. Long legs. Tail: Long with outer feathers tipped white. Eyes: Dark brown. Bill: Black. Female: Paler. Immature: Light brown. Frons speckled white. Belly: Mottled white.

Inquisitive and relatively tame. Often fluffs out breast and belly plumage. Tail is usually cocked when on the ground. Rarely flies.

Habitat: Areas with ample leaf litter. Dense mallee, mulga, dry scrub, heaths, tea-tree thickets.

Feeding: Flicks leaf litter with bill for insects, ants and invertebrates. Snaps flying insects.

Voice: Long, drawn-out whistle contact call.

Status: Uncommon, sedentary. Once common. Habitat lost to wheat cropping.

Breeding: July–Jan. Nest on ground, small depression in debris lined with twigs, rootlets. Often sheltered near shrub or log.

Round Hill NR, NSW. Hattah–Kulkyne NP, Vic. Lincoln NP, Coffin Bay NP and Gluepot NR, SA. Monkey Mia, WA.

Left: Immature. Right: Adult.

Chowchilla

Orthonyx spaldingii

Endemic

270–300mm

Black head with stark blue–white eye ring. Upperparts: Dusky olive-brown. Tail: Olive-brown, spine tipped, strong and used as a prop. Throat, breast, belly: White, remaining underparts olive-brown.
Race *melasmenus*, northern range. Smaller.
Female: Similar, with a brighter orange chin, throat and upper breast. Immature: Duller, browner, mottled cinnamon.

Runs, hops and rarely flies. Territorial. Communal, in family groups.

Habitat: Forest floor in wet tropical rainforests.

Feeding: Leaves circular patches of bare earth after tossing leaves aside with 1 foot, often propped back on its tail, searching for insects, invertebrates, some seeds.

Voice: Loud singing at dusk and dawn. Resonant, repeated 'chow-chowchilla'. Cacophony of notes and mimicry.

Status: Moderately common. Sedentary.

Breeding: May–Nov. Loose, bulky domed nest well hidden near ground / 1 egg.

Lake Barrine, Crater Lakes NP, Mt Hypipamee NP and Mary Cairncross SR, Qld.

Left: Female carrying nesting material. Right: Male.

Australian Logrunner

Orthonyx temminckii

Endemic

180–200mm

Rufous-brown crown. Mid-grey face patch from bill to ear coverts and down sides of neck. Upperparts: Rufous, broadly streaked black-brown, plain rufous rump. Wings: Short, rounded, dusky grey with 2 black bars. Throat, breast: White with black sides. Underparts: White belly, greyish flanks and rufous vent and undertail coverts. Tail: Broad, brown, tipped with short, stiff 'spines'.
Female: Similar, more colourful. Throat, upperbreast: Orange. Immature: Duller, mottled-brown.

Sedentary, communal in small groups. Prefers to hop or run, wings dropped, tail spread. Often propped on tail, it leaves a distinctive circular pattern behind. Infrequent whirring bursts of short flight. Strong feet.

Habitat: Tropical and temperate rainforests.

Feeding: Noisily scratching and tossing leaf litter for insects, snails and other invertebrates from the forest floor.

Voice: Piercing 'qweek-qweek-qweek'.

Status: Uncommon in the south, more common in the north.

Breeding: Apr.–Aug. Domed nest on ground, against rock or tree / 2 eggs.

Lamington NP, Qld.

Left: Female. Right: Male

Quail-thrushes

Whipbirds

Wedgebills

Sittella

Whistlers

Shrike-tit

Shrike-thrushes

Crested Bellbird

A group of three families of medium-sized insectivorous birds with similar foraging and behaviour habits and similar appearance.

Family Cinclosomatidae includes the quail-thrushes, whipbirds and wedgebills.

Quail-thrushes are wary, ground-dwelling birds with beautiful cryptic plumage, named for their quail-like whirring wings and short low flight after bursting from cover when flushed. They are endemic to Australia and are not related to quails or thrushes of the Old World.
The colour of their upperparts varies throughout their range, generally matching the local soil colour found in that habitat.
The Eastern Whipbird, well known for its explosive 'whipcrack' song, is a bird of the damp scrub and rainforests of coastal eastern Australia, while its western counterpart, the Western Whipbird, has adapted to drier coastal thickets and mallee habitats of southwestern Western Australia. Rather than a whipcrack song, the western bird has a rising tinkling song. Land clearing and constant burning of their habitat have decimated populations of Western Whipbirds.
Wedgebills are named for their finch-like short, tapered bills. The two species of wedgebill look almost identical but are separated by their different song and behaviour. Young birds may form large nomadic flocks of up to 100 birds.

Family Neosittidae contains the sittella.
Sittellas are small birds of the upper canopy that forage downwards on trunks and branches for insects. (Treecreepers, on the other hand, spiral up trunks probing for insects.) Communal birds, they are usually found in groups of five to nine. They nest co-operatively, building a beautifully camouflaged nest, and also communally help in feeding the young.

Family Pachycephalidae includes the whistlers, shrike-thrushes, shrike-tit and the Crested Bellbird.
Pachycephalidae means 'thick head' and refers to the generally large, rounded heads and thick necks of this family. It also includes Australia's most accomplished songbirds.
Whistlers are named for their varied, high-pitched, melodious and far-reaching song, sung in a rapid succession of up to 35 notes without a pause.

Shrike-thrushes are noted for their clear melodious song, often in male and female duets. They are not related to the Old World thrushes, after which early settlers named them, but are endemic to Australia, with some species also present on nearby islands.
Shrike-tits are accomplished songbirds and have some resemblance to the true 'tits' of the Old World, but are not related. 'Shrike' refers to their hooked bill, which was thought to resemble the bill of the Old-World shrikes. Shrike-tits have powerful bills to probe for insects and tear loose bark away from tree trunks in search of prey.

The Crested Bellbird is named for the male's ability to raise his crown feathers to form a crest when singing to signal his territory.

Opposite: Male Varied Sittella, race *pilata*, at its beautifully constructed and well-camouflaged nest.

Spotted Quail-thrush

Cinclosoma punctatum

Endemic

260–280mm

3 isolated populations. Nominate race, Qld, NSW, Vic. Largest and plumpest. Face pattern with grey-brown frons, a long white eyebrow and a black patch before and around the eye. Crown, back: Brown-grey, streaked black. Chin, throat: Black with white patches edged black on either side. Neck, breast: Blue-grey. Wings: Black, white-spotted coverts, thin white wing bar, rufous inner feathers. Underparts: Buff-white. Flanks: Streaked black. Eyes: Blue-grey. Race *dovei*, Tas. Female: Black breast band.
Females, immature: Duller. Buff eyebrow, buff-orange neck. Solitary. Short, undulating flight, whirring wings, tail partly fanned displaying white tips.

Habitat: Dry or sometimes wet eucalypt forests with clear understorey. Leaf-littered rocky ridges with short grass clumps.

Feeding: Seeds, insects.

Voice: Repeated, drawn-out, penetrating, musical double whistle from a prominent perch.

Status: Uncommon to rare. Race *anachoreta*, SA. Critically endangered, possibly extinct.

Breeding: July–Feb. Depression under cover / 2–3 eggs.

Dharug NP, NSW.

Left: Female nominate race. Right: Male.

Chestnut Quail-thrush

Cinclosoma castanotum

Endemic

230–260mm

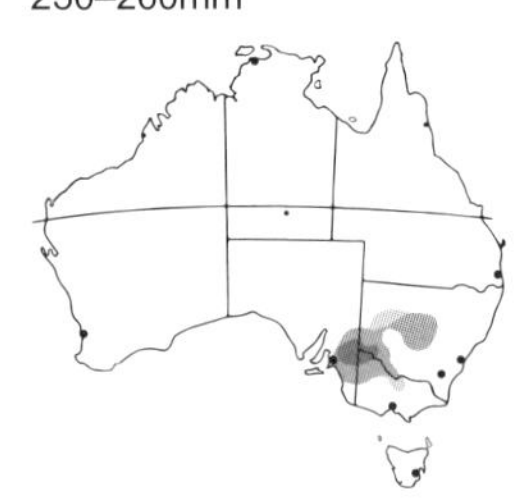

Distinctive deep chestnut lower back and rump, and white eyebrow. Grey-brown head and upper back. Face, throat: Black with broad white streaks on side of throat. Tail: Grey-brown, white tips when fanned. Wings: Chestnut shoulder, dark-grey feathers broadly edged grey and buff. Breast: Black centre, blue-grey sides and flanks. Underparts: White. Undertail coverts: Spotted black.
Female: Smaller, duller. Blue-grey upper breast, brownish throat, buff eyebrow and throat streaks. Upperparts: Chestnut and light brown. Immature: Scalloped breast.

Family groups or pairs. Wary. Flies if startled, wings whirring.

Habitat: Mulga scrub, mallee, semi-arid woodlands, open heaths, spinifex on sandy soils.

Feeding: Insects, seeds from the ground.

Voice: Strong, piping, monotone whistles, repeated rapidly.

Status: Moderately common. Sedentary in south, partially nomadic in desert areas.

Breeding: July–Sept. or Aug.–Dec in desert areas or following rains. Lined depression / 2 eggs.

Left: Female. Right: Male.

Copperback Quail-thrush

Cinclosoma clarum

Endemic

210–260mm

Split from similar Chestnut Quail-thrush (in 2015); more chestnut backed. Male has grey-brown upper back and chestnut shoulder to lower back, forming a wide band to upperparts. White-tipped tail. Face, throat and upper breast are black. White eyebrow and broad white streaks on side of throat.
Female: Paler with grey upper breast, brown face. Immature: Similar to female.
2 races: Nominate, north SA, NT, WA, and race *fordianum*, southwest WA, SA.

Shy, elusive and wary.

Habitat: Dry woodlands, mallee, arid or semi-arid heathlands.

Feeding: Forages on ground, picking up seeds, grasshoppers, beetles, caterpillars, insects, spiders.

Voice: High-pitched, thin, monotone whistles.

Status: Moderately common. Locally nomadic in arid regions. Resident in some regions.

Breeding: July–Sept. or after rain. Depression lined with grass, bark and twigs / 2–3 eggs.

Left: Female. Right: Male.

Cinnamon Quail-thrush

Cinclosoma cinnamomeum

Endemic

280–210mm

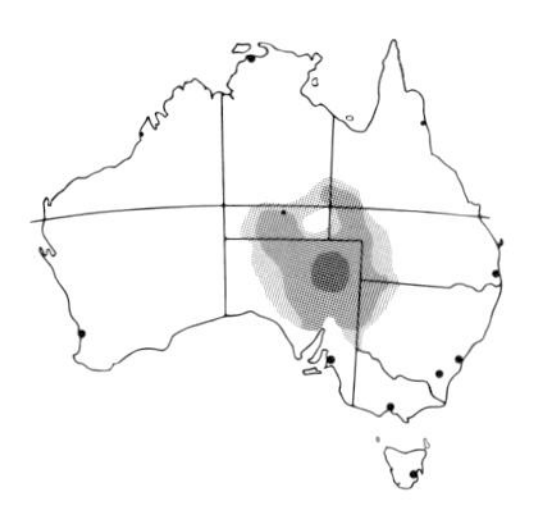

Upperparts are overall chestnut except for a black shoulder patch edged white. Male has a distinctive long buff eyebrow and a black face, with a broad white stripe on the sides of its black throat. Breast: White with cinnamon sides and broad black band. Underparts: White, cinnamon flanks. Back, rump: Pale chestnut. Wings: Dusky, edged rufous, spotted white shoulder. Tail: Pale chestnut centre, outer feathers black, tipped white. Eyes: Red-brown. Female: Paler. Underparts: White. Breast: Grey, lacks black.

Meanders slowly, flicking tail. If flushed, squats under bush or bursting flight on whirring wings. Elusive, shy and wary.

Habitat: Saltbush, bluebush on stony gibber plains and tablelands of Lake Eyre Basin.

Feeding: Ground foraging for seeds and insects, often digging or turning stones with bill.

Voice: High-pitched, thin, monotone whistles.

Status: Rare.

Breeding: July–Sept. or after rains. Lined depression on ground under bush / 2–3 eggs.

Erldunda area, NT. Nullarbor Roadhouse and Mt Lyndhurst, SA.

Left: Female. Right: Male.

Nullarbor Quail-thrush

Cinclosoma alisteri

Endemic

190–210mm

Similar to the Cinnamon Quail-thrush but smaller with richer, brighter colour and distinctive simple black throat and chest. Upperparts: Overall sandy brown with black shoulder patch spotted white. Face: Thin buff eyebrow and broad white stripe on sides of throat. Lower breast, belly: White with a broad black band separating cinnamon flanks. Female: Similar but duller. Black shoulder patch has buff edges. Underparts: White. Breast: Grey, lacks black.

Smallest and shyest quail-thrush. Only endemic bird of the Nullarbor Plain. Elusive, shy and wary. Recently split from Cinnamon Quail-thrush.

Habitat: Saltbush, bluebush sedge lands of Nullarbor Plain.

Feeding: Ground foraging for seeds and insects, often digging or turning stones with bill.

Voice: High-pitched, thin, monotone whistles.

Status: Rare.

Breeding: July–Sept. or after rains. Lined depression on ground under bush / 2–3 eggs.

Opposite: Male.

Chestnut-breasted Quail-thrush

Cinclosoma castaneothorax

Endemic

215–240mm

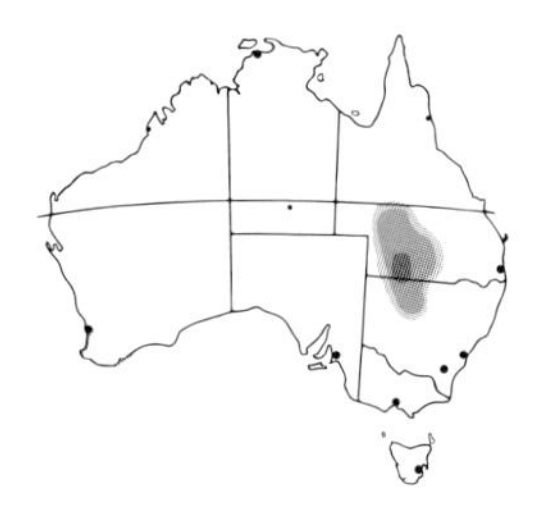

Similar to the Cinnamon Quail-thrush but richer colours. Russet brown head with long white eyebrow and white whisker line. Throat: Black. Breast: Rich yellow-chestnut edged with a black band. Underparts: White with chestnut flanks. Undertail coverts: Mottled brown. Upperparts: Dark rufous. Wings: Black, broadly tipped white. Tail: Centre feathers rufous, outer feathers black, white tipped. Eyes: Red-brown. Female: Paler, lacks black markings. Throat: Buff. Breast: Grey-brown. Immature: Duller. Flecked brown breast.

Shy, squats under cover, runs quickly to cover or bursts 'quail-like' into flight. Solitary or pairs.

Habitat: Open acacia or mallee woodlands on stony ground.

Feeding: Ground foraging, seeds, insects, spiders.

Voice: High-pitched, thin, monotone whistles.

Status: Moderately common. Vulnerable in NSW. Locally nomadic.

Breeding: July–Sept. or after rain. Lined, depression nest / 2–3 eggs.

Bowra, Eulo Bore, Qld.

Left: Male. Right: Female.

Western Quail-thrush

Cinclosoma marginatum

Endemic

230–260mm

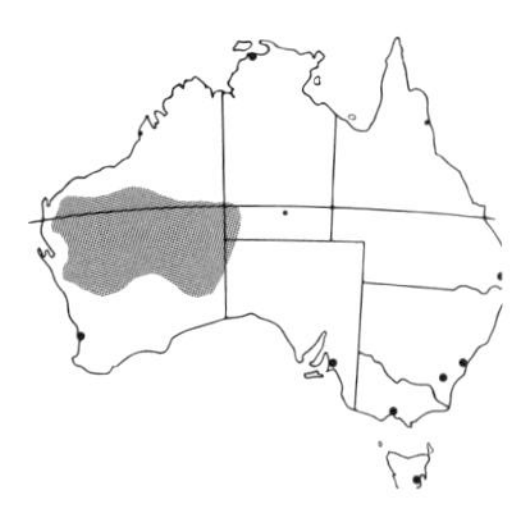

Also called the Mulga Quail-thrush. Similar to the Chestnut-breasted Quail-thrush but with richer brighter colours. Upperparts are overall chestnut except for a black shoulder patch edged white. Long white eyebrow and white whisker line. Throat: Black. Breast: Rich yellow-chestnut edged with cream and a black lower breast band. Underparts: White with chestnut flanks.
Female: Duller than male. Grey breast with rufous wash, pale buff eyebrow and throat.
Immature: Duller. Flecked brown breast.

Pairs or family groups. Shy, runs quickly to cover or bursts 'quail-like' into flight. Solitary or pairs.

Habitat: Arid inland WA. Stony hills with scrubby cover. Open acacia ridges or mallee woodland, spinifex grassland.

Feeding: Ground foraging, walks slowly picking up seeds, insects, spiders.

Voice: High-pitched, thin, monotone whistles.

Status: Moderately common. Locally nomadic.

Breeding: Variable after rain. Lined, depression in ground nest / 2 eggs.

Left: Female. Right: Male.

Eastern Whipbird

Psophodes olivaceus

Endemic

250–300mm

Sexes alike. Mostly olive-green plumage with a black head, black crest and bold white streak down cheek and side of neck. Eyes: Red-brown. Upperparts: Olive green. Tail: Black, broad with white outer tips. Underparts: Olive-black, mottled white centre.
Race *lateralis*, far north Qld. Smaller, olive-brown.
Immature: Similar, duller. Lacks white cheek stripe.

Solitary. Territorial. Weak, short flight. Secretive, moving quietly or dashing through undergrowth.

Habitat: Thickets, gullies in rainforests, heaths, wet sclerophyll forests.

Feeding: Hop-searches through undergrowth for insects, larvae.

Voice: Loud, long-ringing whipcrack call by male, often followed by an instant sharp 'tchew-tchew' reply from female.

Status: Common to moderately common.

Breeding: July–Jan. Shallow cup nest near ground to 4m high / 2 eggs.

Left and right: Adult.

Western Whipbird

Psophodes nigrogularis

Endemic

220–250mm

Sexes alike. Overall olive green with small crest not always erect. Chin, throat: Black, with bold, white streaks. Tail: Outer feathers tipped white.
There are 4 similar but isolated races.
Nominate race, southwestern WA.
Race *oberon*, far southwestern WA. Largest with longest tail.
Race *leucogaster*, Mallee Whipbird, SA, NSW. Pale belly stripe and white streak on throat partially edged black.
Race *lashmari*, Kangaroo Is.
Immature: Olive-brown with rufous throat and neck.

Secretive but curious. Lives in pairs.

Habitat: Coastal sandhills, dense heaths, mallee scrub.

Feeding: Digs, scratches, pokes in debris for insects, small invertebrates, seeds.

Voice: Rising, tinkling, bell-like notes and chattering alarm, without whipcrack call.

Status: Rare to uncommon. Territorial. Sedentary. Vulnerable SA, Vic. Endangered WA.

Breeding: July–Oct. Bulky, domed nest with side entrance, close to ground in a dense shrub / 2 eggs.

Two Peoples Bay NR and Waychinicup NP, WA.

Left: Nominate race. Right: Race *leucogaster*.

Chirruping Wedgebill

Psophodes cristatus

Endemic

190–200mm

Sexes similar. Almost identical to the Chiming Wedgebill but the 2 species differ in behaviour and song and are best identified by location. Mottled sandy brown head and upperparts with a dusky forward-curving crest and black wedge-shaped bill. Wings, tail: Dusky brown edged white. Underparts: Pale grey, indistinct streaks to breast, darker grey flanks. Eyes: Dark brown. Female: Smaller. Immature: Paler, flight feathers edged cinnamon. Bill: Orange-brown.

In colonies of 20 or more. Flat low glides from bush to bush, tail partly spread.

Habitat: Semi-desert areas, open country with sparse acacia, Emu-bush and saltbush.

Feeding: Runs, hops into low shrubs, takes insects, spiders and seeds.

Voice: Male: Sparrow-like 'chirrup' from exposed perch. Female: Descending trill reply. Sings all year, persistently at dawn.

Status: Uncommon. Sedentary. Rare in south of range.

Breeding: Mar.–May., Aug.–Nov. after rains. Shallow loose cup nest close to ground / 2–3 eggs.

Strzelecki Track, SA.

Left: Adult. Right: Immature.

Chiming Wedgebill

Psophodes occidentalis

Endemic

200–220mm

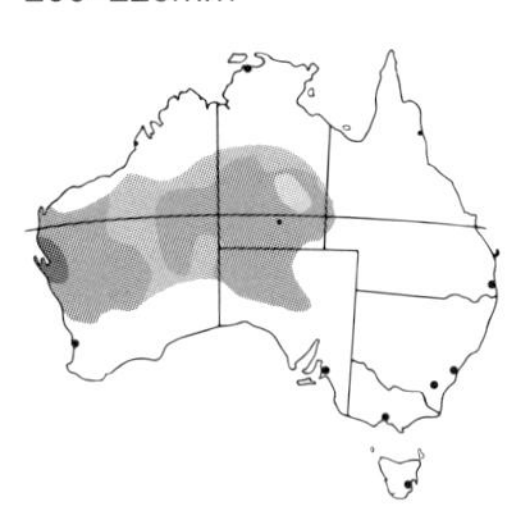

Sexes alike. Similar to and replaces the Chirruping Wedgebill in arid central and western Australia. Glides with tail fanned, displaying white tips. Distinctly longer tail, plainer breast and stubby bill. Immature: Pale, pinkish-brown bill and cinnamon edged flight feathers.

Extremely wary, running quickly to cover if disturbed.

Habitat: Stony ground, thickets of acacia, melaleuca or dune vegetation, Broombush, mallee, spinifex.

Feeding: In low vegetation or on the ground. Insects, seeds.

Voice: Descending chime of 4 notes, repeated and carried far.

Status: Common in west of range. Uncommon inland.

Breeding: Mar.–May, Aug.–Nov. or after rains. Shallow loose cup close to ground in tree or shrub / 2–3 eggs.

Erldunda area, NT. Peron Peninsula and Cape Range NP, WA.

Opposite: Adult.

Varied Sittella

Daphoenositta chrysoptera

100–140mm

Sexes similar. 5 variable races that integrate across range. Nominate race, Orange-winged Sittella, inland Qld and NSW, Vic., southeastern SA. Small, short-tailed bird with a greyish black crown, white throat and bright orange wing bars. Upperparts: Streaked dusky brown. Underparts: Whitish streaked. Bill: Creamish yellow, dark tipped. Eye ring, eyes: Orange. Undertail: Pale with dark barring.
Female has darker head and shorter bill. Immature: Paler.

Foraging at all levels, in pairs or groups of up to 12. Females show a preference for outer foliage; males prefer the tree trunk and often feed lower in the canopy. Undulating flight.

Active and often hard to locate, except for their constant sharp, twittering calls.

Habitat: Woodlands, open forests. Avoids rainforests.

Feeding: Arboreal. Prefers rough-barked eucalypts, zigzagging headfirst down the trunk and along branches, picking, prising, insects and spiders.

Voice: Whistled 'chip' repeated constantly. Call and double-whistle twittery flight song.

Status: Moderately common.

Breeding: June–Apr. Communal group builds a deep nest, well camouflaged with long strips of bark, placed in tree fork / 3 eggs.

Opposite: Nominate race, Orange-winged Sittella, male (left) and female (right).

Race *striata*, Striated Sittella, far north Qld. Body is boldly streaked. Wing band: White. Male: Black crown, striated face. Female: Black helmet. Immature: Plain breast.

Opposite: Race *striata,* Striated Sittella, male (left) and female (right).

Race *pileata*, Black-capped Sittella, southwest Qld, southern NT and WA, SA. Upperparts: Grey-brown. Orange wing band. Underparts: Plain white. Male: Small black cap.
Female: Glossy black crown and face. Immature: Greyish crown.

Opposite: Race *pileata*, Black-capped Sittella, male (left) and female (right).

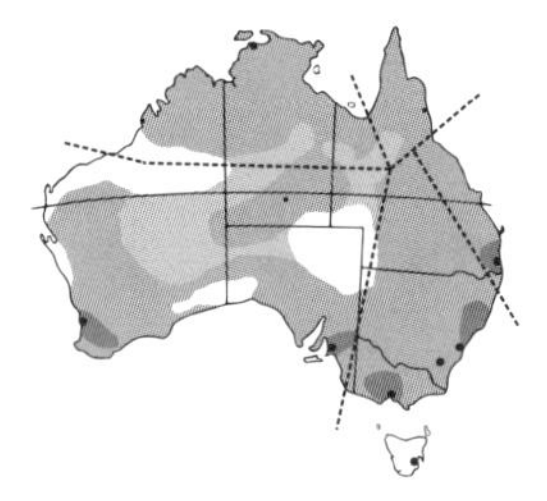

Race *leucoptera*, White-winged Sittella, Top End, Qld, NT, WA. Similar to race *pileata*, but smaller, upperparts darker. Underparts: White. Wing bar: White. Male: Small black cap. Female: Glossy black crown and face.
Immature: Greyish white.

Opposite: Race *leucoptera*, White-winged Sittella male (left) and female (right).

Race *leucocephala*, White-headed Sittella, eastern Qld, far northeast NSW. Both sexes have white head and neck. Upperparts: Streaked dark grey. Wing band: Orange. Immature: Greyish head.

Opposite: Race *leuocephala*, White-headed Sittella, adult.

Red-lored Whistler

Pachycephala rufogularis

Endemic

200–210mm

Sexes similar. Similar to Gilbert's Whistler but larger. Brownish grey upperparts with orange lores and throat. Head: Brownish grey. Wings: Darker grey-brown, edged olive-grey. Breast: Grey. Underparts: Buff-orange. Eyes: Red.
Female: Paler. Immature: Similar to female. Face: Whitish. Eyes: Brown.

Shy and elusive, living in dense cover and difficult to see. Flies infrequently in low undulations.

Habitat: Mallee scrub with dense undergrowth. Native pine and banksia heaths with dense undergrowth.

Feeding: Usually foraging on the ground. Insects, seeds, berries.

Voice: Rich, clear whistle and reedy notes.

Status: Vulnerable. Critically endangered in NSW. Sedentary. Territorial. Solitary.

Breeding: Sept.–Dec. Large cup nest neatly woven around the rim, placed in spinifex or low shrubs / 2–3 eggs.

Gluepot NR, SA. Murray–Sunset NP, Vic.

Left: Male. Right: Female.

Olive Whistler

Pachycephala olivacea

Endemic

200–210mm

Sexes similar. Mid-sized, stocky bird with a dark grey head and white throat, scalloped dark grey. Underparts: Mid-grey breast band, buff-brown below. Back: Dark olive-brown. Wings, tail: Feathers edged lemon. Eyes: Brown. Bill: Black.
Race *macphersoniana*, northern NSW. Lighter crown, less mottled throat.
Female: All are duller with plain whitish throat. Immature: Wings washed rufous, minimal breast band.

Habitat: Northern range: Temperate rainforests of Antarctic Beech above 500m. Southern range: Damp eucalypt forests, heathlands and tea-tree scrub, parks and gardens.

Feeding: Hopping, picking up insects, occasionally in low vegetation.

Voice: Plaintive, sharp, deep whistles. Southern: Whipcrack. Northern: Rising and falling notes.

Status: Uncommon. Sedentary. Overwinters in lower altitudes.

Breeding: Sept.–Jan. Cup nest, low in dense shrub / 2–3 eggs.

Mt Wellington, Tas. Barrington Tops, NSW.

Left: Male. Right: Female.

Gilbert's Whistler

Pachycephala inornata

Endemic

190–200mm

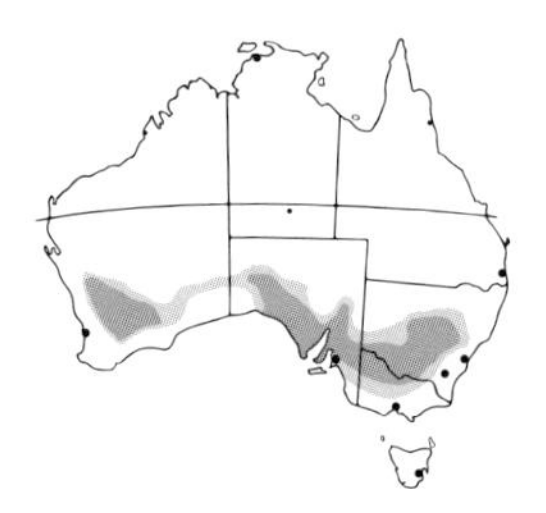

Similar to Red-lored Whistler. Grey upperparts. Black lores. Buff-rufous throat patch. Underparts: Pale grey with rufous wash on belly and under tail. Bill: Black. Eyes: Ruby. Western birds are darker with longer tail.
Female: Pale grey face, throat and pale eye ring. Bill: Grey-brown. Immature: Similar to female, brown eyes.

Usually in pairs. Strong, undulating flight. Unobtrusive.

Habitat: Mallee, mixed thickets mostly with paperbark, and tall, dry scrub.

Feeding: Probes leaf litter or forages low to ground for insects and berries.

Voice: Melodious, deep, whistled sequence of loud far-reaching calls that start low and build to a crescendo.

Status: Moderately common. Locally nomadic.

Breeding: Sept.–Jan. Cup-shaped nest in fork of a low shrub or vines / 2–3 eggs.

Hattah–Kulkyne NP, Vic. Round Hill NR, Back Yamma NR and Gulpa Is. NP, Mathoura, NSW.

Left: Male. Right: Female.

Australian Golden Whistler

Pachycephala pectoralis

160–175mm

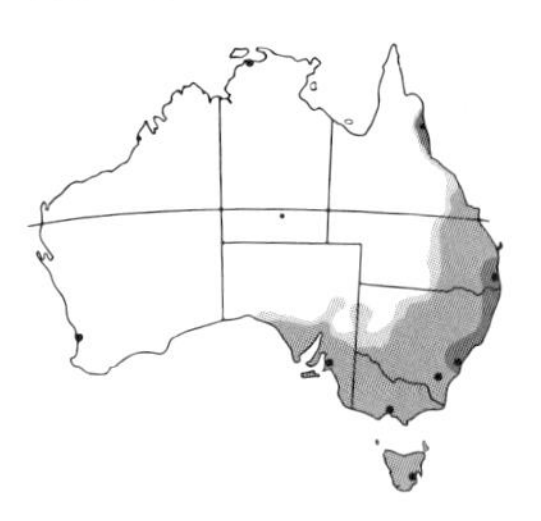

Several races with minor or clinal differences. Black head and breast band contrasting with bright yellow nape and underparts. Chin, throat: White. Upperparts: Olive green. Wings: Edged dark olive. Rump: Dark olive. Eyes: Red-brown. Bill: Black. Tail: Grey, black tipped. Female: Head, upperparts: Mid-grey. Underparts: Whitish grey. Juvenile: Rufous red. Immature: Similar to female, rufous-edged flight feathers, pale bill base.

Solitary or pairs or in mixed flocks after breeding. Undulating flight.

Habitat: Rainforests, dense scrub, eucalypt forests. May move to open forests in winter.

Feeding: In foliage for insects and berries.

Voice: Melodic, clear whistles, often with a whipcrack ending.

Status: Common.

Breeding: Aug.–Jan. Cup nest of bark strips, twigs, grass and spider web, placed low in a shrub or tree fork, 1–4m high / 2–3 eggs.

Left: Female. Centre: Immature. Right: Male

Western Whistler

Pachycephala occidentalis

Endemic

160–175mm

Similar to Golden Whistler, but the male has a paler grey uppertail and dark, narrower terminal band.
Female: Mid-grey upperparts from crown to uppertail. Underparts: Darker buff lower belly and vent than the female Golden Whistler. Immature: Similar to female.

(Previously considered a subspecies of the Golden Whistler, 2015 DNA tests suggested a closer relationship to the Mangrove Golden Whistler and it was made a separate species in 2017.)

Solitary or in pairs after breeding. Undulating flight.

Habitat: From south of Shark Bay, WA, to just east of SA border. Usually in wetter forested areas with a thicker understorey or coastal vegetation, especially with peppermint trees or remnant woodland in wheat belt. Isolated population on Rottnest Island.

Feeding: Usually in foliage, taking insects and berries.

Voice: Melodic, clear whistles, often with a whipcrack ending.

Status: Common to moderately common.

Breeding: Aug.–Jan. Similar to Golden Whistler, 1–4m high / 2–3 eggs.

Left: Female. Right: Male.

Mangrove Golden Whistler

Pachycephala melanura

150–160mm

Generally similar to the Golden Whistler but brighter and smaller, with a broader yellow nape band. Head, breast band: Black. Nape, underparts: Bright yellow. Throat: White. Back, rump: Golden olive. Wings: Blackish, edged grey. Tail: Black. Eyes: Dark red. Female: Head: Grey. Back: Olive-brown. Underparts: Faintly streaked greyish white and yellowish undertail coverts. Bill: Light brown, pale base.
Race *robusta*, Qld, NT, eastern WA. Smaller. Male: Dusky tail. Female: Upperparts: Olive. Breast, underparts: Pale yellow. Immature, both races. Immature: Similar to female, wings grey, edged rufous.

Solitary or in pairs.

Habitat: In the west favours estuarine and tidal mangroves. Tropical north: Rainforest, low tropical scrub, monsoon forests and mangroves.

Feeding: Hops methodically through foliage, gleaning insects or small crabs from mudflats.

Voice: Similar to Golden Whistler but deeper and richer, building to rollicking clear whistles with a final whipcrack.

Status: Moderately common. Sedentary. Territorial.

Breeding: Oct.–Dec. Cup-shaped nest, low in shrubs or tree fork, 1–4m high / 2–3 eggs.

Point Samson and Broome area, WA.

Left: Race *robusta*, female. Right: Nominate male in mangrove.

Grey Whistler

Pachycephala simplex

140–150mm

Sexes alike. 2 races. Nominate race, Brown Whistler, NT. Crown, nape: Grey. Eyebrow: Faint buff with broken eye ring. Upperparts: Grey-brown. Chin, throat: White, faintly streaked. Breast: Pale brown, sometimes dark streaked. Underparts: White.
Race *peninsulae*, Qld. Grey Whistler. Similar to Brown Whistler but upperparts tinged yellow with pale yellow underparts.
Immature, both races: Russet brown first year. Second year, similar to female.
Unobtrusive. Quieter than other Whistlers.

Habitat: Tropical lowland rainforests, tall mangroves and monsoon forests.

Feeding: Quiet. Insects taken from the mid- to upper canopy.

Voice: Complex, melodious whistles.

Status: Moderately common. Sedentary. Solitary or pairs.

Breeding: Sept.–Mar. Cup-shaped nest usually in mangrove fork / 2 eggs.

Iron Range NP and Julatten area, Qld. Middle Arm, Darwin, Fogg Dam CR and Howard Springs, NT.

Left: Race *peninsulae*. Right: Nominate race.

Rufous Whistler

Pachycephala rufiventris

Endemic

160–170mm

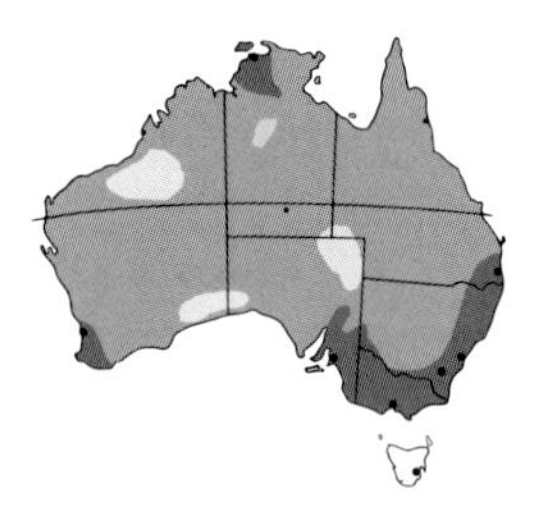

Grey crown and upperparts with a black mask that extends into a broad breast band below a white throat. Underparts: Bright rufous-buff. Eyes: Red. Wings, tail: Dark grey-brown.
Minor geographical variations. Inland and arid desert birds are paler.
Female: Paler, brownish grey. Underparts: Heavily streaked, no black breast band. Immature: Similar to female, pale bill at base, wings edged rufous.

Solitary or pairs. Occasionally hovers.

Habitat: Open forests, woodlands, mallee, mulga, parks, gardens.

Feeding: Mostly insects, sometimes berries.

Voice: Loud, far-reaching variety of ringing, rippling, musical notes with a 'whipcrack-like' ending. Also a penetrating 'joey-joey-joey' incessantly repeated.

Status: Common. Seasonally migratory. Southeastern birds move north in autumn and return in spring. Elsewhere, locally nomadic or sedentary.

Breeding: Sept.–Feb. Cup-shaped nest to 10m high / 2–3 eggs.

Left: Male. Right: Female.

White-breasted Whistler

Pachycephala lanioides

Endemic

180–200mm

Glossy black head with white throat edged black, and rich chestnut breast band extending around nape. Bill: Large, black, slightly hooked. Upperparts: Grey-brown. Tail: Black above, grey below. Underparts: White. Eyes: Red.
Female: Mid-grey to sandy brown face. Underparts: Pale buff with dark streaked throat, breast. Upperparts: Grey-olive.
Nominate race, Kimberley, area WA.
Race *fretorum*, Top End, NT, Qld.
Female: Grey upperparts.
Race *carnarvoni*, Pilbara area, WA.
Female: Brownish upperparts, buff underparts. Immature, all races: Similar to female, darker.
Undulating, low short flights.

Habitat: Low, dense mangroves along estuaries and inlets.

Feeding: Forages in pairs in foliage or on mangrove flats, mostly for crustaceans, insects.

Voice: Similar to Rufous Whistler. Male: Rich, musical 4–5 whistled notes. Scolding alarm calls.

Status: Moderately common. Sedentary. Territorial.

Breeding: Mar.–Nov. Minimal cup-shaped nest in mangroves, 2–10m high / 2 eggs.

Point Sampson and Broome Bird Observatory, WA.

Left: Male in mangrove. Right: Race *fretorum*, female.

Crested Shrike-tit

Falcunculus frontatus

150–190mm

Sexes similar. 3 isolated races. Nominate race, Eastern Shrike-tit. Bold black and white head. Crest, throat: Black. Breast, underparts: Bright yellow. Back, rump: Olive green. Wings, tail: Dusky, edged grey. Bill: Powerful, hooked, black.
Race *leucogaster*, Western Shrike-tit. Similar. Belly: White.
Race *whitei*, Northern Shrike-tit. Similar but smaller with dull yellow wing coverts.
Female: Similar. All have olive green throat. Immature: Similar, with a russet-brown back and pale buff underparts.

Strong flight. Pairs or groups.

Habitat: Open woodlands, eucalypt forests, rainforests.

Feeding: Methodical, upper foliage forager for insects. Hangs from leaves. Uses bill to pull away bark or cut into galls.

Voice: Plaintive piping whistle and chattering.

Status: Moderately common. Sedentary or nomadic. Race *whitei* endangered.

Breeding: Aug.–Jan. Cone-shaped nest up to 30m high / 2–3 eggs.

Chiltern–Mt Pilot NP, Vic. Glen Davis, Gulpa Is. NP, NSW. Stirling Ranges, Dryandra SF, WA.

Left: Nominate female. Right: Race *leucogaster,* male.

Little Shrike-thrush

Colluricincla megarhyncha

170–190mm

Sexes alike. Present in Australia: Race *rufogaster*, Rufous Shrike-thrush and 5 other generally similar races with minor variations in colour and breast streaking: Grey-olive in the southeast; browner, more rufous in the northeast. Underparts: Cinnamon brown with streaked breast.
Race *parvula*, NT, WA. Bill: Dark brown. Upperparts: Grey-brown to dark brown. Underparts: Pale sandy buff, faintly streaked. Throat, lores: Whitish.
Immature: Wing coverts edged brown.

Solitary or in pairs.

Habitat: Dense rainforests, mangroves, monsoon forests.

Feeding: Forages for insects and their larvae from ground to upper canopy. Often tears away bark for prey. Sometimes takes nestlings of other species.

Voice: Varied dialects. Melodious whistles, 'lo-wee-wot-wot' rising, becoming shorter and faster.

Status: Moderately common. Sedentary.

Breeding: Sept.–Feb. Deep cup nest in Lawyer Vine, pandanus, mangroves, low to 10m / 2–3 eggs.

Iron Range NP and Mt Glorious, Qld. Howard Springs NP, NT.

Left: Race *rufogaster.* Right: Race *parvula.*

Bower's Shrike-thrush

Colluricincla boweri

Endemic

190–200mm

Sexes similar. Similar to the Little Shrike-thrush, but with a black bill and darker streaking to the throat and breast. Heavy body, large head and short tail. Crown, upperparts dark grey. Wings: Grey-brown, rufous margins. Face: Grey. Lores, eyebrow: Pale grey. Bill: Black. Eyes: Red-brown, sometimes with pale eye ring. Underparts: Rich rufous. Throat, breast: Fine dark grey streaks.
Female: Rufous eyebrow and lores. Pale bill. Immature: Browner, duller, heavily streaked breast.

Solitary or pairs.

Habitat: Tropical rainforests above 400m altitude.

Feeding: Perch and pounce. Insects, larvae, usually in the mid-canopy, sometimes coming to ground.

Voice: Deep, whistled trills. Iconic tropical summer song.

Status: Moderately common. Overwinters in lower altitudes.

Breeding: Oct.–Jan. Open cup nest in vine or leafy tree fork, 1–10m / 2 eggs.

Lake Eacham, Mt Lewis NP and Paluma Range NP, Qld.

Opposite: Male.

Sandstone Shrike-thrush

Colluricincla woodwardi

Endemic

250mm

Sexes similar. Slender bird with a long tail. Crown: Grey-brown. Upperparts, wings: Olive-brown. Lores: Paler. Eyes: Ruby. Bill: Black. Throat: Buff, lightly streaked. Underparts: Grading to cinnamon buff belly.
Female: Grey brown bill.
Immature: Darker wings, duller breast.
Hops, shelters and nests among rock ledges, outcrops. Occasionally flying to treetops.
Habitat: Remote, rugged sandstone gorges, escarpments.
Feeding: Forages alone among rocks, mostly for grasshoppers, spiders, small lizards.
Voice: Clear, rich, musical, bell-like notes that echo in rocky gorges. Also a 'meow' song.
Status: Moderately common. Sedentary.
Breeding: Nov.–Jan. Nest in spinifex, in rocky fissure / 2–3 eggs.

Boodjamulla NP, Qld. Gunlom Falls, Kakadu NP, NT. Mitchell Falls, Mitchell River NP, WA.

Left: Male. Right: Female.

Grey Shrike-thrush

Colluricincla harmonica

220–240mm

Considered one of Australia's best songsters. Mostly drab grey with a black bill and eye ring. Lores: White. Throat: Light grey streaks. Back: Rich grey-brown. Wings, tail: Grey. Underparts: Light grey.
Female: White eye ring, grey lores, grey-brown bill. Throat, breast: Finely streaked grey.
Immature: Adult plumage reached in 2–3 years. Eyebrow: Rufous. Eye ring: Brown. Underparts: Pale grey, strongly streaked. Wing coverts: Rufous.
5 similar races and various sub-races with subtle or clinal variations. Distinctive races:
Nominate: Qld, NSW, Vic. to eastern SA.
Race *brunnea*, Kimberley area, NT, WA. Upperparts: Light grey-brown. Underparts: Paler.
Race *rufiventris*, Western Shrike-thrush. Upperparts: Grey. Wings, tail: Grey-brown. Undertail coverts: Cinnamon buff.
Race *strigata*, Tas. Female: Eye ring, lores, eyebrow: Whitish. Bill: Grey, paler below. Face, throat, breast: Fine red blackish streaks. Faint eyebrow.

Swift, direct, undulating flight. Easy wingbeats with little flapping.

Habitat: Most habitats, avoids treeless deserts and dense rainforests.
Feeding: Insects, spiders, small reptiles, frogs, nestlings and eggs.
Voice: Clear, rich melodious voice.
Status: Common. Sedentary.
Breeding: July–Feb. Large bowl-shaped nest of twigs, rootlets and grass. Placed in tree or post hollow, on the ground or in a building crevice / 3–4 eggs.

Row 2, left: Nominate male.
Row 2, right: Race *brunnea*.

Row 3, left: Race *rufiventris*.
Row 3, right: Race *strigata*.

Crested Bellbird

Oreoica gutturalis

200–220mm

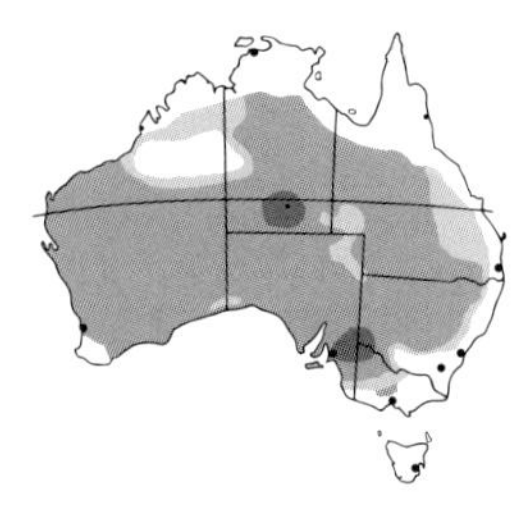

Conspicuous face with a black line from crest through orange-red eyes to black large breast bib. Head: Grey. Frons, chin: White. Back: Grey-brown. Rump: Cinnamon brown. Wings, tail: Dusky grey edged pale brown. Underparts: White. Sides of breast, flanks, vent, undertail coverts: Buff. Bill: Dark grey.
Female: Paler. Crown: Black streak. Face: Grey. Eyes: Brown. Underparts: Buff with white patch on chin, slightly darker breast.
Immature: Similar to female, male has darkish throat bib.
Solitary. Undulating strong flight.
Habitat: Lightly wooded mulga and eucalypt scrub.
Feeding: Hops beneath shrubbery for insects and seeds.
Voice: Deep, melodious 'pan-pan-panella', low, mellow, descending 'plonk-plonk'.
Status: Uncommon. Sedentary. Threatened Vic.
Breeding: Aug.–Jan. or after rains. Cup nest in hollow. Rim often lined with live hairy caterpillars / 3–4 eggs.

Left: Female. Centre: Male showing breast bib. Right: Male.

22

Family Monarchidae, Rhipiduridae, Dicruridae

Flycatchers

Monarchs

Fantails

Willie Wagtail

Boatbill

Magpie-lark

Drongo

A grouping of three families of mostly small to medium-sized insectivorous birds that have an undulating and often twisting flight in pursuit of insect prey.

Family Monarchidae includes flycatchers, monarchs, Boatbill and Magpie-lark. Flycatchers have small crests that are constantly raised and lowered when agitated or when calling and a constantly vertical quivering, flicking or swaying tail when perched. They have broad flattened bills and bristles around their bill that help in catching insects in flight. Flycatchers build beautiful cup-shaped nests of interlaced bark strips and rootlets bound with spider web and decorated on the exterior with camouflaging green moss, lichen or flakes of paperbark.

Monarchs have long bristles around their bill to help scoop up insects in flight. They generally take insects in the outer canopy, sallying out into the open in pursuit of prey.

Frill-necked monarchs are small black and white birds with erectile, frill-like collar feathers that are raised during courtship or in group displays where up to 20 birds participate. Monarch nests are loosely woven twig 'baskets' bound with spider web and hung hammock-like between two vines.

Boatbills are named for their amazingly broad 'boat-shaped' bill. They are brightly plumaged, insect-eating foliage-gleaners found in the lower altitude rainforests. They also have long bristles around their bill to help catch insect prey.

The Magpie-lark forages on the ground, but they show their relationship to flycatchers by their buoyant fluttering flight in pursuit of insect prey. Unlike other family members they build a beautiful bowl-shaped nest constructed of mud pellets reinforced with grass fibre.

Family Rhipiduridae contains fantails. Fantails are specialist aerial feeders named for their large fan-like tails that are usually fanned before takeoff, on landing and during their aerial manoeuviring. They are characterised by their agility in the air, their flight undulating, zigzagging and swooping when snapping at insects. Bristles around the gape help in capturing insects in flight.

The well-loved Willie Wagtail is the largest member of this family. Found throughout Australia, Willie Wagtails become aggressive during the nesting period and use their white eyebrow to signal their mood – expanding the eyebrow in territorial disputes and reducing it when the dispute is settled. Fearless in defence of their nest they will take on all-comers and are often seen chasing away eagles.

Family Dicruridae includes the sole Australian representative, the Spangled Drongo. The Spangled Drongo displays its flycatcher characteristics with its twisting, turning, buoyant flight in pursuit of insect prey and acrobatic courtship display. The bristles around the bill (rictal bristles) help guide prey into the bill. The bird is named for the beautiful blue-green glittering iridescent spots or 'spangles' that are revealed in the sunlight.

Opposite: Female Satin Flycatcher singing its cheerful, rhythmic, whistled song.

Broad-billed Flycatcher

Myiagra ruficollis

150–160mm

Small flycatcher with a distinctively broad and flat-based bill, flanked by long 'whisker-like' bristles. Upperparts: Blue-grey sheen. Crown: Glossy, blue-grey with small erectile hind crest. Eye ring: White, prominent below the eyes. Lores: Pale. Chin, throat, upper breast: Bright rufous-orange. Wings: Grey-brown. Tail: Grey-brown with white edge. Undertail: Dark grey. Underparts: White.
Female: Overall paler, less glossy. Lores, under tail: Whiter. Immature: Buff-tipped wings.

Characteristically quivers its tail when perched.

Habitat: Mostly mangroves, also monsoon forests, paperbark swamps and riverine forests.

Feeding: Solitary. Upper canopy, gleaning or hawking for insects.

Voice: Repeated clear ringing, musical whistling and soft churring.

Status: Moderately common. Sedentary.

Breeding: Oct.–Feb. Shallow nest, often over water in mangroves 1–4m high / 2–3 eggs.

Karumba, Qld. Middle Arm, Darwin and Fogg Dam CR, NT. Derby boat ramp, WA.

Left: Female perched on Sacred Lotus dried seedpod. Right: Male.

Leaden Flycatcher

Myiagra rubecula

150–160mm

Mostly glossy, 'leaden' blue-grey with strongly defined white underparts. Bill: Broad, blue base, black tip. Lores: Black. Eyes: Dark brown. Undertail: Greyish. Legs: Black.
Race *concinna*, Top End, NT, WA. Smaller, narrower bill.
Female: Overall paler upperparts with orange-buff breast and white underparts. Tail: Outer feather edged pale blue-grey. Immature: Similar to female. Breast mottled in males. Wings, tail: Edges buff.

Small crest is constantly raised and lowered. Tail constantly quivering. Upright posture.

Habitat: Eucalypt and paperbark forests and woodlands.

Feeding: Captures insects in flight or gleans from foliage.

Voice: Clear, repeated high whistled notes and twittering.

Status: Moderately common. Sedentary in northern range. Southeastern birds migrate to northern Qld and New Guinea in winter, returning to the same location in spring.

Breeding: Sept.–Oct. in south. Aug.–Feb. in north. Neat cup nest of plant fibre and web, decorated with lichen, 5–25m high / 2–3 eggs.

Cairns area, Qld. Kakadu NP, NT.

Left: Female in nest. Right: Male.

Satin Flycatcher

Myiagra cyanoleuca

170–180mm

Similar to Leaden Flycatcher, but darker with a stubby erectile crest and longer tail. Upperparts, throat: Glossy blue-black. Breast: Black with blue sheen. Underparts: White. Wings, tail: Black. Eyes: Brown. Lores: Black. Bill: Blue-black. Undertail: Grey-black.
Female: Overall paler, greyer and smaller. Throat: Buff-orange. Outer tail feathers: Pale edges. Immature: Similar to female, buff-edged wings.

Tail sways and quivers when perched.

Habitat: Tall, wet eucalypt forest gullies and nearby ranges. Seasonal migration through open woodlands. May visit parks and gardens.

Feeding: Hawks for flying insects from the mid- to upper canopy.

Voice: Sharp, metallic 2-note whistle. Harsh grating.

Status: Moderately common. Seasonal migration to far northeastern Qld and PNG in Mar.–Apr. and returns in Sept.–Oct. to same locality.

Breeding: Oct.–Feb. Cup-shaped nest on horizontal branch 5–25m high / 3 eggs.

Mt Wellington, Tas. Barrington Tops NP, NSW.

Left: Male. Right: Female.

Shining Flycatcher

Myiagra alecto

170–180mm

Overall glossy and iridescent blue-black body with small erectile crest. Eyes: Dark brown. Bill: Blue-grey.
Females have slight variations in colour between races. Nominate: Head, nape: Glossy blue-black. Back, wings, tail: Bright, deep rufous. Underparts: White. Race *wardelli*, east coast. Flanks, undertail covert: Buff-rufous. Immature: Similar to female, duller.

Constantly raises crest and flicks tail upwards.

Habitat: Mangroves, rainforests along watercourses and paperbark swamps.

Feeding: Creeps and flits among mangrove roots and mudflats for small crabs. Hawks for insects in rainforests.

Voice: Diverse repertoire of rapid, clear whistles and rising trilling and frog-like calls.

Status: Common. Sedentary.

Breeding: Aug.–Sept. The nest is constructed by both males and females. Deep cup nest, often in mangroves, usually over water in a sheltered fork, 1–6m high / 2–3 eggs.

Daintree River, Qld. Middle Arm, Howard Spring CR and Kakadu NP, NT.

Left: Male. Right: Female in nest. Decorated with lichen.

Restless Flycatcher

Myiagra inquieta

Endemic

190–210mm

Sexes similar. Glossy blue-black head with a low hind crest. Back: Dark grey-black. Wings: Dark grey-brown. Eyes: Dark brown. Bill: Slate grey. Underparts: White, sometimes with a pale buff-yellow tint to breast.
Female: Dull grey lores.
Immature: Dull grey-black upperparts. Wings: Edged buff or white. Breast, throat: Orange-buff wash.

Solitary or in pairs. Constantly moving. Distinctive hovering motion when plucking insects from foliage. 'Wags' tail. Graceful flight. Noisy and conspicuous.

Habitat: Open eucalypt woodlands, farmlands, inland scrubby areas.

Feeding: Hawking or hovering over foliage for insects. Usually in the middle canopy. Rarely ventures to ground.

Voice: Musical, penetrating, high-pitched, rising, clear whistling. Loud rasping, grating 'schezzp'.

Status: Moderately common. Nomadic.

Breeding: July–Mar. Large, typical nest on horizontal branch 1–20m high. Nest is constructed by both males and females / 3–4 eggs.

Left: Male. Right: Male constructing the large nest of plant fibres and web.

Paperbark Flycatcher

Myiagra nana

170–190mm

Sexes similar. Similar to the Restless Flycatcher, but ranges do not overlap. Slightly smaller and glossier with a slightly wider and shorter bill. Distinctive, glossy blue-black upperparts, forehead, crown and lores. Wings: Dark grey-brown. Eyes: Dark brown. Bill: Slate grey. Underparts: White, sometimes with a buff tint.
Female: Duller. Immature: Duller upperparts. Breast and throat: Rich orange-buff wash.

Solitary or in pairs. Constantly moves and 'wags' tail. Graceful swooping flight. Hovers.

Habitat: Riverine, swampy areas, melaleuca forests, edge of mangroves.

Feeding: Hawks or hovers over foliage for insects. Often hovers near ground when feeding.

Voice: Musical, penetrating, high-pitched 'too whee'. Occasional call of Restless Flycatcher.

Status: Moderately common. Nomadic. Also in southern New Guinea.

Breeding: Nov.–Jan. Small cup-shaped nest of paperbark and spider web on horizontal branch, often over water 1–20m high / 3–4 eggs.

Left: Male singing. Right: Male on Sacred Lotus seedpod.

White-eared Monarch

Carterornis leucotis

Endemic

130–140mm

Sexes similar. Overall small, black and white flycatcher with distinctive facial markings and black head with black erectile frons feathers. Throat: White, speckled black. Upperparts: Dusky black. Wings: Prominent white wing bars. Tail: Dusky black, broadly tipped white outer feathers. Rump: White. Underparts: Whitish grey. Legs: Blue-grey.
Female: Greyer, less white to face and rump. Immature: Brown rather than black markings, buff underparts.

Flits and hovers. Raises crest and tail when alarmed.

Habitat: Rainforest fringes, mangroves and adjoining eucalypt forests.

Feeding: Gleans and snatches insects from above the crown of tall trees down to shrub level.

Voice: Rising–falling repeated sequence of whistles.

Status: Uncommon. Sedentary. Some local movement. Vulnerable in NSW.

Breeding: Aug.–Jan. Deep nest 10–30m high / 2 eggs.

Julatten area, Mt Lewis NP and Iron Range NP, Qld.

Left: Adult male. Right: Immature.

Frill-necked Monarch

Arses lorealis

140–160mm

Sexes similar. 'Frilled' erectile white nape feathers, usually raised when courting. Distinctive cobalt blue eye ring. Head, chin: Glossy black. Upperparts: Black, with distinctive white crescent across back and scapulars. Wings, tail: Blackish brown. Underparts: White.
Female: Lacks black chin. Lores: Pale grey. Immature: Duller, brownish. Grey eye ring.

Noisy and gregarious in aerial pursuits through the rainforests.

Habitat: Rainforests and adjacent eucalypt forests.

Feeding: Flits in outer foliage or spirals up tree trunk like a treecreeper for insects. Mid- to low canopy.

Voice: Soft 'frog-like' calls when feeding. Short series of trilled whistles.

Status: Moderately common. Sedentary.

Breeding: Sept.–Jan. Shallow, camouflaged hung nest between vines and away from foliage, 1–10m high / 2 eggs.

Iron Range NP, Qld.

Left: Male showing white underparts. Opposite: Male, showing 'frilled' erectile nape feathers.

Pied Monarch

Arses kaupi

Endemic

140–160mm

Sexes similar. Similar to the Frill-necked Monarch, but differences include broad black breast band, white chin and thinner blue eye ring. Longer tail. Eyes: Brown. Female has duller, broader black breast band, sides of neck black and shorter erectile collar. Immature: Bill: Yellowish. Eye ring: Bluish. Grey-brown rather than black markings.

Often holds wings half-open and jerks tail downwards. Usually in pairs.

Habitat: Tropical rainforests, vine or palm scrublands near watercourses.

Feeding: Spirals up and down tree trunks and branches in the mid- to upper canopy, flushing insects from bark crevices.

Voice: Short, soft, whistled notes. Harsh call note.

Status: Moderately common. Sedentary.

Breeding: Oct.–Jan. Camouflaged woven basket nest hung between vines and away from foliage, 1–10m high / 2 eggs.

Julatten area and Mt Lewis NP, Qld.

Left and right: Male, showing 'frilled' erectile nape feathers.

Black-faced Monarch

Monarcha melanopsis

160–180mm

Sexes similar. Distinctive black face with black frons, lores, eye ring and throat. Upperparts, crown, breast, tail: Blue-grey. Bill: Hooked, pale blue-grey. Ear coverts, around eye ring: Pale grey. Wings: Darker blue-grey.

Female: Duller. Immature: Similar, without black face markings.
Northern birds are paler and smaller.

Solitary or in pairs.

Habitat: Tropical and temperate rainforests, moist gullies, eucalyptus woodlands. Open woodlands when migrating.

Feeding: Forages for insects in the upper and lower canopy. Sallies for flying insects.

Voice: Clear, rising–falling mellow whistle.

Status: Sedentary and more common in the north. Southern birds migrate northward, some overwintering in PNG.

Breeding: Oct.–Jan. Small, deep, camouflaged nest in shaded fork 1–12m high / 2–3 eggs.

Lamington NP, Qld.

Opposite: Adult male.

Black-winged Monarch

Monarcha frater

170–180mm

Sexes similar. Similar to Black-faced Monarch, but with paler blue-grey plumage and brownish black wings. Upperparts, crown, breast: Pearly grey. Frons, throat patch: Black. Eyes: Black. Ear coverts and around eyes: Pearly light grey. Bill: Pale blue-grey. Tail: Overall dark grey-black. Underparts: Rich rufous-buff. Female: Whitish lores. Immature: Grey-buff rather than black markings.

Completely arboreal, solitary or in pairs.

Habitat: Tropical rainforests, mangroves, adjacent areas.

Feeding: Solitary, hopping through mid- to lower canopy gleaning insects. Rarely hawks.

Voice: Clear, rising–falling whistle, similar to Black-faced Monarch but harsher.

Status: Moderately common. Summer breeding migrant from New Guinea. Arrive: Oct. Depart: Mar.

Breeding: Oct.–Mar. Small, goblet-shaped, camouflaged nest in upright fork 1–10m high / 3 eggs.

Iron Range NP, Qld.

Left and right: Adult male.

Spectacled Monarch

Symposiarchus trivirgatus

150–160mm

Sexes alike. Similar to Black-faced Monarch, but with white belly and rufous-orange cheeks and breast. Tail: Outer feathers broadly tipped white. Upperparts, crown: Blue-grey. Frons, lores, ear coverts form 'spectacles' around black eyes. Throat, chin: Small black patch. Bill: Blue-grey. Underparts: White with merging orange-buff flanks.
Race *albiventris*, White-bellied Flycatcher, Cape York, Qld. Upper breast rufous-orange. Flanks: White.
Race *melanorrhous*, far north Qld. Crown, mantle, wings: Dark grey. Immature, all races: Grey-buff rather than black markings. Lores: Paler.
Solitary or in pairs. Sometimes small groups or in mixed flocks. Active and relatively tame.

Habitat: Coastal rainforests, mangroves, shaded gullies in dense eucalypt forests.

Feeding: Jerky, tumbling, hovering and darting in the mid-canopy to flush insects.

Voice: Lively sequence of squeaky warbling and chattering.

Status: Moderately common. Some southern birds migrate north in winter.

Breeding: Oct.–Feb. Cup-shaped camouflaged nest in tree fork, shrub or vines 1–6m high / 2 eggs.

Iron Range NP, Lamington NP and Mt Glorious, Qld.

Left: Nominate adult. Right: Immature.

Rufous Fantail

Rhipidura rufifrons

Endemic

150–165mm

Sexes alike. Dainty and restless with rufous highlights. Upperparts, wings: Grey-brown. Bill, eyes: Dark brown. Frons, eyebrow, lower back: Orange-rufous. Tail: Grey with orange-rufous base and paler tips. Throat: White. Breast: Black band with black and white mottling below. Belly: White. Flanks, undertail: Buff. Race *intermedia*, Qld, far north NSW. Belly, tail tips: Whiter. Immature: Duller, wings edged rufous. (All.)

Distinctive short, sallying, acrobatic flight, with tail fanned, 1–2m above ground. Constant movement and tail flicking.

Habitat: Densely vegetated areas of temperate rainforests, wet sclerophyll forests and adjacent ranges.

Feeding: Hawks for insects with an audible click of the bill. Gleans insects in the mid- to low canopy.

Voice: Tiny, high-pitched sequence of notes. Similar to Grey Fantail.

Status: Common to moderately common. Moves northward for winter, southern birds to Qld, northern birds to PNG.

Breeding: Oct.–Feb. Compact cup nest with tail, in a shaded fork / 2–3 eggs.

Lamington NP, Qld.

Left: Nominate race. Right: Race *intermedia*.

Arafura Fantail

Rhipidura dryas

150mm

Sexes alike. Similar to the Rufous Fantail, but smaller, duller, with less rufous to tail base and rump. Tail: Longer, broadly tipped white, whiter on outer feathers.

Habitat: Mangroves and fringing scrubland.

Fans tail and flicks wings. Similar to Rufous Fantail.

Feeding: Hawks for insects in the lower foliage of mangroves or sometimes feeds on ground.

Voice: High-pitched falling notes.

Status: Common to moderately common. Sedentary.

Breeding: Sept.–Dec. Small, neat, compact cup nest with a tail, in a shaded fork, usually in mangroves, often over water, low to 2m / 2 eggs.

Fogg Dam CR, NT.

Left: Immature. Right: Adult.

Mangrove Grey Fantail

Rhipidura phasiana

145–150mm

Sexes alike. Similar to Grey Fantail. Upperparts: Pale, mid-grey. Eyebrow: Buff-white. Breast: Thin grey bar. Wings: Wider white bars. Tail: Outer feathers broadly tipped and edged white. Harsh sun and exposure to salts often fades plumage to overall shabby grey and white.
Immature: Duller, browner.

Solitary or in pairs. Sometimes in mixed feeding flocks. Active, constantly fanning tail, flitting from perch, usually returning to same spot. Looping flight.

Habitat: Estuarine mangrove forests.

Feeding: Hawks for insects above tidal mudflats or gleans insects from foliage.

Voice: High-pitched soft, falling, squeaky notes.

Status: Moderately common. Sedentary or wanders locally. More common in west of range.

Breeding: July–Dec. Compact, tiny cup nest with a tail, in a shaded fork, in mangroves over mudflats / 2 eggs.

New Beach and Point Samson, WA.

Left: Adult foraging in mangroves above tidal mudflats. Right: Adult perched on mangrove aerial root.

Grey Fantail

Rhipidura albiscapa

Endemic

150–160mm

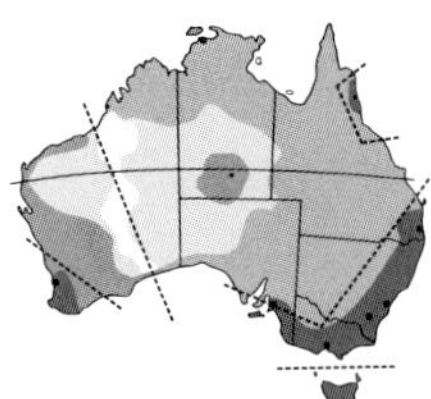

Sexes alike. Mostly grey bird with mid- to blackish grey upperparts. Wings: Dusky grey with 2 narrow bars. Tail: Long, dusky grey, outer feathers broadly tipped and edged white. Eyebrow: White, narrow. Eyes: Black, white mark behind. Throat: White. Breast: Grey, with wide sooty breast band. Underparts: White, fawn or buff.
Subtle variations occur across Australia: Nominate race confined to Tas.
Race *alisteri*, east and southeast Australia. Mid-grey with narrower breast band. Overwintering as far north as northern Qld.
Race *keasti*, tablelands, northeast Qld. Dark sooty plumage. Wide breast band. Sedentary.
Race *preissi*, southwest WA. Larger grey breast band, smaller bill. Moves slightly north or east in winter.
Race *albicauda*, arid interior. Browner with white tail feathers. Erratically nomadic.
Immature: Browner, buffy head and wing markings.

Active, restless and acrobatic flight. Constantly fans tail. Relatively tame. Sometimes in mixed feeding flocks. Usually in singles or pairs, but form small groups in autumn and winter.

Habitat: Most habitats, including tall, open woodlands, parks and gardens.

Feeding: Hawks for and gleans insects from the lower canopy.

Voice: Sweet, musical trilling. Harsh chattering.

Status: Common. Southern birds move slightly northward in winter. Alpine birds disperse north after breeding.

Breeding: Aug.–Jan. Compact, cup-shaped nest with a tail, made from fine grass, bark and spider web, placed in a forked limb, 1–6m high / 2–3 eggs.

Top left: Race *keasti*. NE Qld.

Top right: Race *alisteri* at nest. SE Australia.

Centre left: Race *preissi*. SW WA.

Centre right: Nominate race. Tas.

Northern Fantail

Rhipidura rufiventris

170–180mm

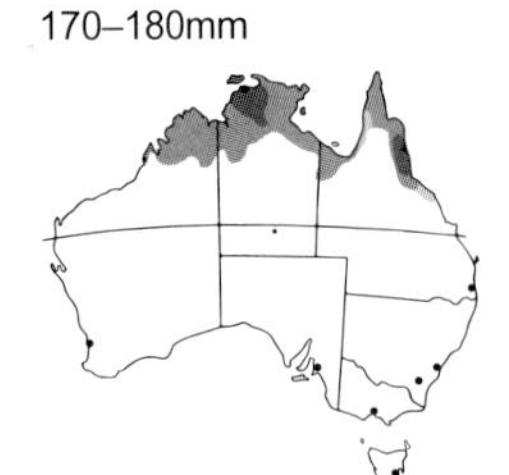

Sexes alike. Similar appearance to Grey Fantail, but larger, stockier with a grey-brown, streaky breast band. Tail: Shorter, less fanned. Belly, underparts: Pale buff. Throat: White. Eyebrow: White, short, thin. Bill: Broader. Upperparts: Dark grey.
Immature: Duller, browner. Face, tail and wing markings: Buff.

Quiet, sedate. Less active than other fantails.

Usually in singles or pairs.

Habitat: Dense monsoon scrub, rainforest margins, vine scrub, mangroves, parks and gardens. Favours paperbark forests.

Feeding: Quietly in upper canopy, hawks insects mid-air.

Voice: Sharp, musical descending notes. Continues after dark.

Status: Moderately common, sedentary.

Breeding: July–Dec. Typical larger nest on fork, 1–6m high / 1–2 eggs.

Left: Adult. Right: Immature.

Willie Wagtail

Rhipidura leucophrys

190–210mm

Sexes alike. Largest fantail, constantly wags, fans tail. Mostly black with glossy black upperparts and sharpy defined white underparts. Wings: Tinged brown. Eyebrow: White, expanded to show aggression, withdrawn completely when submissive. Also white whisker line.
Immature: Similar, with buff-edged feathers.

Fearless defending territory.

Usually pair for life.
Habitat: Lightly timbered country with clearings, parks, gardens.
Feeding: Forages for insects, running or short sallies.
Voice: Distinctive, 'sweet-pretty-creature' song, rattling alarm call. Often sings at night.
Status: Common. Some seasonal movement.
Breeding: June–Feb. Cup nest, 1–15m high / 2–4 eggs.

Left: Male. Right: Female on nest.

Yellow-breasted Boatbill

Machaerirhynchus flaviventer

110–120mm

Distinctive black, wide 'boat-shaped' bill. Black crown, face, lores, ear coverts. Frons, eyebrow: Long, bright yellow stripe. Throat: White. Upperparts: Deep olive green. Wings: Black, white bars, margins. Tail: Black, broad, white tips. Underparts: Bright yellow.
Race *secundus*, south of Cooktown, Qld. Black upperparts. Wings: Yellow bars and margins. Female: Upperparts: Olive. Underparts: Greyish yellow. Immature: Similar to female, buff wing markings.
Solitary or in pairs. Lively, active.
Habitat: Low altitude rainforests.
Feeding: Actively hawks insects in the mid- to upper canopy.
Voice: Clear, penetrating, rising, drawn-out 3–4 whistles, trills, twittering.
Status: Moderately common. Sedentary.
Breeding: Sept.–Mar. Shallow nest 2–25m high / 2 eggs.

Iron Range NP, Qld.

Left: Female. Right: Race *secundus*, Male.

Magpie-lark

Grallina cyanoleuca

260–300mm

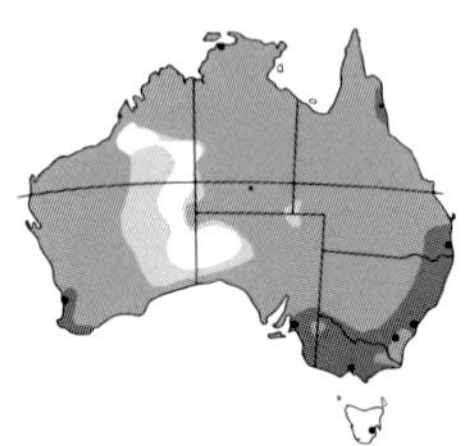

Glossy black and white bird. Black crown, face and bib with bold white eyebrow, cheek patch. Underparts: White. Wings: Bold white bar. Eyes, bill: Whitish. Female: Similar. White face with bold black line linking crown to bib. No eyebrow. Immature: Dark eyes and bill. White throat patch and brow line.

Fearlessly defends territory.

Habitat: Open timbered country near water, wetlands, gardens.
Feeding: Ground foraging for insects.
Voice: Loud, piping calls, often in wing-lifting duets, 'peewee-peewee'.
Status: Common. Sedentary. Non-breeding birds form nomadic flocks.
Breeding: Aug.–Dec. or following rain. Mud and grass, bowl-shaped nest, 4–10m high / 2–4 eggs.

Left: Female at mud bowl nest. Right: Male in flight.

Spangled Drongo

Dicrurus bracteatus

300–320mm

Sexes alike. Long, outward-flared forked tail. Body: Black with blue-green glossy metallic sheen and highly reflective iridescent spots (spangles), to crown, neck, breast. Eyes: Bright red. Bill: Black, heavy, bristles around base. Female: Smaller. Immature: Dusky black. Eyes: Brown. Breast, undertail coverts: Spotted white.

Noisy, active and conspicuous.
Habitat: Coastal open woodlands, rainforests, mangroves and urban areas.
Feeding: Perches patiently, often in the open. Large insects caught in flight or mid-canopy. Tears away bark for spiders, grubs. Also takes fruit and nectar.
Voice: Loud, metallic tearing sounds and harsh, nasal rasping and chattering.
Status: Common. Sedentary in northeast Qld, NT, WA. Nomadic or part-migratory on east coast.
Breeding: Sept.–Mar. Deep hanging basket nest in outer foliage 15–20m high / 3–5 eggs.

Left: Adult taking nectar. Right: Adult on Grass Tree.

23 Family Campephagidae, Oriolidae, Artamidae

Trillers

Cicadabird

Cuckoo-shrikes

Figbird

Orioles

Woodswallows

Magpie

Butcherbirds

Currawongs

A group of birds from three different families, varying in sizes but sharing similarities in appearance.

Family Campephagidae includes the trillers, Cicadabird and cuckoo-shrikes.
Trillers are migratory songbirds named for their loud, melodious, trilling and canary-like warbling in the breeding season, an iconic sound of the Australian bush through spring and early summer. Usually colonial when nesting, their nests are placed only metres apart. However, the breeding pair maintains and defends individual feeding territories that may be up to a kilometre from the nest and up to 15 hectares for each pair.
The Cicadabird is named for its rasping, cicada-like, high-pitched, buzzing call during the breeding season. An elusive bird of the tall forest tree canopy, it is easily disturbed and flies off in a swift undulating flight. The male mostly builds the meagre, shallow, cup-shaped nest of twigs and bark strips stuck together with cobweb and saliva.
Cuckoo-shrikes are named for their cuckoo-like undulating flight and their shrike-like bills. The nest is small, saucer-like and eventually too small for growing nestlings. They are forced to sit precariously on the edge and often fall in a strong wind. Cuckoo-shrikes have spine-like feathers on their lower back and rump that are raised in display and for camouflage when nesting. When a predator appears, the bird lowers its head and tail and raises the spines, forming a hump that blends with the tree branch.

Family Oriolidae includes the orioles and figbird.
Orioles are usually solitary or in pairs, whereas figbirds are gregarious, feeding and moving in search of ripe fruits in small flocks, and often being joined by a few orioles. Orioles are named after their contact call, a musical, bubbling 'orry-ol'. Figbirds are named for their specialised feeding, mostly on fig-tree fruits.

Family Artamidae includes woodswallows, butcherbirds, the Australian Magpie and currawongs.
Woodswallows are small, aerial, insect-catching birds that have a divided, brush-tipped tongue that also allows them to feed on nectar from flowering plants. Unrelated to true swallows, they are one of the few songbirds that soar like swallows. Their range is determined by the extent of open, mostly inland woodlands, giving rise to their name.
The carnivorous butcherbirds are named for their habit of hanging larger prey in a tree fork and then dismembering it with their sharp hook-tipped bill.
Magpies are ground-feeding butcherbirds that have become insectivorous. Early settlers named them after the unrelated European black and white crow, or 'magpie'. Both species have melodious, rich piping calls.
Currawongs are omnivorous scavengers, often robbing the nestlings of other species. They are named after the distinctive, loud, ringing 'cur-ra-wong' call of the Pied Currawong.

Opposite: Male Masked Woodswallow.

White-winged Triller

Lalage tricolor

Endemic

160–180mm

Breeding male: Glossy black cap to below eyes and over upper back. Lower back, rump: Grey. Underparts: White. Wings: Black, with conspicuous white shoulder patch and white-edged feathers. Tail: Black, white corners. Non-breeding: Similar to female. Wings, tail: Black. Rump: Grey. Female: Pale brown upperparts with faint eyebrow and dark line through eyes. Underparts: Pale buff. Breast: Buff, lightly mottled. Immature: Similar to female.

Swift, graceful, undulating flight. Courting males fly over their territory continuously singing.

Habitat: Open forests, acacia woodlands, tree-lined riverbanks.
Feeding: Forages low in foliage or on ground, taking insects, small fruits and seeds.
Voice: Noisy chatter. Melodious trilling, warbling when breeding.
Status: Resident in north and inland. Summer breeding visitor to southern Australia. Wintering in the north or inland. Locally common.
Breeding: Sept.–Feb. Shallow saucer nest 1–12m high in fork or horizontal branch / 2–3 eggs.

Glen Davis, Gulpa Creek, NSW.

Left: Female with well-camouflaged chick below. Right: Breeding male.

Varied Triller

Lalage leucomela

180–200mm

Distinguished from the White-winged Triller by its prominent white eyebrow, long black eye line, cinnamon vent and less black on the wings. Upperparts: Black grading to grey rump. Tail: Black, tipped white. Wings: Black, large white shoulder bar and white edged flight feathers. Underparts: Pale grey, lightly barred flanks.
Female: Similar. Underparts: Barred grey-buff. Upperparts: Barred and streaked.
Race *rufiventris*, NT, WA. Similar. Underparts: Buff, dusky barred lower breast. Pale rufous vent. Female: Similar, brownish back. Immature: Similar to female, flecked white-buff.

Quiet and inconspicuous. Pairs or small groups.

Habitat: Coastal rainforest fringes and adjacent eucalypt or paperbark forests, gardens.
Feeding: Slow and methodical. Mostly in the canopy, but also low shrubbery, taking insects and bite-sized fruit.
Voice: Melodious trilling, taken up by others. Constant churring contact call.
Status: Uncommon. Sedentary.
Breeding: Aug.–Apr. Shallow saucer nest 1–2m high / 1 egg.

Lamington NP, Qld. Howard Springs NP and Holmes Jungle NP Darwin, NT.

Left: Nominate female. Right: Male.

Common Cicadabird

Edolisoma tenuirostris

240–270mm

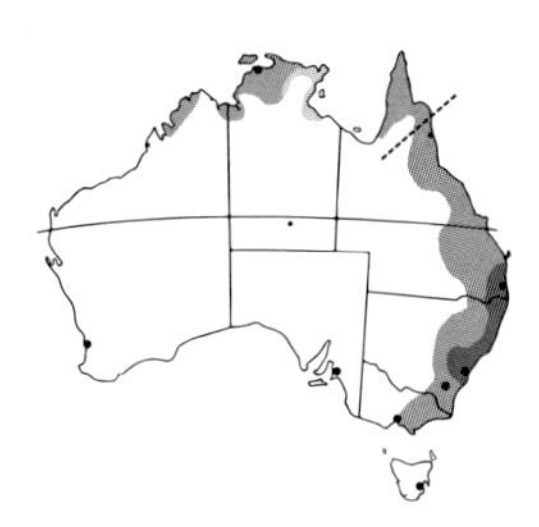

Small cuckoo-shrike, overall deep blue-grey with black streak through dark brown eye. Wings: Black, pale grey feather margins. Tail: Outer feathers black, grey tipped, centre feathers blue-grey, black tipped. Bill, legs: Black. Race *melvillensis*, far north. Similar, behavioural differences. Female, all races: Upperparts: Grey-brown. Underparts: Creamy buff, dark brown barring, whitish throat. Eyes: Dark stripe through brown eye. Buff eyebrow. Bill: Brown, pale base. Immature: Similar, heavier barring. Swift and undulating flight, interspersed with gliding.

Habitat: Eastern range: Eucalypt forests, mangroves, paperbark swamps. Northern range: Monsoon forests, mangroves.
Feeding: Insects, fruit, seeds, taken from tall upper canopy.
Voice: Usually silent. High-pitched, cicada-like buzzing in breeding season. Race *melvillensis*: Low-pitched, slow territorial call.
Status: Moderately common. Sedentary in north. Southern birds winter in Cape York, Qld, PNG.
Breeding: Oct.–Mar. Shallow cup-shaped nest on horizontal branch 3–28m high / 1 egg.

Glen Davis and Royal NP, NSW. Julatten area, Qld. Howard Springs NP, NT.

Left: Male. Right: Female.

Ground Cuckoo-shrike

Coracina maxima

Endemic

330–360mm

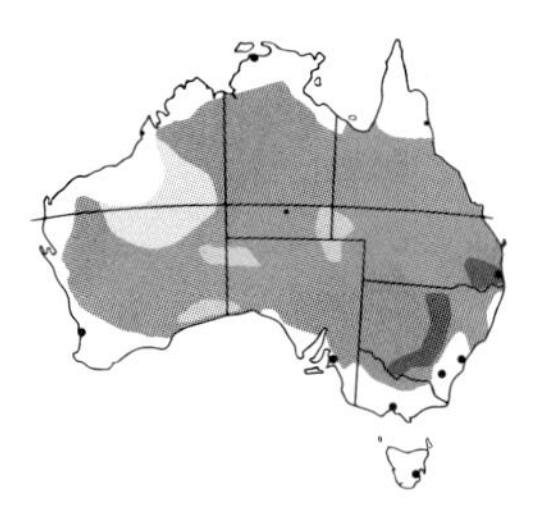

Sexes alike. Largest cuckoo-shrike, graceful, slender and long legged with a pale silvery grey crown and mantle. Face, throat: Mid-grey. Breast, belly, rump: White with fine black scalloped barring. Wings, tail: Black. Eyes: Pale yellow. Bill: Black. Immature: Upperparts finely barred.

Wary. Walks briskly, nodding its head like a pigeon. Usually in groups of 3–4.

Habitat: Sparsely timbered open woodlands, mulga, spinifex and farmland with some trees.

Feeding: Forages mostly on the ground taking grasshoppers, beetles and insects.

Voice: Distinctive, far-carrying metallic drawn-out squeaking.

Status: Uncommon. Nomadic.

Breeding: Aug.–Dec. or following rain. Shallow cup nest of twigs, grass and spider web built into tree fork 3–23m high / 2–3 eggs.

Kunoth Well, Alice Springs, NT. Eulo Bore, Qld.

Opposite: Adult.

White-bellied Cuckoo-shrike

Coracina papuensis

260–280mm

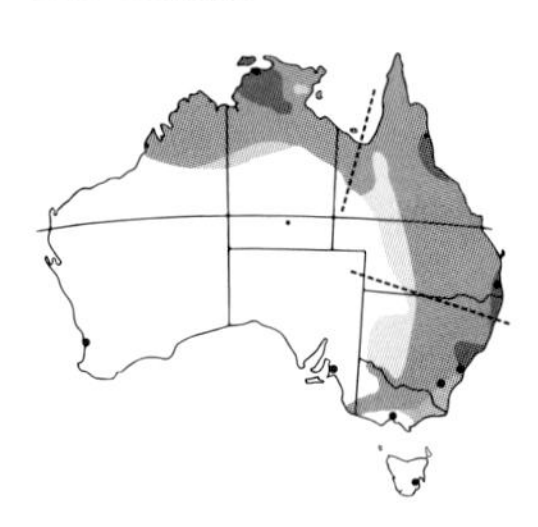

Sexes similar. 2 races found in Australia. Dark morphs are common.
Race *robusta*, southeast Australia. Largest, darkest. Head: Pale grey with bold black eye mask from bill to eyes. Upperparts: Pale grey with dusky edged flight feathers. Breast: Deep grey. Underparts: Mid-grey. Tail: Dusky with outer feathers broadly tipped white.
Dark morph: Head: Dusky black. Lores, face, throat: Black. Breast: Dusky grey mottled and barrred.
Race *hypoleuca*, Top End, NT, WA. Breast: Pure white. Partial, white eye ring behind eyes in small, black mask. Upperparts: Light, mid-grey.
Female, immature: Grey lores and underparts paler.

Undulating, long, gliding flight. When landing, shuffles and folds each wing separately.

Habitat: Coastal eucalypt woodlands, open forests.

Feeding: Forage in understorey shrubs or canopy, diving on insects. Also small fruit.

Voice: Shrill, metallic trills and soft churring.

Status: Moderately common. Locally nomadic, usually family groups.

Breeding: Aug.–Mar. Shallow saucer-like nest high above ground / 2–3 eggs.

Left: Race *hypoleuca*. Centre: Race *robusta*, dark morph. Right: Race *robusta*.

Black-faced Cuckoo-shrike

Coracina novaehollandiae

320–340mm

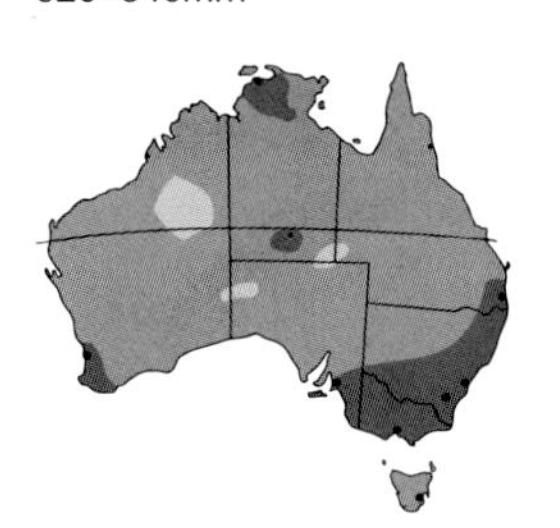

Sexes alike. Conspicuous black face and throat. Back, wings, tail: Blue-grey. Underparts: White. Tail: Tipped white. Eyes: Dark brown. Bill: Black.
Immature: Similar, with grey face, broad black eye stripe, faintly barred breast.
Slight variations in colour and size across range.
Race *subpallida*, Pilbara area, WA. Smaller, with pale silver-grey plumage.

Strong, undulating flight, folds and refolds wings several times when alighting.

Habitat: Open eucalypt forests, woodlands, inland riverine vegetation, scrublands, farms and gardens.

Feeding: Large insects caught mid-air, taken from foliage or less frequently from the ground. Sometimes fruit.

Voice: Metallic, grinding sounds. Loud rollicking and churring, musical notes while shuffling wings.

Status: Moderately common. Sedentary or nomadic.

Breeding: Aug.–Feb. Saucer-shaped nest 8–20m high / 2–3 eggs.

Left: Adult. Right: Immature.

Barred Cuckoo-shrike

Coracina lineata

Endemic

240–260mm

Sexes similar. Distinctively barred black and white lower breast, underparts and underwing coverts. Eyes: Bright light yellow. Head, throat, upper breast: Grey. Lores, base of frons: Black band. Upperparts: Deep grey, paler on rump. Wings: Grey, flight feathers black edged grey. Tail: Black, washed grey base. Bill: Black. Female: Dark grey lores. Immature: Less distinct barring, dark eyes.

In flocks of up to 20. Roosts communally. Fluttering flight, rising–falling low glides over rainforest canopy.

Habitat: Coastal rainforests, less often adjacent eucalypt forests.

Feeding: In the upper canopy, mostly bite-sized fruit and seeds swallowed whole, some insects.

Voice: Sharp, 2–3 notes. Metallic trills. Continuous calling in flight.

Status: Nomadic, following fruiting trees. Southern birds move north for winter. Vulnerable NSW.

Breeding: Oct.–Jan. Small, flat twig nest placed in tree fork, 15–25m high / 2 eggs.

Mt Glorious, Mt Hypipamee NP, Eungella NP and Julatten area, Qld.

Left: Male feeding on Lilly Pilly fruits. Right: Adult male.

Australasian Figbird

Sphecotheres vieilloti

280–270mm

Glossy black head with 'warty' facial skin that becomes redder in excitement. Tail: Black, white tipped, outer feathers almost all white.Nominate race, Green Figbird. Throat, breast, nape: Grey. Underparts: White, green flanks.
Race *flaviventris,* far north Qld. Grey-green with mid-yellow belly. Race *ashbyi*, Yellow Figbird, NT, WA. Throat, breast, belly: Bright-yellow grading to white. Races interbreed.
Female, all races: Olive-brown, dusky streaked. Face skin: Grey. Underparts: White, heavily streaked brown. Immature: Duller, faintly scalloped. Lacks white tail tips. Males foreshadow adult male markings.
Gregarious. Nesting and feeding in small groups.

Habitat: Diverse. Eucalypt forests, rainforest fringes, mangroves, urban trees.

Feeding: Usually with other fruit-eaters. Favours figs, but also insects and nectar.

Voice: Rising–falling, mellow, penetrating musical song. Squeaky notes.

Status: Common. Sedentary. In colonies of 20–40 after breeding. Locally nomadic in large groups, following fruiting trees. Extending range southward.

Breeding: Sept.–Mar. in north. Oct.–Feb in south. Nest hung from horizontal sheltered branch 20m high / 3 eggs.

Left: Female. Centre: Race *ashbyi*, male. Right: Nominate male.

Yellow (Green) Oriole

Oriolus flavocinctus

260–280mm

Sexes similar. Overall golden olive with fine black streaks. Wings: Grey-brown, feathers edged whitish. Tail: Black, tipped yellow or olive. Underparts, undertail coverts: Yellower, less streaked. Throat: Greenish. Eyes: Red. Bill: Large, orange. Female: Paler. Immature: More heavily streaked. Underparts paler. Eyes: Dark. Bill: Brownish-grey.

Elusive, well-camouflaged birds of the rainforest canopy. Entirely arboreal.

Habitat: Wet tropical rainforests, monsoon vine forests, woodlands, parks and gardens.

Feeding: Forages in foliage, taking mostly fruit and insects.

Voice: Melodious, bubbling, warbling song for up to 2 hours at a time, throughout the day. Rich, deep and clear.

Status: Common. Sedentary.

Breeding: Aug.–Jan. Deep nest hung from sheltered horizontal fork 2–15m high / 2–3 eggs.

Left: Female at nest. Right: Adult male.

Olive-backed Oriole

Oriolus sagittatus

250–280mm

Sexes similar. Rich, dull olive green upperparts with fine black streaking. Wings: Dark grey-brown edged white-grey. Bill: Deep-pink. Eyes: Red. Lores: Black. Underparts: Whitish, heavily streaked black with yellowish outer breast, flanks and greenish chin. Tail: Dark grey, white tips.
Race *grisescens*, Cape York, Qld. Greyer and less yellow.
Race *affinis*, Top End, Qld, NT, WA. Less colourful, brownish streaked.
Female all races: Greyer, bill dull pink. Immature: Eyes: Dark. Underparts: White, heavily streaked. Wings: Rufous edged.

Flight is undulating. Singles or in pairs or sometimes small flocks.

Habitat: Open eucalypt forests, woodlands, mallee and tall inland scrub.

Feeding: Usually with other figbirds in the upper canopy. Mostly fruit, also insects.

Voice: Resonating, musical, mellow, bubbly calls.

Status: Moderately common. Sedentary in north. Southeastern birds move north after breeding.

Breeding: Sept.–Jan. Untidy deep nest, hung from horizontal sheltered fork, 2–15m high / 2–3 eggs.

Left and right: Nominate race.

White-breasted Woodswallow

Artamus leucorynchus

170–180mm

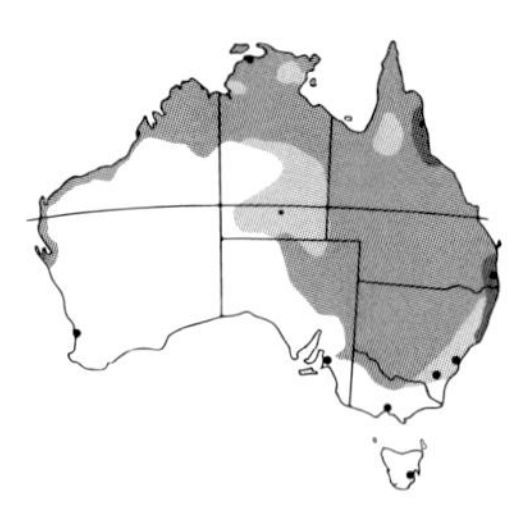

Sexes alike. Distinctive white rump and all-blackish tail. Slate grey hood. Grey-brown mantle and back. White underparts and breast. Bill: Blue-grey tipped black.
Immature: Mottled, buff flecked. Bill: Pale.

Gregarious. Roost together in groups of 10–50 or more in the day and night. Fast flight.

Habitat: Tropical woodlands, mangroves, paperbark swamps and inland along tree-lined rivers.

Feeding: Mostly aerial feeders. Insects caught in open bill, usually over water or treetops. Occasionally nectar from flowers.

Voice: Continuous chattering, soft twittering and mimicry.

Status: Moderately common. Sedentary in north. Summer breeding migrant in southeast Australia.

Breeding: Aug.–Jan. Shallow bowl nest in fork, branch hollow or old Magpie-lark nest 10–30m high / 3–4 eggs.

Left: Adults with immature (on right).
Right: Adult.

Masked Woodswallow

Artamus personatus

Endemic

180–190mm

Distinctive black face mask from lores, around eyes to throat, with white crescent edging. Upperparts: Silvery mid-grey. Underparts; Silvery pale grey. Tail: Outer feathers broadly tipped white. Eyes: Dark brown. Bill: Blue-grey, dark tipped. Female: Grey-black face with faint white edge. Underparts: Washed brown. Immature: Similar to female, with mottled whitish grey head, upperparts.

Often forms large flocks. Most dominant woodswallow in the west of range. In the eastern range in spring, often found with White-browed Woodswallows.

Habitat: Open forests, heaths, farmland with some trees.

Feeding: Insects caught in flight or gleaned from foliage. Some nectar taken.

Voice: Soft twittering, chattering.

Status: Moderately common. Migratory over southern Australia. Vagrant Tas.

Breeding: Aug.–Dec. Cup nest of green grass in a tree hollow or hollow stump, 1–6m high / 2–3 eggs.

Left: Male. Right: Female.

White-browed Woodswallow

Artamus superciliosus

Endemic

190–200mm

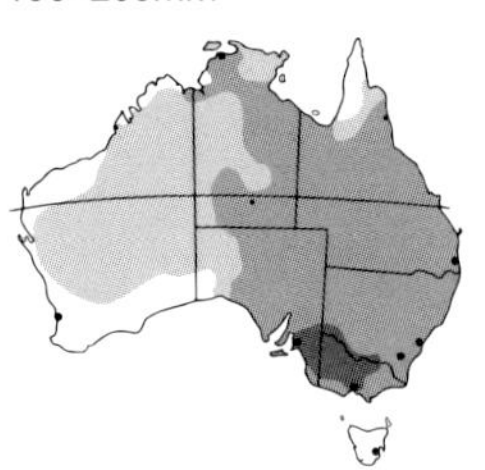

Long, pale blue, black-tipped bill and long, broad, white eyebrow. Upperparts: Blue-grey. Face: Blue-black. Underparts: Rich chestnut. Eyes: Dark brown. Female: Lighter bluish-grey above, cinnamon-buff below. Eyebrow: Muted. Immature: Mottled, similar to female.

Travel and roost together in large flocks of hundreds. Descending in unison.

Habitat: Open eucalypt, acacia woodlands, gardens, mostly west of Great Dividing Range.

Feeding: Insects caught in flight. Some fruit and nectar taken.

Voice: High-pitched, descending twittering. Occasional mimicry.

Status: Common. Nomadic. Migratory, southward in spring and northward in Feb.–Mar.

Breeding: Aug.–Dec. Cup nest in shrub, stump or cavity 1–6m high / 2–3 eggs.

Left: Male. Right: Immature.

Black-faced Woodswallow

Artamus cinereus

180–200mm

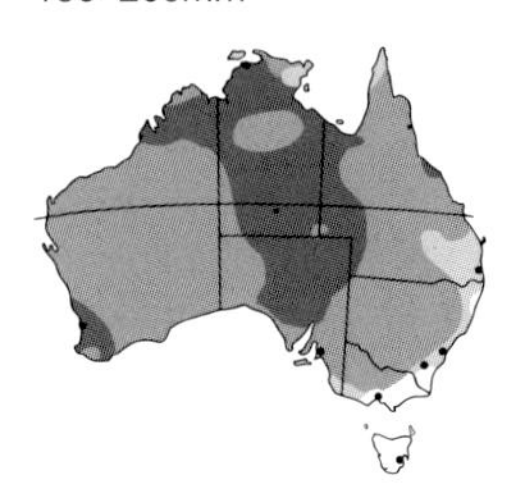

Sexes alike. Smoky grey upperparts and black face mask. Pale blue-grey bill with dark tip. Wings: Broad, pointed, deeper blue-grey. Rump: Black. Tail: Outer feathers broadly tipped white. Underparts: Pale smoky grey. Undertail coverts, vent: Black.
Race *normani*, northeast Qld. White undertail.
Immature: Grey-brown, mottled.

In pairs or small family groups and occasionally large winter flocks. Roost in groups at night.

Habitat: Open woodlands, inland plains, tropical north.

Feeding: Insects caught mostly mid-air. Has a divided brush-tipped tongue for taking nectar.

Voice: Constant twittering. Soft chirping contact calls.

Status: Common, mostly sedentary.

Breeding: Aug.–Jan. Minimal cup nest in tree, stump hollow or cavity 1–5m high / 3–4 eggs.

Opposite: Nominate race, feeding on Pine Banksia (left) and in flight (right).

Dusky Woodswallow

Artamus cyanopterus

Endemic

170–180mm

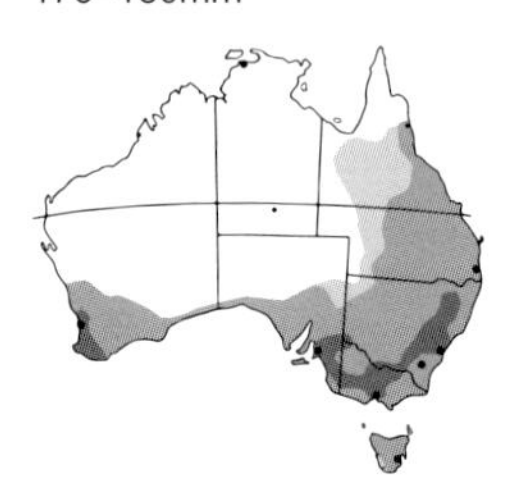

Sexes alike. Upperparts: Dusky brown with distinctive, wide, white edge to slate grey wings. Underwings: Silvery. Underparts: Dark dusky brown. Tail: Dusky black, broadly tipped white. Bill: Blue-grey, black tipped.
Immature: Mottled, brown.

Communal, 10–30 or more birds roost together.

Habitat: Open eucalypt woodlands and forests, Jarrah and Karri forests in the west.

Feeding: Hawks for insects on the wing.

Voice: Constant chattering and churring.

Status: Common. Sedentary or migratory. Summer breeding migrant to southern range.

Breeding: Aug.–Jan. Meagre nest of twigs, rootlets and grass, placed in a tree fork or hollow 1–11m high / 3–4 eggs.

Left: Adult incubating eggs. Right: Adult feeding chicks.

Little Woodswallow

Artamus minor

Endemic

120–140mm

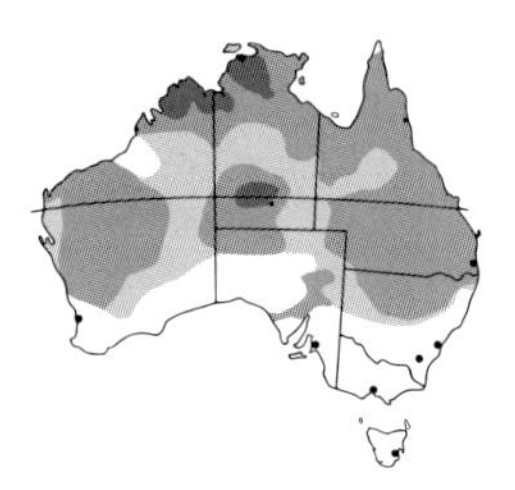

Sexes alike. Similar to Dusky Woodswallow, but smaller and lacks white edge to wings. Body: Dusky brown. Wings: Plain blue-grey. Rump: Grey-black. Tail: Grey-black, outer feathers tipped white. Bill: Blue-grey, black tipped.
Immature: Mottled.

Usually in pairs or small flocks of 6–20 birds. Often soars on thermal currents.

Habitat: Rocky outcrops, gorges near water, open woods, wetlands.

Feeding: Aerial feeders hawking for insects. Occasionally takes nectar.

Voice: Soft chattering.

Status: Common. Sedentary inland and north. Eastern birds migratory.

Breeding: Aug.–Jan. or after rain. Minimal nest in rock crevice or tree stump hollow / 3 eggs.

Left: Immature (on left) and adult. Right: Adults.

Australian Magpie

Cracticus tibicen

380–440mm

Familiar, large black and white bird. Possibly 8 similar races present in Australia. Broadly divided into the Black-backed Magpie and White-backed Magpie groups. Interbreeding occurs where ranges overlap.

Nominate race, Black-backed Magpie, Southeast Qld, NSW, east of Great Dividing Range. Head, throat, underparts: Glossy black. Nape, back: White with broad black centre band. Wings: Black, with white upper and lower wing coverts. Tail: White, broadly tipped black. Bill: Pointed, pale grey, black tipped. Eyes: Red-brown.
Female: Similar, greyish black, pale grey nape. Immature: Mottled, greyer-buff, shorter bill, black eyes.

The Black-backed Magpie group includes the nominate race and the following races:
Race *terraereginae*, inland Qld, inland NSW, adjacent areas in bordering states. Longer bill.
Race *eylandtensis*, Top End Magpie, Kimberley area, WA, NT, Groote Eylandt and across to the Gulf Country, Qld. Smaller. Narrow black tail band and upper-leg white 'trousers'. Male: Large white nape.
Female: Light grey nape.
Race *longirostris*, Long-billed Magpie mid-WA, Pilbara region. Mid-sized. Bill: Long, thin.

White-backed Magpie group: All males have all-white backs with black scapulars.
Females have geographically variable indistinctly mottled, scalloped black-grey backs.
Race *tyrannica*, mostly Vic. and adjacent areas in bordering states. Broad black tail band.
Race *telonocua*, mostly SA. Shorter bill. Paler, smaller. Scapulars: Black.
Race *dorsalis*, Western Magpie, southwest WA. Narrow black tail tip.
Female: Scalloped blackish-brown mantle, back. Scapulars: Black.
Race *hypoleuca*, Tas. and offshore islands. Small, short wings and bill. Scapulars: Black.

Habitat: Open woodlands. Widespread except for very dry desert areas. Well adapted to urban areas.

Feeding: Ground feeding, jabbing with powerful bill for insects and other invertebrates.

Voice: Rich, melodious, complex flute-like carolling and warbling.

Status: Common. Sedentary. Territorial groups of 3–24 birds.

Breeding: June–Dec. Nest is a platform of sticks and twigs placed in outer branches 5–20m high / 1–6 eggs.

Top left: Male Black-backed Magpie, nominate race.

Top right: Female Black-backed Magpie, nominate race.

Centre left: Male White-backed Magpie, race *dorsalis*.

Centre right: Female White-backed Magpie, race *dorsalis*.

Black Butcherbird

Cracticus quoyi

420–440mm

Sexes similar. Large, entirely glossy blue-black with metallic sheen. Eyes: Dark brown. Bill: Blue-grey with broad black tip. Cape York birds are smaller. Female: Smaller. Immature: Blue-black. Bill: Dark grey. Underparts: Buff-brown. Immature rufous-brown morphs occur in race *rufescens*, found in central coastal Qld.

Small groups or solitary. Shy.

Habitat: Coastal rainforests, mangroves, paperbark forests.

Feeding: Hunts from a perch. Insects, frogs, crustaceans, fruit, small birds, eggs.

Voice: Loud 'clonking'. Bubbling, yodelling song, repeated.

Status: Moderately common.

Breeding: Sept.–Feb. Untidy, large nest in upright tree fork, 5–10m high / 2–3 eggs.

Left: Race *rufescens*. Immature rufous-brown morph. Right: Adult with frog prey.

Grey Butcherbird

Cracticus torquatus

Endemic

260–300mm

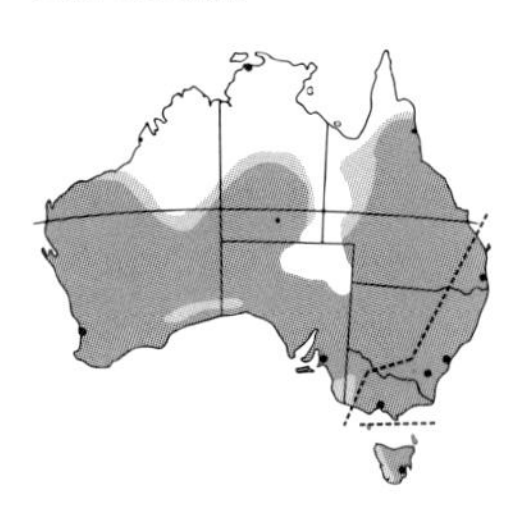

Strong, hooked, blue-grey and black-tipped bill. Head to nape: Black with white lores. Throat, shoulders: White. Back: Grey. Wings: Black with white stripe. Tail: Black, tipped white. Underparts: Pale grey. Nominate race, eastern Qld, NSW, Vic.
Race *cinereus*, Tas. Larger.
Race *leucopterus* elsewhere. Smaller, whiter wings.
Female all races: Similar. Crown, upperparts: Browner. Larger white spot on lores. Immature: Dark brown and buff.
Aggressive predator.

Habitat: Woodlands, open forests, parks and gardens.
Feeding: Perches and pounces, mostly insects, small birds, mice, lizards. Fruit and seeds also taken. Impales prey, including nestlings, often to be eaten later.
Voice: Loud, musical piping whistled song, often in duet.
Status: Common. Solitary. Sedentary.
Breeding: July–Jan. Untidy nest 3–10m / 3–5 eggs.

Opposite: Nominate race, male (left) and female (right).

Silver-backed Butcherbird

Cracticus argenteus

Endemic

240–280mm

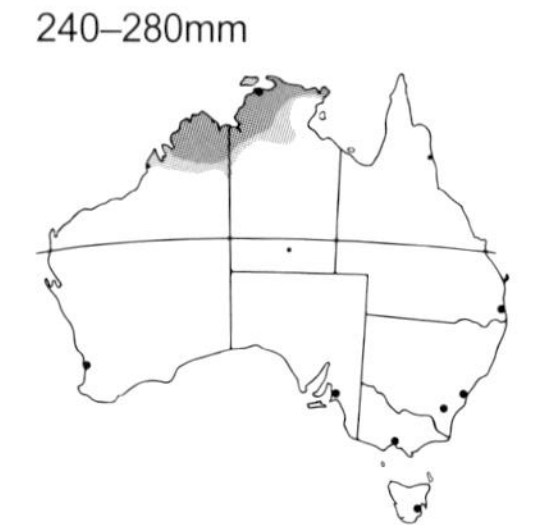

Sexes alike. Similar to the Grey Butcherbird, but smaller, with black lores and a pale silver-grey back. Belly: Silvery white. Bill: Black tipped.
Female: Smaller. Immature: Brown with buff streaking and dark brown head.

Aggressive and fearless. Chases larger birds, dogs or humans away from nesting area.

Habitat: Tall eucalypt woodlands, open forests, edge of rainforests. Urban areas.

Feeding: Impales prey. Perches and pounces, taking mostly insects, mice, lizards. Small birds are snatched in mid-air.
Voice: Similar to Grey Butcherbird but less vocal.
Status: Uncommon. Patchily distributed through range. Sedentary.
Breeding: Aug.–Jan. or after rain. Untidy nest 5–6m / 3–4 eggs.

Left and right: Adult.

Pied Butcherbird

Cracticus nigrogularis

Endemic

320–360mm

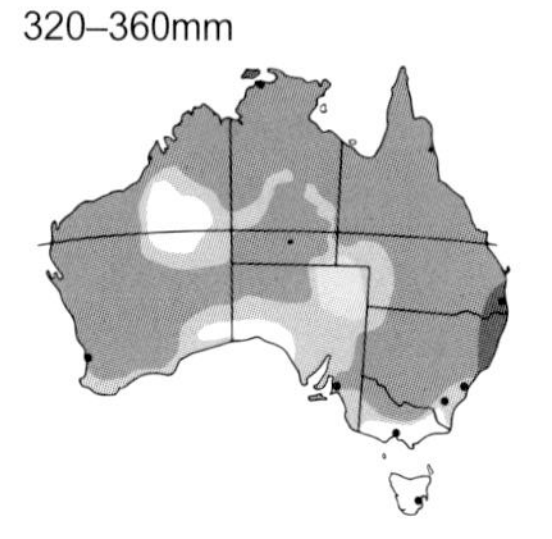

Sexes alike. Black hood and bib. Upperparts: Black, with broad white collar and white rump. Wings: Black, white bars. Tail: Black, outer feathers white tipped. Underparts: White. Immature: Brownish head. Underparts: Buff. Bill: Hooked, pale grey, dark tipped.

Relatively tame. Usually single, in pairs or family groups.

Habitat: Open woodlands and timbered farmland.

Feeding: Pounces from a vantage perch. Snatches mice, snakes, nestlings, insects.
Voice: Noted for its melodious, flute-like warbling, and complex songs, often in duet. Mostly at dawn, but also day and evening.
Status: Common.
Breeding: Aug.–Dec. in south. May–Nov. in tropics. Nest of sticks and twigs, lined with grass, in 3-pronged fork, 5m or more high / 3–5 eggs.

Left: Immature. Right: Adult.

Black-backed Butcherbird

Cracticus mentalis

250–270mm

Sexes alike. Similar to the Grey Butcherbird, but with bold black and white plumage. Upperparts black and white with broad-white partial collar and hooked, grey bill tipped black. Head, chin: Black. Wings: Black with broad white edges. Rump, lower back: White. Tail: Black tipped white. Underparts: White.
Immature: Patterned brown and fawn upperparts. Underparts pale fawn.

Pounces on prey from an open perch.
Habitat: Open forests and woodlands.
Feeding: Mostly in mid-canopy of eucalypts taking large insects, arboreal lizards, small birds. Acacia seeds also taken.
Voice: Piping, whistled carolling.
Status: Common.
Breeding: June–Nov. Untidy nest in tree fork up to 20m high / 3–4 eggs.

Left: Immature. Right: Adult.

Black Currawong

Strepera fuliginosa

Endemic

470–490mm

Sexes alike. Large and overall black with yellow eyes. Wings: Black, white bar to primaries. Tail: Black, white tipped. Bill: Very large, black, pointed. Immature: Similar, yellow gape.

Undulating flight. Usually solitary or in pairs.

Habitat: Diverse. Wet forests, heathlands, high alpine moors, open woodlands, farms, parks and gardens.

Feeding: Omnivorous. Scavenging in foliage, on the ground, for insects, carrion, small vertebrates, berries.

Voice: Rollicking series of wailing, croaking whistles.

Status: Common. Forms flocks in winter and disperses to lowlands.

Breeding: Aug.–Dec. Large, deep nest in fork 5–20m high / 2–4 eggs.

Left and right: Adult.

Pied Currawong

Strepera graculina

Endemic

420–500mm

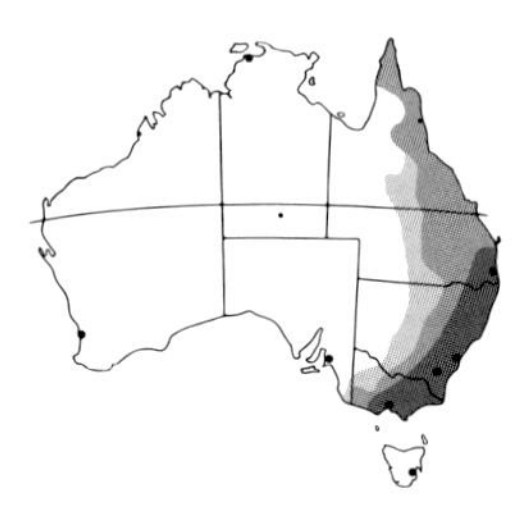

Sexes similar. Possibly 5 races in Australia with minor differences. Overall, large, sooty black birds with yellow eyes. Wings: Black, white band on base of primaries. Tail: Black, tipped white. Underparts: Black. Undertail coverts: White. Bill: Black, large, with fine hook. Far north Qld birds are smaller, with whiter areas on wings and tail. Female: Greyer. Immature: Grey-brown.

Forms flocks in winter and disperses to lowlands.

Habitat: Diverse. Open forests, woodland scrub, alpine areas, farmland, urban areas.

Feeding: Omnivorous scavenger. Insects, especially stick insects, nestlings, snails, carrion, berries.

Voice: Distinctive carolling 'cur-re-wong'. Drawn-out whistles.

Status: Common. Race *ashbyi*, western Vic., endangered.

Breeding: Aug.–Jan. Shallow nest, usually on outer fork of a eucalypt 7–25m high / 3 eggs.

Left: Adult male. Right: Female at nest.

Grey Currawong

Strepera versicolor

Endemic

450–510mm

Sexes similar. 6 subspecies occur in Australia, all with white-tipped tail and white undertail covert. Typical currawong, overall mid-grey, darker around lores and yellow eye. Wings: Grey, white-tipped primaries, white patch underneath. Tail: Grey, broad, white tipped. Underparts: Grey. Undertail covert: White. Bill: Grey, large, straight, pointed. Female: Greyish black. Immature: Grey-brown, Mottled breast, dark eyes.

Race *plumbea,* Squeaker, western SA and mostly southwestern WA. Similar to nominate, but bill is down-curved, thicker and darker. Sometimes plumage is tinged brownish.

Race *arguta*, Clinking Currawong, Tas. Largest, darkest. 'Klink-klank' call.

Race *intermedia*, Brown Currawong, Eyre and Yorke Peninsulas, SA. Smallest, browner.

Race *melanoptera*, Black-winged Currawong, Vic. border, SA, mallee area. Little or no white on wings.

Race *halmaturina*, Kangaroo Is. Dark, with no white on wings and narrow white tail tips.

Rapid undulating flight.

Habitat: Diverse. Coastal open forests, mallee, heathlands, urban areas.

Feeding: Omnivorous scavenger. Probing bark, foraging in foliage or feeding on the ground, taking rodents, frogs, small birds, nestlings, insects, carrion, seed, fruit.

Voice: Ringing, metallic, double-note 'clinking', often in flight.

Status: Common.

Breeding: July–Nov. Large, deep nest in tree fork 3–14m high / 2–3 eggs.

Row 3: Left: Nominate race. Right: Race *plumbea*, Squeaker.

Row 4: Left: Race *arguta*, Clinking Currawong. Centre: Race *intermedia*, Brown Currawong. Right: Race *halmaturina,* with no white on wings.

Family Corvidae, Ptilonorhynchidae, Paradisaeidae, Corcoracidae

Ravens

Crows

White-winged Chough

Apostlebird

Catbirds

Bowerbirds

Trumpet Manucode

Riflebirds

Four family groups that generally share unusual behaviour patterns and/or have exquisite plumage.

Family Corvidae includes ravens and crows.
Australia has 6 species. Most are difficult to identify due to their similarities. Three are ravens; the remainder are crows. All are black. Ravens pair for life and advertise their permanent territory at sunrise, patrolling the boundary and calling. They do not breed until they are over three years old. Crows are generally more common in inland areas, scavenging at rubbish tips and on carrion.

Family Ptilonorhynchidae includes catbirds and bowerbirds.
Catbirds are named for their cat-like wailing. Established pairs maintain a permanent territory and both sexes mark their territory with their distinctive call. Although members of the bowerbird family, they are monogamous and do not construct bowers for courtship.
Bowerbirds are named for the bower built by the male as part of their courtship behaviour. In the breeding season, a male perches above his bower, calling to attract females. When a female approaches, the male commences a display dance, flashing his brightest plumage, and if the female enters the bower copulation occurs. The female leaves and independently builds a nest and raises the young.
Bowers vary in complexity from the simple Tooth-billed Bowerbird's cleared area of forest floor decorated with 8 to 15 large, upturned leaves, to the intricate two 'maypole' towers 1–3m high, bridged by a low display branch, built by Australia's smallest bowerbird, the Golden Bowerbird. Other species build 'avenue' bowers of varying complexity. These range from the simple stick 'mat' display area with two 30cm high, parallel, vertical stick walls 15–20cm long of the Regent Bowerbird, to the largest 'avenue' bower of the Great Bowerbird. This has parallel walls of fine twigs, often overarching, 30–45cm high and up to 1m long, decorated with white objects.

Family Paradisaeidae includes the Trumpet Manucode and riflebirds.
The Trumpet Manucode is named for its far-reaching trumpet-like call and spectacular courtship display that includes the male facing the female, crouching, fluffing feathers and spreading wings while trumpeting.
Riflebirds are named for their green and black plumage that resembles the colours of 19th-century British army riflemen. They are noted for their stunning iridescent plumage, which is gradually obtained between their second and fifth year, and the elaborate courtship display posture and dance of the male, with bill pointed vertically and wings arched upwards.

Family Corcoracidae includes the White-winged Chough and Apostlebird.
White-winged Choughs were incorrectly named after the European Chough, a distant relative. Highly territorial and sociable, they live in family groups of 4 to 20 birds, with one breeding pair and the remainder helping to raise the young. The group eats, sleeps, plays and constructs their beautiful mud 'pudding-bowl' nest communally.
The Apostlebird was believed to live in family groups of 12 birds, giving rise to their name. They have similar communal behaviour to the White-winged Choughs and live in clans of 8 to 16 birds.

Opposite: Male Magnificent Riflebird, in courtship display.

Australian Raven

Corvus coronoides

Endemic

480–540mm

Sexes alike. Overall glossy black with a purple-green sheen. Throat: Shaggy, long, pointed hackle feathers form 'beard'. Eyes: White. Bill: Black, robust. Race *perplexus*, southwest SA, WA. Smaller, shorter hackles. Immature: Eyes: Brown. Underparts: Brownish.

Noisy, conspicuous. Immature and juvenile birds form flocks of up to 50 birds.

Habitat: Open woodlands, farmland, urban areas, rubbish tips, avoids dry desert areas.
Feeding: Omnivore, scavenger. Eats carrion, also insects, grain.
Voice: Strong, high, nasal calling, slow 'ah-ah-ah-aaaah' long wailing last note.
Status: Common. Territorial.
Breeding: July–Oct. Large nest sometimes on the ground, usually in tree fork to 10m high / 3–5 eggs.

Left: Nominate race. Right: Race *perplexus*.

Forest Raven

Corvus tasmanicus

Endemic

520–540mm

Sexes alike. Largest raven. Overall glossy black with purple-green sheen and small feather throat 'frill' when calling. Bill: Black, large. Tail: Relatively short.
Race *boreus*, northern NSW. Smaller bill, longer tail.
Immature: Underparts: Sooty brown. Eyes: Brown.

A breeding pair or family group occupy a territory throughout year. Pairs bond for life.

Habitat: Most habitats.
Feeding: Omnivore, scavenger. Carrion, also insects, fruit, seeds.
Voice: Deep, loud 'korr, korr, korr, korroo'.
Status: Locally common. Nomadic, non-breeding flocks form in winter. Patchy distribution in southeast mainland.
Breeding: Aug.–Nov. Large stick bowl nest, lined with grass, feathers, fur and usually high in a tree / 4 eggs.

Left and right: Nomimate race.

Little Raven

Corvus mellori

Endemic

480–500mm

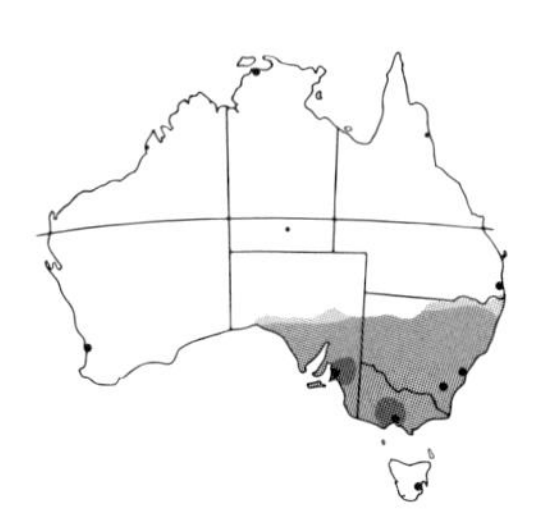

Sexes alike. Similar to, but much smaller than, Australian Raven. Overall black. Bill: Down-curved, less massive. Throat: Small, short hackles.
Immature: Duller. Eyes: Brown, turning hazel.

Strong, direct flight, rapid wingbeats. Wary, but can become relatively tame in urban areas. Synchronised display flights in flocks of hundreds of birds in southeast Australia. Often in colonies of up to 15 birds.

Habitat: Open woodlands, inland plains, farmland with water.

Feeding: Omnivore. Insects, small carrion, vegetable matter.
Voice: Deep, guttural, clipped notes 'kar, kar, kar kar', with distinctive wing shuffle as each note is called.
Status: Common. Nomadic. Breeding pair sedentary if food and water are constant. Alpine birds overwinter in lowlands.
Breeding: Aug.–Sept. Communal, shallow cup nest of twigs in the outer foliage to 10m high / 4–5 eggs.

Left: Adult showing hackles. Right: Adult in flight.

House Crow

Corvus splendens

Introduced

420–440mm

Introduced by shipping arriving from Asia. Invasive and disease carrying. Also known as the Indian Crow. If permitted to establish, it will prey on and displace native species and damage crops.
Mostly black with greyish or brownish breast and nape.

Opportunistic. Feeds mainly on refuse.

Habitat: Dockyards, urban areas near Fremantle, Geelong and Melbourne.
Feeding: Scavenger.
Voice: 'Caw'.
Status: Vagrant. Report sightings to your local State Department of Agriculture.
Breeds: India

Left and right: Adults.

Little Crow

Corvus bennetti

Endemic

450–480mm

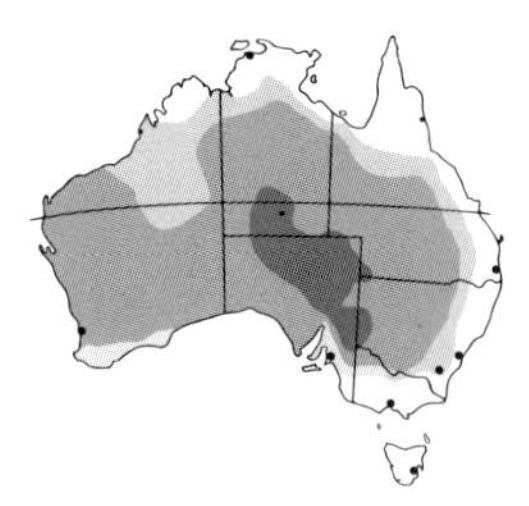

Sexes alike. Similar to the Torresian Crow, but slightly smaller and less scavenging. Overall glossy black with purple-green sheen, no neck hackles. Eye: White. Bill: Relatively short, slender.
Immature: Sooty brown underparts. Eyes: Brown.

Sociable. Large flocks throughout the year. Synchronised display flights.

Habitat: Diverse inland habitats. Arid mulga, mallee, scrublands, inland towns, outback stations and coastal areas in WA.

Feeding: Omnivore, scavenger. Eats large insects, carrion, vegetable matter. Flocks to abundant sources of food such as grasshopper plagues.

Voice: Guttural warbling, nasal calling 'nark, nark, nark, nark'.

Status: Common. Nomadic.

Breeding: July–Oct., or after rain. Sometimes in loose colonies. Small sticks, mud, cup-shaped nest in shrub, tree or telegraph pole, 1–10m high / 3–6 eggs.

Townships along the Strzelecki and Birdsville Tracks, SA.

Left and right: Adults.

Torresian Crow

Corvus orru

480–450mm

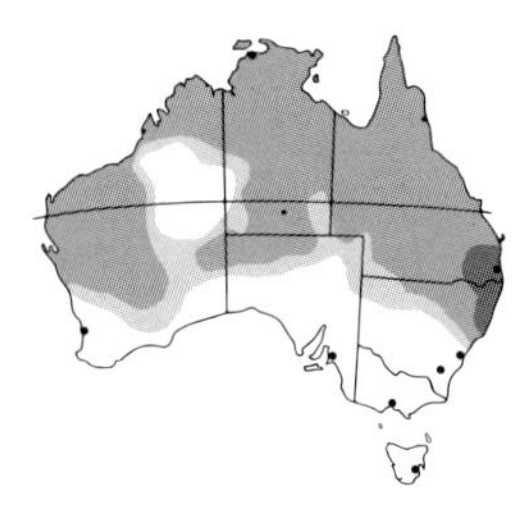

Sexes alike. Race *cecilae* present in Australia. Robust bird with overall glossy black plumage with purple-green sheen and white feather bases revealed when the feathers are ruffled. Throat: Indistinct hackle. Tail: Long, broad, square tipped. Eyes: White with a fine blue eye ring.
Immature: Overall dull black with brown eyes.

Noisy, conspicuous. Usually in small groups. Occasionally flock after breeding. Wary.

Habitat: Tropical farmland, woodlands and along timbered permanent watercourses.

Feeding: Omnivore. In pairs or parties, mostly insects, also carrion and vegetable matter.

Voice: High-pitched, nasal 'uk, uk, uk, uk'.

Status: Moderately common. Adults sedentary, territorial. Non-breeding birds are nomadic.

Breeding: Aug.–Oct. in south. Nov.–Feb. in north. Bowl-shaped nest 10m or more high, usually in a tree / 3–5 eggs.

Left and right: Adults.

White-winged Chough

Corcorax melanorhamphos

Endemic

430–470mm

Sexes alike. Large, sooty black bird with long, strong legs. Bill: Black, slender, curved. Eyes: Red, with orange inner ring. Wings: Black with distinctive white patch, mostly visible in flight. Tail: Long, black.
Immature: Smaller. Eyes: Brown. Tail: Shorter.
Highly sociable. Always seen in family groups of 5–20 birds. Usually 1 breeding pair in a group. Young take 4 years to reach breeding maturity. Young birds help with incubation, preening, feeding and rearing younger birds, as well as helping with nest building and defending the nest against intruders. Group behaves in noisy unison.

Habitat: Forests, woodlands, mallee near water, outer urban areas, parks and gardens.

Feeding: Mostly ground forager, scratching, probing, flicking leaf litter for insects, snails, seeds.

Voice: Mournful, piping descending whistles. Screeching alarm call.

Status: Common, mostly in groups of up to 20.

Breeding: Aug.–Dec. Pudding-bowl-shape mud nest on limb 4–15m high / 3–5 eggs.

Australian National Botanic Gardens, ACT.

Left: Family group. Right: Adult showing white wing patches.

Apostlebird

Struthidea cinerea

Endemic

290–330mm

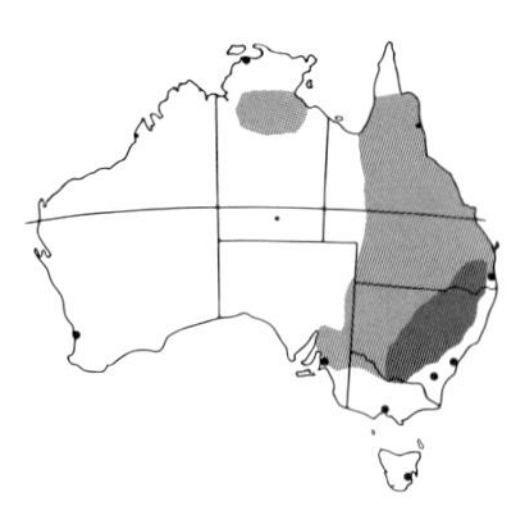

Sexes alike. Overall mid-grey with pale grey–tipped plumage. Longer, pointed feathers on the head and breast give an overall 'shaggy' appearance. Eyes: Grey with pale yellow outer ring. Bill: Stubby, black. Wings: Brownish grey, edges buff-rufous. Tail: Long, blackish.
Immature: Brown eyes.

Highly sociable, communal. Breeding group of about 10 birds with a dominant male and several females with immature young helping with rearing younger birds and nest building. Usually seen on the ground.

Habitat: Near water, open woodlands, mulga scrub, areas with White Cyprus Pine, Brigalow, Lancewood and homestead gardens.

Feeding: Forages on ground for insects, seeds and grasses.

Voice: Continuous variety of harsh chattering, rasping alarm, chorus.

Status: Common.

Breeding: Aug.–Feb. Bowl-shaped mud nest placed on a horizontal limb, 2–17m high / 2–5 eggs.

Eulo Bore, Qld.

Opposite: Adult.

Spotted Catbird

Ailuroedus melanotis

260–300mm

Sexes alike. Prominent bright green back with brownish olive and spotted dull white crown and nape. Wings: Bright olive green with white wing bars. Eyes: Red. Ear coverts: Blackish. Bill: Stubby, whitish. Underparts: Light olive green streaked white. Tail: Light olive, tipped white. Race *joanae*, Cape York, Qld. Similar. Head is darker. Underparts: More heavily spotted.
Immature: Duller. Eyes: Brown.

Usually in singles or pairs. Occasionally small groups of 3–20 in autumn and winter.

Habitat: Tropical rainforests, vine scrubs, paperbark forests.

Feeding: In the upper canopy on fruit, insects, small reptiles and nestlings.

Voice: Nasal, drawn-out cat-like wail, often rising–falling in pitch. Throughout the day, especially in the breeding season.

Status: Moderately common. Sedentary. Moving to lower altitude in winter.

Breeding: Sept.–Jan. Large bulky cup-shaped nest of twigs and vines, usually in a prickly shrub or tree fork, 2–23m high / 1–3 eggs.

Mt Hypipamee NP and Mt Lewis NP, Qld.

Left and right: Nominate race.

Green Catbird

Ailuroedus crassirostris

240–320mm

Sexes alike. Distinguished from the Spotted Catbird by range and lack of black on face. Plumpish green bird with red eyes and strong whitish bill. Back, rump: Emerald green. Wings: Emerald green with bright white spots forming bars. Tail: Olive green, white tipped. Head: Greenish brown, mottled black and buff. Underparts: Light olive green with short white streaks.
Immature: Duller, less streaked.

Bonded pairs. Male feeds the female at the nest and throughout the year.

Habitat: Tropical and subtemperate rainforests and paperbark forests.

Feeding: Forages from ground to upper canopy, mostly fruit, favours figs. Also seeds, insects. In the breeding season takes nestlings, small reptiles, frogs.

Voice: Nasal, drawn-out, cat-like wail. Often rising–falling pitch. Quiet in non-breeding season apart from 'chip' location calls.

Status: Moderately common. Sedentary.

Breeding: Sept.–Feb. Large bulky cup nest of twigs and vines, usually in a prickly shrub, 3–6m but up to 23m high / 1–3 eggs.

Lamington NP, Qld. Royal NP, NSW.

Left and right: Adult.

Satin Bowerbird

Ptilonorhynchus violaceus

Endemic

270–330mm

Overall glossy blue-black plumage. Eyes: Violet-blue. Bill: Short, whitish with yellow-green tinged tip.
Female: Upperparts: Olive green. Wings, tail: Rufous-brown, tinged green. Underparts: Buffy yellow, brown scalloping, heaviest on flanks. Eyes: Blue. Race *minor*, northeast Qld. Smaller and female is greener. Immature, all races: Similar to female. Immature males gradually develop a green and white–streaked throat, green breast and paler bill. These 'green' birds are often seen in groups. Males achieve adult plumage in years 5–6.

Males are solitary. Females, immatures often in groups.
Habitat: Rainforests, wet eucalypt forests and adjacent woodlands. Parks, gardens.
Feeding: Mostly fruit. Insects and leaves taken in breeding season.
Voice: Loud, harsh churrs. Drawn-out whistles. Buzzing and rattling.
Status: Moderately common.
Breeding: Sept.–Jan. Male builds 'avenue' bower adorned with blue objects, painted with saliva and blue berries. Female, builds shallow bowl nest 2–35m high / 2 eggs.

Lamington NP, Qld. Royal NP, NSW.

Left: Male working on his bower. Right: Female.

Regent Bowerbird

Sericulus chrysocephalus

Endemic

240–280mm

Glossy jet-black body with golden yellow crown, mantle and broad wing patch. Crown may have a red-orange patch. Eyes: Yellow.
Female: Upperparts: Olive-brown mottled grey on mantle. Head: Mottled with black patch on crown. Throat: Small black patch. Wings, tail: Olive-brown. Underparts: Fawn, mottled. Eyes: Yellow, brown flecked. Bill: Blackish, yellow gape in breeding season. Immature: Similar to female. Full adult male plumage takes about 5 years.

Habitat: Rainforest margins, coastal scrub, forests.

Feeding: Ground level to upper canopy, mostly fruit, also insects.

Voice: Metallic, harsh churring. Explosive hissing and chattering.

Status: Sedentary. Moves to lowlands in winter.

Breeding: Oct.–Jan. On a shallow base of twigs, male builds small 'avenue' bower decorated with shells, seeds, berries and often paints the walls blue or green with mud-like 'saliva paint' using wads of leaves as a 'paintbrush'. Female builds loosely woven twig 'saucer' nest usually in vines or mistletoe, 3–25m high / 2 eggs.

Lamington NP, Qld.

Left: Male. Right: Female.

Golden Bowerbird

Prionodura newtoniana

Endemic

230–250mm

Smallest bowerbird, with a slender body with glistening golden crest feathers on head and nape. Head, chin, back, wings: Golden olive-brown.
Tail: Long, forked, golden outer feathers, inner feathers washed brown. Eyes: Yellow. Underparts: Golden yellow, glossy highlights.
Female: Upperparts: Olive-brown. Underparts: Ash grey, faintly streaked. Eyes: Yellow. Tail: Shorter. Immature: Similar to female, dark brown eyes.
Habitat: Highland rainforests above 900m.
Groups as little as 20km away have own distinct song routine.
Feeding: Ground level to upper canopy, mostly fruit, also beetles, cicadas, other insects.

Voice: Drawn-out whistles and metallic chattering, mimicry.
Status: Rare.
Breeding: Oct.–Jan. Male builds a 2-towered 'maypole' bower of sticks, 2–3m tall, around 2 slender trunks connected by a fallen branch and skilfully laid sticks. Decorated with flowers, moss and fruits. Males larder fruits to maximise time spent at bower, sometimes stealing prized decorations from other bowers. Female builds a small cup nest of dried leaves lined with twigs, in tree cavity, 1–3m high / 2 eggs.

Mt Lewis NP and Paluma Dam Rd, Qld.

Left: Male at bower. Right: Female feeding.

Spotted Bowerbird

Chlamydera maculata

Endemic

280–300mm

Sexes similar. Upperparts dusky brown, boldly spotted ochre to cream, paler on crown. Small, lilac-pink erectile crest above ash grey lower nape. Underparts: Buff with light brown streaks to throat and breast, mottled barring lower parts.
Female: Much shorter crest, sometimes absent. Immature: Lacks crest.
Solitary when breeding. Flocks of up to 30 may form after breeding.
Habitat: Dry, open sclerophyll woodlands with acacia, eucalypt and fruiting understorey plants. Visits rural gardens.
Feeding: Mostly fruit, flowers and seeds and occasional insects.
Voice: Loud, ringing, metallic calls. Harsh hissing, clicking and mimicry, including the calls of predatory birds, stockwhips, wood chopping and others.
Status: Uncommon NSW. Endangered elsewhere. Sedentary or nomadic.
Breeding: Oct.–Jan. Male builds wide 'avenue' bower of twigs, grass, decorated with white, green, amber, red and shiny objects, adjacent decorated display court for performance. Female nest 2–18m high / 2–3 eggs.

Eulo Bore, Qld.

Opposite: Adult male at bower.

Western Bowerbird

Chlamydera guttata

Endemic

250–270mm

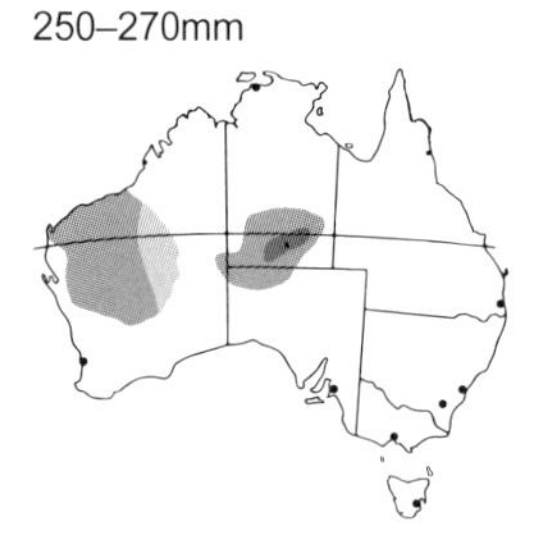

Sexes similar. Similar to Spotted Bowerbird, but smaller, richer and darker colour with no grey patch on lower nape.
Nape: Lilac-pink erectile crest.
Upperparts, throat: Blackish brown with feathers finely tipped and edged orange-buff.
Underparts: Deep buff, brown scalloped. Tail: Brown, whitish tip. Bill: Down-curved, small hooked tip.
Female: Smaller crest. Immature: Paler.
Male performs at bower with tail flicking, churring sounds, mimicry, displaying his lilac-pink crest erect.
Habitat: Rocky gorges in semi-arid inland with figs (especially Desert Fig), close to water. Along watercourses with dense vegetation. Homestead gardens, camping grounds.
Feeding: From ground to canopy. Mostly figs and other fruits, also seeds, nectar, spiders, insects.
Voice: Loud, harsh hissing, clicking and mimicry. Metallic ringing 'advertising' call.
Status: Moderately common. Sedentary.
Breeding: Oct.–Jan. Male builds 'avenue' bower decorated with white and green objects. Female builds twig saucer nest in tree fork, 2–18m high / 2–3 eggs.

Olive Pink Botanic Garden, NT, Cape Range NP, WA.

Opposite: Male decorating bower.

Great Bowerbird

Chlamydera nuchalis

Endemic

330–370mm

Sexes alike. Australia's largest bowerbird. Mid-grey head with silver-flecked crown. Broad, lilac-pink erectile crest to nape.
Upperparts: Grey-brown mottled, scalloped grey-white. Tail: Whitish tip. Underparts: Fawn-grey. Eyes: Dark brown. Bill: Brown, stout, curved.
Immature: Paler, lacks neck patch.
Bounds along ground or between branches. Flight is undulating. Singles or small groups.
Habitat: Tropical woodlands, eucalypt forests, dense undercover, mangroves and tropical gardens.
Feeding: Usually in the canopy, infrequently from ground. Forages mostly for fruit, also insects, spiders.
Voice: Rasping drawn-out call. Mimicry, hissing, churring and grating.
Status: Common. Sedentary. Solitary.
Breeding: Aug.–Feb. Male builds largest 'avenue' bower up to 1m long, 45cm high with walls 20cm apart. Decorated with shells, berries, human objects. Female builds loose, flat nest in a tree fork,10m high / 1–2 eggs.

Nitmiluk NP, NT. Windjana Gorge NP, WA.

Left: Adult. Right: Adult doing bower maintenance.

Fawn-breasted Bowerbird

Chlamydera cerviniventris

260–290mm

Sexes alike. Mostly drab grey-brown bowerbird, spotted buff-white and finely streaked brown head and neck. Lacks crest. Throat: White streaked. Breast to undertail: Fawn. Eyes: Dark brown. Bill: Black, stout, down-curved.
Immature: Similar, but heavier streaking, grey frons.

Solitary or in pairs. Sometimes in small groups.

Habitat: Tropical woodlands, mangrove fringes, vine thickets, paperbark forests.

Feeding: Bounds through foliage, picking figs or other fruits or occasionally insects.

Voice: Grating, rasping, scratching notes and explosive abrupt calls.

Status: Rare. Sedentary.

Breeding: Sept.–Dec. Male builds an avenue bower about 30cm high on top of a 30cm raised stick platform to provide a display area at each end of the bower. Bower is ornately decorated with green fruits and painted green. Female builds a bulky saucer-shaped nest, placed in tree, usually a pandanus,10m high / 1 egg.

Iron Range NP, Qld.

Left and right: Adult.

Tooth-billed Bowerbird

Scenopoeetes dentirostris

Endemic

260–270mm

Sexes alike. Distinctive, heavy, grey-black bill with serrations on the cutting edge of both mandibles for tearing leaves to eat and for pruning leaf decorations for its bower. Upperparts: Olive-brown. Eyes: Brown with pale, bare, narrow eye ring. Underparts: Dark cream, heavily streaked brown.
Immature: Similar. Lighter brown. Bill: Brown. Eyes: Paler.

Solitary or somethimes in small groups. Quiet, unobtrusive.

Habitat: High-altitude tropical rainforests.

Feeding: Usually in the upper canopy, taking fruit, vegetable matter, insects and their larvae.

Voice: Male has melodious, loud, whistles, chatter, mimicry in breeding season from a perch located near display stage. Usually silent at other times.

Status: Uncommon.

Breeding: Nov.–Feb. Male builds a 'stage' of cleared earth 1–3m dia. decorated with fresh, large green leaves upturned to show their pale underside. If a female approaches, he drops to the ground and displays. Old leaves are regularly replaced with fresh leaves. Female builds a flimsy shallow nest, placed in vines or tree fork 3–25m high / 2 eggs.

Mt Hypipamee NP and Mt Lewis NP, Qld.

Left and right: Adult.

Trumpet Manucode

Phonygammus keraudrenii

380–320mm

Sexes similar. Named for its loud, far-reaching trumpet-like call. Overall glossy blue-black with oil-like iridescent sheen, changing to green-purple with light direction. Head: Elongated 'ear tuffs'. Nape: Long, pointed plumes. Eyes: Bright red. Bill: Black.
Female: Smaller, duller.
Immature: Less glossy.

Performs a display dance with chasing, bowing, calling and lifting of wings with plumes fluffed out.

Habitat: Tropical and lowland rainforests, forest fringes, occasionally mangroves.

Feeding: Mostly fruit, favours figs. Also insects from the mid- to upper canopy.

Voice: Loud, deep, guttural trumpet-like call with a resonant quality imparted by a lengthened windpipe just under the breast skin.

Status: Moderately common.

Breeding: Oct.–Jan. Shallow, open basket nest of vine tendrils slung in the upper canopy, often near a Black Butcherbird nest, up to 20m high / 1–2 eggs.

Iron Range NP, Qld.

Left and right: Adult male showing the elongated 'ear tuffs'.

Victoria's Riflebird

Ptiloris victoriae

Endemic

230–250mm

Similar to Paradise Riflebird, but smaller with wider black breast band and smaller, iridescent blue-green triangular-shaped throat patch. Crown: Iridescent, metallic green. Underparts: Black with broad, yellowish scalloped edges. Bill: Long, grey-black, down-curved, yellow mouth. Female: Similar to the female Paradise Riflebird but smaller. Eyebrow: Pale buff. Underparts: Rufous-buff, mottled. Immature: Like female.

Male performs from a display perch, pointing its bill upwards and clapping its arched-up wings.

Habitat: Tropical rainforests above 500m.

Feeding: Mid- to upper canopy. Forages for fruit, insects. Often spiralling up trunks.

Voice: Harsh, rasping, drawn-out call, similar to Paradise Riflebird.

Status: Moderately common. Sedentary.

Breeding: Sept.–Jan. Shallow cup-nest in vines, tree ferns or canopy, 2–3m high / 2 eggs.

Mt Hypipamee NP and Julatten area, Qld.

Left: Female. Right: Male.

Paradise Riflebird

Ptiloris paradiseus

280–300mm

Spectacular when viewed in full sunlight. Long, black, down-curved bill and lime-yellow inside of mouth. Frons, crown, throat, breast: Iridescent green. Back: Black with purple sheen. Wings: Black, rustle in flight (males only). Tail: Short, black, with iridescent green central feathers. Upper breast, throat: Scalloped iridescent blue-green triangular-shaped patch above a broad black breast band. Lower breast to tail coverts: Black with yellow-olive scalloped edges. Female: Olive-brown upperparts. Bill: Fawn, longer. Eyebrow: White. Underparts: Buff, with dark brown crescent barring. Immature: Similar to female.

Usually located by falling debris from feeding birds above. Male performs elaborate courtship displays on a horizontal limb, including wing clapping, tail cocking, swinging upside down and turning side to side.

Habitat: Subtropical, temperate rainforests, dispersing into nearby woodlands after breeding.

Feeding: Probes bark, decaying logs for insects. Forages for fruit.

Voice: Rasping, loud, explosive, drawn out 'yaaas'. Soft churring. Status: Uncommon.

Breeding: Oct.–Jan. Shallow nest in vines or canopy 5–25m high / 2 eggs.

Lamington NP, Qld.

Left: Female. Right: Male.

Magnificent Riflebird

Ptiloris magnificus

Male: 300–320mm
Female: 260–280mm

Iridescent blue-green crown. Velvety black face and neck. Nape, throat, upper breast shield: Iridescent, colour changing with movement with lower yellow border to breast shield. Underparts: Purple-black with fine feathery plumage to flanks. Tail: Short, black, iridescent central feathers. Wings: Short, male feathers rustle loudly. Female: Short, off-white eyebrow and long, off-white whisker line with dark line below. Underparts: Off-white with brown barring. Upperparts: Cinnamon-grey. Wings, tail: Bright rufous. Immature: Like female.
Male courtship display includes suddenly folding out 1 wing at a time, head swaying, shaking silky plumage and pointing bill vertically.

Habitat: Tropical rainforests, swamps, forest edges.

Feeding: Mostly fruit. Tears away bark, foraging for insects.

Voice: Deep, trumpeting in display. Harsh rasping.

Status: Moderately common. Sedentary. Elusive.

Breeding: Sept.–Feb. Deep, loosely woven cup nest 3–15m high / 2 eggs.

Iron Range NP, Qld.

Left: Female. Right: Male.

25 Family Muscicapidae, Sturnidae, Hirundinidae, Zosteropidae, Dicaeidae, Nectariniidae

Thrushes

Starlings

Swallows

Martins

White-eyes

Silvereye

Mistletoebird

Sunbird

A grouping of six families.

Family Muscicapidae includes the thrushes.
There are two similar species of secretive and mostly silent native thrushes best defined by their range: the Bassian Thrush prefers cool highland forests while the Russet-tailed Thrush ocuppies lower elevations. The European Song Thrush and Common Blackbird are introduced species that are renowned for their melodious song.

Family Sturnidae includes starlings.
It is represented in Australia by one native starling, the Metallic Starling, and two introduced species, the Common Myna and Common Starling. The Metallic Starling is noted for its large, noisy nesting colonies containing up to 100 bulky, domed, treetop nests.

Family Hirundinidae includes swallows and martins.
Both groups have short, broad bills to help catch airborne insects, and effortless flight that is aided by their sleek, glossy plumage, which reduces air friction, and long pointed wings. Australia's only endemic swallow, the White-backed Swallow, is noted for its communal behaviour, with up to 80 birds sleeping together in a roosting chamber at the end of a burrow drilled into the side of an embankment. It is uncommon in parts of the inland due to fox predation – foxes dig out the roosting chamber.

Family Zosteropidae includes the widespread Silvereye, with many geographic plumage variations, and the similar white-eyes of northern Australia.
Silvereyes are named for the ring of white feathers surrounding the eyes. After breeding, large groups of Tasmanian and southern Victorian birds gather in flocks and migrate northward along established routes, following coastal heathlands and gardens. Migrating groups travel at night at high altitude, constantly making their whistling chirp contact calls. Not all Tasmanian birds leave and some drift locally.

Family Dicaeidae contains the Mistletoebird, Australia's sole member of the flowerpecker family.
The Mistletoebird has a symbiotic relationship with mistletoe and feeds almost exclusively on the ripe fruits. The bird has a specialised digestive system that allows it to quickly pass the seed with its sticky coating, which helps it to adhere to a branch and grow. Australia has over 60 species of mistletoe. The 'bootee-shaped' nest of the mistletoebird is constructed from plant fibre, densely bound with spider web and suspended from a slender twig, well hidden in the outer foliage.

Family Nectariniidae contains the Olive-backed Sunbird, the sole Australian representative.
A small and fearless, brightly coloured bird, it darts from flower to flower, sucking nectar with its slender tongue that is rolled into a partial tube. It will enter houses for spiders, insects and often in the tropics will nest under verandahs to shelter from storms. The nest is an elongated pendulous dome with a long trailing tail and hooded side entrance. It is suspended and constructed from plant fibre and spider web and decorated with spider cocoons and flakes of paperbark.

Opposite: Tree Martin leaving its nest hollow.

Bassian Thrush

Zoothera lunulata

Endemic

250–290mm

Sexes alike. Heavily scalloped plumage. Upperparts, head, face: Coppery to olive-brown, feathers edged black. Lores, eye ring: Whitish. Wings: Brown flight feathers, pale wing bar in flight. Throat, breast: Pale buff scalloped brown. Lower underparts: White, scalloped black.
2 isolated races.
Race *cuneata*, north Qld. More rufous scalloped breast. Race *halmaturina*, southeast SA, King Is. Pale cinnamon scalloped breast.
Immature: Similar to adult.

Secretive, unobtrusive and well camouflaged in leaf litter. Runs with head down, often freezes if disturbed. Sometimes in loose groups of 10–20.

Habitat: Dense rainforests, wet eucalypt forests, gullies, parks and gardens with thick leaf litter, above 500m altitude.

Feeding: Hopping, scratching for insects, earthworms, snails, fallen fruit.

Voice: Flute-like, rising–falling 3-notes. Whistled trills.

Status: Moderately common. Mostly sedentary.

Breeding: July–Dec. Rounded, deep cup nest, 2–15m high / 2–3 eggs.

Right: Nominate race.

Russet-tailed Thrush

Zoothera heinei

Endemic

260–290mm

Sexes alike. Similar to the Bassian Thrush, but has a more rufous rump and uppertail coverts and finer barring, scalloping. White outer edge to tail, mostly visible in flight. Upperparts: Brighter coppery rufous. Tail: Shorter. Underparts: Whitish, lightly scalloped.
Immature: Similar to adults.

In the field, difficult to distinguish from the Bassian Thrush, particularly as their ranges overlap. Best identified by voice. Altitude can be indicative. Russet-tailed Thrush is usually found in lower altitudes than the Bassian Thrush.

Habitat: Wet, cool rainforests, wet eucalypt forests, lowlands, below 750m altitude.

Feeding: Similar to Bassian Thrush. Ground foragers, hopping, scratching in leaf litter for insects, earthworms, snails, fallen fruit.

Voice: Differs from Bassian. Thrush with a 2-note whistle. Subdued wren-like contact call.

Status: Locally common. Usually sedentary.

Breeding: July–Dec. Rounded, cup-shaped nest covered with moss 2–15m high / 2–3 eggs.

Lamington NP, Qld.

Opposite: Pair at moss-covered nest.

Metallic Starling

Alponis metallica

210–240mm

Sexes similar. Overall glossy black with an iridescent green-purple sheen. Distinctive long tail and orange-red eyes.
Immature: Duller. Underparts: White, streaked black.

Low, fast flying flocks.

Habitat: Coastal lowland tropical rainforests, adjoining woodlands, parks and gardens with fruiting trees. Sometimes mangroves.

Feeding: Mostly fruit, occasionally insects.

Voice: Incessant chatter, harsh wheezing.

Status: Locally common. Breeding migrant, arriving Aug., departing Mar. to New Guinea.

Breeding: Huge noisy breeding colonies. Aug.–Dec. Bulky, woven domed nest with side entrance, crowded together with up to 100 birds in a single large tree / 2–3 eggs.

Centenary Lakes Botanic Gardens Cairns and Cairns Esplanade, Qld.

Left: Adult (foreground) and immature at the nest. Right: Adult.

Song Thrush

Turdus philomelos

230mm

Introduced Melbourne 1853.
Sexes, immature: Similar.
Upperparts: Plain rufous-brown.
Underparts: White with buff tint to breast. Throat to lower belly: Black spotted. Eyes: Dark brown. Bill: Dusky top, yellow-brown below.
Usually solitary or in pairs.
Habitat: Melbourne parks, gardens. Damp woods, forests.

Feeding: Ground foraging. Snails smashed on a rock.
Voice: Male sings continuously in breeding season. Melodious sequence of notes and mimicry.
Status: Uncommon.
Breeding: Sept.–Jan. Mud, grass bowl nest. Low / 4–5 eggs.

Left: Adult. Right: Immature.

Common Blackbird

Turdus merula

250–260mm

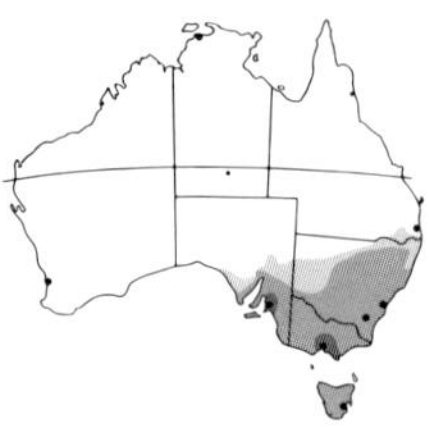

Introduced Melbourne 1850.
Overall black with surface sheen. Orange bill.
Female: Dark grey-brown, darker on wings and tail. Underparts: Light grey, dark streaks.
Immature: Similar to female, paler underparts.
Competes with native birds for food and nesting sites.
Habitat: Urban areas, orchards, parks, sometimes bushland.

Feeding: Ground feeding in damp litter on insects, snails, spiders, seed, fruit.
Voice: Loud, mellow sustained fluting song with high trills.
Status: Common.
Breeding: Sept.–Dec. Mud, grass, cup-shaped nest / 3–5 eggs.

Left: Female. Right: Male.

Common Starling

Sturnus vulgaris

210–230mm

Introduced 1860s. Declared pest. Sexes similar. Glossy, iridescent green-purple sheen, appears black in most light. Pale yellow bill, black in non-breeding. Overall speckled after annual moult.
Immature: Grey-brown. Dark bill. Aggressively competes with native birds for nest hollows, often evicting them and/or killing their young.
Habitat: Urban, farm, bush areas.

Feeding: Insects and orchard fruit in equal measure.
Voice: Nasal, warbling, twittering.
Status: Abundant. Migratory. Nomadic. Displaces native species.
Breeding: Aug.–Jan. Rough, untidy grass cup-shaped nest in roof spaces, natural or other human-made cavity / 4–8 eggs.

Left: Autumn plumage. Right: Spring plumage.

Common Myna

Acridotheres tristis

230–250mm

Native to India. Declared pest. Released in 1860 in a failed attempt to control insects. Sexes similar. Chocolate-brown body. Head, throat, wings, tail: Blackish brown. Eye skin, bill: Yellow. Underwings: Large, white panels seen in flight.
Immature: Duller. Plain brown. Invasive and aggressive pest. Usurps nest hollows.

Habitat: Urban areas.

Feeding: Feral scavenger for food scraps, insects, fruit.
Voice: Whistles, harsh rattling.
Status: Locally abundant. Destroy nests. Trap and dispose of birds where possible.
Breeding: Oct.–Mar. Messy nest, in cavity or thick vegetation / 4–5 eggs.

Left: Adult. Right: Adult taking nest hollow.

Red-whiskered Bulbul

Pycnonotus jocosus

200–220mm

Introduced Sydney in 1890s from India. Sexes similar. Distinctive, pointed black crest. Black head with red tuft behind eyes and white cheeks. Back: Olive-brown. Breast, belly: Pale fawn. Undertail coverts: Red.
Immature: Duller.
Gregarious after breeding season. Roost communally.

Habitat: Shrubbery, urban areas.
Feeding: Insects, fruits including lantana.
Voice: Melodic whistles. Animated chattering.
Status: Common in Sydney, Coffs Harbour, NSW, Mackay, Qld. Rare in Melbourne.
Breeding: Aug.–Feb. Untidy cup nest 2–3m high / 2–4 eggs.

Left and right: Adults.

White-backed Swallow

Cheramoeca leucosterna

Endemic

140–150mm

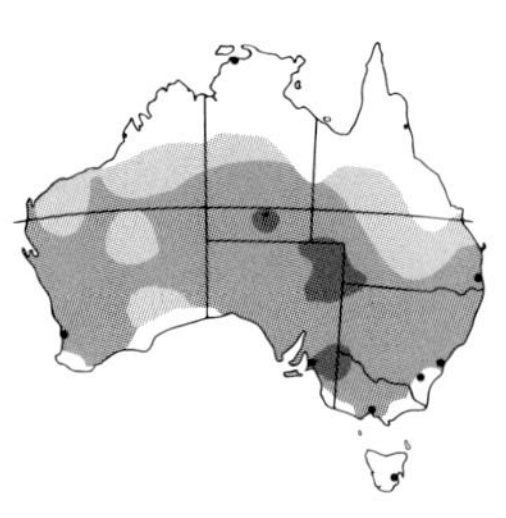

Sexes alike. Distinctive bright white mantle and mottled brown crown. Lores, face: Black. Nape: Deep-brown. Chin to upper breast: Bright white. Underparts: Black. Lower back, rump: Blue-black. Tail: Blue-black, deeply forked, long outer feathers. Immature: Duller.

Usually seen in small flocks. Graceful, fluttering flight.

Habitat: Open country, lightly timbered, inland near water and sandy areas for nesting.

Feeding: Insects caught mid-air or often over water.
Voice: Continuous, twittering.
Status: Moderately common. Sedentary or nomadic. Communal, up to 80 birds nesting and roosting together.
Breeding: Aug.–Dec. Burrow nest dug into sandy banks, severely predated upon by foxes that dig out nesting chamber / 4–6 eggs.

Left and right: Leaving the nest hollow.

Barn Swallow

Hirundo rustica

140–160mm

Sexes similar. Race *gutturalis* present in Australia. Distinctive thin, black broken breast band below chestnut throat. Upperparts: Metallic blue-black. Frons, face, throat: Bright tan. Underparts: White. Tail: Plain central feathers, other feathers edged white, more visible from underneath. Outer feathers extremely long 'streamers'. Underwing coverts: White. Female: Shorter tail 'streamers'. Immature: Variable. Paler. Upperparts: Duller, mottled brown, lacks breast band.

Straight flight, rarely glides.

Habitat: Open country, farmland, urban areas, near water.
Feeding: Forages low, mostly over surface of water. Hawks for insects on the wing.
Voice: High-pitched twittering.
Status: Moderately common. Summer migrant.
Breeding: East Asia.

Left and right: Adult.

Welcome Swallow

Hirundo neoxena

150mm

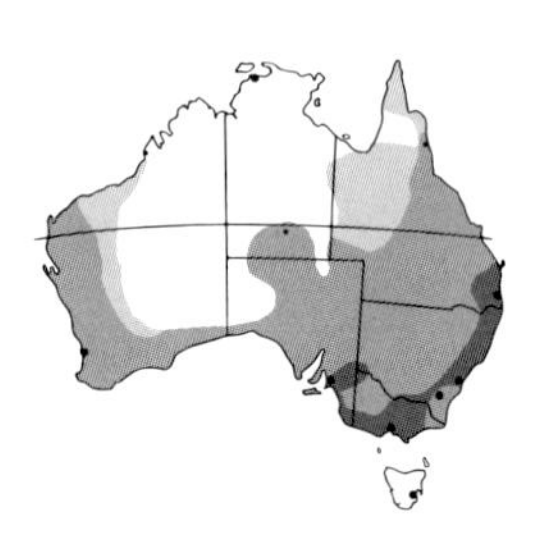

Sexes similar. Distinctive long forked tail. Similar to Barn Swallow, but lacks black breast band and white rump. Tail 'streamers' are shorter. Upperparts: Metallic blue-black. Frons, face, throat: Deep tan. Lores, eye stripe: Black. Underparts: Dull grey. Tail: Deep fork, plain central feathers, other feathers edged white, more visible underneath. Female: Smaller. Immature: Upperparts: Brownish.
Habitat: Open woodlands, wetlands, grasslands, farmland, urban areas.
Feeding: Forages on the wing for small insects, often over water. Fluttering, swooping and diving.
Voice: Musical twittering, squeaky chattering.
Status: Common. Sedentary. Southeastern birds overwinter northward.
Breeding: Aug.–Dec. Mud nest softly lined, often reused for years with same partner. Mostly on man-made structures, or rock shelves, nest shelves / 2–5 eggs.

Left and right: Adult.

Red-rumped Swallow

Cecropis daurica

160–170mm

Sexes alike. Glossy blue-black upperparts, wings, back and crown. Broken chestnut neck collar. Chestnut rump. Tail: Broad, pointed and deeply forked, with long streamers. Underparts: Varies. Off-white to chestnut, heavily or lightly dark streaked.
Immature: Plainer underparts.

Fast flying. Mixes with other swallows.

Habitat: Open areas, coastal grasslands, woodlands.
Feeding: Hawks for insects.
Voice: Warbling twitter. Nasal flight call.
Status: Rare, but regular non-breeding summer migrant Dec.–Feb.
Breeding: Eurasia.

Left and right: Adult.

Fairy Martin

Petrochelidon ariel

Endemic

115–120mm

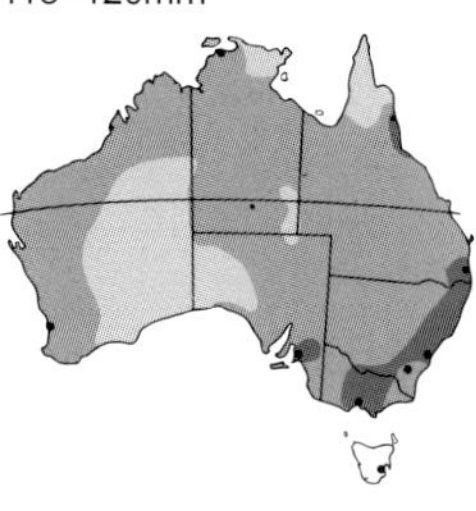

Sexes alike. Rusty-brown head, dull white rump and short square tail. Back, scapulars: Blue-black. Wings, tail: Dusky brown. Underparts: Dull white, finely streaked grey on throat. Eyes: Dark brown. Bill: Black. Birds are darker in WA.
Immature: Similar.
Gregarious, forming large flocks.
Habitat: Open woodlands, open swamps, grasslands, waterways.
Feeding: Hawking high or skimming over water. Often with Welcome Swallows and Tree Martins.
Voice: Chirrup and twittering song, especially at dawn. Distinctive 'drr, drr' churring.
Status: Common. Communal. Flocks of 12–100s. Migratory in winter. Most eastern birds move to northern Australia. Western birds move locally.
Breeding: May–Mar. far north. Aug.–Jan. elsewhere. Mud pellet bottle-shaped nest / 4–5 eggs.

Left: Gathering nesting material. Right: Nests packed closely together.

Tree Martin

Petrochelidon nigricans

120–130mm

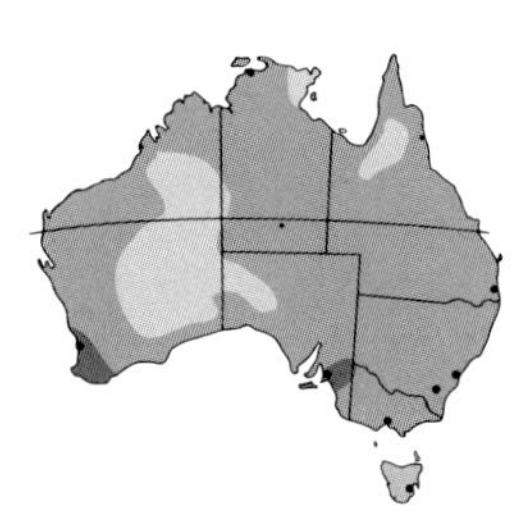

Sexes alike. Distinctive glossy blue-black crown, appears black in flight. Short, slightly forked tail, looks square in flight. Frons: Deep tan patch. Cheeks: Smudged brown-grey. Crown, nape, upper back: Metallic blue-black. Wings: Long, brownish black. Rump: Dull white, streaked brown. Underparts: Dull white, finely streaked brown on throat. Tail: Blackish brown.
Immature: Grey-brown upperparts, small tan frons.
Fast, swivelling flight.
Habitat: Widespread. Eucalypt woodlands near water, wetlands.
Feeding: Aerial foraging, similar to Fairy Martin.
Voice: High-pitched twittering.
Status: Common. Migratory, moving northward Feb.–May gathering in large flocks and returning south July–Oct.
Breeding: July–Jan. In loose colonies. Nests in small tree hollow, mud used to modify opening / 3–5 eggs.

Left: Adult. Right: Adult at nest hollow.

Yellow White-eye

Zosterops luteus

Endemic

95–110mm

Sexes alike. Named for the distinctive white feathers around the eyes and bright yellow underparts. Upperparts: Olive-yellow with dark grey flight feathers. Eyes: Brown. Eye ring: White. Lores: Black. Bill: Slate. Race *balstoni*, coastal WA. Smaller, duller. Upperparts: Greyish yellow.
Immature: Like adults, duller.
Habitat: Mangroves, coastal scrub, paperbark swamps.
Feeding: Insects gleaned from foliage; small fruits; nectar taken with brush-tipped tongue.
Voice: Musical warbling and trills. High-pitched whistled contact call.
Status: Common. Nomadic in groups 5–30.
Breeding: Oct.–Mar. Cup-shaped nest hung in outer foliage usually in mangroves / 3–4 eggs.

Karumba, Qld. Derby boat ramp and New Beach, WA.

Left: Adult in mangroves. Right: Adult.

Ashy-bellied White-eye

Zosterops citrinellus

100–110mm

Sexes alike. Distinctive yellow frons and white eye ring. Frons, chin, throat, rump, undertail coverts: Bright yellow. Upperparts: Greenish-yellow. Wings, tail: Olive-grey, feathers edged yellow. Lower-breast, belly: Whitish. Eyes: Mid-brown. Lores: Black. Bill: Black-grey.
Immature: Paler.
Gregarious. Usually in groups of 3–10.
Habitat: Woodlands and forest with shrubby thickets in coastal far north Qld and Torres Strait Is.
Feeding: Insects, fruits.
Voice: Warbling whistles and trills. Loud contact call.
Status: Moderately common. Sedentary.
Breeding: Dec.–June. Bulky cup-shaped nest suspended from horizontal fork / 2–4 eggs.

Opposite: Adult.

Silvereye

Zosterops lateralis

110–130mm

Sexes alike. Familiar and widespread. There are many plumage variations, but all races have white eye rings, dark lores and olive-green head and nape. Throat: Yellow or olive or white. Mantle: Grey or olive green. Wings, tail: Feathers edged dusky yellow. Underparts: White to grey. Flanks: Variable, browns to white.
Immature: Like adults.

9 generally similar Australian races. Races interbreed where ranges overlap and some races are migratory.
Nominate race, Tas. Head and nape olive green. White throat and light grey to sides of breast. Back is mid-blue-grey. Rich chestnut flanks. Undertail coverts white to greyish white. Rump olive-yellow to olive green. Breeds in Tas. then migrates over southeast and eastern mainland.
Race *pinarochrous*, southwest Vic. to Eyre Peninsula, SA. Olive above, fading into dull grey mantle. Grey breast, deep brown flanks and grey undertail.
Race *chloronotus*, southwest WA. Only Australian silvereye with olive back (other races are grey). Light olive throat. Pale yellow undertail coverts, pale grey breast and pale cinnamon-grey flanks.
Race *westernensis*, southern NSW and eastern Vic. Overall paler. Pale yellow throat, paler brown flanks.
Race *cornwalli*, from near Cairns, Qld, to north of Sydney, NSW. Smaller with bright yellow throat. Light grey breast. White to lemon tint to under tail. Rufous-buff to grey flanks. Off-white belly.
Race *chlorocephalus*, Capricorn Silvereye. Sedentary, restricted to wooded coral cays at the southern end of the Great Barrier Reef. Largest silvereye, with a robust bill and yellow throat. Mid-blue-grey back. Yellow-green head. Pale grey to cream below. Flanks are pale brown. Undertail coverts are yellow.
Race *vegetus*, far northern Qld. Richly coloured, paling south from Cape York. Yellow vent and undertail coverts, flanks washed grey.

Usually seen in small flocks. Group actively moves through feeding area in usison, maintaining contact with constant calling.

Habitat: Widespread. Eucalypt woodlands, coastal heaths, mallee, mangroves, parks and gardens.

Feeding: Forages, usually in low trees and shrubs, taking insects, and fruits. Nectar is lapped up with its brush-tipped tongue.

Voice: Joyful, trilling warbling song. Mournful, high-pitched 'tseep' repeated call.

Status: Common. Most Tas. birds migrate to the mainland for winter, moving as far north as Brisbane, Qld. Southeastern birds move northward. WA birds move locally. Northern birds are sedentary.

Breeding: Aug.–Feb. In good seasons 2–3 clutches may be raised. Nest is a small, tightly woven cup of fine grasses, hair and fine bark fibre, bound with spider web in a horizontal fork, 1–4m high / 3 eggs.

Row 1:
Left: Nominate race, Tas. Right: Race *pinarochrous* SW Vic, SA.

Row 2:
Left and right: Race *chloronotus* SW WA.

Row 3:
Left: Race *westernensis* central NSW, eastern Vic. Right: Race *cornwalli* northern NSW, Qld.

Row 4:
Left: Race *vegetus* far northern Qld. Right: Race *chlorocephalus* Capricorn Silvereye, N Qld.

Mistletoebird

Dicaeum hirundinaceum

Endemic

100–110mm

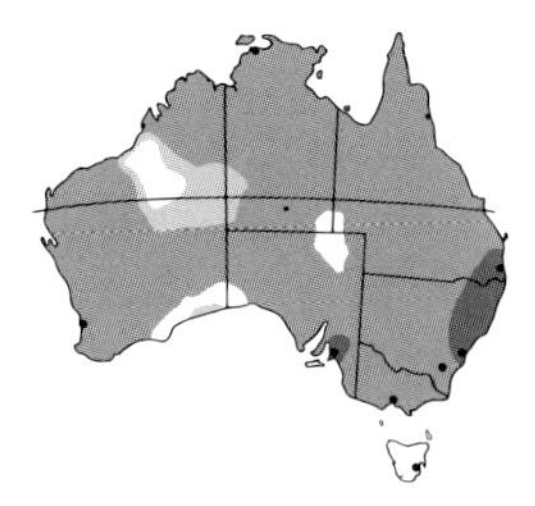

Only Australian member of the flowerpecker family. Glossy blue-black head and upperparts. Chin, breast, undertail coverts: Scarlet. Lower underparts: Greyish with distinctive black central streak. Eyes: Dark brown. Bill: Black. Female: Upperparts: Grey-brown. Underparts: Greyish white with pale centre streak and pale scarlet undertail coverts. Bill: Greyish cream. Immature: Similar to female, yellow gape and paler red markings.

Singularly or in pairs. Usually high in the canopy. Swift, erratic flight.

Habitat: Vegetated areas that support mistletoe.

Feeding: Almost exclusively on mistletoe berries. Rarely other fruit and insects.

Voice: Loud, warbling song and high-pitched flight note.

Status: Moderately common. Locally nomadic, following sequence of ripening mistletoe species.

Breeding: Oct.–Mar. Hanging purse-shaped nest, side entrance, hung from a twig up to 10m high / 3 eggs.

Left: Male. Right: Female.

Olive-backed Sunbird

Cinnyris jugularis

100–115mm

Small, honeyeater-like songbird, with a distinctive long, narrow, down-curved bill, with tubular tongue to suck up nectar. Upperparts: Olive green. Face: Olive-yellow with long dark brown eye stripe and thin bright yellow eyebrow and whisker line. Chin, throat, upper breast: Metallic blue-black with iridescent green-blue and violet sheen. Lower underparts: Golden yellow with orange wash on sides. Female: Similar. Underparts are plain yellow. Immature: Similar to female, duller.

Fast, direct flight. Often hovers while feeding from flowers. Well adapted to humans.

Habitat: Coastal rainforests, mangroves, tropical parks and gardens.

Feeding: Nectar from flowers. Insects, also taken mid-air.

Voice: High-pitched, scratchy, metallic, staccato notes.

Status: Common. Sedentary.

Breeding: Aug.–Mar. Flask-shaped nest, 50–60cm long, with a hooded side entrance and a long 'tail,' suspended from a twig / 2 eggs.

Left: Female at nest. Right: Centre and below, male.

26

Family Locustellidae, Cisticolidae, Alaudidae, Motacillidae, Passeridae, Fringillidae

Reed-warblers

Grassbirds

Songlarks

Spinifexbird

Pipit

Bushlark

Cisticolas

Wagtails

Sparrows

Feral Finches

A grouping of several families that prefer similar open grassland or reedy habitats.

Family Locustellidae, endemic to Australia, includes the Australian Reed-warbler, grassbirds, songlarks and the Spinifexbird.
The Australian Reed-warbler is noted for its continuous melodious song, an iconic sound of summer in wetlands of southern Australia. They build their well-hidden, deep, cup-shaped nests in dense, reedy, grassy vegetation.
The Rufous Songlark and Brown Songlark are not related to larks but have lark-like song flights in the breeding season. The males engage in continuous melodious singing, including song flights as they fly over their territory. The Rufous Songlark song flights are horizontal, while the Brown Songlark rises vertically, flutters and drops to the ground. The female unobtrusively builds a nest and raises the young. The Brown Songlark has one of the greatest variations in size between sexes of any bird species.
Two species of grassbirds are secretive and mostly coastal. They build deep, cup-shaped nests in dense reedbeds or clumps of grass. The Little Grassbird attaches long feathers to the rim that arch over the nest, screening the interior from predators.
The Spinifexbird is found in the arid interior and rarely leaves the cover of spinifex.

Family Cisticolidae has two species in Australia.
They are noted for the males' conspicuous song flights when establishing a breeding territory. The Golden-headed Cisticola, also called the Tailorbird, stitches living leaves together with coarse spider web and fibrous material thread as part of its nest construction. The male passes the thread to the female, who works from inside the nest. Among the birds they are considered to be the finest 'tailors'.

Family Alaudidae includes the larks.
The endemic Horsfield's (Singing) Bushlark is the sole Australian representative of the large genus *Mirafra*, and is noted for its melodious song flights, 10–90m above the ground, quivering its wings while singing.
The Eurasian Skylark was introduced in 1850 because of its superior song. However, it is now widely recognised that the native grassbirds are superior songbirds.

Family Motacillidae includes the pipits and wagtails.
The Australian Pipit is noted for its courtship ritual, including high-altitude swooping dives while continuously singing its melodious song. The migratory wagtails are recent arrivals from the northern hemisphere and are becoming more regularly seen.

Family Passeridae contains introduced sparrows.

Family Fringillidae contains the introduced goldfinch and greenfinch.

Opposite: Australian Reed-Warbler.

Oriental Reed-warbler

Acrocephalus orientalis

180–200mm

Sexes alike. Similar to Australian Reed-warbler, but with a shorter and thicker bill and fine, pale grey streaking to throat and upper breast. Wings are longer and more pointed.

Usually seen warbling from tall reed stem.

Habitat: Dense, coastal reed beds near water, marshes.

Feeding: Forages for insects and invertebrates.

Voice: Warbling phrases. Loud, metallic 'chack'.

Status: Rare, vagrant or migrant to far northern Australia.

Breeding: Central Japan and northeastern China.

Opposite: Adult.

Australian Reed-warbler

Acrocephalus australis

160–170mm

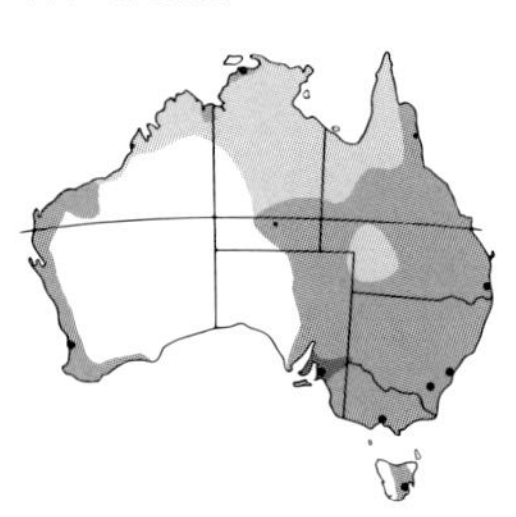

Sexes alike. Mostly plain, tawny brown bird with pale buff eyebrow and long rounded tail. Wings: Pointed, darker. Underparts: Plain pale buff with off-white throat, paler belly and pale cinnamon flanks. Eyes: Brown. Bill: Brown, paler below. Legs: Dusky.
Immature: Darker.

Low, dashing flight between clumps of rushes.

Habitat: Dense reeds, rushes, Cumbungi in freshwater swamps, moist wetlands, inland bore drains, parklands.

Feeding: Clinging to reeds, close to ground or over water, taking insects and small aquatic life.

Voice: Melodious, loud, variable 'chuck-chuck-chuck, dee-dee-dee, kwitchy-kwitchy-kwitchy', ceaseless singing from perch in the breeding season with puffed feathers and small raised crest. Possibly to advertise territory. Usually silent in winter.

Status: Common. Birds arrive in the south in Aug.–Oct to breed, some overwinter. Most depart Mar.–Apr. to northern Qld and northwestern WA.

Breeding: Sept.–Feb. Deep cup-shaped nest in reeds over water / 3–4 eggs.

Left and right: Adult near nest site.

Tawny Grassbird

Megalurus timoriensis

190mm

Sexes alike. Race *alisteri* present in Australia. Rich, tawny rufous crown with pale buff eyebrow, white throat and yellow-brown eyes. Bill: Pinkish brown, paler below. Upperparts: Brown, heavily streaked black with plain rufous rump. Wings, tail: Dark brown edged rufous. Underparts: Plain grey-white with buff washed sides. Legs: Pinkish brown.
Immature: Darker brown with finer streaking.

Conspicuous display song flight and loud continuous singing when breeding. Wings quivering, tail down, flutters and drops to cover. Fluttering flight with tail trailing.

Habitat: Tall grasslands, reeds in floodplains.

Feeding: Hopping, running, gleaning for insects under cover.

Voice: Descending wren-like reels in flight. Scolding calls.

Status: Uncommon. Sedentary, occasionally nomadic.

Breeding: Aug.–Jan. Cup nest of dried swamp grass in dense cover near ground / 3 eggs.

Left and right: Adult.

Rufous Songlark

Megalurus mathewsi

Endemic

Male: 180–190mm
Female: 140–160mm

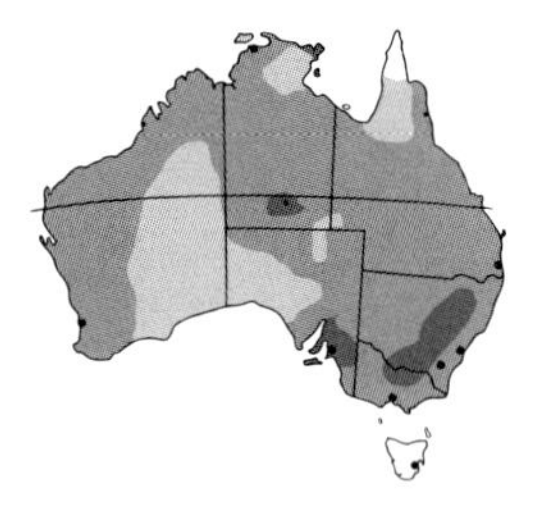

Sexes similar. Streaked brown and dark brown upperparts and crown with a distinctive rufous rump and uppertail coverts. Tail: Dark brown. Wings: Edged pale buff. Underparts: Greyish white, dusky flecked upper breast. Eyes: Brown with pale buff eyebrows and dark line through eyes. Bill: Light brown, black in breeding season.
Female: Smaller with pale brown bill. Immature: Paler, streaked breast.

Continuous melodious song and slow display flight in breeding season.

Habitat: Grassy woodlands and open scrub.

Feeding: Ground feeding on insects and seeds.

Voice: Joyful, rollicking trill followed by explosive 'witchy-weedle' whipcrack. Males sing constantly in breeding season.

Status: Moderately common. Summer breeding visitor to southern states, most migrating to northern Australia for winter.

Breeding: Aug.–Feb. Deep cup-shaped grass nest well hidden on ground / 3–4 eggs.

Left: Non-breeding male. Right: Breeding male.

Brown Songlark

Cincloramphus cruralis

Endemic

Male: 240–260mm
Female: 160–170mm

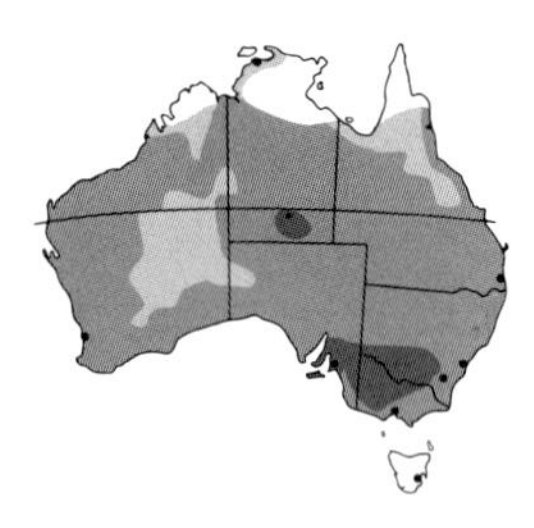

Resembles Rufous Songlark but is larger and lacks rufous rump. Breeding plumage: Mottled dark cinnamon with black lores and bill. Underparts: Paler sooty brown.
Non-breeding: Upperparts: Cinnamon brown with dusky flecking and white eyebrow. Underparts: Pale fawn with dark brown mottled centre strip on belly. Lores: Dark. Wings, tail: Dusky with buff edges. Bill: Pale.
Female: Significantly smaller. Similar to non-breeding male but duller, white eyebrow. Bill: Pale pinkish brown. Breast: Faintly streaked, light buff. Belly: Dark brown centre. Immature: Similar to female.

Spectacular song flights. Upright stance, often with tail cocked.

Habitat: Saltbush plains, treeless plains and farmland.

Feeding: Ground feeding, taking insects and seeds.

Voice: Loud, clear metallic grinding, creaky song followed by musical trill and whipcrack. Male song flights in breeding season.

Status: Moderately common. Nomadic, following rains. Southern birds migrate north or inland after breeding.

Breeding: Sept.–Feb. Nest on ground, hidden under tussocks / 3–4 eggs.

Left: Smaller female. Right: Male in breeding plumage, tail typically cocked.

Little Grassbird

Megalurus gramineus

140mm

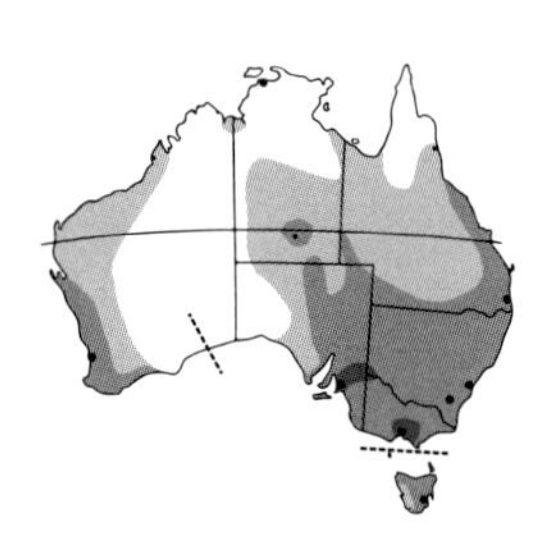

Similar to Tawny Grassbird, but smaller and darker. Crown, upperparts, wings: Grey-brown streaked dusky black. Tail: Brown, long, pointed. Underparts: Off-white with dusky flecks to throat and breast. Flanks: Lightly streaked pale rufous. Eyes: Mid-brown with broad, white eyebrow. Bill: Pale brown. Legs: Brown.
Nominate race found in Tas. Race *thomasi*, WA. Smaller and distinctively darker.
Race *goulburni*, elsewhere. Paler, less streaked.
Sexes alike. Immature: Similar, less streaked.

Secretive. Well camouflaged. Flutters weakly, tail trailing. Hops, runs, usually under cover. Singles, pairs or family groups.

Habitat: Reed beds, wetland grass tussocks, freshwater swamps, saltmarshes.

Feeding: Gleaning insects, often runs on mud, taking aquatic insects, crustaceans.

Voice: Continuous, melancholy 3-note whistling advertising territory. Rapid, harsh chattering.

Status: Common. Sedentary. Sometimes dispersive in winter.

Breeding: Aug.–Jan. or after inland rain. Deep grass cup nest well hidden in reeds / 3–5 eggs.

Left and right: Race *goulburni*.

Spinifexbird

Megalurus carteri

Endemic

150–160mm

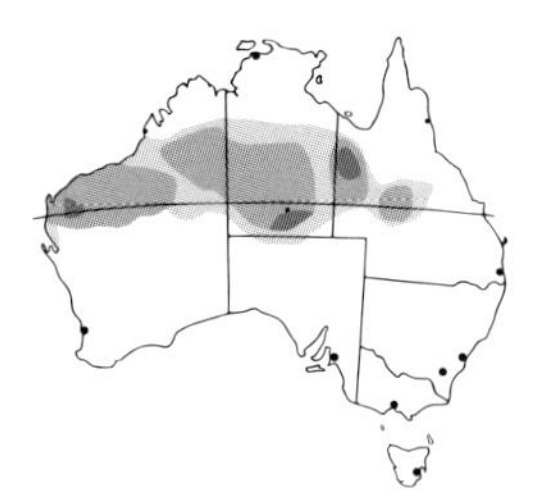

Sexes alike. Mid-sized warbler, best recognised for its preference for spinifex habitat. Rufous frons and crown with buff eyebrow. Upperparts: Rufous-brown, wings and tail darker. Underparts: Fawn-white with white chin and throat and pinkish fawn flanks. Eyes: Brown. Bill: Grey-brown, paler below. Legs: Pinkish grey.
Immature: Paler, slightly streaked.

Weak, low flight with tail trailing. Hopping from 1 spinifex clump to another. Climbing tall stems to observe or advertise territory.

Habitat: Tall clumps of spinifex on stony ground, often among acacia scrub.

Feeding: Forages alone, hopping, tail partly cocked, picking up insects, grasshoppers, seeds.

Voice: Repeated, high-pitched whistled warbling.

Status: Moderately common.

Breeding: Aug.–Nov. Thick grass nest, well hidden in a spinifex clump / 2 eggs.

Barrow Is., WA.

Opposite: Adult perched in spinifex.

Australian Pipit

Anthus australis

160–180mm

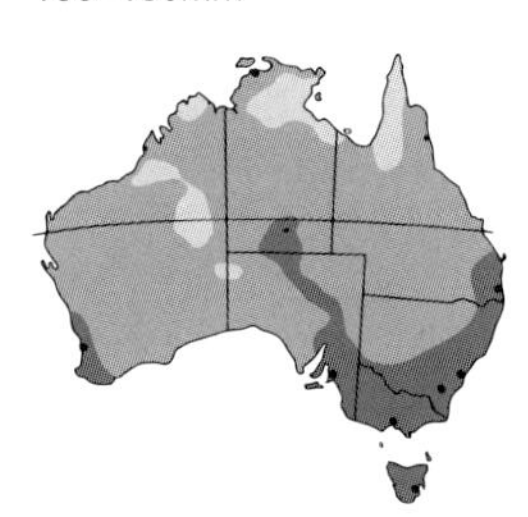

Sexes alike. Also known as Richard's Pipit. Well-camouflaged, ground-dwelling bird, with mid-brown, streaked, dusky and dark brown mottled upperparts and crown. Face: Brown eyes with long, cream-white eyebrow extending to encircle brown ear coverts. Throat: Cream-white with dusky streaked line from bill down either side. Underparts: Cream-white, with dusky speckled breast. Tail: Long, brown, edged white. Bill, legs: Pinkish brown. Immature, similar.

Ground dwelling. Flight is low and swift. Tail is fanned when flushed. Runs with a bobbing tail.

Habitat: Diverse habitat. Open country, grassy woodlands, wet heathlands.

Feeding: Ground foraging for small grasshoppers, ants, other insects, seeds.

Voice: Sparrow-like chirrup and trilling song.

Status: Moderately common. Sedentary, or wanders locally in flocks outside breeding time. Tas. birds overwinter on the mainland, alpine birds move to lowlands.

Breeding: Aug.–Jan. Nest is a deep grass cup, well hidden in a grass tussock / 2–4 eggs.

Left and right: Adult.

Horsfield's Bushlark

Mirafra javanica

Endemic

120–150mm

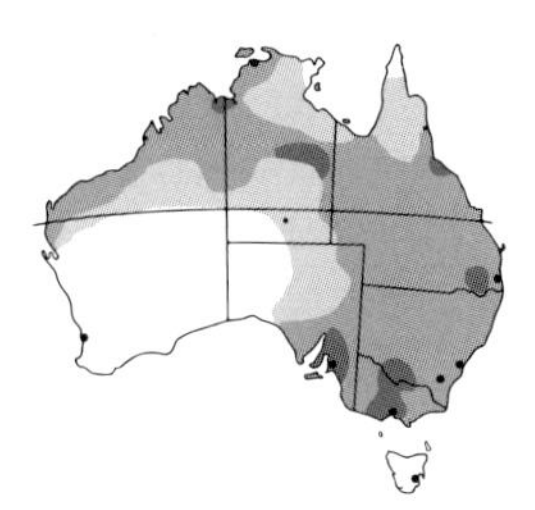

Sexes alike. Upperparts: Brown with heavy dark brown streaking. Underparts: Lighter, with darker mottling to throat and breast. Wings: Brown with rufous margins and rufous patch. Tail: White outer tail feathers. Eyes: Dark with buff eyebrow.
There are 9 races in Australia, with slight plumage variations that reflect local soil colour, from reddish brown to dark brown.
Immature: Paler.

Runs or flutters along the ground with bobbing tail. Soars aloft singing when breeding.

Habitat: Open grasslands, pastures.

Feeding: Fallen seeds, also insects. Feeds while moving.

Voice: Constant singing. Often at night. Rich, varied, melodious trilling, includes some mimicry. Silent in winter.

Status: Common. Southern birds overwinter in northern Australia.

Breeding: Sept.–Jan. in south. After rain in arid areas. Loosely built grass domed nest in a depression under a tussock / 2–4 eggs.

Left and right: Adult.

Eurasian Skylark

Alauda arvensis

170–190mm

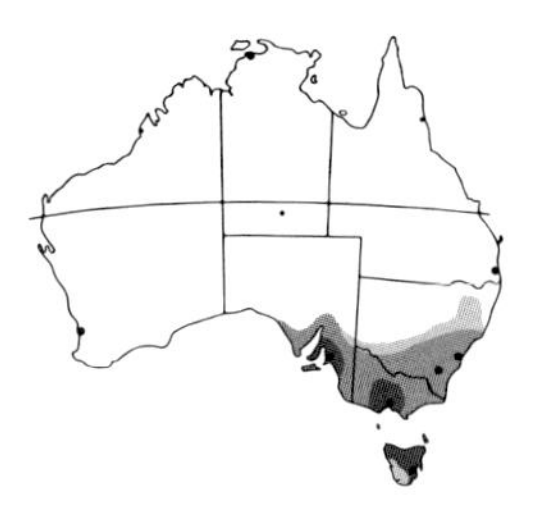

Introduced c.1850. Sexes alike. Similar to but larger than the Horsfield's Bushlark. Does not bob tail. Upperparts: Pale brown, black streaked, small crest. Eyebrow: Buff. Underparts: Pale buff, streaked breast; throat white. Wings: Brown, pale edges. Immature: Paler.

Habitat: Open grasslands, pastures, margins of swamps.

Feeding: Runs across the ground picking seeds, insects.

Voice: Remarkable song flight, rising 100m vertically, singing, then plummeting silently. Continuous, musical high-pitched warbling.

Status: Locally common.

Breeding: Sept.–Jan. Nest on ground under tussocks / 3–5 eggs.

Left: Adult with crest lowered. Right: Adult with crest raised in samphire.

Golden-headed Cisticola

Cisticola exilis

100–110mm

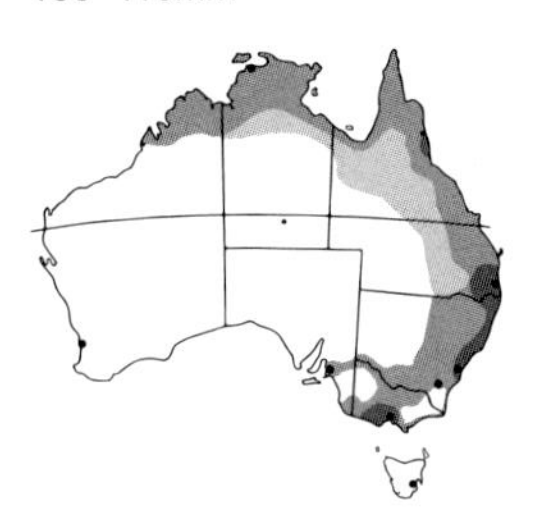

Golden-orange head, crest (raised when calling) and golden buff upperparts with heavily streaked back and wings and plain underparts. Tail: Short, brown, edged and tipped buff. Eyes: Pale brown. Bill: Brown, pinkish brown below.
Non-breeding: Less golden hued. Black streaked crown and back, long, pale-tipped tail. Inland birds are paler, sandy yellow. Western WA birds are deeper rufous.
Female, immature: Similar to non-breeding male. Short tail in breeding season.

Weak, low, undulating flight. Crest raised when calling.

Habitat: Wetlands, tall wet grasses, reed beds, swampy vegetation in open areas.

Feeding: Ground feeding on insects, under cover of tall grass.

Voice: Incessant, metallic buzz, followed by short musical notes. Male advertises territory from a tall grass stem.

Status: Common. Sedentary.

Breeding: Sept.–Mar. Domed nest in shrubs or tussocks / 3–4 eggs.

Left: Breeding. Centre and right: Non-breeding.

Zitting Cisticola

Cisticola juncidis

95–105mm

Resembles the Golden-headed Cisticola in behaviour, size and shape, but is less colourful.
Breeding: Head, back: Black streaked mid-brown. Nape: Dusky streaked. Rump: Tawny, black streaked. Wings: Dusky black and fawn streaked. Tail: Short, dark brown, edged and tipped white, undertail barred black. Lores: White. Underparts: White with tawny flanks. Underwings: Cinnamon 'window'. Eyes: Brown. Bill: Brown, paler below. Legs: Pink-brown.
Non-breeding: Similar to female, but brighter, more heavily streaked, longer tail.

Female: Duller, heavy black streaking to head and upperparts. Immature: Similar.

Habitat: Grassy swamps, coastal plains, saline grasslands, mangrove margins.

Feeding: Insects, plant matter.

Voice: Chattering. Persistent 'zinging' call from breeding male. Song flights to 30m high.

Status: Uncommon. Migratory.

Breeding: Dec.–Apr. Dome nest near ground in grassy clump / 2–6 eggs.

Left: Male, coming into breeding plumage. Right: Breeding male.

Eastern Yellow Wagtail

Motacilla tschutschensis

160–190mm

Race *simillima* most common in Australia. Breeding: Olive-yellow upperparts and bright yellow underparts. Nonbreeding: Crown: Grey. Face: Buff eyebrow, dark ear coverts. Upperparts: Brown-grey with dusky wings, yellow edged. Underparts: Pale yellow. Sexes similar in non-breeding plumage: Immature: Similar to non-breeding, brown underparts.

Graceful flight. Walks briskly, bobbing head and tail.

Habitat: Damp grassland near swamps, mud ponds, wetlands.

Feeding: Often at water's edge; small molluscs, insects, spiders. Often leaping for flying insects.

Voice: High-pitched trilling.

Status: Uncommon, but regular summer migrant Nov.–Apr.

Breeding: Eurasia.

Leanyer Treatment Ponds, NT. Broome Treatment Ponds, WA.

Left: Race *simillima* non-breeding. Right: Race *simillima* breeding.

Green-headed Yellow Wagtail

Motacilla taivana

160–190mm

Similar to Eastern Yellow Wagtail. Bright olive-green head and olive back. Yellow wing-bars. Eyebrow: Yellow, extending behind eyes. Ear coverts: Olive green. Underparts: Bright-yellow. Bill, legs: Slate-grey. Nonbreeding: Sexes similar. Duller with browner upperparts and pale-yellow underparts. Immature: Similar to female with buff throat and breast patch.

Walks briskly, bobbing head, tail.

Habitat: Damp grassland adjoining swamps, grassy wetlands.

Feeding: Often at water's edge. Small molluscs, insects, spiders.

Voice: High-pitched 'pseet' repeated. Harsh alarm call.

Status: Rare, but regular summer migrant.

Breeding: Asia.

Broome oval.

Left: Non-breeding with yellow remnant breeding plumage. Right: Breeding bird beginning moult phase.

Grey Wagtail

Motacilla cinerea

180–190mm

A slender, graceful bird with an exceptionally long tail. Breeding male has grey crown and dark grey face. Eyebrow: White. Throat: Large black patch. Underparts: Yellow. Female and non-breeding male: Upperparts: Mostly Grey. Underparts: White with yellow restricted to between the throat and vent. Throat patch: White.

Incessantly wags tail while walking or running on the ground. Flight is low and undulating.

Habitat: Fresh water streams, sewerage ponds and rainforests.

Feeding: On the ground and surface of shallow water, taking aquatic prey and insects.

Voice: Sharp, metallic call note and trilling song.

Status: Uncommon non-breeding summer visitor mainly to northern Australia. Nov-Apr.

Breeding: Western Europe to Asia.

Left: Non-breeding. Right: Breeding.

White Wagtail

Motacilla alba

180–200mm

Slender, grey, white and black Eurasian species. Several races. Race *leucopsis* most commonly seen here. Upperparts: Black. Face and underparts: White with black bib. Bill: Black. Female and non-breeding male all races have grey upperparts, dull white underparts with a thin black breast band.

Constantly wags long tail and walks with its head bobbing. Often runs in pursuit of insects.

Habitat: Open to semi-open terrain near fresh water. Lawns, wetlands and sewerage ponds.

Feeding: Insectivorous, taking prey from ground or shallow freshwater surface.

Voice: Twittering song. Harsh 'chissick' call.

Status: Scarce migrant. Sightings in WA, Qld, NSW, Vic and NT.

Breeding: Europe, Asia and North Africa.

Left: Race *leucopsis* non-breeding. Right: Race *leucopsis* breeding male.

House Sparrow

Passer domesticus
140–160mm
Introduced c.1860

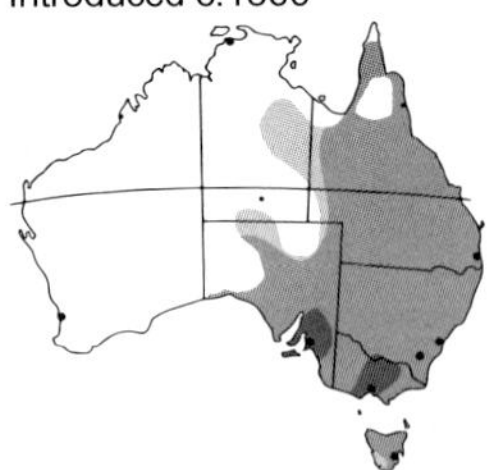

Crown, rump: Grey. Face, throat: Black. Upperparts: Black and chestnut. Underparts: Whitish-grey. Bill: Black.
Non-breeding: Bill: Cream, less black on throat.
Female: Similar, paler, fawn eye stripe. Immature: Similar to female, bill darker.

Habitat: Settled areas, parks and gardens.

Feeding: Food scraps, insects, seeds, flower buds.

Voice: Noisy, constant chirping, harsh twittering.

Status: Common. Pest. Not established in WA.

Breeding: All year, mostly spring. Untidy nest often in roof spaces, or competes with native birds for tree hollows / 3–6 eggs.

Left: Male in summer plumage. Right: Female.

Eurasian Tree Sparrow

Passer montanus
140–150mm
Introduced c.1860

Sexes alike. Similar to the male House Sparrow but smaller. Competes with native birds for tree hollows. Crown, nape: Reddish brown. Cheeks: White with black patch. Immature: Duller.

Habitat: Urban areas with exotic vegetation.

Feeding: Food scraps, insects, seeds.

Voice: Chirping, twittering. Higher pitch than House Sparrow.

Status: Locally common.

Breeding: Sept.–Jan. Compact dome nest, usually in tree hollows, sometimes shrubs or building crevice / 4–6 eggs.

Left and right: Note smaller throat patch than House Sparrow.

European Goldfinch

Carduelis carduelis
120–150mm
Introduced c.1860

Sexes alike. Distinctive bright red face. Crown, shoulders: Black. Wings: Black with large yellow bars and white tips. Tail: Black, tipped white. Back, flanks: Buff-brown. Side of head, rump, throat, belly: White.
Immature: Duller, streaked.
Large feeding flock of up to 200 birds forms after breeding.

Habitat: Settled areas, parks, gardens, roadsides, wasteland, farmland, patches of thistles.

Feeding: Weed seeds, some insects.

Voice: Canary-like twittering. Repeated tinkling in flight.

Status: Locally common.

Breeding: Sept.–Mar. Open cup-shaped nest 2–12m high / 3–6 eggs.

Left: Immature. Right: Adult.

European Greenfinch

Carduelis chloris
140–150mm
Introduced c.1860

Stocky. Upperparts: Olive-green washed yellow. Wings: Bright yellow edges on primaries. Underparts: Greenish yellow. Bill: Cream. Eyes: Brown. Winter plumage: Dull.
Female: Dull brown; less yellow on primaries. Centre belly washed green. Immature: Brown streaked.

Habitat: Areas with exotic vegetation, city parks, gardens, pine plantations.

Feeding: Pinecone seeds, grass seeds, some insects. Small fruits.

Voice: Musical warbling, twittering.

Status: Patchy, locally common.

Breeding: Oct.–Jan. Nest bulky, cup-shaped, 2–12m high / 4–6 eggs.

Left: Male. Right: Female.

27 Family Estrildinae

Finches

Mannikins

A family containing finches and mannikins

There are 19 species of finches and mannikins in Australia, 14 of which are endemic. They are popular with bird lovers because of their striking flashes of colour, spectacular courtship displays and fascinating social bonds. Finches are represented in most parts of Australia, from the widespread, urbanized Double-barred Finch to the huge flocks of Zebra Finch crowded around waterholes in the inland regions, and the extraordinarily colourful Gouldian Finch of northwestern Australia, regarded as the most beautiful of the world's finches.

Finches are highly sociable birds, with most species nesting in loose colonies and moving around in large flocks. Referred to as 'Grass-finches', they have specialised sharp conical bills for cracking small ripe or half-ripe grass seeds. They need to drink regularly because of their dry diet and are never far from a water supply. Some finches need to drink hourly.

Most inland nomadic species have developed the same drinking method used by pigeons and doves, that of 'sucking up' water to allow them to drink quickly while keeping a watch out for possible danger. This method also permits access to water that is inaccessible to most other species. The Diamond Firetail, Star Finch, Gouldian Finch, Black-throated Finch, Zebra Finch, Double-barred Finch and Long-tailed Finch drink this way.

The Painted Finch is the only finch that drinks by scooping water in its bill and tilting its head back to swallow.

Pair bonding among finches is strong, with many species mating for life. Courtship displays include the male parading, dancing and posturing, with quivering tail and a stem of grass held in the bill. The birds typically construct a flask-shaped nest with a side tunnel entrance, from interwoven stiff grass stems and lined with white feathers. The nest is usually placed in a thorny shrub, often near a wasp nest for added protection. At night, pairs sleep in small, unlined spherical grass roosting nests placed high in foliage.

The call of finches is a plaintive musical 'chirrup' and can often be heard when a flock flies overhead, as they softly keep in contact with other members of the flock.

Mannikins are small, brownish-coloured, finch-like birds commonly seen in flocks. Australia has four species, of which two are endemic and one, the Nutmeg Mannikin, is introduced.

Opposite: Male Zebra Finch.

Zebra Finch

Taeniopygia guttata

100mm

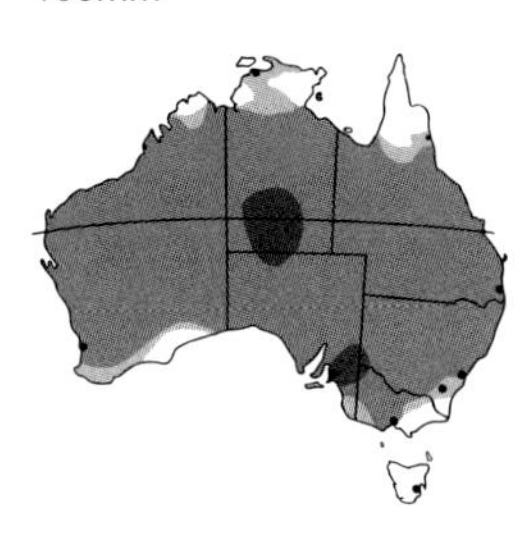

Sexes similar. Distinctive black and white 'zebra' striped tail with a heavy conical red-orange bill. Head, frons: Grey. Face: Orange ear patch and black and white tear streaks before and below eyes. Throat: Fine, black and white barring terminating with a black band on upper chest. Underparts: White, with tan flanks, speckled white. Legs, feet: Yellow-orange. Upperparts: Mid-grey to brown.
Female: Lacks cheek and flank colour. Immature: Similar to female, black bill.

Sociable. Flocks of 10–100s. Pairs form permanent bonds. Flocks congregate at water sources, often in thousands.

Habitat: Varied. Dry wooded grasslands, near water.

Feeding: Ground foraging in flocks on fallen seeds. Plucks ripening grass seeds from stems. Also insects. Drinks frequently.

Voice: Nasal 'tiah' and repetitive nasal trill.

Status: Common to uncommon. Sedentary.

Breeding: Most times of year. Untidy domed grass and twig nest, placed low in prickly shrub, tree hollow or fence post / 4–5 eggs.

Left: Male and female, male drinking.
Right: Male and female pair.

Double-barred Finch

Taeniopygia bichenovii

Endemic

100–110mm

Sexes alike. Long-tailed, owl-faced grass finch. Distinctive white face and throat bordered with a black band and a second black band across the white breast. Eyes: Black. Upperparts: Grey-brown. Wings: Black with white speckles. Rump, upper tail coverts: White. Tail: Black. Belly: Creamy white. Legs: Bluish grey. Undulating flight.
Race *annulosa*, Top End, NT, WA. Rump, uppertail coverts: Black.
Immature: Duller, with indistinct bars.

Forms flocks of up to 40 birds.

Habitat: Open forests, woodlands with grassy understorey and nearby water. Farmland, parks, gardens with low scrubby plants.

Feeding: Fallen grass seeds or half-ripe seeds plucked from grass stems. Also insects. Drinks frequently.

Voice: Nasal 'tiat-tiat' and soft kitten-like meowing.

Status: Nomadic. Common. Range expanding in southeast.

Breeding: June–Nov. in north. Jan.–Mar. in south. Untidy flask-shaped, dry grass nest, lined with feathers and plant down, low in a prickly shrub, near an active wasp nest, 1–5m high / 4–7 eggs.

Left: Nominate race. Right: Race *annulosa.*

Long-tailed Finch

Poephila acuticauda

Endemic

150–160mm

Sexes similar. Distinctive round black bib, orange-red bill and long pointed tail. Head: Pastel blue-grey. Eyes, lores: Black. Back, wings: Fawn-brown. Rump: White. Underparts: Light fawn. Undertail coverts: White. Legs, feet: Orange.
Race *hecki*, Top End, NT. Orange–coral red bill.
Female has smaller bib.
Immature: Duller, bill black.

Bobs head on landing. Sociable, flocks of 30 or more. Strong pair bond.

Habitat: Tropical savannah grasslands, mostly near eucalypt and paperbark woodlands near water.

Feeding: Ground feeders, foraging for half-ripe and ripe grass seeds, also flying insects taken.

Voice: High-pitched, loud whistle, identification call. Flute-like, whistled song.

Status: Common. Mostly sedentary.

Breeding: Jan.–May. Elaborate courtship dance. Large, flask-shaped nest with side entrance tunnel in hollow limb of a eucalypt or pandanus / 4–5 eggs.

Left and right: Adults drinking.

Black-throated Finch

Poephila cincta

Endemic

120–130mm

Sexes similar. Similar to Long-tailed Finch, but with a distinctive black bill, a shorter square cut tail and reddish legs. Bill, lores, throat, upper breast: Black. Head, nape: Pastel blue-grey. Upperparts: Fawn. Belly: Pinkish brown with lower black band. Undertail coverts: White. Rump: White in southern range. Race *atropygialis*, northern Qld. Black rump. Female: Smaller black bib. Immature: Duller.

Bobs head when landing. Drinks by sucking up water. Sociable, flocks of 10–30 birds drink, bathe, rest and preen communally.

Habitat: Savannah woodlands, open timbered grasslands near water.

Feeding: Feeds on the ground on ripe and part-ripe grass seeds, also occasional insects, ants, spiders, flying termites taken. Drinks regularly.

Voice: Harsh, loud whistle. Soft flute-like song.

Status: Uncommon. Rare, possibly extinct in NSW.

Breeding: Sept.–Jan. in south. After monsoon season, Feb. in the north. Typical flask-shaped nest 1–15m high / 4–5 eggs.

Opposite: Race *atropygialis*, with black rump.

Masked Finch

Poephila personata

Endemic

120–130mm

Sexes similar. Distinctive, heavy, deep yellow bill with a black face mask and chin. Upperparts: Cinnamon. Rump: White. Underparts: Light cinnamon with black flanks. Tail: Black, pointed and constantly flicked. Race *leucotis*, White-eared Finch, Cape York, Qld. Pale cheek patch and underparts.

Female: Smaller face mask. Immature: Duller grey-brown. Lacks mask, black bill.

Sociable, usually in pairs or small flocks of 10–20. Sometimes in mixed flocks with Long-tailed and Black-throated Finches.

Habitat: Grassy tropical eucalypt or paperbark woodlands, and wooded grasslands scattered with shrubs near water.

Feeding: Mostly fallen grass seeds. Flocks move to water in mornings and evenings.

Voice: Rapid, loud, toneless, nasal 'twaat-twaat' notes.

Status: Sedentary. Moderately common.

Breeding: Mar.–Jun. Globular nest of grass strips lined with feathers and plant down and charcoal, placed in long grass or shrub to 8m high / 4–6 eggs.

Gunlom Falls, Kakadu NP, NT. Wyndham, WA.

Left: Race *leucotis*. Right: Nominate race.

Plum-headed Finch

Neochmia modesta

Endemic

110–120mm

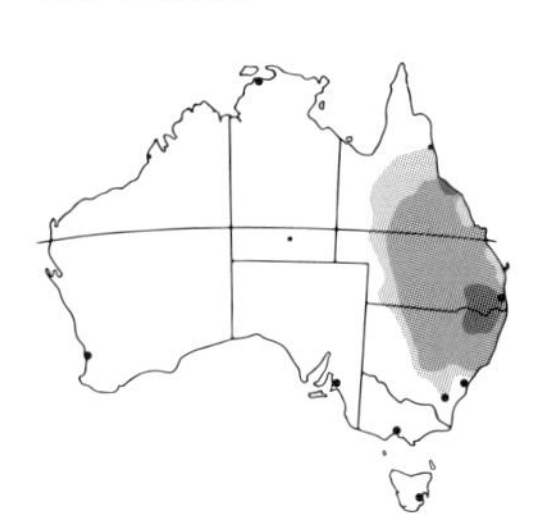

Sexes similar. Black eyes, bill, tail and lores and dark claret crown, frons and chin. Upperparts: Mid-brown. Ear coverts: White, with brown streaks. Breast: White with fine brown barring. Underparts: White, with fine brown barring on flanks. Female: Duller, whitish chin and throat. Immature: Duller.

Relatively tame. Usually in pairs or small flocks. Forms large flocks in drought conditions.

Habitat: Tall grasses, reed beds near swamps or watercourses. Parks and gardens.

Feeding: Fallen seed. Climbs stems to reach ripe and half-ripe grass seeds.

Voice: Tinkling whistle call. Trilling, chirping song.

Status: Uncommon. Rare in north. Nomadic.

Breeding: Aug.–Mar. in north. Sept.–Jan. in south. Small, elliptical nest with side entrance, of green grass strips lined with feathers and plant down interwoven into living tall grass clump or in shrub / 4–7 eggs.

Glen Davis area, NSW. Eulo Bore and Girraween NP, Qld.

Left: Male. Right: Female.

Red-browed Finch

Neochmia temporalis

Endemic

110–120mm

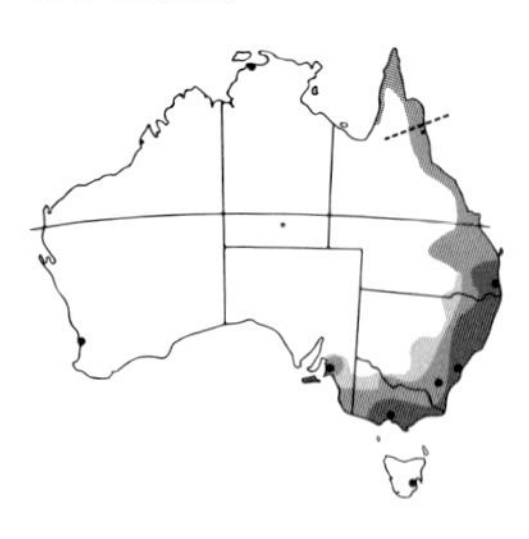

Sexes similar. Easily recognised by its red eyebrow, bill, rump and uppertail coverts. Upperparts: Olive-yellow with yellow side patches each side of neck Underparts: Buff-grey.
Race *minor*, far north Qld. Brighter olive-yellow back. White throat, belly and wider eyebrow. Black tail.
Female: Thinner eyebrow.
Immature: Browner, lacks red eyebrow, dark bill.

Large flocks form after breeding.

Habitat: Wet, semi-open woodlands, grassy areas with adjoining dense vegetation. Parks, gardens near water.

Feeding: Usually ground feeding on ripe or half-ripe grass seeds, small fruits, insects. Frequently visits water.

Voice: 'Switt-switt', high-pitched, repeated.

Status: Common, sedentary, sometimes nomadic. Small introduced population occurs around Perth, WA.

Breeding: Sept.–Nov. in south. Jan.–Apr. in north. A large grass domed nest with a side entrance tunnel, placed in a low prickly shrub such as Blackthorn or Silky Hakea / 4–6 eggs.

Opposite: Adults, nominate race.

Star Finch

Neochmia ruficauda

Endemic

110–120mm

Sexes similar. Red face and chin. Upperparts: Dark olive. Underparts: Yellow-olive. Ear coverts, cheeks: Spotted white. Breast, flanks, rump: Lightly spotted white. Tail: Purple to dull-red hued with large white spots on uppertail coverts.
Nominate race, Qld.
Race *clarescens*, far north Qld. Greyer upperparts.
Race *subclarescens*, NT, WA. Brighter olive upperparts. Richer yellow underparts. Fine spotting. Females in NT, WA and far north Qld have less red on face.
Immature: Paler, black bill.

Usually small flocks of up to 20.

Habitat: Grasslands, grassy woodlands near fresh water, reedy swamps, mangroves.

Feeding: Ripe or half-ripe grass seeds, including spinifex, also insects. Far north, Rice Grass is a critical food source.

Voice: Loud, high-pitched 'sseet' contact call. Toneless twittering.

Status: Rare. Eastern birds are endangered. Mostly sedentary.

Breeding: Dec.–Aug. Domed grass nest lined with feathers, placed in tall grass or shrub to 6m / 3–6 eggs.

Left: Nominate race. Right: Race *subclarescens*, female.

Crimson Finch

Neochmia phaeton

120mm

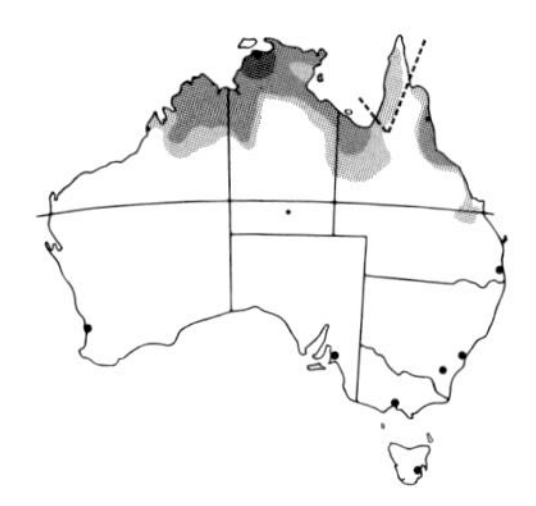

Sexes similar. Crimson face, bill and breast and crimson flanks, speckled white. Eyes: Brown. Crown, upperparts: Grey-brown, with crimson wash on wing. Undertail coverts: Black. Legs, feet: Yellowish.
Race *evangelinae*, west Cape York. White belly and undertail coverts.
Female: Duller, grey underparts.
Immature: Duller, black bill.

Pairs or small family groups. Constantly flicks its long crimson tapered tail up and down. Flits between clumps of grass.

Habitat: Close to water in swampy grasslands with scattered pandanus. Eucalypt and paperbark woodlands with long grass, margins of canefields, homesteads.

Feeding: Grass and pandanus seeds, also insects. Rarely feeds on the ground.

Voice: Penetrating chatter. 'Chee-chee' call. Soft, rasping, descending-note song.

Status: Moderately common. Mostly sedentary.

Breeding: Sept.–May. Low, domed, bulky nest usually in pandanus, sometimes under eaves, inside building crevices / 5–8 eggs.

Fogg Dam, NT.

Opposite: Nominate race, female (left) and male (right).

Painted Finch

Emblema pictum

Endemic

110–120mm

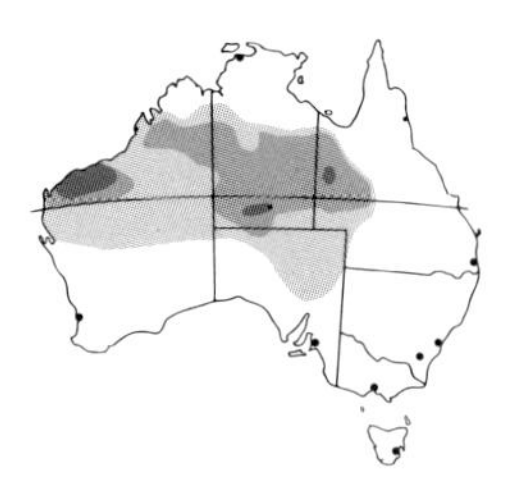

Sexes similar. Brown back with bright scarlet face, throat and rump. Underparts: Black with bold white spotted and barred sides of breast and flanks and a crimson belly streak. Bill: Long, blue-black base, red tipped more prominently on lower bill. Female: Duller. Red lores, eyebrow and cheeks. No red on throat. Immature: Lacks red face. Rump: Red. Bill: Black, red tipped.

Small flocks of 5–20 birds. Well camouflaged.

Habitat: Close to water. Rocky gorges, stony hills with spinifex, acacia scrub.

Feeding: Mostly fallen grass seeds. Drinks frequently.

Voice: Wheezy chattering. Male in courtship and solo, loud, harsh 'trut'. Long, repeated whistled song.

Status: Moderately common. Highly nomadic.

Breeding: Dependent on rain. Domed nest, near to ground. Usually in spinifex, sometimes on a twig, bark, or stone platform / 3–4 eggs.

Mica Creek, Mt Isa, Qld. Tennant Creek, NT. Cape Range NP, WA.

Left: Male. Right: Female.

Diamond Firetail

Stagonopleura guttata

Endemic

120mm

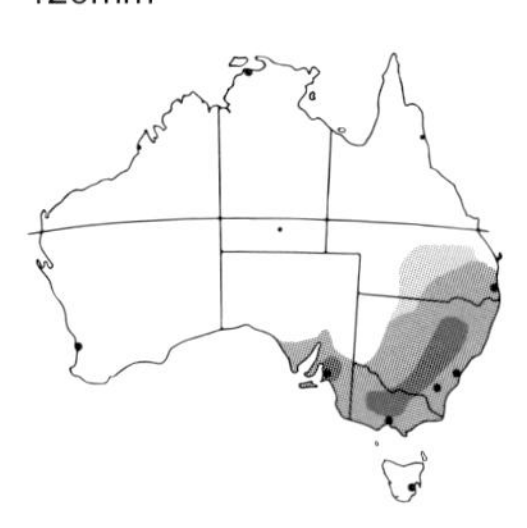

Sexes alike. Named for the strikingly well defined, white, diamond-like markings set into the black flanks. Breast band, lores, tail: Black. Rump, uppertail coverts: Crimson. Crown, frons, neck: Grey. Upperparts: Grey-brown. Chin, underparts: White. Immature: Olive-greyish where adult is black, barred breast and flanks. Rump: Red.

Usually seen feeding on the ground in flocks of 5–40 birds. Male courtship display with long grass stem.

Habitat: Open eucalypt forests, heaths, mallee fringes, grassy woodlands, near water.

Feeding: Ground foraging, usually close to cover. Ripe to half-ripe seeds, sometimes insects. Drinks frequently.

Voice: Plaintive, repeated 'oowee'. Harsh alarm call.

Status: Moderately common. Vulnerable NSW. Sedentary or locally nomadic.

Breeding: Aug.–Jan. In small colonies. Bulky flask-shaped nest of grasses with entrance tunnel, in dense vegetation, mistletoe clumps or eucalypt trees up to 10m high / 4–7 eggs.

Glen Davis and Back Yamma SF, NSW. You Yangs RP, Vic.

Left: Adult. Right: Immature.

Red-eared Firetail

Stagonopleura oculata

Endemic

110–120mm

Sexes similar. Distinctively black belly with bold white spots and a brilliant crimson rump revealed in flight. Upperparts: Brown-grey, finely barred black. Bill: Scarlet. Upper ear covert: Scarlet stripe. Lores: Black. Eye ring: Light blue. Uppertail coverts: Crimson. Throat: Grey-brown, finely barred black.
Female: Pale ear patch when breeding. Immature: Paler, black bill, no ear patch.

Usually shy, unobtrusive. Sedentary in small territory, in mated pairs or family groups. Male courtship display with long grass stem and bobbing.

Habitat: Dense coastal forests, scrub and dunes, paperbark thickets and swampy areas.

Feeding: Forages without tail flicking, mostly for grass and sedge seeds. Seeds also picked from trees and shrubs. Insects taken.

Voice: Piercing, mournful, whistled 'oou'.

Status: Uncommon.

Breeding: Sept.–Jan. Large bulky, flask-shaped nest of green grass stems, with an entrance tunnel and cup-shaped vestibule, placed in a shrub or tree 1–15m high / 4–6 eggs.

Two Peoples Bay NR, WA.

Left: Adult. Right: Immature.

Beautiful Firetail

Stagonopleura bella

Endemic

110–120mm

Sexes similar. Small, plump finch with a scarlet bill, rump and undertail coverts. Upperparts are olive-brown with fine black barring. Lores: Black, extending around the eyes. Eye ring: Light blue. Underparts: Light grey with fine black barring. Undertail coverts, centre belly: Black. Female: Lacks black to undertail coverts, belly.
Immature: Duller, faint barring, dark bill.

Usually in pairs or family groups. Male courtship display with long grass stem held in flight or while bobbing head.

Habitat: Thick coastal heaths, swampy grasslands, tea-tree scrub, casuarina stands.

Feeding: Forages on the ground under cover. Preferring seeds of grass, casuarina, tea-tree. Also insects. Frequently drinks and bathes.

Voice: Penetrating, mournful whistle.

Status: Uncommon Tas. and offshore islands. Rare on mainland.

Breeding: Sept.–Jan. Untidy dome nest of grass and twigs with a long entrance tunnel, placed in dense foliage up to 2m high / 4–8 eggs.

Strahan area, Tas. Barren Grounds NR, NSW.

Opposite: Male, note black belly.

Blue-faced Parrot-finch

Erythrura trichroa

120mm

Sexes similar. Distinctive bright green upperparts and cobalt blue face. Bill: Light grey. Underparts: Bright green with olive wash. Rump, uppertail coverts: Dull red. Tail: Long, rust coloured. Female: Duller, less blue on face. Immature: Dull, no blue on face.

Solitary or in small groups. Wary, shy and elusive.

Habitat: Grasslands edged by coastal rainforests, mangroves.

Feeding: Forages in dense vegetation, clinging to grass stems, mostly seeds and some insects.

Voice: High-pitched 'chirp'. Low trills and high trill calls.

Status: Uncommon to common within its restricted range.

Breeding: Nov.–Apr. Domed, pear-shaped nest of moss and vines in a rainforest tree up to 25m high, 3–4 eggs.

Mt Lewis NP, Qld.

Opposite: Adult male.

Gouldian Finch

Erythrura gouldiae

Endemic Endangered

140mm

Sexes similar. Colourful finch with 3 colour morphs. Face is usually black but sometimes red or, rarely, yellow. Upperparts: Bright green with a bright yellow belly. Breast: Purple. Bill: Creamy white with red tip. Tail: Black, finely pointed.
Female: Duller, paler, all morphs. Immature: Dark olive upperparts, light buff below. Bill: Light grey.

Highly sociable in groups of 10–15 or occasionally large flocks.

Habitat: Mangrove fringes, tropical open timbered grasslands and woodlands, near water.

Feeding: In groups, climbing stems to pick ripe or half-ripe seeds, especially Native Sorghum. In breeding season relies almost exclusively on insects. Rarely feeds from the ground.

Voice: Usually silent. Call 'ssitt'. High-pitched whistling, hissing song.

Status: Endangered, mostly due to large-scale illegal trapping. Partly migratory, following food sources. Moves to the coast in the dry season.

Breeding: Jan.–Apr. In small colonies. Always nests in tree hollows or termite mounds. May share a nest hollow / 4–8 eggs.

Wyndham area, WA. In the late dry season near water.

Left: Red-faced morph. Right: The more common black-faced morph.

Chestnut-breasted Mannikin

Lonchura castaneothorax

100mm

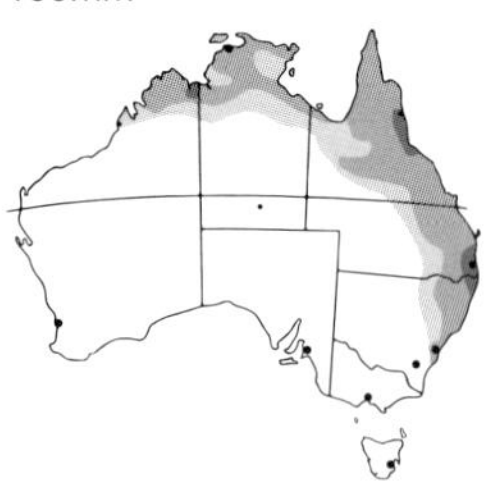

Sexes similar. Grey crown. Black face mask. Chestnut brown breast with narrow black band and white belly. Small black bars edging flanks. Crown, nape, back, wings: Mottled grey-brown. Bill: Blue-grey. Rump, tail: Orange-brown. Undertail coverts: Black.
Female: Paler. Immature: Olive-brown upperparts. Underparts: Pale buff.
Sociable, in small flocks of 5–30 or after breeding in flocks of several hundred.

Habitat: Reed beds, long grassy areas.
Feeding: Ripe and half-ripe grass seeds. Occasionally insects.
Voice: Sharp, bell-like 'teet'.
Status: Common. Locally nomadic in the north. Sedentary in the coastal east.
Breeding: Anytime, mostly Sept.–Feb. Nests in colonies. Domed nest in grassy clumps to 2m high / 4–6 eggs.

Left: Adult male. Right: Top young adult male and female.

Yellow-rumped Mannikin

Lonchura flaviprymna

Endemic

100mm

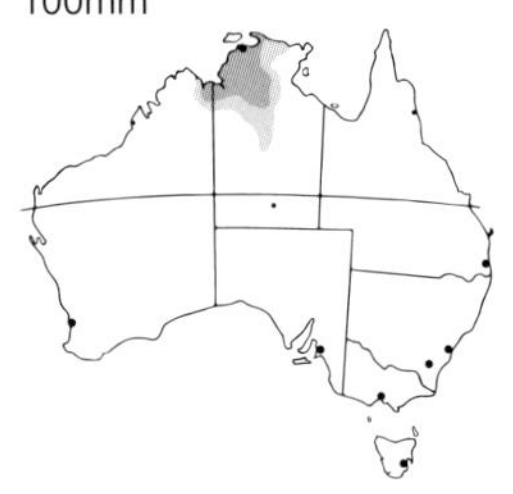

Sexes alike. Pale grey head with a thick grey bill. Upperparts: Cinnamon brown. Underparts: Creamy buff with black undertail coverts.
Immature: Dark bill. Olive-brown above, paler below.
Sociable, flocks to 100 birds, usually of mixed finches.
Habitat: Moist grasslands, flood plains, mangroves, irrigated areas.
Feeding: Ripe to half-ripe grass and weed seeds.
Voice: Sharp, ringing calls. Similar to Chestnut-breasted Mannikin.
Status: Rare. Sedentary. Local seasonal movement, moving towards coastal regions in the dry season.
Breeding: Jan.–Mar. Similar to Chestnut-breasted Mannikin / 4–5 eggs.

Ivanhoe irrigation area, Kununurra, WA.

Opposite: Adult drinking.

Pictorella Mannikin

Heteromunia pectoralis

Endemic

110mm

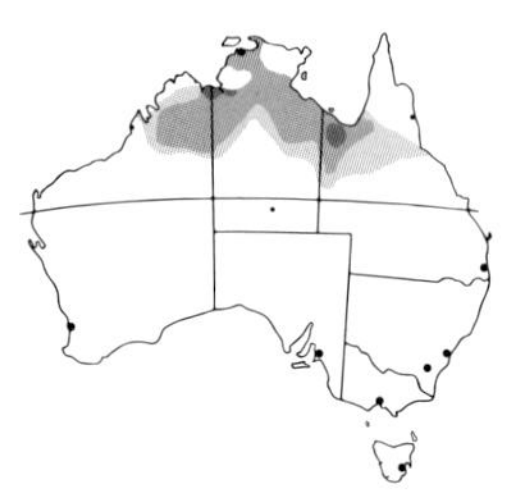

Sexes similar. Distinctive glossy black face mask, edged around ear coverts in rosy fawn. Underparts: Rosy fawn. Bill: Thick blue-grey. Upperparts: Brown-grey with white spots to wing coverts. Breast: Mostly white with some scaly black.
Female: Duller. Breast black, finely scalloped white. Pale flecked flanks. Immature: Plain brown-grey upperparts, paler below.
Usually in small flocks.
Habitat: Dry acacia, spinifex, tropical grasslands, near water.
Feeding: Forages, climbing among grass stems or hopping along the ground, mostly insects, small seeds. Drinks in rapid sips, once or twice a day.
Voice: Vocal groups around water source. Sparrow-like 'chip' call.
Status: Locally common. Nomadic in dry conditions. Moves away from the coast in the wet season.
Breeding: Jan.–Apr. in west. Apr.–May in east. Flask-shaped nest of grass lined with feathers, in a spinifex clump or shrub / 4–6 eggs.

Left: Male. Right: Female.

Nutmeg Mannikin

Lonchura punctulata

Introduced

110mm

Introduced from Asia. Sexes alike. Rich brown upperparts and white underparts with pronounced dark brown scalloping to breast, flanks and undertail coverts. Bill: Blue-grey.
Immature: Plain brownish upperparts, plain buff below.
Usually in small flocks. Constantly flicks its dull brown tail.
Habitat: Tall, grassy areas along riverbanks, roads, cropland; scavenger around dumps, farms.
Feeding: Mostly grass and herb seeds. Sometimes insects. Competes with native species.
Voice: High-pitched 'chip', 'kit-tee'.
Status: Common to uncommon.
Breeding: Sept.–Feb. in south. All year in north. Spherical, green grass and bark nest placed in a shrub / 4–8 eggs.

Left and right: Adult.

Glossary

adult: bird of breeding age.
arboreal: living in trees.
arthropod: invertebrate animal with a hard exoskeleton, segmented body and many jointed legs; includes insects, spiders, crustaceans, millipedes, scorpions.
breast-shield: scaly or iridescent feathers on the breast.
carpal joint: bend in the wing.
casque: French word for helmet, referring to an enlargement on the upper mandible of the bill or crown.
cere: waxy, fleshy structure at the base of the bill. Usually bare, sometimes brightly coloured.
chevrons: pointed or V-shaped markings forming a contrasting pattern.
clinal: a gradual change in inherited characteristics across the geographic range of a species (not identifiable in the field).
colonial: roosting or nesting together.
communal breeding: group containing adult breeding pair and other members co-operatively helping to raise the young.
conspecific: members of the same species.
continental shelf: part of the landmass submerged under the ocean.
cosmopolitan: worldwide distribution, but not necessarily polar regions.
coverts: feathers covering the bases of the main flight feathers of the wings and tail.
dewlap: fold of loose skin hanging from the throat.
diagnostic: key features that confirm a bird's identity.
diurnal: active by day.
dorsal: upper surface or back.
eclipse plumage: phase between breeding and non-breeding plumage.
endemic: confined to a region or country.
frons: forehead.
frugivore: fruit eating.
gape: join of the upper and lower mandibles.
glean: picking food from the surface of leaves, etc.
gorget: patch of distinctive feathering on the breast.
gregarious: living in groups.
gular: pertaining to the throat.
honeydew: rich sugary secretion mostly from aphids and some scale insects.
hybridisation: crossbreeding between different species.
immature: fledged bird, free flying, that has not reached adulthood.
insectivore: a diet of mostly insects.
juvenile: young bird with its first true feather covering, before free-flying.
lamellae: comb-like membrane on the inner edge of the bill used as a sieve for food.
lanceolate: spear-shaped.
lerp: crystallised honeydew produced by larvae of psyllid insects as a protective cover.
lores: area on either side of a bird's face between the eyes and the base of the bill.
manna: sugary sap that exudes from eucalypt leaves when attacked by psyllid or sap-sucking insects. The sugary sap later crystallizes into white blobs or 'manna'.
mollymawk: medium sized albatross found only in the southern hemisphere.
morph: consistent colour variations within a species.
moult: seasonal replacement of feathers
nape: the back of the neck
nocturnal: active at night.
nomadic: irregular wandering.
nominate race: where there is more than one race in a species, the race that has the scientific specific name identical to the species is called the nominate race.
nuchal: nape crest.
nuptial plumage: breeding plumage.
opportunistic: capable of living in and adapting quickly to changing conditions.
passerine: bird of the order Passeriformes; perching birds.
pelagic: living on food from the upper level of oceanic waters. Usually beyond the continental shelf.
pied: black and white colours.
primaries: outer flight feathers.
psyllids: tiny sap-sucking insects.
race: another term for subspecies. Where more than one form of the species occurs that share common characteristics that differ from the nominate race.
resident: living in one place. Another term for sedentary.
riparian: relating to or situated on the banks of a river.
roost: resting or sleeping place.
saltmarsh: plant community of salt-tolerant plants including grasses, sedges, saltbush, samphire.
samphire: succulent, salt-tolerant, perennial shrub.
savannah (savanna): generally grassland with well-spaced trees.
scapulars: shoulder area.
sedentary: living in one area. Not nomadic or migratory. Another term for resident.
speculum: distinctive plumage colour in the wing.
subspecies: where more than one form of the species occurs that share common characteristics that differ from the nominate race. Another term for race.
terrestrial: ground dwelling.
tominal tooth: notched bill. Not a true tooth.
ventral: underside of the bird.
whipstick mallee: multi-stemmed eucalypt community.
window: translucent area in the wing, viewed in flight.
zooplankton: tiny drifting aquatic invertebrates.

Botanical Plant Names

Acorn Banksia (*Banksia prionotes*)
An-binik (*Allosyncarpia ternata*)
Antarctic Beech (*Nothofagus* sp.)
Bangalow Palm (*Archontophoenix Cunninghamiana*)
Beach Daisy (*Arctotheca populifolia*) – introduced weed.
Belah (*Casuarina cristata*)
Belah woodland – where *Casuarina cristata* predominate.
Berrigan (*Eremophila longifolia*)
Black Box (*Eucalyptus largiflorens*)
Blackseed Samphire (*Tecticornia pergranulata)*
Blackthorn (*Bursaria spinosa*)
Black Wattle (*Callicoma serratifolia*)
Bluebush (*Maireana* sp.)
Blue Tussock Grass (*Poa poiformis*)
Blue Waterlily (*Nymphaea caerulea*)
Box-ironbark – forest community found in central Vic.
Brigalow – habitat where *Acacia harpophylla* is dominant.
Brittle Gum (*Eucalyptus mannifera*)
Broombush (*Melaleuca uncinata*)
Brown Stringybark (*Eucalyptus baxteri*)
Bulkuru Sedge (*Eleocharis dulcis*)
Bull Banksia (*Banksia grandis*)
Cliff Bottlebrush (*Melaleuca comboynensis*)
Coachwood (*Ceratopetalum apetalum*)
Coast Beard-heath (Leucopogon parviflorus)
Coastal Wattle (*Acacia sophorae*)
Common Boobialla (*Myroporium insulare*)
Cumbungi (*Typha* sp.)
Cypress Pine (*Callitris* sp.)
Darwin Woollybutt (*Eucalyptus miniata*)
Desert Bloodwood (*Corymbia opaca*)
Desert Fig (*Ficus platypoda*)
Emu-bush (*Eremophila* sp.)
Figs (*Ficus* sp.)
Firewood Banksia (*Banksia menziesii*)
Gorge Hakea (Hakea fraseri)
Grass-leaf Hakea (*Hakea multilineata*)
Grass Tree (*Xanthorrhoea australis*)
Grey mangrove (*Avicennia marina*)
Heath-leafed Banksia (*Banksia ericifolia*)
Hoop Pine (*Araucaria cunninghamii*)
Jarrah (*Eucalyptus marginata*)
Karri (*Eucalyptus diversicolor*)
Knobthorn Acacia (*Acacia nigricans*)
Lacebark (*Brachychiton discolour*)
Lancewood (*Acacia shirleyii*)
Laurels, native (*Cryptocarya* sp.)
Lawyer Vine (*Smilax australis*)
Lignum (*Muehlenbeckia florulenta*)
Manna Gum (*Eucalyptus vimminalis*)
Manuka Tea-tree (*Leptospermum scoparium*)
Marri (*Corymbia calophylla*)
Mistletoe – a diverse group of parastic plants, with over 90 mostly endemic species.
Monga Waratah (*Telopea mongaensis*)
Mountain Ash (*Eucalyptus regnans*)
Mount Morgan Wattle (*Acacia podalyriifolia*)
Mountain Swamp Gum (*Eucalyptus camphora*)
Mulga (*Acacia aneura*)
Native Cherry (*Exocarpos cupressiformis*)
Native Laurel (*Anopterus glandulosus*)
Native Raspberry (*Rubus moluccanus*)
Native Sorghum (*Sorghum leiocladum*)
Nepean Spider Flower (*Grevillea arenaria*)
Netbush (*Calothamnus quadrifidus*)
Northern Paperbark (*Melaleuca leucadendra*)
Old Man Banksia (*Banksia serrata*)
Parrot Bush (*Banksia sessillis*)
Pear-fruited Mallee (*Eucalyptus pyriformis*)
Pigface (*Carpobrotus* sp.)
Pine Banksia (*Banksia tricuspis*)
Platypus Gum (*Eucalyptus platypus*)
Rat's Tail Grass (*Sehima nervosum*)
Red Bloodwood (*Corymbia gummifera*)
Red-flowering Gum (*Corymbia ficifolia*)
Red-flowered Yellow Gum (*Eucalyptus leucoxylon rosea*)
Red Ironbark (*Eucalyptus tricarpa*)
Red Spider Flower (*Grevillea speciosa*)
Red Swamp Banksia (*Banksia occidentalis*)
River Red Gum (*Eucalyptus camaldulensis*)
Rock Correa (*Correa glabra*)
Sacred Lotus (*Nelumbo nucifera*)
Salmon Gum (*Eucalyptus salmonophloia*)
Saltbush (*Atriplex* sp.)
Samphire (*Sarcocornia quinqueflora*)
Sandhill (Dune) Canegrass (*Zygochloa paradoxa*)
Sandpaper Fig (*Ficus coronata*)
Sassafras (*Doryphora sassafras*)
Scarlet Honey Myrtle (*Melaleuca fulgens*)
Screwpine (*Pandanus spiralis*)
Sheoak (*Casuarina* sp.)
Silky Hakea (*Hakea sericea*)
Silky Oak (*Grevillea robusta*)
Silver Saltbush (*Atriplex bunburyana*)
Smithton Peppermint (*Eucalyptus nitida*)
Spinifex (*Triodia* sp.)
Stringybark – includes several eucalypt species that have stringy, thick, fibrous bark.
Sturt's Desert Pea (*Swainsona formosa*)
Swamp Mahogany (*Eucalyptus robusta*)
Sword Grass (*Gahnia sieberiana*)
Sydney Golden Wattle (*Acacia longifolia*)
Tasmanian Blue Gum (*Eucalyptus globulus*)
Tea-tree (*Leptospermum* sp.)
Telegraph Rush (*Mesomelaena stygia*)
Wandoo (*Eucalyptus wandoo*)
Waratah Banksia (*Banksia coccinea*)
Water Snowflake (*Nymphoides indica*)
Weeping Rice-grass (*Microlaena stipoides*)
Whipstick Mallee (*Eucalyptus multicaulis*)
White Cyprus Pine (*Callitris glaucophylla*)
White-flowered Black Mangrove (*Lumnitzera racemosa*)
White-flowered Bottlebrush (*Callistemon salignus*)
Wild Rice (*Oryza meridionalis*)
Yellow Box (*Eucalyptus melliodora*)

Associations and Societies

Australian Conservation Foundation

Australian National Botanical Gardens

Australian Native Plant Society

BirdLife Australia

Environmental Justice Australia

Environmental Defenders Office ACT (EDO ACT)

Environmental Defenders Office New South Wales (EDO NSW)

Environmental Defenders Office (EDO Northern Territory)

Environmental Defenders Office Queensland (EDO Qld)

Environmental Defenders Office South Australia (EDO SA)

Environmental Defenders Office Tasmania (EDO Tasmania)

Environmental Defenders Office Western Australia (EDO WA)

Friends of Grasslands

La Trobe Wildlife Sanctuary
To purchase nestboxes visit the website or visit their nursery.

WIRES (Wildlife Information and Rescue Service): 1300 WIRES or 1300 094 737.

Australian Birding Organisations

New South Wales

Birds Australia, Olympic Park

BirdLife:
- Echuca
- Mildura
- Northern NSW
- Southern NSW/ACT

Bird Observer Clubs:
- Blue Mountains
- Cumberland
- Far South Coast Birdwatchers
- Hastings
- Hunter
- Illawarra Birders Inc.
- Ovens and Murray
- Shoalhaven
- Southern Highlands
- Tweed

Murrumbidgee Field Naturalists Inc.
NSW Field Ornithologist Club Inc.

Victoria

BirdLife:
- Bass Coast
- Bayside
- Bellarine Peninsula

The Field Naturalists Club (FNCV)

Queensland

BirdLife:
- Bundaberg
- Capricornia
- Mackay
- Northern Queensland
- Southern Queensland
- Townsvillle

Tasmania

BirdLife Tasmania

South Australia

BirdLife:
- Gluepot Station
- Kangaroo Island
- South East SA

Fleurieu Birdwatchers

Northern Territory

BirdLife Central Australia

Western Australia

BirdLife WA
WA Recent Bird Sightings
Leewin Current Birding

Australian Capital Territory

BirdLife Southern NSW/ACT
Canberra Ornithologist Group

Acknowledgements

The author wishes to thank the following people:

Corinne Mansell, Errol and Ethel Newlyn, Andrea McNamara, Ann Wilson, Julian Mole, Bronwyn Sweeney, Catherine Dunk, Penguin Books staff especially Catherine Hill, Izzy Yates, Ben Ball and Brandon VanOver. Matt Hermitage, Thirdegree, BirdLife Australia, Bush Heritage Australia, Australian Broadcasting Commission – Alice Springs.

I would also like to thank the designer of this book, my wife Dianne, and my children, Elizabeth, Franchesca, George, Benjamin, David and Mary and their extended families and our friends for putting up with years of temperamental moods, late nights and strange dinner table conversations about not much else but birds, taxonomy splits, bird behaviour and birdlife histories and how if they were good, Santa might one day bring them a copy of the completed book.

The author would like to acknowledge the generous contribution the photographers have made to this book, and especially thank Colin Cock, Michael Schmid, Eric Sohn Joo Tan, Duade Paton, Mark Sanders, John Anderson, Alwyn Simple, Peter Jacobs, Andrew Bell, Tony Ashton, Nolan Caldwell, Chris Wiley, Alan Danks, John Irvine, Maureen Goninan and Marlene Lyelle.

Photographic Acknowledgements

George Adams Author. All drawings. 12r4, 15r3, 41r2R, 83r2C, 105r2L, 203r1L, 203r2R, 203r3L, 205r2L, 205r1L, 205r3, 211r2LR, 213r1LR, 215r2R, 215r3L, 217r1R, 217r3L, 221r2R, 223r1R 225r3R, 227r3L, 229r1R, 229r3R, 231r1LR, 235r3r, 241r2LR, 255r2R, 287r2L, 289r1R, 293r3C, 297r2, 305r2R, 309r1LR, 355, 357r2R, 357r3R, 359r3L, 363r2R, 365r1R, 367r2L, 387r2LR, 479r1R, 485r1L, 487r3L, 491r1R, 501r1, 529r2L
Stuart Amer 157r2L, 157r3L
Graham Anderson Holmview, Qld. Wildlife photographer and flight studies using high-speed flash. 285r2LR
John Anderson Began photographing birds in UK at the age of 15 with a second hand Zenith E and Soligor lens, has now progressed to natural history photography in Western Australia with Nikon equipment. 33r2R, 33r3L, 83r2L, 83r3R, 103r3L, 115r3R, 133r1R, 135r1L, 143r1L, 153r1R, 155r2R, 173r2, 191r3R, 193r4L, 223r2R, 245r2R, 249r2, 253r2L, 265r3LR, 293r1R, 303r1L, 313r3R, 315r2L, 339r1R, 345r4L, 357r2L, 359r1R, 379r2L, 385r3R, 391r2L, 393r2L, 399r3L, 419r3LR, 421r1LR, 437r3L, 441r2LR, 469r3L, 475r2L, 493r2, 509r2L, 515r2R, 535r3L
Michael J Anderson 87r2R
Diane Armbrust 215r1R, 367r3
Thibaud Aronson 335r3R
Tony Ashton Northern Qld. Birding obsessive; former Kiwi journo; better birdwatcher than word botcher. Read Tony's blog at tytotony. 21r1R, 25r1R, 25r2R, 37r2R, 91r2L, 91r3L, 93r3L, 133r2, 141r1L, 155r2R, 177r2R, 177r3R, 263r2R, 277r1LR, 277r3R, 315r1LR, 339r3R, 369r1R, 369r2L, 373r3L, 373r2R, 375r2R, 395r3R, 397r3R, 421r3R, 423r1LR, 445r2L, 455r3LR, 461r3L, 471r1R, 495r2R, 501r3L, 513
Nick Athanas Avid bird photographer as well as a tour leader for Tropical Birding Tours. While Nick currently resides in the US, he lived for 16 years in Ecuador and is the lead author of *Birds of Western Ecuador: A Photographic Guide*. 409r3L, 459r1R, 463r2R
Steve Axford Northern Rivers NSW. More of his work can be found at steveaxford on smugmug. 41r3R, 99r3L, 111r3R, 161r1R, 199r1R, 277r2L, 417r3R, 433r2R
Nick Baldwin 361r1L
Stephen Michael Barnett 11r3
Brent Barrett 257r2
Mike Barth UK. Bird photography is my passion. See more of Mike's work at mikebarthphotography. 141r2L, 241r3R
Gill Basnett 13r4, 14r3
Tim Bawden 45r1R, 55r1R, 271r3LR
Doug Beckers 14r4
Andrew Bell Medical practitioner who has lived with his family in Katherine NT for the last 23 years working predominantly with remote communities. He has a long-standing interest in the natural environment and has been a keen photographer and birdwatcher since he was a teenager. 31r3R, 39r1R, 105r2R, 127, 143r1R, 145r1L, 145r3, 147r1L, 147r2R, 149r1L, 149r2L, 151r1LR, 155r1R, 155r3R, 157r1R, 167r2L, 167r3LR, 169r1L, 177r1, 177r3L, 179r1R, 193r2R, 197r1L, 199r2R, 221r2L, 231r2R, 243r3R, 245r2L, 287r2R, 305r3R, 369r3R, 379r1, 399r2R, 487r2R, 519r1, 529r2R, 531r2R, 537r3R
Jim Bendon 113r3R, 221r1R, 293r3R, 411r2C, 475r2R
David Bertram 457r2R
Leo Berzins Merimbula NSW. Now retired, Leo is a keen amateur nature photographer whose favourite subject is birds. 61r4L, 67r3R, 79r3R, 97r1R, 141r1R, 381r3L, 395r1L
S J Bond ACT. I have always had a passion for birds and the natural world. I'm interested in how the power of photography can generate interest in our fascinating and unique wildlife, and hopefully foster a deeper ecological understanding of our environment. 107, 183r3, 191r4L, 235r1L, 239r1R, 347r2L, 447r1L, 453r2L, 461r3R, 481r2R
Michael Bouette 433r1L
Adrian Boyle wildlifeimages.com.au 181r2
Winfried Bruenken (Amrum) [CC BY-SA 2.5 (https://creativecommons.org/licenses/by-sa/2.5)], via Wikimedia Commons 79r2
Daniel Burgas Finland. His photography is an omnivorous approach, which largely gravitates around those details from nature that are often overlooked; either taken in the backyard or in the antipodes, either from tiny creatures or from impressive landscapes. Birds are often present in his images, as he has been birding since an early age. More of his work can be found at danielburgas.com 49r3L, 53r3L, 65r1L, 65r4R, 67r1L, 69r1LC
Julie Burgher 321r1L
Tod Burrows Ecologist, birder and wildlife photographer based on the Gold Coast since 2001. 11r4, 16r3, 73r2L, 179r3R, 183r1R, 211r3L, 281r3, 297r3, 299r1R, 425r3L, 431r3L, 491r1L
Wayne Butterworth 37r1R, 39r1L, 39r2, 41r1L, 43r1R, 83r2R, 93r3R, 123r2L, 125r2L, 133r1L, 161r2R, 263r2L, 269, 445r1L, 481r3L
Pablo Caceres Valparaiso Chile. Architect and birding enthusiast, particularly pelagic birds. 55r2L, 63r2R, 63r3R, 63r4LR, 67r2R, 71r4R
Nolan Caldwell Darwin, the capital of 'Nature Territory', he has been seriously taking photos for about the last fifteen years of the wonderful nature that the Top End has to offer. Travelling throughout the Top End and Kimberley recording birds, flora and fauna along with landscapes and the wonderful storm activity that abounds in this part of Australia. Nolan's images have been used by various environmental groups including Birds in Backyards and have featured in *Australian Geographic*. More of his work can be found at encimages.com.au 23r1R, 29r1L, 85r2L, 95r2R, 103r1R, 105r1L, 111r1L, 113r1R, 113r2L, 113r3L, 123r2R, 123r3LR, 175r2L, 207r2L, 207r3, 213r2L, 225r2L, 251r1L, 273r1R, 289r2R, 291r1R, 293r2L, 313r3L, 401r3L, 411r2R, 451r2R, 493r3L, 533r3R, 570
Anna Calvert 115r3L
Sandy Carroll 287r3L, 371r2R
Martin Casemore UK. Follow Martin's birding blog at ploddingbirder.blogspot 55r2C, 57r2LR, 57r3L, 59r1R, 59r2L, 59r3R, 61r1RL, 63r2L, 97r3R, 187r1L, 485r4L

Graeme Chapman Basin View NSW. Spent most of his life as professional ornithologist and photographer with CSIRO Division of Wildlife and Research, and has one of the biggest collections of wildlife images in Australia. 247r1R, 249r3, 267r1R, 311r3R, 319r1, 349r2R, 389r2, 389r3, 431r2, 437r3R, 443r1R, 517r2R
Timmy Chen 285r1LR
Katarina Christenson ACT. Hobby wildlife photographer mostly interested in bird and insect or spider macro shots. 41r2L, 85r1R, 119r4L, 289r1L, 365r1L, 463r3R, 469r2C, 477r1L, 487r3R, 503r3R, 509r1L, 509r3L, 515r2L
Jon Clark Nightcliff, NT. I love travelling and photographing wildlife. My aim with photography is to capture a little of the amazing sights I'm lucky enough to see around home and beyond. 143r3, 209r1R, 289r2L, 471r2C, 493r3R
Phil Clarkson Photography 497r2R
Ron and Alenka Co Tasmania, near Colebrook in a basically blue-gum dry forest area. Moving here from SA about 8 years ago we have endeavoured to make our patch of forest as wildlife and bird friendly as possible. We now feed Bennett's pademelons and Brushtail possums and are visited by quolls and a family of Tasmanian native hens has taken up residence. I took up birding after moving here and have recorded about 80 species in my garden. 21r2R 351r2R, 383r1L, 387r1R, 439r2L, 461r2R, 485r2LR
Colin Cock SA. I spent much of my youth exploring the creeks and valleys around the foothills around Adelaide's southern suburbs, or off fishing and diving. Always fascinated by the variety of creatures that abounded in the native bush region and seas at that time; much of the region is now dominated by urban sprawl. A great lover of animals and often finding comfort in their presence led to my long-time employment with Royal Society for the Prevention of Cruelty to Animals and now various volunteer conservation groups. As an extension of my love and need to be in the great outdoors I became a passionate nature photographer, both as a hobby and for educational purposes. Finding joy in using the camera to enlighten others about our wonderful and diverse flora and fauna: 'Conservation through Education'. 21r2L, 23r3, 25r2L, 37r1L, 93r2L, 95r3L, 101r3L, 105r3L, 109r2LR, 115r1R, 121r3L, 123r2C, 125r2R, 125r3R, 129r1R, 131r1, 131r3, 157r1L, 161r3L, 171r3LR, 173r3LR, 175r1R, 195r2L, 211r1, 217r1L, 225r1R, 225r2R, 233r1R, 233r2R, 243r1L, 245r3R, 253r3, 255r3L, 265r1L, 283, 287r1L, 293r1L, 311r2R, 315r2R, 315r3L, 317r1LR, 317r3LR, 327r1LR, 337r1LR, 343r3R, 345r2L, 349r1L, 353r2L, 373r2L, 377r3LR, 379r2R, 379r3L, 391r1LR, 403r2LR, 403r3L, 405r2LR, 405r3L, 409r1R, 413r1R, 413r2LR, 417r1R, 419r2, 425r1LR, 427, 429r2LR, 435r1LR, 447r4R, 463r1R, 469r3R, 509r1R, 517r1R, 521r2LC, 527, 529r1R, 535r1L, 539r1R, 549, 566, 568, 572
Fred Coles Perth WA. I enjoy hiking, camping, and hosting dinner parties with friends. In my spare time, I enjoy photographing the birds of Western Australia and as much wildlife as possible, with a few landscape shots here and there. I live in Perth with my wife. 35r2R, 119r2R, 227r2, 279r3L, 313r2R, 345r1L, 349r3L, 391r2R, 475r4R
Ian Colley 29r2L, 35r3R, 203r2L, 217r3R
Anne Collins Hobart Tas. Following on from a life-long interest in nature, photography seemed a natural progression. My preference is bird photography, with an emphasis on capturing interesting behaviour or demonstrating the habitat of a particular species. 137r3R, 289r3R, 313r1L, 349r3R, 405r3C, 437r2LR, 479r2L, 529r1L, 209r1L, 209r2L, 275r1R
Dave Curtis UK. Passionate birder for over 50 years. Usually with his wife Alison, out in the field or enjoying the birds in their garden or on guided bird tours abroad. 23r2R, 117r2L, 153r3L, 367r1LR, 367r2R, 395r1R, 479r2R, 481r1L, 481r2L, 503r4L, 525r3LR
Marc Dalmulder https://creativecommons.org/licenses/by/2.0/ 15r1
Alan Danks 295, 299r2LR
John Dart 43r1L, 103r3R, 137r2L, 175r2R, 195r1L, 205r1R, 205r2R, 217r2R, 245r1LR, 247r2LR, 265r1R, 277r3L, 481r3R
Arron Davies 333r3
Simon Davies 523r2L
Peter Deasy of Stan-Dea Photographics 235r2L
Garry Deering 439r1L
Andy Deighton 109r3R
Judith Deland ACT. Completed a Masters in Biological Anthropology at ANU and in about 2008 I picked up a camera to try and record fungi on Brown Mountain when we thought we were going to lose the old growth there. My interests include wildlife and walking and using photography to foster our care of the environment. As a 'newbie' to photography I was privileged to be included in Natural Forests Australia's wilderness coast exhibition.12r3, 15r4, 19, 239r1L, 277r2C, 507r2R
Bram Demeulemeester Birdguiding Philippines 305r1R
Jim Denny 75r4L
Drew Douglas 237r2L
Ed Dunens CC by 2.0 73r1
Larry Dunis Late of Queensland. Larry passed away Sep. 2010 aged 54 years. His major interests were his family and photography, particularly nature and especially birds. 83r1L, 181r1L, 231r3R, 311r3C, 331r4L, 353r4L, 503r4R, 519r2R, 525r1L
Vic Dunis Arrived in Australia from England in 1970 at age 15 and fell in love with the bush. I spend much of my spare time travelling in remote parts of Australia, observing and photographing the local animals and plants. 63r3L, 99r2R, 109r3L, 193r3R, 207r1R, 321r1R, 321r3L, 331r3R, 341r1R, 369r3L, 391r3R, 509r4L, 525r2L
Graham Ekins UK. Semi-retired ecologist with an interest in single species studies, having spent 11 years working on tree-nesting cormorants and also studying the dispersal characteristics of Essex Greenfinches. I am a seabird and cetacean surveyor for the UK MARINElife charity. Evie and I are keen travellers and we try to focus on a country's endemic species and obtain images for Avibase, Cornell Institute, OBC and NBC websites. I have also donated seabird images to the RSPB (UK) for their Albatross Taskforce programme. Other natural history interests include Lepidoptera, Odonata, Orthoptera and recently mycology. 51r1LR, 53r1CR, 53r2C, 57r1L, 65r3R, 67r4R, 71r2L, 81r1R, 81r3L 191r2, 263r3R, 523r2R
Tony Enticknap Dorset, UK. Keen amateur wildlife and nature photographer with a passion for travel. 81r2C, 87r1LR, 101r1R, 145r2LR, 153r3R, 159r2L, 167r2R, 169r3R, 189r4LR, 191r1LR, 503r1L
Peter Ericsson 141r2R
Malcolm Fackender 233r2L
Eric Gofreed 339r3L, 361r2L, 385r1R, 397r3L, 401r1L
Maureen Goninan I'm addicted to bird photography, an addiction which is easy to understand if one really appreciates the exquisite beauty of birds! 29r3R, 33r1R, 39r3R, 41r1R, 79r1, 81r3R, 97r2L, 97r3L, 99r1LR, 103r2, 163r1R, 187r3R, 193r4R, 199r2L, 221r3R, 227r1LR, 365r2R, 387r3L, 471r3L, 473r2L, 487r2L, 489r1, 529r3R
Brian Gratwicke 11r1, 21r1L, 197r3R, 235r3L
Mark Greatorex 13r2
Ralph Green Vic. I have had a long interest in photography, mainly landscapes, which started with the purchase of my first camera in the early 1960s. Now in semi-retirement and with more time and resources available I have been able to extend that interest into bird

photography. Both interests now provide an opportunity to see more of Australia and appreciate the remarkable scenery and diversity of the birdlife in this country. 14r2, 15r2, 27, 29r2R, 31r2R, 79r3L, 83r1R, 83r3L, 95r2L, 135r3R, 137r2R, 223r2L, 237r2R, 287r1R, 329r4L, 373r2R, 375r3R, 393r3L, 477r1R, 481r4R, 503r3L

Mehd Halaouate 237r1L

Knut Hansen 55r1L, 65r4L, 159r2R

Mark Harper 433r3L

JJ Harrison I joined Wikipedia in 2006. Outside of uni and wiki related activities, I like riding my bike and bird watching. Wikimedia Australia has supported my work in the form of a number of small grants and travel grants. 69r4L, 75r2R, 217r2L

Mark Helle Hoppers Crossing Vic. Interests are travel, photography, bushwalking and the challenge and unpredictability of photographing wildlife. These days all my spare time is spent with my daughters and I am looking forward to taking them bushwalking when they are old enough. 39r3L, 197r3L, 207r2C, 243r3L, 353r3L, 485r3L, 535r1R

Rachel Hopper 191r4R

Robert Horl 393r2R

Bruce Hosken NSW. After graduating from the University of WA and following conscription and a lengthy period overseas, I went on a trip through the Sandy Desert with my brother-in-law, Prof. M Tarburton, and Don Hadden, a brilliant wildlife photographer, and I was just amazed at the birdlife in Australia. Soon I was looking at cameras with big lenses and addicted to capturing birds in their native environment. Member of Hunter BOC. 99r3R, 117r2R, 123r1LR, 229r1L, 229r3L, 243r1R, 333r1, 347r3L, 351r3L, 459r1L, 489r2R

Lee Hunter I took up bird photography as a post-retirement avocation. My wife and I have travelled to some of the world's most exciting birding countries, trying to come home with the best photos possible. Many more places to go and birds to see. 35r1L, 87r3L, 289r3L, 417r3L, 489r2L, 497r1R, 501r3R

Robert Hutchinson 319r2

Jon Irvine Formerly resident in Sydney, NSW but now based back in Cornwall, UK. My main interest is in birds and bird photography with raptors, seabirds and waders being particularly important. I am currently studying the nesting behaviour of White-bellied Sea-eagles at Sydney Olympic Park using remote CCTV footage from the nest. 16r1, 47, 51r2R, 67r2L, 67r4L, 87r4L, 101r2R, 113r1L, 117r1L, 129r3L, 229r2L, 251r2R, 267r2L, 303r3L, 425r2R, 471r3R, 477r3L, 479r4R, 491r3R, 511r1L

Richard Jackson 273r2L, 275r3L

Peter Jacobs Waikerie SA. Sadly, Peter passed away in February 2016, but his amazing photography lives on as an everlasting legacy. Peter was a science, maths and biology teacher, before returning to his family's citrus and wine grape property. Always interested in natural history, his skill lay in his deep knowledge of bird behaviour and his ability to anticipate where birds will land for a good shot, always considering the light and the background. Many of his shots were taken in his home garden. Diagnosed with pancreatic cancer, he maintained his passion for nature and took great comfort in his garden, birds and environment, which were an integral part of his being. 65r1R, 101r2L, 239r2, 241r3L, 251r2L, 253r2R, 291r3LR, 301r1L, 303r3R, 309r3R, 321r2L, 325, 335r1L, 337r3, 343r3L, 345r3L, 351r2L, 353r1R, 359r2L, 365r3LR, 381r3R, 383r3L, 407, 409r2, 411r1, 413r3R, 439r3LR, 447r3L, 447r4LC, 465, 473r3L, 475r1LR, 499, 507r1L, 507r2L, 517r3L

Fred Jacq 65r2L

Honey Jeffs 35r3L, 329r4R

David Jenkins Melbourne Vic. Natural history photo-illustrator and photo tutor. I've been involved in wildlife photography all my life. Until recently I shot illustrations for a number of magazines, and I've been teaching photography for about 15 years. Now, slowing down a bit and enjoying the birds. David's blog: Birds as poetry. 33r3R, 95r3R, 115r1L, 119r3R, 121r2, 125r1LR, 195r1R, 223r1L, 249r1, 263r1R, 271r2L, 337r2, 383r3R, 413r3L, 415r1LR, 415r2L, 415r3L, 441r1C, 453r2R, 475r3LR, 521r1L, 521r2R

Darryl Jones WA. Vietnam veteran who took up bird photography as a way of dealing with life after Vietnam. Sadly, waited too long and only began the photography 8 years ago. Married happily to a lovely lady who shares my passion for the birds and the outdoors. I have been interested in birds and wildlife for as long as I can remember. 43r4, 73r2R, 111r2L, 121r1L, 129R1L, 163r3, 177r2L, 207r2R, 261r1R, 275r1L, 309r2L, 359r1L, 373r3R, 441r3R, 505r1LR, 535r3R

Somchai Kanchanasut Thailand. Retired, now acting as a freelance bird photographer. My inspiration may be like 'Retirement is a free life as the bird without captivity.' 37r2L, 97r2R, 147r1R, 149r3R, 155r2L, 157r2R, 157r3R, 169r3L, 171r1L, 523r4R

Patrick Kavanagh Vic. Passionate photographer of nature. He particularly enjoys exploring the beauty and wonder of the wildlife with whom he shares a bush property in central Victoria. Birds and invertebrates (via macrophotography) feature heavily in his photographs. 345r2R, 347r2R, 385r2R, 435r3LR

David King Gold Coast, Qld. Most admired photographers: Ansel Adams and Graeme Chapman. Philosophy: Photography is about establishing a relationship with the subject – a bird, a plant, a landscape, another person – and letting others share that experience. 21r3LR, 29r1R, 41r3L, 93r2R, 359r2R, 373r1, 375r3L, 457r3R, 473r2R, 515r3R

Gary Knight Life-long interest in photography, particularly orchids, but only recently hooked on bird photography. I joined the local Birdlife Australia group and as my knowledge of birds grew so did my appreciation of getting out in the bush and enjoying the peace and tranquillity. I soon developed a great deal of patience and began to love the stillness of the wait. Maybe it's that deep-down hunting instinct that drives me to see how close I can get to my subject to get that great shot, and I often use one of my shorter lenses to make it just that little bit harder. 91r2R, 131r2, 261r2LC, 385r1L, 399r3R, 401r1R, 451r1R, 455r1L, 457r3L, 531r3R

Martijn de Kool 33r2L, 85r1L, 105r3R, 121r1R, 137r1L, 251r3, 281r1, 327r4L, 361r2R, 393r1R, 395r2L, 421r3L, 443r2R, 467r1L, 481r4C, 505r3L, 507r1R

Daisuke Kudo 285r4L

Davis Kwan 331r1, 347r1R

Tim Lenz 49r4L, 81r1L, 189r1R

Sheau Torng Lim I love travelling, birds and photography, so this is a great hobby for me. 25r3LR, 115r2R, 175r3L, 231r2L, 273r3R, 291r1L, 503r5R, 505r4L, 515r1LR, 519r3L, 521r3R, 523r1LR, 523r3R, 529r3L, 539r2, 539r4L, 569

David Lochlin 271r1LR

Marlene Lyell Bendigo Vic. on the Campaspe River. I am fortunate to be surrounded by birdlife. Our property list is 146 species, of which I've photographed nearly 50 at the 11 birdbaths around the garden. Got into birding rather late in life after a trip around Australia in 1979. Been photographing birds since then. All aspects of the natural world inspire me. My greatest joy is sitting and watching the birds and their antics. Photography offers the opportunity to study the nuances in the colour and detail of the bird's plumage, which the eye often fails to see. Beautiful! 53r2R, 85r3R, 91r1R, 117r1R, 135r2, 179r2R, 183r2L, 199r3R, 225r3L, 233r1L, 303r2, 329r3L, 343r4L, 345r1R, 353r1L, 353r3R, 365r2L, 381r1, 385r2L, 413r1L, 453r3L, 479r3L, 481r1R, 481r4L, 485r3R, 514r1LR, 505r1L, 521r1R

Stewart MacDonald 271r2C

John Mansell Born in Yorkshire UK, and it was there that at an early age I developed an interest in birds. When I was 15, the family migrated and we settled in Brisbane where I have spent most of my life. I am married with 3 children and 4 grandchildren. My interest in birds was rekindled while I was working in a remote part of central Queensland after drought-breaking rains. I was amazed at the huge numbers and variety of birds gathered around the waterholes, and at the fact that I could not identify any of them. On returning to Brisbane, I bought my first field guide, and soon bird watching was added to my love of photography and bushwalking. 165, 277r2R, 339r4R, 363r1R, 399r1, 451r2L
Petter Zahl Marki 397r2L
Greg McLachlan Sydney NSW. An Australian birder who also enjoys photographing the many varied birds of Australia and beyond as well as recording their song and call. I have always been fascinated by wildlife, and have developed the urge to use photography to share that fascination with others. I enjoy trying to reveal the delights and surprises of nature. I hope my photography will help people appreciate the need to conserve what we have left, while we still can. 59r2R, 67r1R, 73r4R, 101r3R, 331r2L, 341r3R, 361r3LR, 379r3R, 467r1R, 509r3R
Jim McLean Sydney. I have been fascinated by wildlife since I was a child. I'm drawn to share that passion in the form of photographs, partly for fun, and partly in the hope that I will help raise community awareness of the importance of preserving biodiversity. 23r1L, 31r2L, 219, 537r1
Angus McNab Ecologist, birder and wildlife photographer based on the Gold Coast since 2001. 69r1R, 181r3R, 511r2L, 255r1L, 257r1L
Ellis McNamara www.ardea.com 299r1L
Shanta McPherson 13r3, 393r1L
Robert McRobbie Sydney. He is passionate about wildlife and often takes his family on excursions to remote parts of Australia with a secret desire to capture new animals. Through the challenge of bird photography Robert has discovered several thriving habitats around Sydney and takes every chance to wander softly through the bush listening for new species to add to his collection of images. 163r1L
Greg B Miles Central Coast, NSW. Recently retired wholesale nurseryman. My lifetime love of birds started early, when I found my father's edition of Cayley's *What Bird Is That?* on the bookshelf. I still have it. I have been close to birds most days of my life and photographing them is as much about retaining a memory of a birding experience as it is obtaining the best possible image of them. 77, 179r3L, 183r2R, 215r1L, 339r2L, 263r3L, 351r1, 387r1L, 403r1R, 419r1L, 443r1L, 467r3L, 495r3L
Michael Minter 433r2L
Ross Monks My wife Thea and I love being in the bush; chasing beautiful and elusive birds gives us an excuse and a challenge at the same time. Bird photography has introduced us to some wonderful friends and spectacular places all around the world. 99r2L, 129r3R, 149r1R, 151r3R, 273r1L, 181r1R, 405r1L, 429r1LR, 441r1L, 467r3R, 517r3R
Ian Montgomery Based near Townsville Qld. Follow Ian at Birdway. 51r3LR, 81r2R, 87r2L, 91r1L, 93r1R, 189r3L, 197r4L, 215r2L, 279r1R 289r3R, 437r1L, 461r1L, 471r2L, 533r2L
Rob Morris Grew up with a father who was a mad keen birder. He starting birding in the UK at the age of 4 and hasn't stopped for the last 40 years. His passion for birds and wildlife led him to become a bird bander from 16. He then did a BSc (Hons) in Ecology followed by a Masters in Environmental Assessment where he looked at the treatment of birds in over 100 Environmental Impact Statements. Today he heads a leading Environmental Consultancy that specialises in understanding the impacts of major mining and oil and gas projects. 71r3LR, 73r3CR, 75r4R 323r1LR, 343r2LR, 371r1L, 429r3LR, 433r1R
Colin Mount 159r1
Greg Muir Perth WA. Enjoys photography, mostly birds. 443r3L
Dr Steve Murphy 257r3
Troy Mutton Sydney NSW.
A birder since his childhood and bird photographer since the mid 2000s, Troy is passionate about Australian birds, particularly pelagic seabirds. 51r2L, 97r1L, 179r2L, 261r2R, 377r1L, 519r2L
Craig Nieminski NT. Photography and nature are a great combination. Taking photographs and then identifying what I have captured has taught me a great deal about the world I live in and choose to explore. Being in nature keeps me centred in a 'here and now' world, well away from yesterdays or tomorrows. As I have learned about its habits and occupants, my insertion in it becomes more meaningful, yet usually results in far more questions and absolutely more appreciation. 13r1, 105r1R, 237r3, 301r3R, 451r1L
Bianca Nogrady 415r2R
Tom Oliver 111r3L, 471r2R
Organic Maven 12r1 https://creativecommons.org/licenses/by-nd/2.0/
Hrafn Óskarsson 187r4
Tony Palliser UK, moved to New Zealand in 1973 and then to Australia in 1980. My Interest started at an early age and when I retired from the IT industry, I started up my own photographic business, to pursue wildlife photography. I have pioneered many expeditions to all corners of the continent and am probably best known for the work done coordinating pelagic trips around Australia and for being the chairman of the Australian Rarities Committee since 1995. Presently I am working on a personal project to see all the bird families of the world. 59r4R, 69r3L, 187r3L, 285r3C, 301r2LR, 329r1L, 347r3R, 357r3L, 375r2L, 497r1L, 507r3R, 507r4
Kay Parkin Adelaide SA. Obtained a Bachelor of Applied Science in Conservation in 1995 and travels extensively with the aim to observe and photograph every bird in Australia. Hobbies include travel, photography and birding. Read Kay's birding blog, kay parkin birding. 25r1L, 61r3L, 81r4L, 87r3R, 209r3R, 343r1LR, 349r2L, 439r1R
Duade Paton South Coast, NSW. I took up photography in November 2011 and have developed a passion (some may say addiction) for it. I love getting outdoors and finding and watching birds. To capture a photo of that moment is what motivates me and I hope to be able to photograph as many species as I can over the coming years. 53r3R, 55r1C, 55r3LR, 57r1R, 61r2LCR, 61r4R, 63r1C, 65r3L, 69r4R, 85r2R, 111r1R, 113r2R, 117r3R, 121r3R, 125R3L, 135r3L, 139, 143r2, 147r3R, 153r2L, 155r1L, 161r1L, 161r2L, 161r3R, 167r1LR, 169r1R, 169r2LR, 173r1, 195r2R, 197r2L, 199r3L, 255r3R, 291r2L, 293r2R, 311r3L, 315r3R, 327r2LR, 329r1R, 339r2R, 341r3L, 377r2L, 431r3R, 441r1R, 447r2L, 453r3R, 489r3LR
Aaron Payne 507r3L
Shelley Pearson Secret Harbour WA. I am a crazy bird photographer and love nothing better than being outside exploring, looking for that bird. I am passionate about conservation of our environment and particularly concerned about man's progress and the effect it is having on our wildlife. 43r2, 109r1R, 199r1L, Front cover, 559
Grant Penrhyn 69r3R
Christina Port Central Coast NSW. Passionate birder and photographer, and most days will see her out exploring with her camera.

She is also a bird guide. 33r1L, 73r4L, 89, 91r3R, 103r1L, 145r1R, 185, 227r3R, 247r3, 259, 307, 453r1LR, 461r1R, 571
Public domain 11r2, 43r3, 503r5L
Liam Quinn Software developer, Canada. Dabbles in photography on his vacations and when he spots interesting wildlife in his backyard. 45r2, 45r4, 51r2C, 59r1C, 187r2R
Belinda Rafton Wide Bay, Burnett region of Qld. Recently moved for a change of pace and to pursue her interests in nature and photography. She has travelled widely within Australia and a bonus, Belinda says of being a birdwatcher, is the wonderful places and people you come across in the pursuit. 23r2L, 45r1L, 213r2R, 223r3, 233r3R, 253r1L, 281r2, 341r1L, 347r4R, 375r1
Brett Ramsey 263r4L, 339r4L, 403r1L
John D Reynolds 487r1R
Neil Robertson 45r3L
Julian Robinson 49r2R
Matthew Rodgers 81r4R
Neon Tomas Buenaflor Rosell II Philippines. Geologist, with an avid interest in bird-watching and photography. His work took him to all parts of Australia. With his camera gear always packed, his spare time was spent viewing and photographing the local wildlife. 35r1Rtop & below, 141r3, 179r1L, 363r1L, 505r2LR, 511r2 top, 523r3L, 525r1R, 525r2R, 533r3L
Bill Rosenthal Photography 55r2R
Chris Ross 311r1R, 313r2L
Mark Sanders Qld. Professional ecologist and accomplished photographer. His interests are wide and varied, extending from birds to butterflies, frogs to gliders and snakes to bats. He takes every opportunity to return to the 'wild', and now works as a professional ecologist travelling the length and breadth of the continent in search of the next adventure. 37r3, 305r3L, 319r3, 321r3R, 323r2, 323r3, 327r2r, 329r2LR, 341r4L, 395r2R, 395r3L, 397r2R, 425r2L, 431r1R, 457r1, 459r2L
Roman Sandoz 297r1
Michael Schmid Townsville Qld. Passion for birding began at age 25 when Neville Cayley's *What Bird Is That* sparked an enduring passion. Now retired he is a self-taught bird photographer who enjoys the challenge of improving his skill, knowledge and experience and most of all travelling to beautiful places with his wife Christie in pursuit of this passion. Michael's work has been published in books and magazines and includes a butterfly on an Australian stamp. Follow Mick's blog, 'A Good Look Around'. 31r3L, 67r3L, 85r3L, 93r1L, 95r1LR, 101r1L, 115r2L, 137r3L, 193r3L, 203r3R, 211r3R, 221r1L, 229r2R, 235r2R, 237r1R, 241r1LR, 245r3L, 261r3, 265r2L, 267r2R, 285r4R, 291r2R, 301r1R, 311r1L, 311r2L, 317r2L, 327r4R, 333r2R, 335r2, 343r4R, 345r3R, 347r1L, 369r1L, 369r2R, 371r3L, 383r1R, 391r3L, 395r3L, 397r1, 401r2L, 411r3L, 415r3R, 417r2, 419r1R, 421r2L, 423r2, 423r3R, 425r3R, 437r1R, 443r2L, 443r3R, 455r2L, 463r2L, 463r4R, 467r2L, 475r4L, 479r3R, 479r4L, 491r2LR, 491r3L, 495r3R, 497r3L, 511 below right, 515r3L, 521r3L, 531r1, 531r2L, 531r3L, 533r1, 539r3L, 539r4R, 573
Seeboundy [CC BY-SA 3.0 (https://creativecommons.org/licenses/by-sa/3.0)], from Wikimedia Commons 12r2
Merrilyn Serong 377r2R
Ross Silcock 195r4LR
Alwyn Simple Started taking photos while still at school using his Mother's old Kodak Box Camera. He advanced to a 35mm film and slide camera back in the late 1950s. With the introduction of the 'Digital Age' he purchased his first digital camera. He has always used Nikon cameras and updates them each time Nikon release a new professional camera. Alwyn was a bushwalker, canyoner and abseiler until his knees gave way in 2007. Needing a challenge, he took up bird photography and purchased the equipment to meet his needs. He has travelled to all seven continents and has photographed more than 600 Australian bird species in high-quality images. 119r1LR, 151r2LR, 171r1R, 207r1L, 213r3, 233r3L, 265r2R, 273r3L, 275r2LR, 279r2R, 305r2L, 331r2R, 331r3L, 331r4R, 333r2L, 335r3L, 353r4R, 357r1L, 363r3, 383r2LR, 389r1LR, 409r1L, 423r3L, 431r1L, 435r2, 437r4, 439r2R, 495r1R, 497r2L, 497r3R, 517r1L, 517r2L, 535r2LR, 537r2, 537r3L
David Sinclair 63r2C, 73r3L
Eleanor Sobey 109r1L
Rafal Soroczynski 247r1L
John Stirling Melbourne Vic. I took up birdwatching (i.e. keeping lists) in 1982 and photographing birds in 2004. It was inevitable really as both my parents were keen amateur ornithologists and botanists. I've now covered quite a bit of Australia in my quest for new birds and photographic opportunities, but I don't think I'll ever run out of places to go in a continent as vast and varied as ours. The main challenge these days is trying to get a better photo of a particular species than the last time I photographed it. One of the great things about it is that it's a lifetime's work. 49r2L, 255r2L, 371r3R, 519r3R
Peter Struik 239r3L
Max Sutcliffe Has had a keen interest in bird photography for over forty years. Many of his photographs have been published in books, magazines and calendars both in Australia and overseas. Additionally many of his photographs appear on local tourist attraction information signage. 117r3L, 193r1R, 193r2L, 199r4R, 203r1R, 221r3L, 225r1L, 253r1R, 299r3LR, 329r3R, 363r2L, 473r1R, 477r3R, 479r1L
Phil Swanson 49r3R
Eric Sohn Joo Tan Rural General Practitioner by profession, with a deep passion for adventure and a love for nature and wilderness. He has been to some of the most remote parts of the world, and his pursuits include mountaineering, ice climbing, caving, canyoning, mountain biking, and photography. He combines his passion for the outdoors and photography with his love for nature, specialising in avian photography. 49r4R, 75r1L, 111r2R, 119r2LC, 119r3LC, 149r2R, 163r2R, 171r2LCR, 175r3R, 189r1L, 189r2R, 197r2R, 209r2R, 215r3R, 257r1R, 267r1C, 277r3C, 293r3L, 317r2R, 327r3L, 339r2R, 347r4L, 351r3R, 357r1R, 361r1R, 381r2R, 401r3R, 403r3R, 405r1CR, 409r3R, 411r2L, 445r1R, 445r2R, 447r2R, 447r3R, 449, 451r3LR, 459r3LR, 469r2L, 473r3R, 487r1L, 493r1, 509r2R, 509r4R, 525r4L, 533r2R
Simon Tan 71r4L
Tom Tarrant 75r3R, 243r2R, 261r1L 267r1L, 267r3LCR, 285r3LR, 341r2R, 341r4R, 441r3L, 455r1R, 463r4L, 467r2R, 469r1, 469r2R, 473r1L, 539r1L
Pete Taylor 445r3
Dion Thompson 263r1L
Serena Thompson 433r3R
Bruce Thomson 239r3R
Jon Thornton 279r3R, 309r3L, 321r2R, 335r1R
Michael Todd 401r2R, 411r3R

Frank Vassen 503r1R https://creativecommons.org/licenses/by/2.0/

Nigel Voaden Somerset UK. Part-time, amateur photographer whose passion is all things birds and birding. I've travelled all seven continents in search of birds and my interest and drive is the rarely seen and rarely photographed. I try and contribute as much as possible to conservation projects for endangered species by donating photographs and proceeds from any sales I make. 57r3R, 59r3L, 61r3R, 75r1R, 75r3L, 147r3L, 149r3L, 151r3L, 153r1L, 155r3L, 187r1R, 187r2L, 189r3R, 495r1L, 503r2LR, 525r4R

Peter Waanders Avid birdwatcher since age 10. Born and raised in the Netherlands, Peter obtained a B Sc. degree in Environmental Management. He has travelled extensively to over 40 countries in pursuit of his main interests: birdwatching and nature photography (bird species currently totals 2,200). In 1996 Peter moved permanently to to Australia and in 2002 started conducting professional bird guiding trips showing visitors the rare and hard-to-see species of southern Australia. Bellbird Birding Tours / SABirding. 371r1R, 405r3R, 455r2R

Paul Walbridge 59r4L

Geoff Walker In 2007, when he purchased his first Digital SLR camera, he discovered he loved photography and took up the challenge of photographing birds in 2008, when he purchased a 100–400 zoom lens. He has enjoyed finding and photographing nature ever since, visiting places as diverse as Iron Range National Park and Macquarie Island. Some of Geoff's photos can be found at bushpea.com. 129r2, 133r3, 197r4R, 209r3L, 243r2L, 263r4R, 279r2L, 371r2L, 471r1L, 495r2L, 501r2, 505r4R

Ian Wallace I have a passion for observing, recording and learning about the behaviour of wildlife, especially birds. With a digital camera, it is possible to take a series of photos of a bird as it goes about its business and you can often capture things that the eye is not quick enough to pick up. I dabbled in photography back in the 1960s but my interest was re-awakened in 2009 when I purchased a DSLR camera and lens kit including a zoom lens. Looking for a use for the zoom lens we visited a local lake and photographed the birds. We identified and learned about the birds from field guides and our interest and our knowledge just exploded. We are still learning and expanding our activities by camping in the field and photographing in areas that are new to us. I can't imagine a more absorbing activity to occupy our time. 16r4, 29r3L, 35r2L, 135r1R, 137r1R, 163r2L, 191r3L, 193r1L, 309r2R, 339r1L, 345r4R, 349r1R, 353r2R, 387r3R, 417r1L, 461r2L, 463r1L, 477r2LR, 485r1R, 505r3R

Dillon Wan 201, 463r3L

Laura West 175r1L, 385r3L

Chris Wiley Qld. My interest in birding started in my childhood, and has only grown stronger over the years. This passion has taken me to all of the world's continents, including years spent living in the USA and Sweden. Today, I work as an ecological consultant, based in Brisbane. I bought my first camera four years ago, and naturally birds became the primary subject of my photography. While I consider myself very much an amateur in this regard, I'm immensely enjoying this new dimension to birding. 71r1LR, 87r4R, 147r2L, 153r2R, 159r3LR, 189r2L, 195r3LR, 197r1R, 235r1R, 251r1R, 255r1R, 287r3R, 301r3L, 303r1R, 313r1R, 341r2L, 349r4, 377r1R, 381r2L, 393r3R, 399r2L, 421r2R, 447r1R, 457r2L, 459r2R, 511r1R, 539r3R

Martin Willis 483

Graham Winterflood Cairns, Qld. I have been an observer of bird-life in Australia for around 40 years, then in 2014 I purchased my first camera with a zoom lens, a Canon SX50. That same day I took a photo of a dragonfly that ended up on the cover of *Wildlife Australia* magazine. 31r1, 279r1L

James Wood 53r1L

John Woods 275r3R

Lip Kee Yap 485r4R, 523r4L https://www.flickr.com/photos/lipkee/2216110201. https://www.// creativecommons.org

Neil Zoglauer 181r3L

Stephen Zozaya North-eastern Queensland. PhD student, environmental consultant, and wildlife photographer. I strive to create quality images that are both aesthetically pleasing as well as telling about the animal's biology. 305r1L, 14r1, 16r2, 183r1L, 231r3L, 271r2R, 273r2R

Kirk Zufelt I am driven by the love of the open sea, remote islands and lands free from civilization. The study of the adaptations and lives of seabirds and the other ocean dwellers is endlessly fascinating for me. The wind, waves and rough weather is a perfect combination to keep the masses on shore. Follow Kirk's pelagicodyssey blog. 45r3R, 49r1LR, 53r2L, 59r1L, 63r1LR, 65r2R, 69r2LR, 71r2R, 75r2L, 81r2L

Brown Falcon

Index of Common Names

Orange Chat

Index of Scientific Names

Chestnut-rumped Thornbill
exiting its nest hollow.

Quick Index

Male White-throated Treecreeper.

Female Magpie-lark on its deep mud bowl nest, plastered together with grass and plant fibre and generously lined with grass, feathers and fur.

Male Australian Bustard. One of Australia's largest birds. In flights it has measured wingbeats on broad wings.

Female Shining Flycatcher on its closely woven nest decorated with lichen.

Male Scarlet Robin.

Australian Ringneck. Race macgillivrayi 'Cloncurry Parrot'.

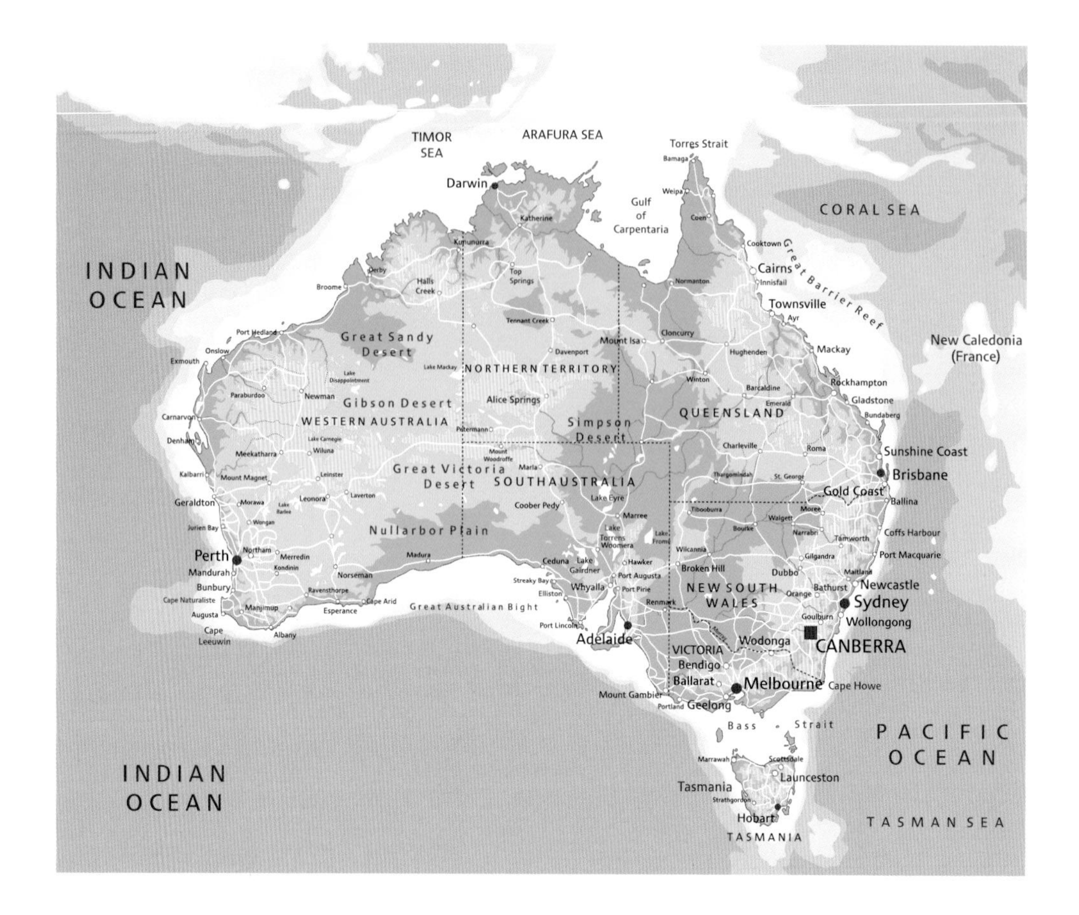

Australian Birding Hot-spots

Queensland

Atherton Tablelands, Birdsville Track, Bowra Station, Cairns Esplanade, Cape York, Cumberland Dam, Daintree NP, Hasties Swamp, Iron Range NP, Julatten, Lamington NP, Mt Glorious, Mt. Hypipamee, Mt Isa, Mt Lewis NP, Townsville

New South Wales

Back Yamma SF, Barren Grounds NR, Capertree Valley, Dorrigo NP, Hunter Wetlands, Lake Cargelligo TP, Lord Howe Is, Macquarie Marshes, Menindee Lakes, Munghorn Gap NR, Pooncarie, Richmond Woodlands, Sydney Olympic Park, The Rock NR

Victoria

Banyule Flats Reserve, Chiltern area, Dandenong Ranges, Gipsy Point, Grampians NP, Great Otway NP, Kooyoora SP, Hattah-Kulkyne NP, Healesville Sanctuary, Little Desert NP, Phillip Is, Terrick Terrick NP, Werribee TP

Tasmania

Bruny Is, Eaglehawk Nest, Melaleuca, Narawntapu NP

South Australia

Adelaide River Crossing, Birdsville Track, Gluepot Reserve, Kalbarri NP, Kangaroo Is, Lake Alexandrina, Lake Gilles CP, Strzelecki Track, Yorke Peninsula

Western Australia

Broome Bird Observatory, Canning Stock Route, Cheynes Beach, Dryandra Woodland, Houtman Abrolhos, Kununurra NP, Lake Argyle, Parrys Lagoon, Porongurup NP, Roebuck Bay, Stirling Range NP, Two Peoples Bay

Northern Territory

Alice Springs TP, Barkley Tablelands, Buffalo Creek, Kakadu NP, Knuckleys Lagoon, Fogg Dam CR, Pine Creek, Simpsons Gap, Timber Creek, Uluru-Kata Tjuta NP, Victoria River Crossing